HOW TO BE A QUANTITATIVE ECOLOGIST

HOW TO BE A QUANTITATIVE ECOLOGIST

THE 'A TO R' OF GREEN MATHEMATICS AND STATISTICS

Jason Matthiopoulos

University of St Andrews, Scotland, UK

WILEY

A John Wiley & Sons, Ltd., Publication

This edition first published 2011

© 2011 John Wiley & Sons, Ltd

Registered office
John Wiley & Sons Ltd, The Atrium, Southern Gate, Chichester, West Sussex, PO19 8SQ, United Kingdom

For details of our global editorial offices, for customer services and for information about how to apply for permission to reuse the copyright material in this book please see our website at www.wiley.com.

Library of Congress Cataloging-in-Publication Data

Matthiopoulos, Jason.
     How to be a quantitative ecologist : the 'A to R' of green mathematics and statistics / Jason Matthiopoulos.
          p. cm.
     Summary: "The book will comprise two equal parts on mathematics and statistics with emphasis on quantitative skills" – Provided by publisher.
     Includes bibliographical references and index.
     ISBN 978-0-470-69978-2 (hardback) – ISBN 978-0-470-69979-9 (paper)
     1.   Ecology – Mathematics.   2.   Ecology – Vocational guidance.   3.   Ecology – Research.
4.   Quantitative analysts.   5.   Quantitative research.   6.   Mathematics – Vocational guidance.
I.   Title.
     QH323.5.M387 2011
     577.0285'5133 – dc22

                                                                                                                2010051191

A catalogue record for this book is available from the British Library.

ISBN HB: 978-0-470-69978-2
ISBN PB: 978-0-470-69979-2
ePDF ISBN: 978-1-119-99158-8
oBook ISBN: 978-1-119-99159-5
ePub ISBN: 978-1-119-99172-4

Typeset in 9/11pt PalatinoLinotype by Laserwords Private Limited, Chennai, India
Printed in Malaysia by Ho Printing (M) Sdn Bhd

Στους αγαπημένους μου γονείς,
Έφη και Σπύρο

How to be a Quantitative Ecologist

The ***A to R*** *of green*
mathematics &
statistics

How I chose to write this book, and why you might choose to read it   **xvii**

Preface

0. How to start a meaningful relationship with your computer   **1**

Introduction to R

1. How to make mathematical statements   15

Numbers, equations and functions

2. How to describe regular shapes and patterns 67

Geometry and trigonometry

3. How to change things, one step at a time 107

Sequences, difference equations and logarithms

4. How to change things, continuously   137

Derivatives and their applications

5. How to work with accumulated change   177

Integrals and their applications

6. How to keep stuff organised in tables   213

Matrices and their applications

9. How to identify different kinds of randomness 299

Probability distributions

10. How to see the forest from the trees   345

Estimation and testing

11. How to separate the signal from the noise   **381**

Statistical modelling

12. How to measure similarity   **425**

Multivariate methods

How I chose to write this book, and why you might choose to read it
(Preface)

'Many of our biology students are refugees from high-school mathematics'
John Ollason, thinker, cynic and the first quantitative ecologist I ever met.

The evolution of most languages is driven by ease of use and the need for fast information exchange. In this sense, phrases like the 'language of mathematics' and 'computer program-ming language' are cruel euphemisms (arguably, even seasoned mathematicians find it harder to read through an unfamiliar equation than an unfamiliar piece of text). These languages are driven by the desire to eliminate ambiguity, at the expense of user-friendliness. Because of our excellent ability to perceive context in everyday situations, few of us feel the need to communicate unambiguously. However, scientific research happens outside the comfort zone of well-understood phenomena. For this reason, modern ecologists need to be trained in quantitative methods but find the process painful. Although there is also a great deal of pleasure in using mathematics and computers in science, it is sometimes hard to keep sight of it during the early period of training. Assuming you have decided that you need quantitative skills for your scientific career (a good call), there are five tricks that can make your learning experience less arduous:

❶ Begin from material that you already know well and work your way up to the harder stuff. Try not to be impatient: often, you will discover that you didn't know the basic stuff as well as you thought.
❷ Pick material that contains a good mix of equations and words. The idea of using text around equations is to give an intuitive understanding of their meaning, but you cannot implement the concepts with words alone. So, do not skip the equations. It defeats the purpose.

❸ If you are (or training to be) a research scientist, then quantitative techniques are just a means to an end. So, look for a book that contains lots of examples from your own area of expertise. The mathematical concepts may be the same, but there is no doubting the motivational power of examples that are interesting to you.

❹ You are unlikely to know, in advance, which techniques will come in handy later on in your research. Who knows? Maybe you will need to combine two apparently unrelated techniques to achieve your objective. So, early on, look for breadth rather than depth.

❺ Don't be fazed by notation and terminology. Most terms and symbols have a common-sense, plain-language interpretation. I would hazard that only about 1 in 10 new terms requires repeated readings to digest. If you are struggling with every single term you encounter, you need to go back to a simpler text.

This book is geared towards these requirements. It's the sort of textbook that I wish I had as an ecology student. I have tried to give it a logical structure that nevertheless doesn't make the narrative so linear as to be boring. Each section prepares you for future sections, each chapter builds on previous ones and the entire book prepares you for more advanced texts that are certainly out there. Almost all chapters follow a classical entry to their subject matter and develop to more contemporary themes towards the end. This should re-animate faint high-school skills and then smoothly carry you to what you need to know for your research today.

Focusing on ecology was a selfish choice but it has allowed a happier co-existence between elementary and more advanced material. Basic maths and stats should not be seen as the bitter pill that has to be swallowed in anticipation of the good stuff. Therefore, while the later examples aim to elucidate abstract mathematical concepts by couching them in an ecological context, the earlier examples hopefully show that even basic high-school techniques can be used to address some interesting ecological questions. Indeed, the priority throughout is to convince you that quite a lot of useful quantitative ecology can be done with a modicum of technical knowledge and that some rather fuzzy ecological concepts can readily be recast and understood in formal mathematical language.

Having said that, I have not tried to pretend that quantitative ecology is easy – it is not. Making this admission means that I do not have to skip over the easier bits by pretending that they are trivial and neither do I need to hide the harder bits under the carpet. There are, therefore, few black boxes in the presentation of the theory. I realise that this is a risky decision because most ecologists don't particularly want to know what goes into the methods they are using, but perhaps this is the mentality that we should be working hard to change. As a result, this text contains hundreds of equations, but one of the earlier ones is $1 + 1 = 2$.

The inclusion of computing with R is a double blessing: from your point of view as a reader, it is useful to see how to implement complicated numerical solutions in practice with minimum effort. From my point of view, the extensive R libraries meant that I could include advanced techniques by explaining their theoretical basis but not their exact implementation. This trick makes for a lower page count and stretches some of the chapters towards the cutting edge of the discipline.

It is not the aim of this book to be an exhaustive guide to either R or the science of ecology, but it is most definitely intended as a comprehensive introduction to maths and stats for green scientists. Similarly, I am not hoping to convert ecologists into modellers (although using this book for a structured, two-semester course would go a long way towards this objective). Quite honestly, this material represents the minimum level of quantitative skills currently required of practising ecologists. The choice of topics is broad for two reasons: first, modern ecology is a melting pot of different quantitative concepts and techniques. The best advances in the primary literature seem to come from cross-fertilisation of ideas. Second, the biggest obstacle faced by a neophyte theoretician is psychological: lack of familiarity with the basic terminology and scope of applications means that ecologists lack the confidence to tackle more technical papers and they are left looking for a 'way in' among textbooks that are either far too basic or way too advanced. As a result, many good hunches remain unsatisfactorily verbal (and unpublished) because they are conceived by colleagues who lack the overview of quantitative theory needed to either formalise their own notions or contact the correct specialist. If this book helps a few of these ideas make it into the primary literature, then writing it will have been worth my while. I hope you will find reading it just as rewarding, especially if one of those ideas belongs to you.

Supplementary material for this book (Exercises, Computer projects, R code, etc.) can be found on the online resource : www.wiley.com/go/quantitative_ecologist

Jason Matthiopoulos
St Andrews
20 May 2010

Thank you to . . .

. . .those who toiled

Popi Gkikopoulou, John Harwood, Helen Heyes, Debbie Russell, Gayatri Shanker, Debbie Steele, Students of St Andrews MRes in Environmental Biology and in Marine Mammal Science (classes 2001–2010), Steve Smart.

. . .those who advised

Geert Aarts, Christian Asseburg, Nicole Augustin, Mike Begon, Tim Benton, Luca Borger, Steve Buckland, Peter Corrigan, Will Cresswell, Carl Donovan, Ann Farrow, John Fieberg, Marie Guilpin, John Halley, Sonja Heinrich, Monique Mackenzie, Marc Mangel, Juan Morales, Dave Moretti, Robert Moss, Leslie New, Theoni Photopoulou, Sophie Smout, Matthew Spencer, Simon Wood, Mark Woolhouse.

. . .those who provided

Susan Barclay, Richard Davies, Heather Kay, Ilaria Meliconi, Sheila Russell, Prachi Sinha-Sahay, University of St Andrews, Wiley publishing house, the R development team.

. . .my loved ones

Spyros Matthiopoulos, Spyros Phevos Matthiopoulos, Effie Matthiopoulou, Valia Tavoularie-Matthiopoulou

0

How to start a meaningful relationship with your computer
(Introduction to R)

'Part of the inhumanity of the computer is that, once it is competently programmed and working smoothly, it is completely honest'

Isaac Asimov (1920–1992), author of science and fiction

This chapter looks and feels different to the rest of the book. It is short, contains no ecology and simply aims to familiarise you with the language used by scientific programmers and the particular conventions of R. It is not exhaustive, so all further R skills will be presented as needed in later chapters, in their appropriate mathematical, statistical and ecological context. The essential questions of what R is, why I chose to burden you with it and what it feels like to use it are covered in Sections 0.1–0.3. In Sections 0.4–0.7 you will find out where to obtain R and some of its valuable accessories, how to set them up in your computer and where to find help when you need it. I also outline the typesetting conventions that I will use to explain R code in this book. The last three sections (0.8–0.10) explain the basics of R usage and how to import data from other software into **data frames**.

0.1.   What is R?

R is an open-source software package developed by a core team of academics and continually augmented by a large list of contributors. It is a numerical environment with a particular

How to be a Quantitative Ecologist: The 'A to R' of Green Mathematics and Statistics, First Edition. Jason Matthiopoulos.
© 2011 John Wiley & Sons, Ltd. Published 2011 by John Wiley & Sons, Ltd.

bias towards statistical analysis and modelling. To some extent, it is what you make of it. It may be used interactively to interrogate a data set or as a programming language to construct simulations and automate complicated tasks. Despite being free to academic users, R compares favourably with other data-analysis and modelling software. For example, it can do considerably more than basic proprietary software such as SPSS or MiniTab and it competes well with very expensive software such as SAS, MATHEMATICA and MATLAB.

0.2.   Why use R for this book?

It is generally better to teach scientific computing using real rather than pseudo-code. It is much better to understand the lofty concepts of programming through a particular language, any language. It is then easier to cross over to another if it is better suited to your purposes. There are several accomplished environments for data analysis and scientific programming but there are several reasons why the choice of R for this book is particularly sound.

The first is its overall suitability to the workflow of ecologists. In the field of quantitative software, packages have historically belonged to one of three camps:

❶ traditional programming languages like Pascal, Fortran and C with basic numerical libraries;
❷ mathematical software like MATHEMATICA and MAPLE with extensive libraries for symbolic analysis;
❸ statistical software such as S, R and SAS with extensive libraries for data analysis.

Any one of these would be suitable, but since the majority of quantitative ecologists spend most of their careers analysing data sets and running simulation models (or solving analytical models numerically), software from the third category seems to fit best.

Another important reason for choosing a software tool, particularly considering the time and effort required to become proficient in it, is longevity. R has a respectable pedigree (its foundations were laid in the software S that has existed since 1976) and it also has considerable potential. Currently, the momentum behind R shows no signs of abating and this augurs well for its future. This momentum guarantees the continuous supply of contributed packages to do almost any imaginable task, books at various levels of specialisation, online resources and text editors for programming (see Section 0.6). Many of these tools and textbooks are aimed at, or motivated by, ecological applications.

Finally, R is freely distributed under the Gnu public licence. There are great ideological reasons for supporting a piece of software developed by publicly-funded academics who then freely distribute their work, placing it at the service of the worldwide academic community. The fact that it is also open-source means that the number of good brains working to improve it is likely to exceed those employed by a private software company.

0.3.   Computing with a scientific package like R

Most of the tasks that an ecologist would care to do on other specialised packages (e.g. geographical information systems, spreadsheets, databases etc.) can also be done in R, with one crucial difference: because it is a programming language, R is considerably more flexible and customisable. Using the built-in commands and the additional packages that can readily be downloaded from the CRAN website, you can write computer code for any imaginable task. This comes at a price to user-friendliness: as with any large tool-box, you need to know what the tools are for, how to use them and in what order. For larger tasks that need to be done several times over, you will need to bundle together several tools in a well-defined sequence. This is called **programming**. If, in your career so far, you have only dealt interactively with a computer (ask a question, get back an answer, then ask another question, possibly based on the previous answer, and so on) then you might find that you need to shift your way of thinking about computers somewhat. Specifying complex tasks for a computer to do is an unforgiving and frequently frustrating job. Not only do you first need to perform the task manually (at least once) to make sure you know what you want done, but you then need to explain it to the computer unambiguously, in a language that looks nothing like written speech. Once this is done, you will often spend long hours looking at the screen wondering why on earth your apparently perfect piece of code comes up with an incomprehensible error message. The problem may be a tiny typo, a missing bracket or a fundamental logical inconsistency. Invariably, you will have to swallow the humiliation that, whatever the mistake was, it was yours and not the computer's. Even when these errors (or **bugs**) have been detected and fixed, there is always the possibility that the computer flawlessly performs a task other than the one you want. For example, a computer program may obligingly allow biological populations to recover from extinction long after their size has become negative. So, the process of **debugging** requires you to be untrusting and critical towards your own creation. This is probably one of the best life-lessons that your computer can teach you.

Once you have adjusted your expectations of how long it takes to develop a piece of code, things can only get better. You may start to enjoy the hunt for bugs, the creative process of constructing a functioning tool out of nothing, the rewarding feeling of uncovering the secrets of your data. Crucially, you will get better as a scientific programmer. You may even savour the rare occasion when code works perfectly the very first time you run it.

0.4.   Installing and interacting with R

Day-to-day work with R involves the **R base package**, additional R packages as required, a good text editor and a quick-reference document of your liking. I explain what each of these is and how to obtain them.

The R base package contains the functionality required by most users including the basic user interface, mathematical, graphical and programming functions and all essential statistical tests and models. It can be downloaded from the Comprehensive R Archive Network (or CRAN for short) at http://cran.r-project.org/. You need to select the appropriate version for your operating system (Linux, Windows or Mac). You then need to follow the link to the base package and download the current version (v2.9.2 at the time of writing this book). Once prompted by a dialogue box, ask for the executable to be run and follow the default options in the various prompts of the setup program. When the program installs, start it up (e.g. by clicking at the desktop shortcut), you should see a screen like the one in Figure 0.1.

R uses a **command-line interface**, meaning that all the interesting stuff isn't done through the drop-down menu in the RGui window, but by typing commands next to the prompt ($>$) in the **R Console window**. When using it interactively, you type something at the prompt which can cause R to give you an output. Try typing something, say a numerical calculation, and then press return:

```
> 1+1
```

The response from R is

```
[1]  2
```

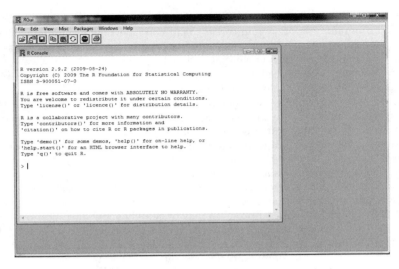

Figure 0.1: Start-up screen of the R command-line interface.

The serial number in square brackets indicates the order of a particular output line resulting from the previous user input. For example, to get R to print a list of all the years from my birth until writing this paragraph, the input and output would be:

```
> 1970:2009
 [1] 1970 1971 1972 1973 1974 1975 1976 1977 1978 1979 1980
[12] 1981 1982 1983 1984 1985 1986 1987 1988 1989 1990 1991
[23] 1992 1993 1994 1995 1996 1997 1998 1999 2000 2001 2002
[34] 2003 2004 2005 2006 2007 2008 2009
```

The colon (:) always indicates a range of values. Depending on the width of your screen, R will generate the list in as few lines as possible, quoting the serial number of the next item within square brackets, at the start of each line.

When your input is not understood by R, you will get an error message:

```
> this_input_is_rubbish
Error: object 'this_input_is_rubbish' not found
```

The flow of printed information in the R Console is always downwards. Although you can scroll up to view your previous workings, and you can use your mouse to highlight and copy bits of printed input/output, you cannot navigate up to edit any part of your previous inputs. You may, however, use the up arrow key on your keyboard to quickly copy a previously typed line of code onto the currently active prompt line. Working with such one-liners is not a problem within the R Console but it can get cumbersome when you need to input several lines of code together. In these cases, a better alternative is to type your code in a text editor (Section 0.6) and copy/paste it into the R Console.

Assignments in R can be done with the arrow symbols (<-, ->). For example, to give a name (say, years) to the above list of years, you would type

```
> years<-1970:2009
```

An assignment prompts no response from R. The information is simply stored under the name `years`, ready for later use. If you want to inspect the information, type `years` and press return.

R is **vectorised**, meaning that operations can be applied to entire collections of things with the same ease as applying them to single items. For example, my entire list of birthday anniversaries can be calculated (and filed under the name `ages`) as

```
> ages<-years-1970
```

While the R Console handles the alphanumeric input and output, the **R Graphics Device** deals with images. This window appears separately within the RGui window. For example, a plot of ages versus years can be created by typing

```
> plot(years, ages)
```

Pressing return brings up the graphics window on screen (Figure 0.2). Feel free to rearrange and reposition this within your workspace.

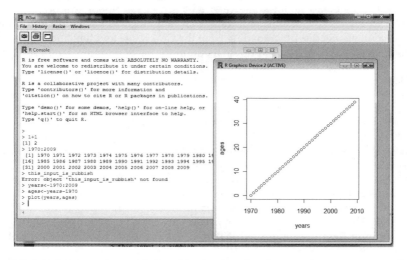

Figure 0.2: R Gui, R Console and R Graphics Device: the three main components of the R interface.

0.5.   *Style conventions*

When editing your code outside of R, it is a good idea to use monospaced fonts (such as `Courier`), because, unlike proportional fonts (like Times), they allocate equal space to all characters. This retains the spacing of some tabular forms of output and makes code easier to debug. In this chapter, I have placed R input and output on a grey background. In the rest of the book, I use a grey background for the entire R boxes to make the computing sections more obvious among the rest of the text. When describing interactive use of R, I precede user input by the prompt (>) and use boldface for the output:

```
> 1+1
[1] 2
```

Larger pieces of code, of a length that might be typed up in a text editor and then pasted into R, are presented without the prompts, accompanied by short comments in English for most lines. Such detailed annotation is good practice when programming and not just for the benefit of others: I am always surprised by how hard it is to understand my own uncommented code a mere few weeks after writing it. The special character # tells R to ignore the remainder of that line so that when copy/pasting code, the comments do not interfere with the computation. In the book, such comments are shown in italics. For example:

```
# This plots age v calendar year for a person born in 1970
years<-1970:2009      # A list of years since 1970
ages<-years-1970      # A list of ages since birth
plot(years,ages)      # Generates the plot
```

0.6. Valuable R accessories

The functionality of R can be expanded by installing additional packages. A **package** is a collection of additional functions, example data sets and documentation. Whenever this happens in this book, you will be informed which package to get, but you need to know how. There are two types of packages: those that only need to be loaded into R and those that require a full install (i.e. downloading from the CRAN site and then loading into R). Both can be done via the 'Packages' drop-down menu in the RGui window. Upon selecting 'Load package...' you will be presented with a selection of about 28 packages. Simply select the one you need and press OK. The package is then loaded and can be used by your current R session. Alternatively, packages not on this list can be obtained by selecting 'Install package(s)...', again from the 'Packages' drop-down menu. You may be asked to select a CRAN mirror site. Just pick the one that is closest geographically to you. This list of packages is considerably larger. Pick your package and wait until it downloads and expands. Installed packages are saved on the hard disc and stay with your computer even after your R session ends. You do, however, need to load them into R when you start a fresh session. Go through 'Packages-> Load package...' as before. You will notice that your recently installed package has made its appearance in this list. An alternative to using the drop-down menu to load a package is to do it using the commands require() or library(). For example, require(MASS) will load the package MASS. If your code requires a package, then place the require() command right at the beginning, so that the package is loaded before the rest of the code is executed.

The constant introduction of new R commands throughout the book might leave you overwhelmed by the apparent arbitrariness of their names. Rest assured you are not alone. Each computer language may use different names for different purposes and no programmer can remember more than a small vocabulary. Navigating help files is therefore an essential skill (see Section 0.7) but equally important is a good quick-reference guide of the names and syntax of the most frequently used R commands. My personal favourite was created by Tom Short and can be found at http://cran.r-project.org/doc/contrib/Short-refcard.pdf.

Finally, there is the delicate issue of the text editor to be used for developing longer pieces of code. You do need one, but the choice is a matter of taste (Figure 0.3). Word-processors are to be avoided, because their spell-checking and slow searching facilities tend to get in the way. You may, instead, use a fast and simple text editor such as TextPad (freely available from http://www.textpad.com/products/textpad/index.html). Better still, you can download a text editor that has been developed specifically for R programming. More information on editors can be found at http://www.sciviews.org/_rgui/projects/Editors.html.

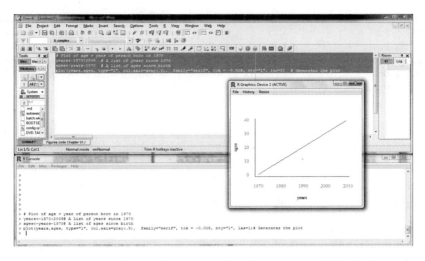

Figure 0.3: The combination of a specialist text editor with R can greatly facilitate programming work. Here, R is seen running together with Tinn-R, a great editor for the Windows operating system.

0.7. Getting help

There are three sources of information on various aspects of R. Printed and online manuals can inform you about the capabilities of the program and offer detailed, worked examples. Several pdf manuals come together with the R base package and you can access them from the 'Help' drop-down menu in the R Console. A list of further references can be found in Sections 0.13 and 0.14.

By working through books (such as the one you are holding), you will become aware of what is required of you as a programmer and what R can do for you as a scientist. The quick-reference guide mentioned in the previous section will keep you right regarding the syntax of commands for common tasks. However, you will regularly need to read more about the syntax and details of particular commands by researching the R help files. You can access these in two ways from within R. Go to the 'Help' drop-down menu and select 'Html help'. This will launch your web browser to display a page for searching and browsing keywords. The contents of this page are stored on your hard disc, so don't worry if you happen to be working off-line.

Alternatively, within the R Console, you can type a question-mark followed by the R command you want help on. For example, `?plot` will bring up, in a new window, the help file for the basic plotting command. You will initially find R help files somewhat...unhelpful. The way they present information takes some getting used to but, thankfully, surmising what you need becomes easier with practice. Each help file usually contains a section summarising the purpose of a command ('Description'), its correct syntax ('Usage'), its inputs ('Arguments') and outputs ('Value'). Most importantly, towards the bottom, all help files have examples of usage ('Examples') and hyperlinks to relevant commands ('See also'). If you are unsure of the name of the command for which you want help but you vaguely remember that it pertains to or contains some keyword, try typing `??keyword`. This performs a search through the help files for related commands. If the command cannot be found by the help searches, then it is possible that you are trying to read the help files for a package that you have not yet installed into R.

If all else fails you may seek information from the web. There are several searchable help archives on the CRAN website and Google will usually come up with the goods within its first page of search results. For questions on the base package, the forum **R-help** is a good starting point. The R community has a great record of responding to questions but before sending a question to the group, make sure that your question has not previously been answered in the archives.

0.8.   Basic R usage

At its very simplest, R can be used as a glorified calculator. For example, the expression $\sqrt{(20 - 1.5)^2 + (20 - 5)^2}$, can be calculated as

```
> sqrt((20-1.5)^2+(20-5)^2)
[1] 23.81701
```

Note the use of bracketing to specify which parts of the expression are to be squared and which are to be square-rooted. Brackets are used to specify the priority of operations. Hence, the expression

```
> (1+2)*4
[1] 12
```

does not give the same result as

```
> 1+2*4
[1] 9
```

Bracketing is also used for R commands. The following generates a pretty plot of the age v calendar year data

```
plot(years, ages, type="l", col.axis=gray(.5), family="serif",
tck=-0.008, bty="l", las=1)
```

Since this syntax is representative of most R commands, it is useful to elaborate: The name of the command (in this case, `plot`) comes first, followed by a pair of brackets containing the input to be used by the command and the options specifying how to use it. In this example, the inputs are two lists, for the x- and y-axis data (`years` and `ages`) followed by a total of six options. The name of the option comes first and its assigned value follows the equality sign (=). For example, `type = "l"` specifies the type of plot as a line plot (rather than a scatter plot), `col.axis = gray(.5)` specifies that the text used for the tick mark labels will be a medium shade of grey and `las = 1` that these labels will be horizontal (rather than parallel to the two axes). Two things must be noted: first, option assignments are not done by arrow signs (`<-`, `->`), if you do, R will respond with an error message. Second, specifying options is *optional* because all options are set by the R developers to some default value. For example, the command `plot()` has about 70 possible options (type `?plot` and follow the link to `par` to see the complete list of graphical parameters). You can use as many of them as you like to specify a plot to your exact standards, but omitting all the options and simply typing `plot(years, ages)` will generate a perfectly decent plot.

Other types of brackets are also employed by R. As we will see in later chapters, square brackets [ ] are used to modify and extract elements from data sets and curly brackets { } are used in programming to package together multiple lines of commands.

0.9. *Importing data from a spreadsheet*

Ecologists spend long days painstakingly collecting data in the field and long nights analysing them in front of a computer. It would be a shame if the analysis was spoiled simply because measurements of animal body weight were accidentally imported into the column for vegetation cover. One way to ensure this doesn't happen is to standardise the protocol for importing data.

The main R command for data import is `read.table()`. This requires information on the location of the file that holds the data and its specific format. Consider an Excel spreadsheet containing a column of years and a column of ages, as shown in Table 0.1.

Table 0.1

Years	*Age*
1970	0
1971	1
1972	2
1973	3
1974	4
1975	5
1976	6
1977	7
1978	8
1979	9
1980	10
1981	11

Because Excel spreadsheets have multiple sheets, it is easier to export a single sheet by saving it as a text file. In Excel, go to 'File→Save as. . .' and save the current sheet as 'Text (Tab-delimited)'. Assuming that the full path to the file is C:\My documents\Data\YearsAge.txt, then it can be imported into R by typing:

```
read.table("C:/My Documents/Data/YearsAge.txt", header=TRUE)
```

If the option `header` is set to `TRUE`, the first row of the spreadsheet is interpreted as a header, meaning, in this example, that R will recognise the first column by the name `Years` and the second by the name `Ages`. Note that the file path needs to be enclosed in double quotes and specified in terms of double backslashes. If you are likely to be importing different files every time, you may want to consider the following version of the command which launches a browsing dialogue box:

```
read.table(file.choose(), header=TRUE)
```

If you already have a nonstandard text file (one that is not tab-delimited) you can adapt the `read.table()` command so that R can read it. Options available for importing data can be found by typing `?read.table`. If you are still having problems, refer to the R Console manual 'R Data Import/Export'. You will find this through the 'Help' drop-down menu, under the option 'Manuals (in PDF)'.

0.10.   Storing data in data frames

A **data frame** is a two-dimensional tabular object used for storing different types of data. The columns of a data frame store qualitatively different types of measurements and its rows correspond to sampling units or replicates. The data frame shown in Table 0.2 contains 16 observations from five animals sighted in different habitats, performing different behaviours. Although for the purposes of statistical analysis, it is debatable whether the appropriate sampling unit is the observation or the individual, the data frame must contain the full information of the data. Hence, the definition of replicate used for the rows of the data frame must represent the data in its most resolved form.

Table 0.2

ID	Individual ID	Observation ID	Habitat	Behaviour
1	1	1	Grass	Foraging
2	1	2	Forest	Socialising
3	1	3	Forest	Socialising
4	1	4	Grass	Foraging
5	1	5	Rock	Transiting
6	2	1	Grass	Foraging
7	2	2	Grass	Socialising
8	2	3	Grass	Foraging
9	2	4	Grass	Foraging
10	2	5	Grass	Socialising
11	3	1	Forest	Transiting
12	3	2	Forest	Socialising
13	4	1	Forest	Socialising
14	4	2	Forest	Socialising
15	4	3	Grass	Transiting
16	5	1	Rock	Transiting

The command `read.table()` automatically imports data into a data frame, so all you need to do is name it. For example, if the above data set is saved by Excel as a tab-delimited text file ('`Sights.txt`'), the data frame is created at import

```
data<-read.table("C:/My Documents/Data/Sights.txt", header=TRUE)
```

It is good to check that the data sheet imported has the right number of rows and columns

```
> nrow(data)
[1] 16
> ncol(data)
[1] 5
```

R has a wealth of commands for manipulating data frames. To see the column names of the data frame type

```
> names(data)
[1] "ID" "Individual" "Indiv_ID" "Habitat" "Behaviour"
```

To extract any one of your columns, you may call it by using the following syntax

```
> data$Habitat
 [1] Grass   Forest Forest Grass   Rock    Grass   Grass   Grass   Grass
[10] Grass   Forest Forest Forest Forest Grass   Rock
Levels: Forest Grass Rock
```

Here, R has identified that the variable Habitat takes a discrete number of values. Statisticians call such variables **factors** and R has taken it upon itself to identify the three values (or **levels**) that this variable has taken. An even easier way to access the content of the data frame is to attach it to the R search path. You only need to do this once in any one session and it allows data frame columns to be called directly by name.

```
> attach(data)
> Habitat
 [1] Grass   Forest Forest Grass   Rock    Grass   Grass   Grass   Grass
[10] Grass   Forest Forest Forest Forest Grass   Rock
Levels: Forest Grass Rock
```

This leaves open the possibility for naming conflicts: e.g. there may be more than one attached data frame with a column called `Habitat`. R will give you warnings if one attached data frame is about to mask the contents of another. To avoid these issues, you can `detach()` a data frame when you are finished with it.

Specific segments of the data frame can be extracted by specifying row and column numbers (in that order) inside square brackets. I will talk a lot more about these conventions in Chapters 1 and 6 but here are a few examples. A particular row (say the fifth), can be extracted as follows:

```
> data[5,]
   ID Individual Indiv_ID Habitat  Behaviour
5  5          1        5    Rock Transiting
```

Here, the column reference is left vacant after the comma in the square brackets, meaning that all column values for that row are required. A range of rows (say, all observations from the first animal) can also be obtained using the colon notation:

```
> data[1:5,]
   ID Individual Indiv_ID Habitat   Behaviour
1  1          1        1   Grass    Foraging
2  2          1        2  Forest Socialising
3  3          1        3  Forest Socialising
4  4          1        4   Grass    Foraging
5  5          1        5    Rock  Transiting
```

The data frame can also be queried using conditioning. For example, all the rows that refer to the second animal can be found as follows:

```
> data[Individual==2,]
   ID Individual Indiv_ID Habitat   Behaviour
6   6          2        1   Grass     Foraging
7   7          2        2   Grass Socialising
8   8          2        3   Grass     Foraging
9   9          2        4   Grass     Foraging
10 10          2        5   Grass Socialising
```

0.11. Exporting data from R

Some analyses take a long time to design. Others take a long time to run. Their numerical results may therefore be valuable enough to store on the hard disc via the command `write.table()`. At its simplest, you are required to specify the data frame to be saved to file and the full path. For example, the line,

```
> write.table(data, file="C:/My Documents/Data/SightsE.txt",
col.names=TRUE)
```

will write the data into the file `SightsE.txt`. If you would like to browse for the target location of your new file type

```
> write.table(data, file.choose(), col.names=TRUE)
```

During a new session the data can be re-acquired by typing

```
> data<-read.table("C:/My Documents/Data/SightsE.txt",
header=TRUE)
```

Communicating with other users, not all of whom may use R, will require you to export your data in formats readable by other packages. You can do this by customising the options in `write.table()`. For example, here is how to create an Excel file called `SightsE.xls`

```
> write.table(data, "C:/My Documents/Data/SightsE.xls",
sep="\t", row.names=FALSE, col.names=TRUE)
```

The special character \t uses tab as a column separator and the appropriate file extension (here, .xls) ensures that, once double-clicked, the file will be opened by Excel.

0.12. Quitting R

You can end your R session either by closing the R Console, or by typing `q()`. You will be asked if you want to save your workspace image. This will store all the imported data and variable names so that you find yourself exactly where you were when you next open R. This sounds appealing but it is best avoided because R then tends to accumulate a lot of definitions from previous analyses that take up memory and may lead to conflicts. If you have saved your previous sessions and want to wipe the slate clean you can do it by typing

```
rm(list=ls())
```

Further reading

The core developers of R have, over the years, produced several introductory references. Prime examples are Venables and Smith (2002) and Dalgaard (2008). Other, general references are Crawley (2005, 2007) and Zuur *et al.* (2009). Recently, several example-based books have appeared (Everitt and Hothorn, 2006; Braun and Murdoch, 2007) that are great for people who like to adapt worked solutions to their own needs. If you are looking for R books with an ecological bias, check out Bolker (2008) and Stevens (2009).

References

Bolker, B. (2008) *Ecological Models and Data in R*. Princeton University Press. 408pp.

Braun, W.J. and Murdoch, D.J. (2007) *A First Course in Statistical Programming with R*. Cambridge University Press, Cambridge. 163pp.

Crawley, M.J. (2005) *Statistics: An Introduction using R*. John Wiley & Sons, Ltd, Chichester. 342pp.

Crawley, M.J. (2007) *The R Book*. John Wiley & Sons, Ltd, Chichester. 942pp.

Dalgaard, P. (2008) *Introductory statistics with R*. Springer. 364pp.

Everitt, B.S. and Hothorn, T. (2006) *A Handbook of Statistical Analyses using R*. Chapman & Hall/CRC, Boca Raton. 275pp.

Stevens, M.H.H. (2009) *A Primer of Ecology with R*. Springer. 388pp.

Venables, W.N. and Smith, D.M. (2002) *An Introduction to R*. Network Theory Ltd. 156pp.

Zuur, A.F., Ieno, E.N. and Meesters, E.H.W.G. (2009) *A Beginner's Guide to R*. Springer. 220pp.

1

How to make mathematical statements
(Numbers, equations and functions)

'In science there is only physics and stamp collecting'
Ernest Rutherford (1871–1937), the father of nuclear physics.

'I have hardly ever known a mathematician who was capable of reasoning'
Plato (428–348 BC), the father of all science.

One of the exciting challenges of quantitative ecology is to examine whether a set of observations that have been classified by name can be **ordered** along a continuum. Therefore, this chapter begins with a discussion of nominal and ordinal scales (Section 1.1). Although there is still a valuable role for **nominal classification** (see Chapter 12), the deceptively simple act of comparing two, apparently different, individuals, species or communities along one or more quantitative scales, propels us forward from natural history to modern ecology. This transition is mediated by **numbers** (Sections 1.2 and 1.17). **Symbols** (Section 1.3) are often used instead of numbers either to cope with ignorance or to make general statements. Mathematical **operators** (Sections 1.4 and 1.5) are used to connect different (known or unknown) quantities into algebraic **expressions. Algebra** is the set of rules dictating how these expressions may be manipulated (Sections 1.7–1.9). The two main scientific applications of mathematics are in formalising known facts or assertions as **equations** or **inequalities** (Sections 1.10–1.15) and expressing **relationships** between variables (Sections 1.18–1.25).

How to be a Quantitative Ecologist: The 'A to R' of Green Mathematics and Statistics, First Edition. Jason Matthiopoulos.
© 2011 John Wiley & Sons, Ltd. Published 2011 by John Wiley & Sons, Ltd.

1.1. Qualitative and quantitative scales

Data are called **qualitative** if they cannot be compared using some measure of magnitude. For example, **nominal** observations can only be compared in a rudimentary way, by checking for 'sameness'. If they are not the same, one nominal observation cannot readily be said to be greater than another. In contrast, **quantitative** data can be ordered and the degree of dissimilarity between them can be evaluated objectively. This rudimentary taxonomy of data will be elaborated in Chapter 7. For now, it is sufficient to say that the distinction between quality and quantity is not always clear. Often, observations that appear to be nominal can be ordered by means of their attributes, as in Example 1.1.

Example 1.1: Habitat classifications

Fern frond

We can easily distinguish between marine and terrestrial habitats. In the marine environment there are polar, upwelling, shelf, open-ocean and coral habitat types. In the terrestrial environment, examples include the boreal, tundra, tropical, temperate, desert and montane habitat types. The definitions of these are generally vague but suffice for most applied purposes. However, studies in spatial ecology (Manly *et al.* 2002; Aarts *et al.* 2008) have increasingly found that it is more useful to describe the distribution of plants and animals in terms of individual habitat characteristics such as temperature and precipitation (measured on a quantitative scale) rather than using arbitrary – and occasionally anthropocentric – habitat types (Figure 1.1).

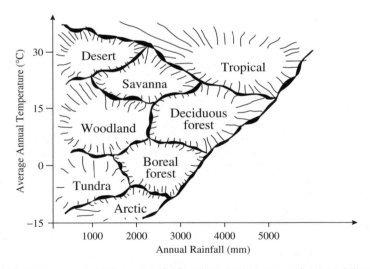

Figure 1.1: Habitat types are arbitrary subdivisions imposed on an environmental continuum.

ℝ 1.1: Declaring nominal categories

To create a simple computerised taxonomic scheme involving the categories of Animals, Plants, Fungi, Protoctista, Archaea and Monera, it is first necessary to tell R that these labels are to be treated as text, so that it doesn't expect a numeric value for them. This is done by enclosing the labels in quotation marks:

```
"An", "Pl", "Fn", "Pr", "Ar", "Mo"
```

The labels can be collected together using the **concatenation** command c ():

```
c("An", "Pl", "Fn", "Pr", "Ar", "Mo")
```

and the taxonomy is declared using the command factor() which says to R that a collection of specimens can be classified according to this scheme of labels (more on factors in Chapter 7):

```
factor(c("An", "Pl", "Fn", "Pr", "Ar", "Mo"))
```

so, to classify a collection of organisms according to kingdom, each specimen needs to be associated with one of the six categories in this factor.

1.2. Numbers

Numbers are certainly useful for counting, but not all measurable quantities can be counted. Thankfully, the different types of numbers used for measurement are both countable and few; all-in-all there are only five. Each type is a **set**, an imaginary container that may enclose (or be itself enclosed in) other sets (Figure 1.2).

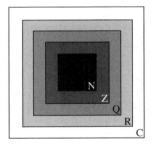

Figure 1.2: There are five types of numbers, usually represented as a hierarchy of nested sets.

The first set of numbers, both historically and in order of simplicity, are the **naturals** (collectively denoted by $\mathbb{N}$). These are the numbers 1, 2, 3, 4, etc., that you would use to count whole items, such as the number of animals in a population or the number of species in a community. If we use curly brackets to enclose the elements of a set, then we have

$$N = \{1, 2, 3, 4, 5, 6, 7, 8, 9, 10, ...\} \tag{1.1}$$

The three dots at the end of the sequence imply an infinite continuation of the pattern already expressed by the preceding numbers.

The second set of numbers are the **integers**, collectively denoted by $\mathbb{Z}$. They are also known as the **signed numbers** because they are preceded by a minus or a plus

$$\mathbb{Z} = \{\ldots, -4, -3, -2, -1, 0, +1, +2, +3, +4, \ldots,\} \tag{1.2}$$

Zero represents the absence of any magnitude and the plus signs of the positive numbers are usually implied,

$$\mathbb{Z} = \{\ldots, -4, -3, -2, -1, 0, \underbrace{1, 2, 3, 4, \ldots}_{\mathbb{N}}\} \tag{1.3}$$

Compare (1.1) with (1.3) and note that the set of naturals is a **subset** of the integers (i.e. $\mathbb{N}$ is contained in $\mathbb{Z}$). In mathematical notation, this is written $\mathbb{N} \subset \mathbb{Z}$.

The third set of numbers are the **rationals**, denoted by $\mathbb{Q}$. They are the numbers produced from the **ratio** or division of any two integers n, m, assuming m is not zero. In mathematical notation:

$$\mathbb{Q} = \left\{ \frac{n}{m} \forall n, m \in \mathbb{Z}, m \neq 0 \right\} \tag{1.4}$$

Try not to panic when you see an expression like this. Mathematical notation is admittedly unfriendly but it makes up for it by being both precise and brief. Often, even the most intimidating expressions have a plain-language translation. In Equation (1.4) the symbols $\forall$ and $\in$ are mathematical shorthand, meaning 'for every' and 'belonging to', respectively. So, the whole expression says: The rationals are the numbers that can be obtained by dividing two integers n over m, excluding the value zero for the denominator m.

Example 1.2

$6/3 = 2$, $\quad 1/2 = 0.5$, $\quad 7/4 = 1.75$ and $\ -10/2 = -5$ are rational numbers

---

All integers can be produced as the ratio of other integers so that all integers are also rationals. However, not all noninteger numbers can be produced as ratios of integers. This surplus set of numbers are, quite appropriately, termed the **irrationals**. We will encounter examples later on (e.g. square root of 2) but, for now, it is useful to note that irrational numbers have an infinity of nonrepeating decimals. The combined set of rationals and irrationals gives us the fourth type of numbers called **real numbers**. The set of reals (denoted by $\mathbb{R}$) is used when we need to measure continuous quantities, such as length, density or mass.

The fifth and final type are known as the **complex numbers**, denoted by $\mathbb{C}$. A more detailed presentation of complex numbers is left until Section 1.17, but it is worth noting here that the set of complex numbers is a **superset** of the reals. So, we can represent Figure 1.2 in mathematical terms by:

$$\mathbb{N} \subset \mathbb{Z} \subset \mathbb{Q} \subset \mathbb{R} \subset \mathbb{C} \tag{1.5}$$

Example 1.3: Observations of spatial abundance

There is a correspondence between the different types of ecological measurement and the sets of numbers in Equation (1.5). Consider the measurements that might typically be used to describe the distribution of a plant population along a linear study site, such as a stretch of river, which is 1 km long and has been subdivided into ten segments (Figure 1.3). The easiest description is in terms of **occupancy**, the presence or absence of the species from any particular segment. Although occupancy can be thought of as a qualitative trait, it is readily made quantitative by attributing the value 0 to absence and the value 1 to presence (Figure 1.3(a)). If, in addition, there are data on the number of occurrences in each segment, then the plant distribution can be described by a series of counts which take their values from the set of non-negative integers, $\{0, \mathbb{N}\}$ (Figure 1.3(b)). These count data can be readily converted to densities by dividing each count by the size of the segment – in this case 100 m (Figure 1.3(c)). Standardised density (or relative abundance) can be obtained by dividing the count in each segment by the total number of observations. This conveys the proportion of the total count occurring in any given segment (Figure 1.3(d)). Both density and relative abundance are rational numbers. Finally, we may want to compare the distribution of the species with that of some environmental covariate that could be used as a proxy at other unsurveyed sites (a covariate is a quantity that is closely related to the measure of interest). For example, there may be a gradient in soil pH along the study site (pH measurements are real numbers). A look at Figure 1.3(e) suggests that the plant has a preference for soil pH around 6.

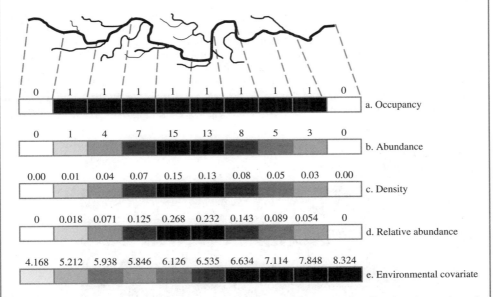

Figure 1.3: Measurements belonging to different sets of numbers naturally occur in ecology. In quantifying the spatial distribution of a species, we may use (a) occupancy (nominal data), (b) counts of abundance (non-negative integers), (c) density, (d) relative abundance (both rational numbers) or (e) environmental covariates, such as soil pH (real numbers).

Ⓡ **1.2: Declaring simple sets of numbers**

As we saw in R.1.1, a set is declared by the concatenation command c(). The set of the first nine natural numbers can be declared as c(1,2,3,4,5,6,7,8,9). A quicker alternative is to specify these as a range using a colon c(1:9). These two types of declaration can be used in combination. Here are three different ways of declaring the set of the first ten non-negative integers:

```
c(0,1,2,3,4,5,6,7,8,9) or c(0,1:9) or c(0:9)
```

1.3.   Symbols

A time-honoured mathematical trick is to use symbols to represent unknown quantities. This is done for two reasons. The first is practical: an unknown quantity of interest needs a name (say, x, y or z) while its value is being deduced. As we will see in Sections 1.9–1.13, this kind of speculation involves formulating mathematical statements (equations or functions) that combine known facts with unknown quantities. One might then try to manipulate (i.e. solve or plot) these statements to find out more about the unknown quantities. The second reason is more important: one of the primary endeavours of all scientists is to establish **generality** from their results. In most cases, you won't want to limit your mathematical arguments to specific numbers and symbols facilitate this task. You can fix some of the properties of a symbol and let the others loose. You can, for example, stipulate that x is any integer number. Alternatively, you may specify a narrower range for it, or leave it completely unspecified. Hence, when biological first principles dictate limitations for the values of symbols (e.g. biological populations cannot be negative), *you can avoid investigating biologically meaningless scenarios*.

You can treat a symbol just like a number and operate on it using the rules of algebra (Section 1.8). However, remember that once you've defined your symbol (say, as an integer) you can't treat it as anything that defies that definition (e.g., by talking about its fifth decimal). Of course, the end result of a calculation using symbols may also be partly symbolic (as a result of partial ignorance or the desire for generality).

When forming a mathematical statement, it is important to distinguish between **variables** (quantities that change over time, space or in response to other inputs) and **constants** (quantities that don't change). This distinction is purely pragmatic, and may vary between different systems or applications.

Example 1.4: Population size and carrying capacity

Figure 1.4: Highly-packed colonial breeders. If available space is the only factor limiting the density of these birds, then we can interpret peak densities as a colony's carrying capacity (© Sheilla Russell).

In population models, the size of a population is represented by a variable, say P. If the population size is restricted by limited resources, then the environment can only sustain a limited number of individuals (Figure 1.4). This is called the environment's **carrying capacity** (K). If the environment does not change, carrying capacity should remain constant. So, although P and K are both symbols representing population size, the first is a variable and the second a constant. More complicated models may acknowledge the fact that carrying capacity is subject to environmental change, hence treating it as a variable.

1.3: Naming conventions

When introducing a new symbol in an R session, make sure that you will understand the meaning of the symbol when you look again at your code at a later date. This can be done by using entire words rather than single letters, which also ensures an infinite supply of names, certainly more than the combined Latin and Greek alphabets used in maths notation. However, bear in mind that you will be typing some symbols many times, so try to be sensible with the length of names.

1.4.   Logical operations

Logical operators can be used to express equality ($=$), inequality ($\neq$), directed inequality ($>$, $<$) or inequality with the possibility of equality ($\geq$, $\leq$). These enable us to make comparisons, or to make assertions about numbers and symbols.

Example 1.5

The following are true: $1 = 1$, $2 < 3$, $4 \neq -4$

The following are false: $1 \neq 1$, $2 > 3$, $4 = -4$

The following are assertions: $x = 1$, $x \geq y$

1.4: Assigning values to symbols

The general operator for making assignments in R is the arrow `->` or `<-`. The value 2 can be assigned to the symbol x, in any one of two ways: `x<-2` or `2->x`. If both sides of the assignment are symbolic, then the information flows according to the arrow, assuming the symbol at the base of the arrow already has a value. For example:

```
> x<-2
> y<-x
> y
[1] 2
```

1.5.   Algebraic operations

Addition, subtraction, multiplication and **division** are the fundamental algebraic operations that bind together numbers and symbols to create mathematical expressions, such as $12 + 45$, x/y, $x - 5$.

The result of addition is called a **sum**. The number zero is also known as the **identity operator** in addition and subtraction because it leaves the result unaffected ($a \pm 0 = a$). The result of subtraction is called a **difference** and is sometimes denoted by an upper case Greek delta prefix (Δ). Sometimes, it is convenient to interpret subtraction as the addition of a negative number because, numerically, the result is exactly the same.

Example 1.6: Size matters in male garter snakes

Garter snake, genus Thamnophis

Mating success in male garter snakes appears to depend on body length (Shine *et al.*, 2000). The longest (l_{Long}) and shortest (l_{Short}) adult males recorded in a population were, respectively, 50 and 45 cm. Therefore, the difference between a successful and an unsuccessful male is $\Delta l = l_{Long} - l_{Short} = 50 \; cm - 45 \; cm = 5 \; cm$. Although body length can only increase with time, an individual's weight w may go either up or down. We can express the overall change in w during a 28-day period as the sum of weekly measurements of weight change (gain or loss). For example:

$$\Delta w_{Total} = \Delta w_1 + \Delta w_2 + \Delta w_3 + \Delta w_4 = 5 + 7 + (-4) + (-2) = 6g \qquad (1.6)$$

The result of multiplication is called a **product** and the result of division is called a **ratio**. Remember the various incarnations of multiplication. If a and b are defined as numbers then, in mathematical notation, $a \times b, a \cdot b, ab$ are all the same. The **identity operator** in multiplication is 1. Multiplication is allowed between any two numbers, but division by zero is not. Products and ratios can be combined in all possible ways (Table 1.1):

Example 1.7

Table 1.1

Operation	Example	Simplifies to
Product of products	$(1 \times 2) \times (3 \times 4)$	$1 \times 2 \times 3 \times 4$
Product of ratios	$\dfrac{1}{2} \times \dfrac{3}{4}$	$\dfrac{1 \times 3}{2 \times 4}$
Ratio of products	$\dfrac{1 \times 3}{2 \times 4}$	$\dfrac{3}{8}$
Ratio of ratios	$\dfrac{1/2}{3/4}$	$\dfrac{1 \times 4}{2 \times 3}$

Furthermore, ratios with the same denominator can be added. Any two ratios can be added by first forcing them to have the same denominator.

Example 1.8

$$\frac{2}{3} + \frac{1}{5} = \left(\frac{2}{3} \times \frac{5}{5}\right) + \left(\frac{1}{5} \times \frac{3}{3}\right) = \frac{10}{15} + \frac{3}{15} = \frac{10 + 3}{15} = \frac{13}{15} \qquad (1.7)$$

Multiplication gives rise to powers. Hence, multiplying the same number (a) a total of n times gives

$$\underbrace{a \times a \times a \times a \ldots}_{n \text{ times}} = a^n \tag{1.8}$$

In this notation, a is called the **base** and n the **exponent**. An exponent of 1 returns the base and is therefore the **identity operator for powers**.

$$a^1 = a \tag{1.9}$$

Multiplication of powers with the same base translates to addition of their exponents. Here's why:

$$a^n a^m = \underbrace{(a \times a \times a \ldots)}_{n \text{ times}} \times \underbrace{(a \times a \times a \ldots)}_{m \text{ times}} = \underbrace{a \times a \times a \ldots}_{n+m \text{ times}} = a^{n+m} \tag{1.10}$$

Powers with negative exponents have the following interpretation:

$$a^{-n} = \frac{1}{a^n} \tag{1.11}$$

To see why this is useful, consider the following example.

Example 1.9

$$\frac{a^3}{a^2} = \frac{a \times a \times a}{a \times a} = \frac{a}{a} \times \frac{a}{a} \times a = 1 \times 1 \times a = a \tag{1.12}$$

or, alternatively

$$\frac{a^3}{a^2} = a^3 a^{-2} = a^{3-2} = a \tag{1.13}$$

An exponent of 0 indicates that the base should not even appear once and yields 1.

$$a^0 = 1 \tag{1.14}$$

To see why, consider an exponent of zero as the result of two equal and opposite exponents (say 1 and -1). This gives

$$a^0 = a^{1-1} = a^1 \times a^{-1} = \frac{a}{a} = 1 \tag{1.15}$$

Therefore, Equation (1.14) does not hold for $a = 0$, because $0/0$ is not defined.

Roots of numbers are a by-product of powers. The nth root of a number b (written $\sqrt[n]{b}$) can be thought of as the number a that was raised to the power n to give b. The alternative notation for roots uses fractional powers. In short,

$$\text{If } a^n = b \text{ then } \sqrt[n]{b} = a \text{ and } b^{\frac{1}{n}} = a \tag{1.16}$$

Example 1.10

The shorthand for the square root of four is $\sqrt{4}$. This can also be written $\sqrt[2]{4}$ or $4^{\frac{1}{2}}$. Note that 4 can result from a power of 2 in one of two ways, either 2^2 or $(-2)^2$. Hence, mathematically, the square root of 4 is written ± 2, because it can be either 2 or -2. This applies to all even roots. In some cases, we may be able to use biological first principles to specify the result further. For example, negative numbers are not plausible values for variables like body length or population size.

Finally, a particularly useful operator, the **absolute value**, is used to turn any real number into a positive number with the same magnitude. Formally,

$$|a| = \begin{cases} a & \text{if } a \geq 0 \\ -a & \text{otherwise} \end{cases} \tag{1.17}$$

Example 1.11
If $x = -3$, then $|x| = |-3| = 3$. If x is an unknown negative number, then $|x| = -x$, which, perversely enough, is a positive number. So, given an expression like $-x$, don't be misled into thinking of it as negative.

There is an agreed order to how the different parts of a complicated algebraic expression should be calculated. The priority of operations is as follows: powers, divisions and multiplications, additions and subtractions. Brackets can be used to override this order as required. Also, the notations for absolute values ($|\ |$) and roots ($\sqrt{\ }$) operate as brackets, signifying the extent of their application within the expression.

Example 1.12
The expression $\frac{1}{2} \times 4^2 + 1$ has the value 9. Alternative priorities are possible and can be specified using brackets: $\frac{1}{2} \times (4^2 + 1) = 8.5$ or $(\frac{1}{2} \times 4)^2 + 1 = 5$.

The expression $\sqrt{1+2+1} + 5$ has the value 3. The scope of the square root can be modified by the length of the overhanging line: $\sqrt{1+2+1} + 5 = 7$. A clearer way of writing these two expressions uses brackets and fractional powers: $(1+2+1+5)^{\frac{1}{2}} = 3, (1+2+1)^{\frac{1}{2}} + 5 = 7$.

Finally, the expression $|-1-1|$ is not the same as $|-1| - 1$.

ⓡ   1.5: Encoding algebraic expressions
Algebraic operations are denoted as follows in R:

Addition and subtraction	`+   -`
Multiplication and division	`*   /`
Power	`^`
Square root	`sqrt()`
Absolute value	`abs()`

For example, the mathematical expression $\sqrt{5^2 - 2^2}/(2 \times 3)$ would be encoded as `sqrt(5^2-2^2)/(2*3)`. Note that, in R, as in any other computer language, only parentheses () can be used for specifying priority in algebraic expressions. Other types of bracketing, such as [] or {}, are reserved for different purposes.

1.6. Manipulating numbers

Here is a trick question: if the number 1.30719 is printed at the end of a calculation, do you interpret it as rational or irrational? Strictly speaking, the rational number 1.30719 can be generated as the ratio of two integers (200/153) but an irrational number beginning 1.30719 . . . and having infinite digits may also be curtailed to 1.30719. This is an example of **numerical approximation**, also known as **rounding off**. The number of **decimal places** in a numerical result is the number of digits after the decimal point. Hence, 1.30719 has five decimal places. We can round it off to four decimal places to get 1.3072. Similarly, rounding off to an integer gives 1. Rounding off the more ambiguous case of 1.5 can be treated in several different ways. We may, for example, decide to round off to the nearest integer up, to get the result 2. However, this introduces a consistent tendency to obtain numerically larger results. Alternatively, we may toss a coin every time we need to make a decision, but doing this mentally is likely not to be random. A more systematic approach (implemented by R) is to round off to the nearest even number.

Example 1.13

The rational numbers 1.5, 2.5, −1.5, −2.5, 0.5 can be rounded off to the integers 2, 2, −2, −2, 0.

Rounding off a number leads to some loss of information, known as **rounding error**. The accumulation of even small rounding errors can lead to inaccurate results, a process that can afflict both manual and computerised calculations (Chartier, 2005).

The information content in a particular number can be characterised by its **significant figures**. All nonzero digits in a number are significant. In contrast, 'padding' a number with zeroes doesn't add to the information content of a reported result. If an instrument gives you measurements to one decimal point, reporting a measurement of 1.3 as 1.3000 adds no information. Similarly, multiplying the reported unit of measurement doesn't add any new information to it (1.3 cm is the same as 0.013 m). So, **leading zeroes** are considered nonsignificant and so are **trailing zeroes** before the decimal point (1.3 m = 130 cm). If a number is reported with trailing zeroes after the decimal point, this is taken to indicate the accuracy of the measurement and these zeroes are considered significant.

Example 1.14

The numbers 123, 1.23, 0.123 and 0.0123 all have three significant figures. A reported value of 123.0 has four significant figures.

To simplify the reported results when the number of decimal digits is much greater than the significant figures we can use a combination of significant figures and **orders of magnitude**, expressed as powers of ten.

Example 1.15

The numbers 123, 12.3, 0.123 and 0.0123 can, respectively, be written as $1.23 \times 10^2, 1.23 \times 10^1, 1.23 \times 10^{-1}$ and 1.23×10^{-2}. Calculations with orders of magnitude use the properties of powers. For example, it is rather tricky to see without a calculator that $(0.003 \times 300)/0.0000009$

is equal to a million. But the calculation is greatly simplified using orders of magnitude:

$$\frac{0.003 \times 300}{0.0000009} = \frac{3/1000 + 3 \times 100}{9/10,000,000} = \frac{(3 \times 10^{-3}) \times (3 \times 10^{2})}{9 \times 10^{-7}}$$

$$= \left(\frac{3 \times 3}{9}\right) \times \left(\frac{10^{-3}10^{2}}{10^{-7}}\right) = 1 \times 10^{-3+2+7} = 10^{6} \qquad (1.18)$$

ℝ 1.6: Orders of magnitude and rounding off

R decides automatically when it needs to report numbers using orders of magnitude. It does so by using the 'e notation'. Typing `0.000123` gives the output `1.23e-5`, which stands for 1.23×10^{-5}.

Several commands can be used to deal with decimal digits in subtly different ways. The command `round(x, digits = n)` will round the number x to n decimal digits. The command `floor(x)` rounds the number x to the nearest integer down (effectively chopping off all the decimal digits) and the command `ceiling(x)` rounds the number x to the nearest integer up. The command `signif(x, digits = n)` gives x to n significant digits. For example, `round(1.2345,3)` gives `1.234` but `signif(1.2345,3)` gives `1.23`. Finally, the command `trunc(x)` rounds both positive and negative numbers towards zero. The difference between `trunc(x)` and `floor(x)` can be illustrated by the following examples:

```
> floor(1.3)
[1] 1
> trunc(1.3)
[1] 1
> floor(-1.3)
[1] -2
> trunc(-1.3)
[1] -1
```

1.7. Manipulating units

Many pencil-and-paper calculations are accompanied by measurement units. Most of us think of units as a nuisance along the way to a numerical result, and it's very tempting to drop them and get on with the numerical calculation. This can backfire in two ways: first, in our haste to number-crunch, we may forget to 'homogenise' the units of measurement to the same unit set. A calculation that simultaneously involves centimetres, metres and inches will yield nonsense. Second, units are a means of verifying the physical relevance of the answer. If a calculation about daily energetic costs yields a result in joules instead of joules per day, it is just possible that the numerical result is also wrong. Units can be treated as symbols and manipulated algebraically alongside the numerical part of the calculation. Example 1.19 will provide an illustration of how this works.

1.8. Manipulating expressions

The presence of symbols in a mathematical expression may make it impossible to reduce it to a single numerical result. Similarly, if we are after a general answer to a problem, we may prefer to express it symbolically. The objective is then to simplify these expressions as much as possible, by moving things around or collecting similar things together. Sometimes, an expression may first need to be expanded before it can be simplified. To carry out these manipulations it is useful to brush up on the basic rules of algebra. The first type of rule deals with the order of terms within a sum or product. The **commutative** rule states that, when adding or multiplying any two terms a and b, the order of the terms is reversible

$$
\begin{aligned}
a + b &= b + a \quad &\text{(commutative addition)}\\
ab &= ba \quad &\text{(commutative multiplication)}
\end{aligned}
\tag{1.19}
$$

The **associative** rule says that, when adding or multiplying any three terms a, b and c, the priority with which the operation is performed is not important

$$
\begin{aligned}
(a + b) + c &= a + (b + c) = (a + c) + b \quad &\text{(associative addition)}\\
(ab)c &= a(bc) = b(ac) \quad &\text{(associative multiplication)}
\end{aligned}
\tag{1.20}
$$

The second set of rules deals with collecting similar things together in sums and products. The sum of identical symbols can be expressed as a multiple of the same symbol. This readily extends to sums of identical expressions or powers:

Example 1.16

$x + x + x = 3x$	Identical symbols
$3x + 8x = 11x$	Multiples of identical symbols
$3x^2 + 8x^2 = 11x^2$	Multiples of identical powers of symbols
$3(a + b) + 8(a + b) = 11(a + b)$	Multiples of identical expressions
$3x^2 + 8x^2 - 3x - 8x = 11x^2 - 11x$	Sum involving different exponents cannot be simplified

Simplifying products is achieved by collecting powers of the same base.

Example 1.17

$xxx = x^3$	Identical base
$x^2 x^3 = x^5$	Powers of the same base
$x^{\frac{1}{2}} x^3 = x^{\frac{1}{2}+3} = x^{\frac{1+6}{2}} = x^{\frac{7}{2}}$	Powers of the same base
$(a + b)^2 (a + b)^3 = (a + b)^5$	Powers of the same base
$a^2 b^3$	Product of powers with different bases cannot be simplified

Sometimes, sums within brackets need to be multiplied out if this eventually leads to a simplification of the overall expression. This is done by the **distributive** rule which expresses the product of sums as the sum of pairwise products,

$$
a(b + c) = ab + ac
\tag{1.21}
$$

Example 1.18

The expression $3x(x + 1)(x - 2) + 4x^3 + x(2x + 1)$ can be simplified as follows:

$$
\begin{aligned}
3x(x^2 - 2x + x - 2) + 4x^3 + (2x^2 + x) &= \text{(Distributive rule)} \\
3x^3 - 6x^2 + 3x^2 - 6x + 4x^3 + 2x^2 + x &= \text{(Distributive rule)} \\
3x^3 + 4x^3 - 6x^2 + 3x^2 + 2x^2 - 6x + x &= \text{(Commutative rule for sums)} \\
7x^3 - x^2 - 5x & \qquad \text{(Collecting similar terms)}
\end{aligned}
$$

We can employ the same set of rules for manipulating numbers (Section 1.6), units (Section 1.7) and expressions (this section) to find the answer in the following example:

Example 1.19: Energy acquisition in voles

Vole, genus Microtus

A species of vole spends 70% of its time foraging and the remainder of its time resting. While resting, its metabolic rate (in joules per day) is $E_r = 30000 \, \text{J} \cdot \text{d}^{-1}$. This increases by 20% during foraging. Females defend territories of about 100 m². Each cm² in the territory produces 0.0005 g of grass that could be entirely consumed by the vole if it spent 100% of its time foraging. The vole gains $\varepsilon = 40.6$ cal from a single gram of grass. We want to calculate the daily rate of energetic gain (or loss) of the animal.

This is a complicated calculation, made particularly messy by the incongruence between the units employed to report physical quantities. For example, area is reported both in m² and cm². The first task is to sketch a symbolic solution and simplify it by algebraic manipulation. Daily change in energy is the difference between foraging intake (E_f) and metabolic expenditure (E_c),

$$\Delta E = E_f - E_c \tag{1.22}$$

First, we deal with metabolic costs: If p is the proportion of its time that a vole spends foraging, then its complement, $1 - p$, is the proportion of time that it spends resting. If E_a and E_r are the active and resting metabolic rates, energetic cost can be written as

$$E_c = pE_a + (1 - p)E_r \tag{1.23}$$

This can be simplified somewhat by using the fact that $E_a = 1.2E_r$ (because active metabolic rate exceeds resting metabolic rate by 20%).

$$E_c = p(1.2E_r) + (1 - p)E_r = (1.2p + 1 - p)E_r = (1 + 0.2p)E_r \tag{1.24}$$

Second, we deal with foraging intake: the territory of the vole has area A and each unit of area produces m units of food daily. The vole exploits its territory efficiently, so it consumes a mass mA for every complete day spent foraging. The effective energetic value of this to the animal is

$$E_f = \varepsilon m p A \tag{1.25}$$

Placing Equations (1.24) and (1.25) back into Equation (1.22) we get the symbolic answer

$$\Delta E = \varepsilon m p A - (1 + 0.2p)E_r \tag{1.26}$$

We must now put some numbers into this in the same units (I will use g for mass, d for time, cm² for area and J for energy). Some conversions are required: 1 cal = 4.184 J, $1\,\text{m}^2 = (100\,\text{cm})^2 = 10^4\,\text{cm}^2$. We can use these to obtain a complete list of required values in a consistent system of units (Table 1.2).

Table 1.2

Quantity	Unconverted	Converted
p	70%	0.7
E_r	$3000\,\text{J·d}^{-1}$	$3 \times 10^4\,\text{J·d}^{-1}$
A	$100\,\text{m}^2$	$10^6\,\text{cm}^2$
m	$0.0005\,\text{g·d}^{-1}\text{·cm}^{-2}$	$5 \times 10^{-4}\,\text{g·d}^{-1}.\,\text{cm}^{-2}$
ε	$40.6\,\text{cal·g}^{-1}$	$1.7 \times 10^2\,\text{J·g}^{-1}$

We are now ready to do the calculation in Equation (1.26):

$$\Delta E = (1.7 \times 10^2\,\text{J} \cdot \text{g}^{-1})(5 \times 10^{-4}\,\text{g} \cdot \text{d}^{-1} \cdot \text{cm}^{-2})(0.7)(10^6\,\text{cm}^2) - 1.14(3 \times 10^4\,\text{J} \cdot \text{d}^{-1}) \quad (1.27)$$

For best book-keeping, we can separate the numbers, orders of magnitude and units and simplify as follows:

$$\begin{aligned} \Delta E &= (1.7 \times 5 \times 0.7)(10^2 10^{-4} 10^6)(\text{J} \cdot \text{g}^{-1} \cdot \text{g} \cdot \text{d}^{-1} \cdot \text{cm}^{-2} \cdot \text{cm}^2) - (1.14 \times 3)10^4(\text{J} \cdot \text{d}^{-1}) \\ &= 5.95 \times 10^4\,\text{J} \cdot \text{d}^{-1} - 3.42 \times 10^4\,\text{J} \cdot \text{d}^{-1} \\ &= 2.53 \times 10^4\,\text{J} \cdot \text{d}^{-1} \end{aligned} \quad (1.28)$$

The subtraction leading to the final result was only possible because the two quantities had the same order of magnitude (10^4) and units (J·d^{-1}). By way of verification, note that the final result is an energetic rate correctly expressed in J·d^{-1}.

 1.7: Symbolic manipulations

R is primarily intended for data manipulation and consequently has only limited capabilities for symbolic mathematics (but see Chapter 4). Proprietary packages, such as MAPLE and MATHEMATICA, specialise in symbolic calculations and are well worth investigating if you are planning to do complicated algebra.

1.9. Polynomials

Polynomials are a class of algebraic expressions that occur naturally in many ecological problems (Example 1.21). Their generality is matched by their ease of use. Polynomials are the best-behaved mathematical expressions, partly due to their repetitive structure. By way of introduction, consider how you would summarise the following set of four expressions in words

$$\{3x, -5x, 0.33421x, 1298x\} \quad (1.29)$$

You may have noticed that each expression is the product of a real number and the same symbol. Mathematically, these algebraic expressions have the form $a_i x$, where i and a_i have the values shown in Table 1.3.

Here, the symbol a is used to represent a constant number. The subscript i locates a_i in the set and declares that it is not necessarily the same as the others. So, $a_i x$ conveys the *pattern* of the set rather than its detail, and it works just as well with four thousand terms as with four.

Table 1.3

$i =$	$a_i =$
1	3
2	−5
3	0.33421
4	1298

This notation can now be applied to something more complicated. A **polynomial** is an algebraic expression of the form

$$a_n x^n + a_{n-1} x^{n-1} + \cdots + a_1 x + a_0 \tag{1.30}$$

where the as are indexed constants and n is a non-negative integer called the **order of the polynomial**. Equation (1.30) contains some clever tricks of notation. It is a sum of $n + 1$ terms. To avoid writing them all out, we interrupt listing them once we have conveyed the pattern, write three dots to indicate that the sequence continues, and then pick it up again to show where the sequence ends. Can you see that all the terms in Equation (1.30) are of exactly the same form? Remember, that $x^1 = x$ and $x^0 = 1$ (Equations (1.9) and (1.14)). So, Equation (1.30) can be written

$$a_n x^n + a_{n-1} x^{n-1} + \cdots + a_1 x^1 + a_0 x^0 \tag{1.31}$$

where each term of the sum is the product between a constant (known as a **coefficient**) and the variable x raised to a non-negative, integer power.

In each term of the polynomial, the subscript of the coefficient and the superscript of the variable are the same but, while the subscript is a book-keeping convention to distinguish between coefficients, the superscript is an exponent.

Example 1.20

Let's have a look at some specimen polynomials. First, one that looks like Equation (1.30)

$$2x^5 + 4x^4 + 6x^3 + 8x^2 + 3x + 5 \tag{1.32}$$

The highest power is 5, so this is a polynomial of order 5. The coefficients don't have to be positive or integers. The following is also a polynomial of order 5 in which the coefficients of the fourth and second order terms are −4 and −1.

$$2x^5 - 4x^4 + 2.6x^3 - x^2 + 0.001x + 5 \tag{1.33}$$

The coefficients may also be zero, so that an expression like $4x^5 - 6$ is also a fifth order polynomial.

Example 1.21: The law of mass action in epidemiology

Bacterial, airborne pathogen

A theoretical abstraction that has contributed greatly to the study of epidemics is the law of **mass action**. It asserts (quite unrealistically) that all individuals in a population are identical and interact randomly. The second of these assumptions is known as **perfect**

mixing because it implies that spatial proximity does not influence transmission between any two individuals in a population. To investigate the implications of these assumptions, consider a small population of ten animals, four of which (black discs in Figure 1.5(a)) carry an infectious disease. Because the population is so small, every individual comes into contact with every other individual with approximately the same frequency (say, three times a day). As a result, each infected individual is equally likely to pass the infection to each of the susceptible individuals. The daily number of transmission opportunities between a particular carrier and a susceptible individual can readily be tabulated (Figure 1.5(b)).

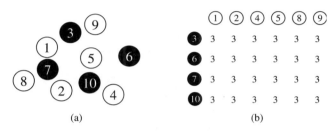

(a)   (b)

Figure 1.5: (a) A population of ten individuals in which four (shown in black) carry the disease; (b) the number of opportunities of transmission in one day between any two individuals.

The total number of interactions in Figure 1.5(b) is $3 \times 4 \times 6$. If only one out of a thousand interactions leads to transmission, then the number of expected transmissions is $\frac{1}{1000} \times 3 \times 4 \times 6 = 0.072$.

More generally, if K is total population size, N is the number of carriers, b is the rate of interactions between two individuals and c is the proportion of interactions that lead to transmission, then the rate of transmission (γ) is

$$\gamma = cbN(K - N) \tag{1.34}$$

You can read more on the law of mass action in Heesterbeek (2005). For now, note that if we multiply out the right-hand side of Equation (1.34) we get a second order polynomial in the variable N, the number of infected animals

$$-cbN^2 + cbKN \tag{1.35}$$

You can check that this is a typical polynomial of the form $a_2N^2 + a_1N + a_0$ by setting $a_2 = -cb, a_1 = cbK$ and $a_0 = 0$.

1.8: Set indexing and element extraction

We saw in this section how indexing the members of a set makes it easier to identify them individually. Consider the set of numbers A, defined as

```
A<-c(2,4,1,6,0,5)
```

The documentation for R sometimes refers to such sets as **vectors** (more on vectors in Chapter 6). The nth element of A, in R syntax, is A[n]. In the example above, if you type A[1], R will return 2, typing A[3] will give you 1. In addition to this, there are several other ways of using indexing to get specific subsets of a set, as shown in Table 1.4.

Table 1.4

If you want...	...the R syntax is	Example input	Result
The nth element of A	A[n]	A[5]	0
All but the nth element	A[-n]	A[-5]	2 4 1 6 5
The first n elements	A[1:n]	A[1:5]	2 4 1 6 0
The last n elements	A[-(1:n)]	A[-(1:3)]	6 0 5
Specific elements from A	A[c()]	A[c(1,3)]	2 1
Elements above a value a	A[A>a]	A[A>2]	4 6 5
Elements between a and b	A[A>a & A<b]	A[A>2 & A<6]	4 5
Elements between a and b, inclusive	A[A> = a & A< = b]	A[A> = 2 & A< = 6]	2 4 6 5

1.10. Equations

An equation consists of two parts, separated by an equality sign. It states that these two expressions have the same value. Verifying this may be just a matter of numerical calculation. Such equations are called **trivial**.

Example 1.22

Here's a trivial equation

$$1 + 1 = 2 \tag{1.36}$$

The following is also a trivial equation, because it contains no symbols,

$$\frac{\sqrt{\frac{1295.78}{1987.43}} \times (9.32^2)^2}{9569.86} = 0.637 \tag{1.37}$$

Nontrivial equations come in two flavours: an **identity** is valid for all values of its variables (a refresher of basic algebraic identities can be found in Section A1.3 in the appendix). A **conditional equation** is valid only when the variables take specific values (called the **solutions** or **roots** of the equation).

Example 1.23

By trying out different values for x you can convince yourself that the following is an identity

$$(x + 4)^2 = x^2 + 8x + 16 \tag{1.38}$$

whereas

$$x^2 + 8x + 16 = 36 \tag{1.39}$$

is a conditional equation because it is only valid when $x = 2$ or $x = -10$.

Conditional equations may be solved; that is, we can try to find the value(s) of x that satisfy the equation. In a simple equation, this can be done by trial and error but, in general, it is best achieved by systematically isolating the variable on one side of the equality sign and all the known elements (numbers or symbols with known or assumed known values) on the other. Note, however, that some equations cannot readily be solved. For example, there is no real number that can satisfy the equation $1 + x^2 = 0$.

At this point, I owe you an apology for a particular misuse of terminology: cross-references to numbered mathematical expressions are incorrectly labelled as 'equations', throughout the book. Hence, references like 'Equation (1.30)' are commonplace, even though that particular expression may not strictly be an equation as defined in this section. Unfortunately, for the sake of brevity and uniformity, almost all books and journals commit this intentional error.

1.11.   First order polynomial equations

A polynomial can be used to construct an equation, which can always be written as

$$a_n x^n + a_{n-1} x^{n-1} + \cdots + a_1 x + a_0 = 0 \tag{1.40}$$

But polynomials don't always appear in this form and you may need to carry out some manipulation before they can be identified.

Example 1.24

The following expression is a third order polynomial equation. Can you see why?

$$\frac{3}{2} + x + \frac{1}{x} = x(3x + 2) - 1 \tag{1.41}$$

To get a first order polynomial equation from the general definition in Equation (1.40) we set the higher-order coefficients to zero (mathematically, $a_i = 0, \forall i > 1$)

$$a_1 x + a_0 = 0 \tag{1.42}$$

The equation is solved by isolating x. To maintain the equality sign, exactly the same things must be done to both sides of it. Subtracting a_0 from both sides

$$a_1 x = -a_0 \tag{1.43}$$

and then dividing both sides by a_1

$$x = -\frac{a_0}{a_1} \tag{1.44}$$

gives the general solution of a first order polynomial equation. It is general because it gives the root of any equation of the form of Equation (1.42) for any values of a_0 and a_1.

Example 1.25: Population size and composition

Sexual dimorphism in redstarts (Phoenicurus phoenicurus)

In a population of tree-nesting birds there are two males for every three females. During the breeding season nests are highly cryptic, but it is known that all females breed and each produces five eggs, of which only three survive to become fledglings. Although an accurate, independent estimate of population size can be obtained after the breeding season, by that time it's no longer possible to distinguish juveniles from adults. Nevertheless, we want to know the number of males in a population of 70 animals counted just after the breeding season.

Let's denote males by M, females by F, juveniles by J and population size by P. If we assume that no deaths of adults occur during the breeding season, then all four of these quantities refer to a point in time just after the breeding season so that

$$M + F + J = P \tag{1.45}$$

The number of unknown quantities can be reduced by using the information on sex ratio and chick-rearing success. Specifically, we can use the following two facts

$$J = 3F \text{ and } \frac{M}{F} = \frac{2}{3} \tag{1.46}$$

to reduce the four symbols in Equation (1.45) to two. There are three ways to do this:

$$M + 1.5M + 3(1.5M) = P \tag{1.47}$$

$$\frac{2}{3}F + F + 3F = P \tag{1.48}$$

$$\frac{2}{3}\left(\frac{1}{3}J\right) + \frac{1}{3}J + J = P \tag{1.49}$$

Since we are interested in the number of males, Equation (1.47) is the more appropriate. We can simplify it further and solve it for M

$$M(1 + 1.5 + 4.5) = P$$
$$M = \frac{P}{7} = 10 \tag{1.50}$$

1.12.   Proportionality and scaling: a special kind of first order polynomial equation

A scale drawing is one that maintains the proportions but not the size of its subject. Decreasing or increasing a picture while maintaining its relative proportions is called **scaling** (Figure 1.6). You will probably have encountered scales on maps, where an inscription of, say, 1:100 000 indicates that a single unit of length on the map corresponds to 10^5 units of length on the ground. Formally, a scaling operation is given by a first order polynomial equation with $a_0 = 0$. If we use y to denote the size of the scaled object and x for the size of the real object, then

$$y = ax \tag{1.51}$$

where a is called the **proportionality constant**. If the proportionality constant is greater than 1, then the scaling operation is a **magnification**, if it is smaller than 1 it is called a **contraction**. If we don't know the value of a, but we know that y and x are proportional, we state this using the notation $y \propto x$.

Any two dimensions (x_1, x_2) of an object (e.g. height and width in Figure 1.6) are related to the scaled dimensions by the same constant $y_1 = ax_1$ and $y_2 = ax_2$. Scaling maintains the

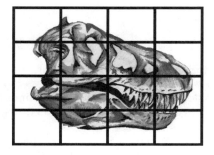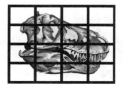

Figure 1.6: The second drawing is smaller than the first but it maintains the same ratio of height over width.

relationship between any two of the object's dimensions so that

$$\frac{y_1}{y_2} = \frac{x_1}{x_2} \tag{1.52}$$

Equation (1.52) is an alternative way of stating proportionality. A number of scaling problems can be solved using either Equation (1.51) or Equation (1.52).

Example 1.26: Simple mark-recapture

Ringing is a traditional mark-recapture field technique for bird species

Mark-recapture methods are used to estimate the size of animal populations and vital rates, such as survival. Simple mark-recapture assumes that the population has a constant size. This implies that there is no migration into or out of the population (in which case the population is said to be **closed**), and that the study happens quickly enough to exclude the possibility of new births and deaths. Suppose we make two visits to the population. In the first, a number of animals n are captured and marked. In the second, we capture M animals, and find that m of these had already been marked (i.e. they are **recaptures**). If animals mix freely and the event of capture does not affect the chance of recapture, then we can use proportionality to estimate the (unknown) size of the population (N): the proportion of marked animals in the population should be the same as the proportion of recaptures in the second sample

$$\frac{n}{N} = \frac{m}{M} \tag{1.53}$$

We can now solve this for N

$$N = \frac{M}{m}n \tag{1.54}$$

The proportionality constant linking N and n is M/m. This constant cannot be less than 1, because the number of recaptures cannot exceed the total number of marked animals. If the proportionality constant is close to 1, this implies that almost all of the animals in the second sample were found to be marked or, equivalently, that almost the entire population was caught in the first visit.

Example 1.27: Converting density to population size

Regular quadrats, used to sample plant diversity

It is usually difficult to make a complete inspection (a **census**) of a field site to count the number of occurrences of a species. In plant ecology, this is approached by randomly placing quadrats of known size (a) at the site and counting the number of occurrences of the species within each. To estimate the population size N in a field of area A, we may count the number of occurrences n in a total of m quadrats, each of size a. Assuming that the quadrats are representative of the field, then the density of plants in the quadrat ($n/(ma)$) should be the same as the overall density of plants in the field (N/A), giving the estimate

$$N = \frac{A}{ma}n \tag{1.55}$$

1.13. Second and higher order polynomial equations

As a starting point for this discussion consider the following example:

Example 1.28

$$(3x - 2)(x - 4) = 0 \tag{1.56}$$

Setting $x = 4$ gives $(12 - 2)(4 - 4) = 0$, hence 4 is a solution (or root) of this equation. There is also a second root: $2/3$.

Equation (1.56) has the general form

$$(s_1 x - r_1)(s_2 x - r_2) = 0 \tag{1.57}$$

where s_1, s_2, r_1 and r_2 are constants. Look at the two parts of the product on the left-hand side. They are first order polynomials. A product of zero can only arise if either or both of the following equations are true:

$$s_1 x - r_1 = 0 \text{ or } s_2 x - r_2 = 0 \tag{1.58}$$

The two roots of Equation (1.58), and hence the solutions of Equation (1.57), are

$$x = \frac{r_1}{s_1} \text{ and } x = \frac{r_2}{s_2} \tag{1.59}$$

However, if the left-hand side of Equation (1.57) is multiplied out prior to solving, it is

$$\begin{aligned}(s_1 x - r_1)(s_2 x - r_2) &= s_1 s_2 x^2 - s_1 r_2 x - s_2 r_1 x + r_1 r_2 \\ &= s_1 s_2 x^2 + (-s_1 r_2 - s_2 r_1)x + r_1 r_2\end{aligned} \tag{1.60}$$

The constant components of the result can be renamed as follows:

$$a_2 = s_1 s_2, a_1 = -s_1 r_2 - s_2 r_1, a_0 = r_1 r_2 \tag{1.61}$$

Using these definitions we can rewrite Equation (1.57) as

$$a_2 x^2 + a_1 x + a_0 = 0 \tag{1.62}$$

This is a second order polynomial equation, which also goes by the name **quadratic equation**. Although it is the same as Equation (1.57), it is impossible to tell at a glance what its roots are. Unfortunately, second order polynomial equations appear in the form of Equation (1.62) more often than in the form of Equation (1.57) and we need a reliable way of solving them. The roots (x_1 and x_2) of a quadratic equation are given by the **quadratic formula**

$$x_1, x_2 = \frac{-a_1 \pm \sqrt{a_1^2 - 4a_2a_0}}{2a_2} \tag{1.63}$$

The plus-or-minus sign ($\pm$) means that there are a maximum of two roots obtained by respectively adding and subtracting the entire expression $\sqrt{a_1^2 - 4a_2a_0}$ in the numerator. The expression within the square root is called the **discriminant**. It has a special name because it gives information about how many roots the quadratic equation has. If the discriminant is positive, its square root can be added and subtracted from a_1 to give two roots. If the discriminant is zero, it makes no difference in the numerator and the equation has a single root. Finally, if the discriminant is negative, the two roots are complex numbers (see Section 1.17).

Example 1.29

The roots of the quadratic equation $2x^2 - 11x - 21 = 0$ are found as follows

$$\frac{-(-11) \pm \sqrt{11^2 - 4 \cdot 2 \cdot (-21)}}{2 \cdot 2} = \frac{11 \pm \sqrt{289}}{4} = \begin{cases} 7 \\ -3/2 \end{cases} \tag{1.64}$$

Example 1.30: Estimating the number of infected animals from the rate of infection

In a perfectly mixed population of size $K = 1080$ animals, a total of $\gamma = 3$ new cases of an infectious disease appeared today. The daily rate of encounters between animals is $b = 0.03$ and a proportion $c = 0.001$ of encounters lead to transmission. The disease is not lethal but individuals do not recover within the timescale of a single disease outbreak. We would like to estimate the total number of infected animals (N) in the population. To do this, we can rearrange the model in Equation (1.34) as follows:

$$(cb)N^2 - (cbK)N + \gamma = 0 \tag{1.65}$$

Placing some of the constants in parentheses highlights the fact that this is a second order polynomial equation. It can have a maximum of two solutions (say, N_1, N_2) that can be found by applying the quadratic formula in Equation (1.63)

$$N_1, N_2 = \frac{cbK \pm \sqrt{(-cbK)^2 - 4cb\gamma}}{2cb} \tag{1.66}$$

Replacing the symbols with numbers and doing the arithmetic,

$$N_1, N_2 = \frac{0.001 \times 0.03 \times 1080 \pm \sqrt{(0.001 \times 0.03 \times 1080)^2 - 4 \times 0.001 \times 0.03 \times 3}}{2 \times 0.001 \times 0.03}$$

$$= \frac{3.24 \times 10^{-2} \pm 2.63 \times 10^{-2}}{0.006 \times 10^{-2}} = \frac{3.24 \pm 2.63}{0.006} \cong \begin{cases} 978 \\ 102 \end{cases} \tag{1.67}$$

The answers have been rounded off to the nearest integer animal and so the symbol $\cong$ expresses approximate equality. The appearance of two answers that are so different (relative to the total population size of 1080) may seem biologically unrealistic. However, both values are feasible because the (relatively low) rate of three new cases per day can occur either when there are only a few infected individuals (giving $N_1 = 102$) or a few uninfected individuals (giving $N_2 = 978$).

Higher order polynomial equations are easy to write, but not so easy to solve. Analytic, general solutions such as those presented in Equations (1.44) and (1.63) exist only for equations up to fourth order. However, it may be possible to simplify the equation into lower-order components.

Example 1.31
Strictly speaking, $a_6 x^6 + a_5 x^5 + a_4 x^4 = 0$ is a sixth order equation but, since the terms below fourth order are missing, it can be written as the product of two lower-order components

$$a_6 x^6 + a_5 x^5 + a_4 x^4 = x^4 (a_6 x^2 + a_5 x + a_4) \tag{1.68}$$

When such simplifications are not possible, the easiest way to find the roots is **numerical approximation**, a group of computationally efficient methods for finding numerical answers by trial-and-error. In doing this, it is useful to know that the maximum number of solutions of a polynomial equation is equal to the order of the equation. R can handle equations up to order 49 in this way.

1.9: Numerical solution of higher order polynomial equations
The command for numerically solving polynomial equations is `polyroot()` which requires a list of the polynomial's coefficients in increasing order. To find the roots of the polynomial equation $2x^3 + 19x^2 - 150x - 100 = 0$ type

```
> polyroot(c(-100,-150,19,2))
[1] -0.6210104+0i  5.5716520+0i -14.4506415-0i
```

These are the three solutions of the third order equation (namely, $-0.621, 5.572$ and -14.451). The $\pm 0i$ parts are of no relevance to the answer because they are effectively zero. Their significance should become clearer in Section 1. 17.

1.14. Systems of polynomial equations

The equations we have looked at so far contain a single variable but this does not have to be so.

Example 1.32
The equation $3x - 4y + 8 = 0$ contains two variables, x and y.

To get numerical results for multiple variables we need just as many *unique* equations. Complementing the equation in Example 1.32 with the equation $1.5x - 2y + 4 = 0$ does not lead to two unique equations because the second equation is simply a scaling of the first. When more than one variable are entangled in more than one equation, we call these equations **simultaneous** or **coupled**. A list of coupled equations forms a **system of equations**.

Example 1.33

A system of two coupled equations involving the variables x and y is

$$3x - 4y + 8 = 0$$
$$x + 2y - 4 = 0 \tag{1.69}$$

Solving systems of polynomial equations relies on the same ideas used for single, self-contained equations, although more work is required. One approach, called **solution by substitution** is to pick any equation from the system and solve it with respect to one of the variables it contains. The solution is phrased in terms of the remaining variables. The next step is to choose another equation from the system and replace the first variable with the expression that was obtained in the previous step. This procedure is repeated until an expression is obtained containing only one variable. Then, if all has gone well, a numerical answer for that variable can be obtained. Finally, these steps need to be retraced backwards, replacing variables with their values in the process.

Example 1.34

The system in Example 1.33 can be solved as follows:

Step 1: Take the first equation and solve it for one of the two variables, say x, expressing it in terms of y.

$$x = \frac{4y - 8}{3} \tag{1.70}$$

Step 2: Substitute Equation (1.70) into the second equation of the system

$$\frac{4y - 8}{3} + 2y - 4 = 0 \tag{1.71}$$

x has now vanished and we are left with one equation containing only y.

Step 3: Solve the second equation for the remaining variable

$$4y - 8 + 6y - 12 = 0 \tag{1.72}$$
$$10y - 20 = 0$$
$$y = 2$$

Step 4: Retrace the steps replacing variables by values. Substituting y by 2 in Equation (1.70) gives $x = 0$.

The complexity of a system of equations is potentially unlimited. We could, for example, have a lot more variables (and equations), or the equations themselves could be more complicated. In Chapter 6, we will discuss more efficient ways of solving large systems. For now, here is a biological example:

Example 1.35: Deriving population structure from data on population size

The size of a particular bird population has been accurately estimated to be 18 050 just before the breeding season and 20 130 at some time after it. We know that 10% of males and 5% of females died in the interval between the two counts. It is also known that each female produced 0.3 juveniles that survived (as adults) until the post-breeding count. We would like to calculate the number of males and females at the beginning of the breeding season. Using intuitive notation we have the system

$$M + F = 18050$$
$$0.9M + 0.95F + 0.3F = 20130 \qquad (1.73)$$

This can be solved to give $M = 6950, F = 11100$

1.15. Inequalities

Often, the available biological information can only be specified up to a range of values rather than a countable set of solutions, leading to **inequalities**.

Example 1.36

Table 1.5 lists some verbal statements and the corresponding inequalities.

Table 1.5

Statement	Corresponding inequalities
x is greater than 3	$x > 3$
x is at least 3	$x \geqslant 3$
x is no larger than 3 and no smaller than 1	$1 < x < 3$
x is outside the range 1 to 3	$x < 1, x > 3$

The fourth statement requires two separate inequalities. The comma between them implies that either $x < 1$ or $x > 3$ will be true for any given value taken by x. An expression like $1 > x > 3$ is not meaningful because both inequalities cannot be true for any given value of x.

Example 1.37: Minimum energetic requirements in voles

The voles from Example 1.19 feed on two types of plant during part of the year, each with a different energetic value, ε_1 and ε_2 (in some appropriate unit, say J·g^{-1}). The amounts (say, x and y) consumed by any given vole on any given day may vary but, in total, each vole requires EJ·day^{-1} to survive. Voles that satisfy their energetic requirements are described by the following inequality

$$x\varepsilon_1 + y\varepsilon_2 \geq E \qquad (1.74)$$

An inequality such as Equation (1.74) may be solved with respect to a single variable. As with equations, this is achieved by using the basic rules of algebra (see Section 1.8) to rearrange the terms on either side of the inequality and applying the same algebraic operations to both sides to move things from one side to the other. For inequalities, the second of these two actions requires particular care.

Example 1.38

Table 1.6 considers different algebraic operations applied to both sides of the inequality $4 < 5$.

Table 1.6

Operation	Outcome
$+1$	$5 < 6$
-1	$3 < 4$
$\times 1$	$4 < 5$
$\times(-1)$	$-4 > -5$

Notice that the first three operations maintain the direction of the inequality but multiplication (or division) by a negative number reverses it.

In Section 1.5 we saw how absolute values and even powers can turn negative numbers into positive ones. Therefore, inequalities involving their variable in even powers or absolute values contain some ambiguity that needs to be expressed when solving for the variable.

Example 1.39

Table 7.1 provides the interpretation of four inequalities involving powers and absolute values.

Table 1.7

If we are given	......then this implies
$x^2 < 9$	$-3 < x < 3$
$x^2 > 9$	$x < -3$ or $x > 3$
$|x| < 3$	$-3 < x < 3$
$|x| > 3$	$x < -3$ or $x > 3$

Example 1.40

The inequality $x^2 - 4x - 1 > (x - 3)x$ can be solved for x as follows:

❶ Expand the right-hand side $x^2 - 4x - 1 > x^2 - 3x$
❷ Subtract x^2 from both sides $-4x - 1 > -3x$
❸ Add $3x$ to both sides $-x - 1 > 0$
❹ Add 1 to both sides $-x > 1$
❺ Multiply both sides by -1 $x < -1$

(remembering that this reverses the inequality).

ℝ 1.10: Comparisons and the logical TRUE and FALSE

It is often useful to be able to test for equality or inequality between two quantities. For example, consider the following assignments (the semicolon allows us to put multiple statements in one line):

```
a<-2; b<- -4; c<-1; d<-8
p<-(a+b)/(c-d)
q<-(a-b)/(c+d)
```

It is not immediately clear if p is equal to, smaller than or greater than q. We can query R to tell us using any one of six logical operators $= =, \, ! =, \, >, \, > =, \, <, \, < =$. The response can either be TRUE or FALSE with numerical values 1 and 0, respectively. Table 1.8 shows some examples of queries and their responses for the above example.

Table 1.8

The query...	...translates to...	...and R responds with...
Is p the same as q?	p==q	**FALSE**
Is p different from q?	p!=q	**TRUE**
Is p greater than q?	p>q	**FALSE**
Is p smaller than or equal to q?	p<=q	**TRUE**

1.16.   Coordinate systems

Comparison between two real numbers m and n can have one of three outcomes ($m > n, m = n, m < n$). Such comparisons enable us to order numbers.

Example 1.41
The small set of integers $\mathbf{S} = \{5, 3, 9, 4, -4, 1, 7, 10, -2\}$ can be rewritten in ordered form as $\mathbf{S} = \{-4, -2, 1, 3, 4, 5, 7, 9, 10\}$.

The entire ordered set of real numbers can't be represented in the form of Example 1.41 because there is an infinity of real numbers before, after and between any two real numbers. We therefore need a different way to visualise such noncountable sets. If there are infinite real numbers between any two real numbers, then we need something that is so small that an infinity of it would fit in the gap. We envisage such a dimensionless object and call it a **point**. We arrange an infinity of points along the **line of real numbers** to visualise the ordered set of real numbers (Figure 1.7). We think of it as having an origin (a point corresponding to the number 0) and extending indefinitely to the left and to the right. Viewing numbers on this line is more informative than ranking them because we can also show how far apart they are from each other.

Example 1.42

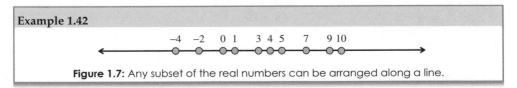

Figure 1.7: Any subset of the real numbers can be arranged along a line.

Such one-to-one correspondence between numbers and points along a line (or **axis**) creates a **coordinate system on the line**. However, we often want to characterise things according to more than one of their attributes. We may, for example, want to characterise a species of animal according to both its life-span and its metabolic rate. We therefore need to visualise pairs of numbers on a **coordinate system on the plane**. Mathematicians most frequently use the Cartesian system. This consists of two axes intersecting at right angles at their origin (O). The same unit is used to place numbers along both axes of a Cartesian coordinate system. Just like coordinate systems on the line, systems on the plane are visualisation tools. Each pair of numbers must be ordered so that we know which one is to be mapped on which axis. Conventionally, the first number in the ordered pair is used as a coordinate for the horizontal axis and the second for the vertical. To represent a point on a planar system of coordinates, you need to mark the two numbers on their corresponding axes and from there draw two lines parallel to the other axes. The representation of the ordered pair is the point of intersection of these two lines.

Example 1.43: Non-Cartesian map projections
The ordered pair (2,4) can be plotted on the Cartesian system of coordinates (Figure 1.8(a)) but the approach to plotting would be the same even if the coordinate system was not Cartesian; for example, if the axes were not perpendicular and the unit of length was different along each axis (Figure 1.8(b),1.8(c))

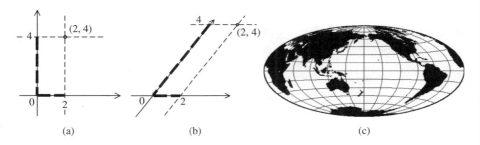

(a)   (b)   (c)

Figure 1.8: The same ordered pair of coordinates in a Cartesian (a) and non-Cartesian system (b). Projections of the globe onto two-dimensional maps (c) are a familiar example of non-Cartesian coordinates.

® 1.11: Basic plotting
As part of the larger body of literature motivated by the 'rate of living' hypothesis (Pearl, 1928), a study was undertaken to look at the relationship between mass-specific, daily energy expenditure and body mass in small mammals (Speakman *et al.*, 2002). I will ignore which indexes were used for measuring the two variables (to avoid specifying units), but simply seek to visualise the data. The body mass and energy expenditure of 12 species were

```
> bodM<-c(3.9, 3.1, 2.0, 8.3, 5.8, 5.1, 3.7, 6.6, 3.2, 4.3, 2.9, 7.4)
> enExp<-c(-1.1, -0.5, 2.5, -2.2, -0.7, -0.1, 0.4, -0.4, 0.8, 0.7,
0.4, -1.5)
> plot(bodM, enExp, xlab="Body Mass", ylab="Energy expenditure")
```

Lines 1 and 2 assign the data to the symbols `bodM` and `enExp`. The order of the numbers matters because these are paired measurements for each species. The command `plot` places the variables `bodM` and `enExp` on a system of coordinates. The axes are labelled as instructed by the options `xlab` and `ylab` (remember that options within a command are assigned using `=`, not the arrow `<-`). The graph is shown in Figure 1.9.

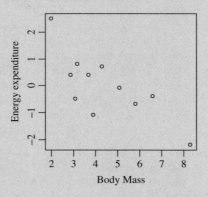

Figure 1.9: Graphical output of the data on body mass and energy expenditure.

1.17.   Complex numbers

Although mathematics is renowned as a system of rigorous and formal rules, progress has very often come about by acts of rule-breaking. Complex numbers are a good example. Their introduction was motivated by the need to make sense of quadratic equations with a negative discriminant (see Section 1.13). The elementary maths approach at this point is to report no solution. However, the lack of real roots does not necessarily imply a lack of useful information. To work around this problem, a new number i is introduced. This is called an **imaginary** number and is defined as

$$i^2 = -1 \text{ or, equivalently } i = \sqrt{-1} \tag{1.75}$$

Example 1.44

The square root of any negative number can now be 'calculated'. For example, $\sqrt{-9} = \sqrt{-1 \times 9} = \sqrt{-1} \times \sqrt{9} = 3i$.

Mixing real with imaginary numbers gives rise to the set of complex numbers, already mentioned in Section 1.2. A complex number z is always of the form

$$z = a + bi \tag{1.76}$$

where both a and b are real numbers. This section presents the ground rules for how to work with complex numbers graphically and algebraically.

Unlike any other type of number, complex numbers cannot readily be compared. So, we cannot say whether $3 + 2i$ is greater or smaller than $2 + 3i$ (despite the fact that both of them are numbers and they are clearly not the same). Instead, we can visualise them in a two-dimensional system of coordinates by plotting the real part on one axis and the imaginary part on another (Figure 1.10).

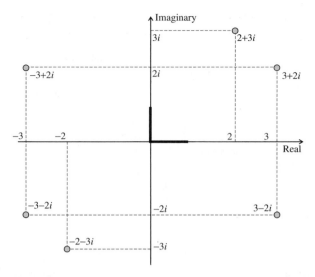

Figure 1.10: The complex plane, with the real component plotted along the horizontal axis and the imaginary part on the vertical axis. In this example there are two conjugate pairs, $3 \pm 2i$ and $-3 \pm 2i$.

In the complex plane, numbers that are symmetric around the horizontal axis (the line of real numbers) are called **complex conjugates**. Such numbers are usually highlighted with an over-bar notation. So, the conjugate of any complex number $z = a + bi$ is $\bar{z} = a - bi$.

For any two complex numbers, say $z_1 = a_1 + b_1 i$ and $z_2 = a_2 + b_2 i$, addition and subtraction are done separately for the real and imaginary parts

$$z_1 + z_2 = (a_1 + a_2) + (b_1 + b_2)i$$
$$z_1 - z_2 = (a_1 - a_2) + (b_1 - b_2)i \tag{1.77}$$

Multiplication is equally common-sense if we think of the imaginary number i as just any other symbol

$$z_1 z_2 = (a_1 + b_1 i)(a_2 + b_2 i)$$
$$= a_1 a_2 + a_1 b_2 i + a_2 b_1 i + b_1 b_2 i^2 \tag{1.78}$$
$$= a_1 a_2 + (a_1 b_2 + a_2 b_1)i - b_1 b_2$$

Note that the last step in Equation (1.78) uses the definition of i in Equation (1.75) to get $b_1 b_2 i^2 = -b_1 b_2$. A particularly important version of Equation (1.78) arises when the two complex numbers being multiplied are conjugate,

$$z\bar{z} = (a + bi)(a - bi)$$
$$= a^2 + abi - abi - b^2 i^2 \tag{1.79}$$
$$= a^2 + b^2$$

So, in general, the product of two complex conjugates is a real number.

Division of a complex number by a noncomplex number is a straightforward division of the real and imaginary parts

$$\frac{z}{c} = \frac{a + bi}{c} = \underbrace{\left(\frac{a}{c}\right)}_{\text{Real}} + \underbrace{\left(\frac{b}{c}\right)}_{\text{Imaginary}} i \tag{1.80}$$

However, division of two complex numbers is a bit tougher. Feel free to avoid the following derivation if you are not in the mood for mental gymnastics, and skip to the result in Equation (1.87).

We can think of division between two complex numbers as the multiplication of one with the inverse of the other

$$\frac{z_1}{z_2} = z_1 \frac{1}{z_2} \tag{1.81}$$

Unfortunately, we don't really know how to calculate the inverse of a complex number. We therefore start by assuming that it exists, and try to calculate it. In general, if $z = a + bi$, then its inverse may also be a complex number, say $1/z = a' + b'i$. To find a' and b' we begin by stating a requirement: for one number to be the inverse of the other, their product must equal 1

$$z\frac{1}{z} = 1 \tag{1.82}$$

We can expand this to see where it leads

$$(a + bi)(a' + b'i) = 1 \\ (aa' - bb') + (ab' + ba')i = 1 \tag{1.83}$$

This says that the complex number with real part $(aa' - bb')$ and imaginary part $(ab' + ba')$ must equal 1. Since 1 is a real number, this statement can only be true if the coefficient $(ab' + ba')$ of the complex part is zero, implying

$$ab' + ba' = 0 \\ aa' - bb' = 1 \tag{1.84}$$

The only two unknown quantities in Equations (1.84) are a' and b'. So, we have a system of two equations in two unknowns that can be solved (see Section 1.14) to get

$$a' = \frac{a}{a^2 + b^2}, \ b' = -\frac{b}{a^2 + b^2} \tag{1.85}$$

Putting everything together, the inverse of a complex number is

$$\frac{1}{z} = \frac{\bar{z}}{z\bar{z}} \tag{1.86}$$

Which, in turn, yields a rule for dividing two complex numbers z_1 and z_2

$$\frac{z_1}{z_2} = \frac{z_1 \bar{z}_2}{z_2 \bar{z}_2} \tag{1.87}$$

Note that the denominator on the right-hand side $(z_2 \bar{z}_2)$ is the product of complex conjugates and therefore, according to Equation (1.79), just a real number.

Example 1.45

Here are the four operations applied to the numbers $z_1 = -2 + i$ and $z_2 = 3 - 2i$:

$$z_1 + z_2 = (-2 + 3) + (1 - 2)i = 1 - i$$

$$z_1 - z_2 = (-2 - 3) + (1 + 2)i = -5 + 3i$$

$$z_1 z_2 = -6 + 4i + 3i + 2 = -4 + 7i$$

$$\frac{z_1}{z_2} = \frac{(-2 + i)(3 + 2i)}{(3)^2 + (-2)^2} = \frac{-6 - 4i + 3i - 2}{13} = -\frac{8}{13} - \frac{1}{13}i$$

1.12: Complex numbers in R

If you type something like `(-2)^0.5` or `sqrt(-2)`, R will return the value NaN meaning 'not a number'. So, despite the fact that some of the commands in R will deal with complex numbers (see R1.9), R is generally reluctant to perform complex algebra. Nevertheless, given a complex number z, you can extract its real and imaginary parts by typing `Re(z)` and `Im(z)`.

1.18.   Relations and functions

Consider two sets A and B containing n and m members respectively. We can create as many as $n \times m$ pairs from these two sets, implying a total of $n \times m$ possible associations. Any collection of such associations is called a **relation**.

Example 1.46: Food webs

Adélie penguin (Pygoscelis adeliae)

Trophic interactions in ecological communities relate predators to their prey. If A is the set of predator species and B the set of prey species, we may be interested in the linkages between members of A and B. In most ecological communities, some species may act as both predators and prey. Therefore, with no loss of generality, we may assume that the sets A and B have the same membership. However, generally not all predators are also prey and not all prey are also predators. The food web can be illustrated as a network of interconnected nodes (Figure 1.11(a)), or as a regular grid of all conceivable associations (Figure 1.11(b)).

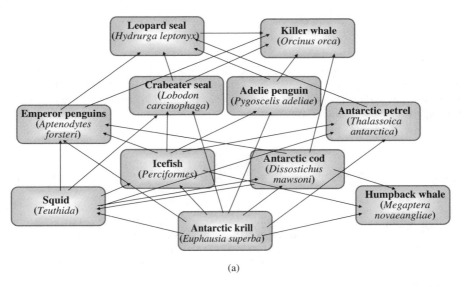

(a)

Figure 1.11: (a) A food-web representation of part of an Antarctic marine community and (b) the corresponding graph of associations.

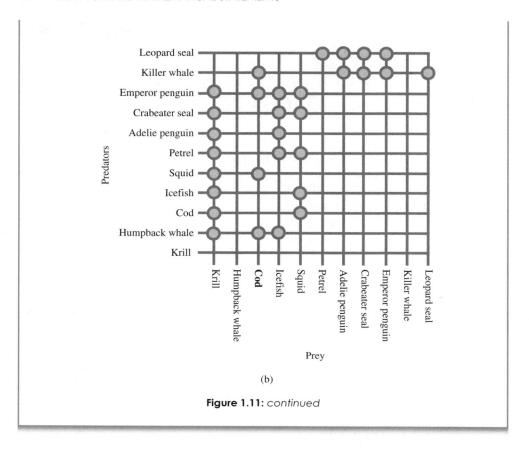

(b)

Figure 1.11: *continued*

In Figure 1.11(b), it is clear that a species can eat more than one prey and be eaten by more than one predator. Other types of relations may present simpler associations. One particularly simple case is when every member of the set A is associated with exactly one member of the set B; such a relation is called a **function**.

Example 1.47: Mating systems in animals

The set of reproductively active males and females in a sexually reproducing species can be associated in any one of four mating schemes (monogamous, polygynous, polyandrous, promiscuous). Figure 1.12 shows examples of associations between females and males. Although all of them are relations, only two are functions, because in the other two more than one arrow leaves certain members of the set of females. If we were associating males with females, then the situation would change: the polygynous system would cease to be a function and the polyandrous system would become one.

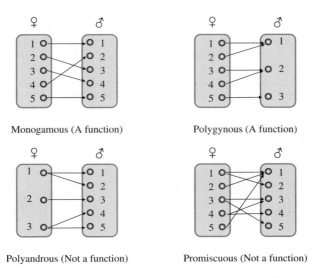

Figure 1.12: Four types of animal mating system represented as associations between females and males.

Algebraic operations can be used on numbers and symbols to create algebraic expressions. The variables contained in these expressions can be as widely or as narrowly defined as we require. We envisage a set of values that such a variable can take and call this the **domain** of the variable.

Example 1.48

Consider the expression $5x + 7$ and let the set $A = \{1, 2, 3, 4\}$ be the domain of the variable x. Now imagine that we pick any value from the domain of the variable x (say, 1) and calculate our algebraic expression (to get 12). If we repeat this process for all the other values, we notice that the expression $5x + 7$ transforms the set A to a new set $B = \{12, 17, 22, 27\}$.

In mathematical terminology we say that the expression in Example 1.48 **maps** the set A onto the set B. The term 'map' originates from cartography where people would identify points in real space as points on a map. The set generated by operating on the function's domain is called the function's **range**.

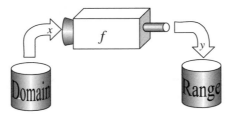

Figure 1.13: A function f operates on an input x, selected from the function's domain. This operation produces an output y that belongs to the function's range.

Like most things in mathematics, functions can be given symbolic names (conventionally letters like f or g). The input of the function is called an **independent variable**. It is convenient to think of the function as **operating** on the independent variable (Figure 1.13). For a function f with independent variable x, the operation is denoted by $f(x)$. The result of the operation $f(x)$ generates an output, say y. The function's output is also a variable whose value depends on the choice of the input value x, so it is called the **dependent** variable.

 1.13: Defining computer functions

In computing, the term 'function' has a somewhat broader definition than in mathematics, although computer functions can be used to encode mathematical functions. In general, a computer function is a self-consistent piece of code that performs a particular task. Just like a mathematical function, it usually requires some input, upon which it operates to produce an output. However, unlike mathematical functions, the output generated by a particular input does not have to be unique. Four components are needed to fully specify a computer function:

❶ its name
❷ a temporary name for its input (to be used internally by the function);
❸ a description of how the function performs its task (sometimes called its 'main body');
❹ a command to return the function's output.

The syntax looks a bit like this:

```
function.name<-function(input)
    {
    # insert function's main body here
    return(output)
    }
```

The main body of the function may comprise several lines of code. As we will see below, the ability to replace several lines by a `function.name` is the greatest advantage of 'functional' computing. To illustrate, consider the simple task described in Example 1.48. The easiest way to perform this is by first entering a value for the input and then operating on it directly:

```
> x<-4
> y<-5*x+7
> y
[1] 27
```

For such a simple task, this approach is perfectly acceptable. However, more complex tasks, which will need to be performed several times for different inputs, are best encoded as functions. To achieve the above task functionally, type the following in your text editor and then paste it into R:

```
f<-function(x)
    {
    y<-5*x + 7
    return(y)
    }
```

No output is produced because no specific input has been provided. R knows that it should expect an input (here called x) and it knows what to do with that input once received, but it is waiting for the function to be 'called' by typing its name and specifying the input (e.g. 4):

```
> f(4)
[1] 27
```

The entire mapping of Example 1.48 can be generated as

```
> f(1:4)
[1] 12 17 22 27
```

1.19.   The graph of a function

The input x and output y of a function are an ordered pair of coordinates that can be visualised on the plane. Every number from the function's domain generates a new ordered pair and hence a new point on the plane. Sweeping through the entire domain yields a collection of points called the **graph of the function**.

Example 1.49: Two aspects of vole energetics

In Example 1.19, an equation was derived for energy acquisition in voles in terms of physical quantities such as territory size (A) and the proportion of a day spent foraging (p)

$$\Delta E = \varepsilon m p A - (1 + 0.2p)E_r \qquad (1.88)$$

To see how changes in these quantities affect ΔE we can interpret Equation (1.91) as a function with dependent variable ΔE and independent variables A and p. Calculating the value of ΔE for 20 example values of A and p gives two graphs (Figure 1.14) that hint at linear responses.

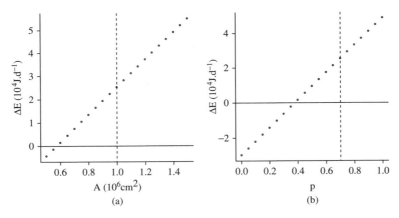

Figure 1.14: Plotting combinations of the independent and dependent variables can help visualise how changes in one affect the other. The above is a graphical representation of the common-sense fact that energy gain will increase if the vole exploits a greater home range with the same amount of effort (a), or if it increases its daily foraging effort but maintains the same home range (b). Note that, at low values of the independent variables, the vole will suffer net energy losses (signified by negative values of the dependent variable). The dashed lines indicate the fixed values of the independent variables used in Example 1.19.

1.14: Plotting the graphs of functions

R offers an overwhelming variety of options for customising graphs. These allow the production of publication-quality figures but require considerable input from the user. However, plotting a simple function isn't too challenging. The bare minimum required is to generate a set of paired data for the independent and dependent variables, plot them and label the axes in a meaningful way. Here is an example for the function $f(x) = 5x + 7$, using the integers between -10 and 10 as the domain.

```
x<-seq(-10,10,by=1)
y<-5*x+7
plot(x,y)
```

The first line introduces the new command `seq()` that is here used to produce a sequence of values from -10 to 10 in steps of 1. The second line takes this entire list (now called x) and applies to it the function. The third line plots the data. You will notice that R automatically uses the names of the two variables to label the axes. If you want to specify your own labels you can do it as follows:

```
plot(x,y, xlab="My own x label", ylab="My own y label")
```

There is a subtle difference between the true graph of a function and that generated by a computer because many functions have infinite or noncountable domains. Computers always plot functions using only a finite number of values from a function's domain. The resulting output is therefore only a discrete approximation of the true graph. To give the impression of continuity, the dots can be joined up by asking for a line plot via the `type` command:

```
plot(x,y, type="l")
```

However, it is strictly not certain that the graph follows a straight line between any two successive dots. Much of this uncertainty can be dispelled by knowing roughly what the graph of a function should look like. This is the topic of the next section.

1.20.   *First order polynomial functions*

Polynomials can be used to construct functions of the general form

$$f(x) = a_n x^n + a_{n-1} x^{n-1} + \cdots + a_1 x + a_0 \tag{1.89}$$

The simplest polynomial function is

$$f(x) = b \tag{1.90}$$

for $x \in \mathbb{R}$ and for some constant b. The graph of such a constant function is a straight line parallel to the x-axis. The line meets the y-axis at b. Technically, Equation (1.90) is a function of x whose domain includes all the real numbers. However, its output and range are limited to a single value: b.

Example 1.50: Population stability in a time series

Let $N(t)$ describe the number of lizards in an island population as a function of time (t, measured in days). We know that at time $t = 3.2$, the population $N(3.2)$ was 503 and that during the interval $3.2 \leq t \leq 9.2$, no lizards were born and none died. We can express this mathematically using the constant function: $N(t) = 503, \forall t \in [3.2, 9.2]$.

The second simplest polynomial function, the **identity function**, maps its domain exactly onto its range.

$$f(x) = x \tag{1.91}$$

Its graph is a straight line that goes through the origin (the point 0,0) because $f(0) = 0$, and it forms a $45°$ angle with the x-axis. We can interpret such a function as having no impact on its input.

Example 1.51: Population stability and population change

If we use a unit of one day to measure time, then we may be interested in plotting how today's population, N(today), determines tomorrow's population, N(tomorrow). We can simplify this notation by using the symbols N_t and N_{t+1}. Future populations depend on current populations so we can envisage a function that describes this transition for any current population size. The independent and dependent variables of such a function are the population size at time t and the population size at $t + 1$

$$N_{t+1} = f(N_t) \tag{1.92}$$

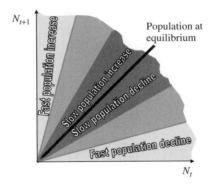

Figure 1.15: A population remains constant if its size is the same from any time instant to the next ($N_{t+1} = N_t$). This situation is represented by the identity function, the graph of which (labelled 'Population at equilibrium') splits the plane into two parts. The lower part corresponds to population decline ($N_{t+1} < N_t$) and the upper part to population increase ($N_{t+1} > N_t$).

In Chapter 3 we will learn how to construct mathematical functions that describe such transitions. For now, it is useful to be able to interpret the biological meaning of different regions of a graph of this function (Figure 1.15). It is particularly important to recognise that if $N_t = f(N_t)$ (compare this with Equation (1.91)), the input and output of the function are identical and population size remains constant.

Example 1.52: Visualising goodness-of-fit

 Scientific rigour should not necessarily lead to boring science. Contrary to popular belief, good scientists are not averse to leaps of faith. They do, however, need to put these wild guesses to the test. The best way to challenge a hypothesis is to use it to make predictions. Mathematical modelling simply extends our scientific reach by enabling us to examine the quantitative consequences of ever-more-complicated hypotheses (so, in a sense, modelling leads to more interesting science...). To do this, we need to know how well our predictions fit the data (not the other way around!). As you work your way through this book, you will see increasingly elaborate ways of doing this. However, when predictions involve a single variable, there is a simple graphical method using the identity function.

Consider, a model that aims to predict the density of a plant species in different parts of its range, using several types of information (soil characteristics, climate characteristics, density of grazers, density of competitors, etc.). For each set of observed conditions, we can obtain a single prediction of plant density. For the same set of conditions, we can provide a corresponding measurement of observed density at the same scale. Inspecting the cloud of paired values in a plot can be immensely informative about the quality of a model (Figure 1.16).

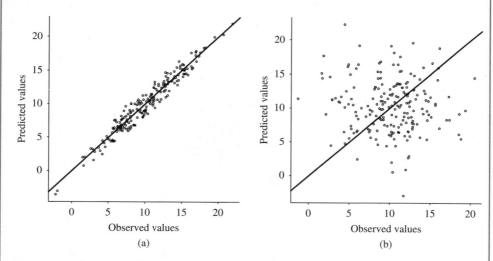

Figure 1.16: (a) The holy grail of every modeller who works with real data; (b) a more typical graph.

We can easily convert the identity function to a scaling relationship (see Section 1.12) by multiplying the right-hand side of Equation (1.91) by a constant a, to get

$$f(x) = ax \tag{1.93}$$

The graph of Equation (1.93) is a straight line that goes through the origin, but the angle it forms with the x-axis depends on the value of a. Adding another constant (b) to Equation (1.93) (i.e., combining Equations (1.90) and (1.93)) gives

$$f(x) = ax + b \tag{1.94}$$

This function also has a straight-line graph. Note that if we set $a = 0$ we recover Equation (1.90); setting $b = 0$ we recover the scaling function in Equation (1.93); and setting $a = 1$ and $b = 0$ we recover the identity function (1.91). In short, Equation (1.94) is the most general function with a linear graph, we will call it a **linear function**. Two important pieces of information about Equation (1.94) and its linear graph are that the constant a determines the **slope** (the inclination) of the line while the constant b is its **intercept** with the vertical (y) axis. More about slopes in Chapter 4.

ℝ 1.15: Adding lines to plots

Graphs of linear functions can be generated using the techniques of R.1.14. However, you may want to add further lines to a plot, perhaps to indicate features such as goodness of fit, or reference axes. The command `abline()` can be used for this purpose in three ways:

`abline(b,a)`   draws a line of slope a and intercept b
`abline(h=y)`   draws a horizontal line that intercepts the y-axis at the value y
`abline(v=x)`   draws a vertical line that meets the x-axis at the value x

For example, the following code first plots the graph of the function $f(x) = 1.6x + 2$ (lines 1–3) and then adds the Cartesian axes (lines 4–5) and the graph of the identity function (line 6).

```
x<-seq(-10,10)
f<-1.6*x+2
plot(x,f,type="l",xlab="x",ylab="f(x)",xlim=c(-8,8),ylim=c(-8,8))
abline(v=0)   # y-axis
abline(h=0)   # x-axis
abline(0,1, lty=2) # Identity function
```

Options `xlim` and `ylim` in line 3 specify the ranges of values to be plotted in the x and y axes, and the option `type = "l"` indicates that a line graph is required. The option `lty = 2` in line 6 specifies a dashed line. Line styles can be specified by name: `"solid"`, `"dashed"`, `"dotted"`, `"dotdash"`, `"longdash"` and `"twodash"`. The numerical codes 1, 2, 3, 4, 5 and 6 can also be used in place of these named options.

1.21. Higher order polynomial functions

The simplest second order polynomial function is

$$f(x) = x^2 \tag{1.95}$$

It is a bit harder to imagine what the graph of Equation (1.95) looks like, but trying out some values for x should help (Figure 1.17). This characteristic cup-like shape is called a **parabola** and it is common to the graphs of all second order polynomial functions.

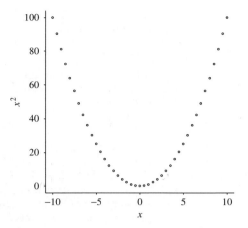

Figure 1.17: This graph was generated by trying out all values of x between -10 and 10, at intervals of 0.5.

Let's examine the effect on the graph of introducing a coefficient to the second order term:

$$f(x) = ax^2 \tag{1.96}$$

We can treat Equation (1.95) as a special case of Equation (1.96), obtained by setting $a = 1$. To explore the effect of other values of a we can use values smaller or greater than 1.

Example 1.53

Figure 1.18 shows the graphs of Equation (1.96) with six different values (± 0.5, ± 1 and ± 2) for the parameter a. The absolute value of the coefficient regulates the spread of the parabola (the closer to zero it is, the wider the parabola) and its sign determines the parabola's orientation (a negative coefficient leads to a downward-pointing cup).

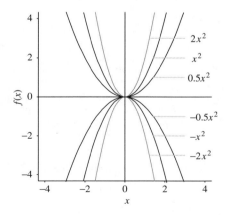

Figure 1.18: The graphs of six different versions of the function $f(x) = ax^2$.

The graph of the more general polynomials $f(x) = a_2x^2 + a_1x + a_0$ is still a parabola that intercepts the y-axis at a_0. The term a_1x shifts the graph horizontally.

Example 1.54

To illustrate the effect of the sign of different polynomial coefficients, we look at an example $f(x) = a_2x^2 + a_1x + a_0$ with $|a_2| = 0.1$, $|a_1| = 0.4$ and $|a_0| = 1$ (Figure 1.19).

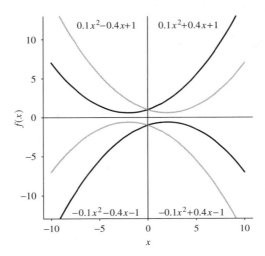

Figure 1.19: The effect of the signs of different polynomial terms.

As with the simpler second order polynomials in Example 1.53, the sign of a_2 determines the orientation of the parabola in relation to the x-axis (if it's positive, the parabola is upward-pointing). The sign of a_1 determines the position of the parabola's peak (or trough) in relation to the y-axis (if it's positive, the peak/trough is to the left of the y-axis). Finally, the sign of a_0 determines whether the intercept will be above/below the x-axis.

1.22.   The relationship between equations and functions

In Section 1.19 we saw how an equation could be turned into a function by allowing a constant to become a variable. The converse is also useful. To plot the graph of a function, we repeatedly fix its independent variable to a particular value and then calculate the dependent variable. Similarly, we can fix the dependent variable and try to solve the resulting equation for the independent variable. The most frequent use of this trick is to decide where the graph of a function will cross the x-axis. For any function $f(x)$ this is equivalent to solving the equation $f(x) = 0$.

Example 1.55

To find if and where the graphs of the three quadratic functions $f(x) = x^2 - 2, g(x) = x^2$ and $h(x) = x^2 + 2$ cross the x-axis, we set each function to zero and attempt to solve the resulting equation. From this, we find that $f(x)$ crosses the x-axis at $\pm\sqrt{2}$ (the equation has two solutions). The graph of $g(x)$ doesn't cross the x-axis but it touches it at 0 (the equation has one solution). The graph of $h(x)$ is an upward-pointing parabola with y-intercept equal to 2, so it can't cross the x-axis (indicated by the fact that the equation $x^2 + 2 = 0$ has no real solutions).

Example 1.56: Extent of an epidemic when the transmission rate exceeds a critical value

Consider a situation in which the rate of transmission of a disease in a population of K animals exceeds a critical value γ_c. We want to find what this implies for the total number N of infected individuals. Assuming perfect mixing, we can rewrite Equation (1.34) in Example 1.21 as an inequality:

$$cbN(K - N) > \gamma_c \qquad (1.97)$$

where N are the infected animals. Since $cb > 0$, this can be manipulated into

$$-N^2 + KN - \frac{\gamma_c}{cb} > 0 \qquad (1.98)$$

The graph of the function $f(N) = -N^2 + KN - \gamma_c/cb$ (Figure 1.20) can help visualise the range of values of N over which this might occur. In the particular example seen in Figure 1.20 we can identify that, for the given transmission threshold, the population size of infected individuals must lie in the range $N_1 < N < N_2$. These values can be found from the quadratic formula

$$N_1, N_2 = \frac{K \pm \sqrt{K^2 - 4\gamma_c/cb}}{2} \qquad (1.99)$$

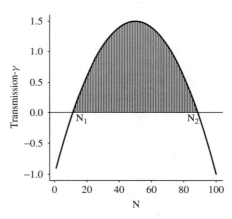

Figure 1.20: The graph of the function $f(N) = -N^2 + KN - \gamma_c/cb$, which gives the difference between actual transmission rate and the critical value of interest γ_c as a function of infected population size. In the shaded region the transmission rate exceeds the critical value γ_c. In this example, the parameter values are $K = 100, cb = 0.001$ and $\gamma_c = 1$.

1.23.   Other useful functions

My initial focus on polynomials was entirely due to their flexibility and ease of use. However, they are not suited to describing certain ecological relationships. Chapter 2 discusses the fact that seasonal phenomena require a different class of (trigonometric) functions, and in Chapter 3, population growth gives rise to yet others (exponential and logarithmic). This section aims to broaden your concept of functions beyond polynomials by illustrating how we can use the basic algebraic operators to expand our toolbox of functions.

Example 1.57

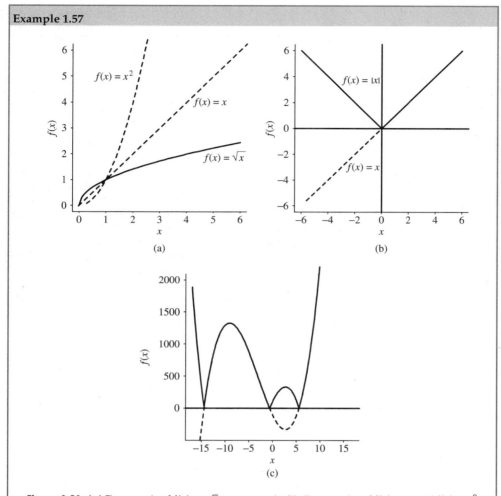

Figure 1.21: (a) The graph of $f(x) = \sqrt{x}$ compared with the graphs of $f(x) = x$ and $f(x) = x^2$ (dashed curves); (b) the graph of $f(x) = |x|$; (c) the graph of $f(x) = |2x^3 + 19x^2 - 150x - 100|$, the dashed curve shows the segments of the original graph that have been reflected upwards by taking its absolute value.

Roots and absolute values (see Section 1.5) can be used to construct functions. The simplest ones are $f(x) = x^{1/a}$ and $f(x) = |x|$. You may remember that functions using roots are not real-valued if the expression under the root takes negative values. So, the function $f(x) = x^{1/a}$ must take (and also give) non-negative values. Its graph increases sharply and then slows down (Figure 1.21(a)), although it never levels off. The function shows an interesting symmetry with the function $f(x) = x^a$ around the $45°$ line (Figure 1.21(a)), and the two functions cross at $x = 1$. The graph of the absolute value of a function resembles the graph of the function itself (see Figure 1.21(b) and 1.21(c)). However, values that fall below the x-axis are reflected upwards in the graph of its absolute value.

One particular biological phenomenon that cannot easily be described using a polynomial is 'plateauing' or saturation behaviour. For example, there is usually an upper limit on the number of prey items that an individual predator can take per day, no matter how many prey are available to it. There are several different formulations that will achieve this effect; one of the least elaborate is given in Example 1.58.

Example 1.58

A useful function for modelling saturation is the **rectangular hyperbola**,

$$f(x) = \frac{ax}{b+x} \tag{1.100}$$

For most biological applications of saturation, the independent variable (e.g. prey density, enzyme or pollutant concentration) can only take non-negative values. The function starts from the origin ($f(0) = 0$) and increases with x. It approaches, but never attains, a value known as the function's **asymptote** (a Greek word meaning 'uncrossable'), equal to the constant a. Because the asymptote is never attained, the value of half-saturation ($f(x) = a/2$) is used to describe how fast the value of the function increases. Half-saturation occurs when $x = b$ (check this by calculating $f(b)$ in Equation (1.100)). The properties of Equation (1.100) are summarised graphically in Figure 1.22. The rectangular hyperbola appears throughout the biological literature, often with a different name. For example, in foraging ecology (see Example 4.14 in Chapter 4), it is called a **Holling type II, or hyperbolic, functional response**, in enzyme kinetics it is known as the **Michaelis–Menten equation** and in medical, behavioural and epidemiological applications, it is a type of **dose–response curve**.

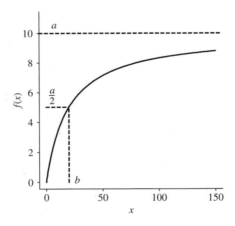

Figure 1.22: The rectangular hyperbola with $a = 10$ and $b = 20$.

1.24. Inverse functions

In Section 1.18, functions were described as processing machines $f(x) = y$. The operation of the function f on x gives the output y. However, given the output of a known function, is it possible to guess what the input was? The answer is yes, if each output in the function's range can only be produced by using a single input x from the function's domain. These are called **one-to-one** functions.

Example 1.59

We are given the function

$$f(x) = 2x - 4 \qquad (1.101)$$

and three of its outputs: $y_1 = 2, y_2 = -4, y_3 = -1$. To find the corresponding inputs x_1, x_2, x_3 we must solve the general equation $2x - 4 = y$ for x,

$$x = \frac{y + 4}{2} \qquad (1.102)$$

If we treat this as a function of y and we denote it by f^{-1} we get

$$f^{-1}(y) = \frac{y + 4}{2} = x \qquad (1.103)$$

Putting in the values of y_1, y_2, y_3 we get $x_1 = 3, x_2 = 0, x_3 = 1.5$. We have thus **inverted** the effect of the function f by introducing the new function f^{-1} (Figure 1.23).

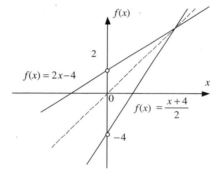

Figure 1.23: Graphs of a simple linear function and its inverse.

Such inverses can be defined for all one-to-one functions (Figure 1.24). Hence, Chapter 2 will introduce the inverses of trigonometric functions (within a constrained part of their range) and Chapter 3 will present the logarithm as the inverse of the exponential function.

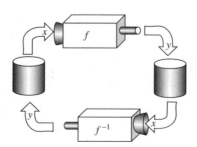

Figure 1.24: We can now complete Figure 1.12. What serves as the domain of f is the range of f^{-1} and vice versa.

1.25. Functions of more than one variable

A function need not be constrained to just one input. Most ecological phenomena are affected simultaneously by many influences and it would be very limiting if we could only examine them one at a time. The notation is similar to what we have encountered before: we need a name for the function (say, f) and names for its independent variables (we can use heuristic symbols as in Example 1.60 below, or indexing, $x_1, x_2, \ldots$). The simplest function in two independent variables is linear

$$f(x_1, x_2) = a + bx_1 + cx_2 \tag{1.104}$$

The graphs of such functions are constrained by the three dimensions of physical space but functions of two independent variables can be readily represented in three axes. The graph of Equation (1.104) is a plane (Figure 1.25(a)). Nonlinear functions give more complicated surfaces (Figure 1.25(b)).

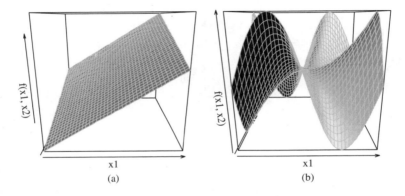

Figure 1.25: (a) The graph of the simple linear function $f(x_1, x_2) = -10 + 3x_1 + 5x_2$; (b) the graph of the nonlinear function $f(x_1, x_2) = x_1 x_2 (x_1^2 - x_2^2)/(x_1^2 + x_2^2)$.

Example 1.60: Two aspects of vole energetics

More/less active voles may hold large/small territories. The variables affecting energy acquisition are the daily proportion spent foraging (p) and territory size (A). In Example 1.49, their impact on ΔE was plotted separately for each variable but, in truth, the value of ΔE will result from the values of both independent variables. There are infinite combinations (A, p), each yielding a value for ΔE under the model

$$\Delta E = \varepsilon m p A - (1 + 0.2p)E_r \tag{1.105}$$

An overview of the behaviour of Equation (1.105) can be obtained by plotting it (Figure 1.26). This reveals new aspects of the model's behaviour. Compared to Figure 1.13, which gave the impression of linearity, the plots in Figure 1.26 show how the two independent variables can interact to give higher energy gains for active animals with large territories.

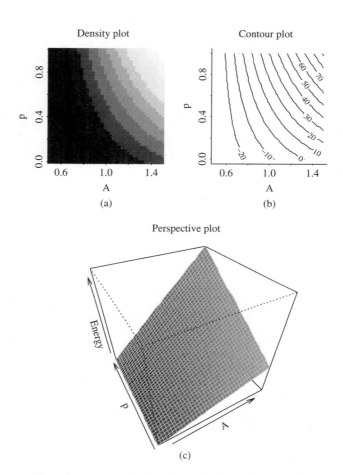

Figure 1.26: Three different ways to plot a function with two independent variables. The first two, (a) and (b), are plotted in two dimensions and use implicit representations of the third. In the case of the density plot (a), the third dimension is represented by shading or colour. A contour plot (b) uses isopleths for this purpose. A perspective plot (c) tries to emulate three dimensions.

1.16: Perspective, image and contour plots

Generating 3D plots requires some more preparation than 2D ones because the function of interest needs to be calculated for values on a grid. The first step, therefore, is to prepare the grid by specifying the plotting range and number of subdivisions to be used for each of the two independent variables:

```
#Plotting parameters
steps<-5 # Number of points in each dimension of the grid
x1r<-c(1,5) # Plotting range for 1st indep. variable
x2r<-c(1,5) # Plotting range for 2nd indep. variable
```

These are used to generate the labels for the two axes. The command `seq()` that we previously saw in R1.14, is here used with the option `len`, which specifies the total desired number of members in the sequence.

```
x1lab<-seq(x1r[1], x1r[2], len=steps) # Labels for x1
x2lab<-seq(x2r[1], x2r[2], len=steps) # Labels for x2
```

We now need to construct *all* pair-wise combinations of values of `x1` and `x2` on the grid. One way to achieve this is, again, by use of `seq()`

```
# Repeats the entire sequence for 1st variable a number of times
# equal to steps
x1<-rep(x1lab, steps)

# Repeats each value of the 2nd variable a number of times
# equal to steps
x2<-rep(x2lab, 1, each=steps)
```

In this example, `x1` and `x2` are:

```
> x1
 [1] 1 2 3 4 5 1 2 3 4 5 1 2 3 4 5 1 2 3 4 5 1 2 3 4 5
> x2
 [1] 1 1 1 1 1 2 2 2 2 2 3 3 3 3 3 4 4 4 4 4 5 5 5 5 5
```

We are now ready to calculate the function and then arrange the values in a tabular form using the command `matrix()` (more on matrices in Chapter 6).

```
fxy<- x1*x2*(x1^2-x2^2)/(x1^2+x2^2) # Type your own function here
f<-matrix(fxy,steps,steps,byrow=T) # Matrix of values for plotting
```

A perspective plot can now be generated with a single line

```
persp(x1lab,x2lab,f, xlab="x1", ylab="x2", zlab="f(x1,x2)")
```

A density plot (command `image()`) and a contour plot can be generated as follows:

```
image(x1lab,x2lab,f, xlab="x1", ylab="x2")
contour(x1lab,x2lab,f, xlab="x1", ylab="x2")
```

A combination between these two plots is the `filled.contour()` which colours the intervals between the contours. These four plots can be endlessly customised according to taste. The help files in R provide more information on graphing options.

Further reading

Most of the mathematical concepts covered in this chapter are also covered in Cann (2003, pp. 1–34), Foster (1998, pp. 3–48, 59–72), Harris *et al.* (2005, pp. 2–46) and Mackenzie (2005, pp. 1–9, 25–28, 46–53). The presentation is briefer with less relevance to ecology but the mathematical problems will provide additional practice. A slightly more old-fashioned, but very thoughtful, treatment of this chapter's topics is given by Batschelet (1979, pp. 1–14, 17–35, 59–109). Extensive coverage of scaling in ecology can be found in Schneider (2009).

References

Aarts, G., Mackenzie, M.L., McConnell, B.J., Fedak, M.A. and Matthiopoulos, J. (2008) Estimating space use and environmental preference from wildlife telemetry data. *Ecography*, **31**, 140–160.

Batschelet, E. (1979) *Introduction to Mathematics for Life Scientists*. Springer, Berlin.

Cann, A.J. (2003) *Maths from Scratch for Biologists*. John Wiley & Sons, Ltd, Chichester.

Chartier, T. (2005) Devastating roundoff error. *Math. Horizons*, **13**, 11.

Foster, P.C. (1998) *Easy Mathematics for Biologists*. Harwood Academic Publishers, Amsterdam.

Harris, M., Taylor, G. and Taylor, J. (2005) *CatchUp Maths & Stats For the Life and Medical Sciences*. Scion, Kent.

Heesterbeek, H. (2005) The law of mass-action in epidemiology: A historical perspective. In *Ecological Paradigms Lost: Routes of Theory Change* (Eds K. Cuddington and B. Beisner). Elsevier, Amsterdam, pp 81–106.

Mackenzie, A. (2005) *Instant Notes: Mathematics and Statistics for Life Scientists*. Taylor and Francis, New York.

Manly, B.F.J., McDonald, L.L., Thomas, D.L., McDonald, T.L. and Erickson, W.P. (2002) *Resource Selection by Animals*. Kluwer Academic Publishers, Dordrecht, The Netherlands, 221 pp.

Pearl, R. (1928) *The Rate of Living*. University of London Press, UK

Schneider, D.C. (2009) *Quantitative ecology: Measurement, models and scaling*. Academic press. 432 pp.

2

How to describe regular shapes and patterns
(Geometry and trigonometry)

'... and when they were up, they were up;
and when they were down, they were down.
But when they were only halfway up,
they were neither up nor down.'
English nursery rhyme

Quantifying shapes and patterns is useful for studying organism morphology, survey design, telemetry triangulation, modelling animal movement, periodic population dynamics and phylogeny. This chapter first covers **geometry**, the field of mathematics that deals with shapes, their relative sizes and relative positions. Its name (Greek for 'earth measurement') betrays its origins: the first geometers in Egypt and Babylon were surveyors, architects and engineers. Yet 'measurement' is only a minor part of geometry and practitioners during its golden age (around 300 BC) used unmarked rulers, solely to construct and compare shapes, not to measure them. Geometry is the most intuitive of all areas of elementary mathematics and the one that best captures the deductive reasoning required for higher mathematics and rigorous scientific thinking in general. The focus here is on **Euclidean geometry**, the variety that simultaneously occupies pages in high-school textbooks and treatises on special relativity. Euclid, a Greek mathematician living in Egypt during the 4th century BC, wrote nine books, only five of which have survived to the present day. His book *Elements*, consists of thirteen volumes. Only segments from the first volume are considered here.

How to be a Quantitative Ecologist: The 'A to R' of Green Mathematics and Statistics, First Edition. Jason Matthiopoulos.
© 2011 John Wiley & Sons, Ltd. Published 2011 by John Wiley & Sons, Ltd.

Euclid's geometry is known as **axiomatic**: it aims to reduce the number of unproven statements in our scientific description of the world by deducing as many logical conclusions as possible from a small set of self-evident empirical observations. It may, therefore, be argued that Euclid's aims are common to all modern quantitative disciplines, including ecology. However, many of the empirical observations used as ecological ground-truths are not so self-evident. Euclid used three types of statement: **Definitions** (Section 2.1) and **axioms** (Section 2.2) set out the properties of primitive elements such as the point, the line and the plane. **Propositions** are initially proved on the strength of definitions and axioms but, once proven, they are accepted as truth and can be used to prove others. This process is briefly traced in Section 2.3 by proving some useful propositions.

A brief refresher of measures of distances, areas, volumes and angles (Sections 2.4–2.6) leads to a discussion of **trigonometry**, this chapter's second main component. The name 'trigonometry', Greek for 'triangle measurement', understates its utility in describing **periodic phenomena** (those that occur repeatedly and regularly). The main trigonometric functions are introduced using a graphical device known as the **trigonometric circle** (Sections 2.7–2.8) and are then used to construct **polar coordinates** (Section 2.9), an alternative to the system of Cartesian coordinates discussed in Chapter 1. Polar coordinates offer a more natural way of posing questions about animal movement and orientation. The next few sections cover the graphs of trigonometric functions (Section 2.10) and techniques for manipulating and solving trigonometric equations (Sections 2.11–2.13). Further techniques for shifting, stretching and combining the basic trigonometric plots (Sections 2.14–2.15) enable the mathematical description of more general periodic patterns. Even more complex natural phenomena can be approached with methods like **spectral analysis**, used to extract periodic patterns from noisy time series (Section 2.16) and **fractal geometry**, which can succinctly summarise highly complicated natural shapes (Section 2.17).

2.1.   Primitive elements

The raw materials of geometry are its **primitive elements**. To understand these, we need to appeal to our everyday experience of space. A **point** is an (almost metaphysical) shape with no dimensions. A **curve** has length but no breadth or depth. A **surface** has breadth and length but no depth. A **straight line** is a curve and so is a circle. A **plane** is a flat surface. A **line segment** is a line of finite length, and the figures enclosed in flat shapes (such as a circle, a triangle or a square) are finite, planar surfaces. An **angle** is defined as the figure formed by two line segments that extend from a common point, or two planes that cross at a common line (Figure 2.1). These 'definitions' merely introduce the names of the primitive elements rather than rigorously specifying their properties, a task better achieved by the geometrical axioms.

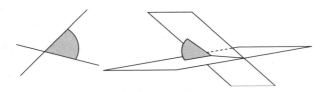

Figure 2.1: Angles in two- and three-dimensional space.

2.2. Axioms of Euclidean geometry

Axioms are self-evident statements, taken to be true without proof. Just like its definitions, the axioms of Euclidean geometry are based on everyday experience. Here are some examples:

❶ Only one straight line passes through two points.
❷ Only one plane passes through three points.
❸ Two or more parallel lines have no point in common.
❹ If one shape can be moved to coincide with another, then the two are equal.
❺ Given a line L_1 and a point P that is not on L_1, there is only one line L_2 parallel to L_1 that goes through P.

Euclid also used axiomatic properties, which he called common notions, that were equivalent to those of modern algebra (Chapter 1). For example, 'If equals are added to equals, then the sums equal each other.'

Note how axioms add to our understanding of a straight line and a plane, hence complementing their definitions. However, it seems that Euclid anchored his entire conceptual masterpiece on nothing but intuition: why should two parallel lines never cross? Why can't any three points define a line? If this most elegant of mathematical constructs ultimately relies on fallible common sense, then why should it be accepted as an objective and ultimate truth? The answer is that it should not; Euclid's reality is just one of many possible realities explored by mathematics. The presentation of axioms as a set of unprovable beliefs, poses some wider epistemological questions. How complex a set of explanations should we need in science? Can we prove that life on Earth has evolved by natural selection or that natural selection can give rise to life forms such as us? Should the discovery of a working model be taken as proof that a mechanism implied by the model operates in nature?

Example 2.1: Suicidal lemmings, parsimony, evidence and proof

Norwegian lemming, Lemmus lemmus

Scandinavian lemmings are well known for their spatially extensive and temporally regular population fluctuations, known as cycles. This phenomenon has attracted a great deal of popular and scientific interest (not to mention the use of creative licence from the occasional 1950s documentary film crew – just type 'white wilderness' in YouTube), but its definitive explanation is still elusive. Since the early work of Charles Elton (1924), several biological explanations have been proposed, from the, now discredited, suicidal emigration hypothesis to the currently favoured specialist predation hypothesis. In 1993, Batzli enumerated twenty-two different explanations for population cycles (e.g. weather, social stress, disease, food quantity and quality). Several, or all, of these mechanisms could act in combination to produce the observed dynamics. Since Batzli's list seems to be near-exhaustive, we might be tempted to bracket the truth by invoking all explanations simultaneously. However, it is generally a good idea to work up from as simple an explanation as possible, or, alternatively, prune back a complicated explanation to the bare minimum that can account for the observations. This is known as the principle of **parsimony** or **Ockham's razor**, one of scientists' best guards against the accruement of superstition. We could try to reduce the list by refuting each hypothesis, one-by-one, using experimentation. However, very often the process leading from the experimental data to their interpretation is complicated and unreliable, so we may find it hard to decide if we have actually achieved a refutation. Instead, it might be easier to operate on the relative weight of evidence for each hypothesis. An alternative route to (at least partial) refutation is to construct a mathematical model. A model is a formalisation of a set of assumptions about how the real world functions. 'Assumption' is an alternative word for 'axiom' so, in yet another manifestation of the principle of parsimony, it is a good idea to keep these to a minimum. For the model to be credible, all of the assumptions except those pertaining to the hypothesis at hand need either to be based on common sense (e.g. only female lemmings give birth) or supported by evidence. Given such a model, it may be possible to apply mathematical wizardry to prove that it can or cannot replicate the observed behaviour. However, proving that a set of assumptions has a set of logical (and mathematical) consequences does not also prove that the hypothesis operates in nature. The fact that any effect may have alternative causes was called the 'philosophical vulnerability' of modelling by Nisbet and Gurney (1982).

2.3.   Propositions

This section reviews some useful propositions from the first volume of Euclid's *Elements*. The aim is to present geometrical results for use in this and later chapters but also to offer a flavour of deductive reasoning (otherwise known as **proof**). The very first propositions are proved by **construction**. Quite literally, we prove that something is possible by doing it.

Proposition 1: It is possible to construct an equilateral triangle on a given line segment AB.

Proof: A circle is a shape whose every point is equally distant from a particular point called the circle's centre. This characteristic distance is called a circle's **radius**. We assume that we are able to draw circles of any desired radius (e.g. using a compass). We draw a circle with centre A and another with centre B, both with a radius equal to the length of the segment AB (Figure 2.2).

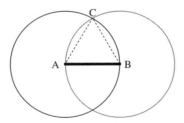

Figure 2.2: Drawing an equilateral triangle on a segment AB.

These circles will meet at two points (strictly speaking, we should also prove this). We choose one of these and call it C. Since C belongs to both circles we have

$$AB = BC = AC \tag{2.1}$$

Which proves that the triangle $\triangle ABC$ is an equilateral.

Example 2.2

A **Koch arc** is created by repeatedly applying the rule: replace the middle third of a line segment by two line segments, to form an equilateral triangle (Figures 2.3 and 2.4).

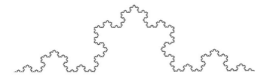

Figure 2.3: Koch arc, first iteration.

Using ruler and compass, this construction has two stages: the first requires us to trisect a line segment. We don't examine this more challenging problem but it can be solved by construction in several different ways (Hartshorne, 2000). The second stage is the construction of the equilateral triangle seen in Proposition 1.

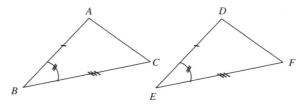

Figure 2.4: Koch arc after five iterations.

Proposition 2: If two triangles have two sides equal to two sides respectively and have the angles contained by the equal sides also equal, then the two triangles are equal.

Proof: Consider two triangles $\triangle ABC$ and $\triangle DEF$ (Figure 2.5) with equality between two of their sides $AB = DE, BC = EF$ and the containing angle $A\hat{B}C = D\hat{E}F$.

Figure 2.5: The tally lines indicate equality between different features of the figures.

If we superimpose the two triangles so that the point A coincides with the point D and the line segment AB coincides with the line segment DE, then, since $AB = DE$, the point B coincides with the point E. Furthermore, since the angle $A\hat{B}C$ equals the angle $D\hat{E}F$, the segment BC will coincide with the segment EF. Hence, the points C and F will also coincide. Therefore, the three vertices of the triangles coincide, the triangles are equal and so are all of their characteristics (including the remaining side and two angles).

*Proposition 3: In **isosceles** triangles (those that have two equal sides), the angles at the base equal one another.*

Proof: Let $\triangle\,ABC$ be an isosceles triangle, such that $AB = AC$ (Figure 2.6(a)).

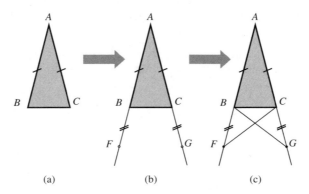

(a)   (b)   (c)

Figure 2.6: This proof requires some additional construction.

Extend the segments AB and AC and take two arbitrary points F and G on these extensions such that $BF = CG$ (Figure 2.6(b)). Note that

$$AB + BF = AC + CG \quad \text{hence} \quad AF = AG \tag{2.2}$$

The triangles $\triangle\,AFC$ and $\triangle\,AGB$ (Figure 2.6(c)) have two equal sides and an angle in common (by construction). Therefore, from Proposition 2,

$$\triangle\,AFC = \triangle\,AGB \tag{2.3}$$

The triangles $\triangle\,BCF$ and $\triangle\,BCG$ have $BF = CG$ (by construction), $BG = FC$ and $B\hat{F}C = B\hat{G}C$ (from Equation (2.3)). Therefore, $\triangle\,BCF = \triangle\,BCG$, which means that

$$F\hat{B}C = B\hat{C}G \tag{2.4}$$

Now, note that the angles $A\hat{B}F$ and $A\hat{C}G$ are both on straight lines and are therefore equal. This means that the sum of their parts will be equal

$$F\hat{B}C + A\hat{B}C = G\hat{C}B + A\hat{C}B \tag{2.5}$$

Combining Equations (2.4) and (2.5) implies $A\hat{B}C = A\hat{C}B$, which is the required result. Although not proven here, the reverse of this proposition is true: if the angles at the base are equal then the triangle is an isosceles.

I will use the symbol π to represent an angle whose sides lie on a straight line. The next two propositions (presented without proof) refer to **right angles**.

Proposition 4: If a line segment stands on a straight line then it makes either two right angles or two angles whose sum equals two right angles (Figure 2.7). This implies that the magnitude of a right angle is $\pi/2$.

Figure 2.7: Right angles are shown by a square. All other angles are depicted by an arc.

Proposition 5: The sum of the angles inside any triangle equals two right angles (Figure 2.8).

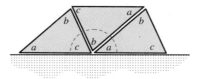

Figure 2.8: A graphical illustration of Proposition 5. By drawing the same triangle three times and rotating the copies to bring the three angles next to each other, we can rest the arrangement along a straight line (an angle of size π).

Proposition 6: If two straight lines cut one another, then they make the vertical angles equal to one another (Figure 2.9).

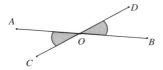

Figure 2.9: There are two sets of vertical angles in this diagram. The proposition can be proved for either set, and it is therefore true for both.

Proof: In Figure 2.9, $A\hat{O}D + D\hat{O}B = \pi$ and $A\hat{O}D + A\hat{O}C = \pi$. These facts imply,

$$D\hat{O}B = A\hat{O}C = \pi - A\hat{O}D \tag{2.6}$$

which is the required result.

Proposition 7: In an isosceles triangle, the line drawn at right angles from the base to the opposite vertex bisects the triangle's base and the angle opposite to the base.

Proof: From Proposition 3, we know that the angles at the base of an isosceles triangle are equal, so $A\hat{B}C = A\hat{C}B$. From Proposition 4, we know that the angles formed by the line segment AD that is perpendicular to the base BC are both right angles. Proposition 5 gives the following relationships for the angles of the two triangles $\triangle ABD$ and $\triangle ACD$.

$$\begin{aligned} B\hat{A}D + A\hat{D}B + D\hat{B}A &= \pi \\ C\hat{A}D + A\hat{D}C + D\hat{C}A &= \pi \end{aligned} \tag{2.7}$$

$$\begin{aligned} B\hat{A}D &= \pi - A\hat{D}B - D\hat{B}A \\ C\hat{A}D &= \pi - A\hat{D}C - D\hat{C}A \end{aligned} \tag{2.8}$$

The right-hand sides of these expressions are equal, meaning that the angle at the triangle's apex is bisected by the dashed line segment in Figure 2.10. In notation,

$$B\hat{A}D = C\hat{A}D = \frac{B\hat{A}C}{2} \tag{2.9}$$

The two triangles $\triangle ABD$ and $\triangle ABD$ have two sides equal ($AB = AC$), one side in common (AD) and the angle contained by these sides is the same (from Equation (2.9)). Therefore, Proposition 2 can be used to prove that the triangle's base is also bisected ($BD = DC$). Following similar logic, it may be shown that a line drawn perpendicularly from the middle of an isosceles's base, will go through its apex.

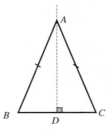

Figure 2.10: This proof focuses on the equality of the two smaller triangles formed by dropping a line from the apex (the topmost vertex), perpendicular to the base.

Example 2.3: Radio-tracking of terrestrial animals

Bengal tiger, Panthera tigris

A tiger expedition uses very high frequency (VHF) transmitters to track the animals. The reception equipment comprises two identical receivers mounted on a rotating bracket at distance d from each other. What sort of spatial information can be obtained from this arrangement? Because of variable interference, we cannot use the absolute strength of the signal to estimate distance from the animal. We may, however, be able to use the fact that the intensity of the signal at two different receivers will be identical if the tiger is the same distance from both of them. When this happens, we know that the tiger lies at the apex of one of many possible isosceles triangles, whose base is the line segment between the two receivers (Figure 2.11).

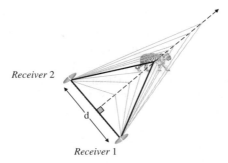

Figure 2.11: When the signal at the two receivers is the same, the tiger is at an equal distance from both.

Proposition 7 says that a line drawn perpendicular to the line joining the two receivers, starting from a point halfway between them, will go through any and all of the apexes of the isosceles triangles in Figure 2.11. Therefore, the tiger lies in the direction of that line. This simple arrangement reveals the tiger's **bearing** (see also Example 2.7).

Proposition 8: **(Pythagoras's theorem)**: *In right-angled triangles, the squared length of the side opposite the right angle (the **hypotenuse**) equals the sum of the squared lengths of the other sides.*

Proof: We can prove this theorem by a combination of diagrams and algebra. Its graphical interpretation is shown in Figure 2.12(a). The construction described in Figure 2.12(b) shows that the four triangles fit neatly in the largest of the squares because their hypotenuses match with the sides of the square. Also, the combination of the tips of the triangles make up right angles (you can use Proposition 5 and the fact that these are right-angled triangles to convince yourself of this). Similar arguments can be used to show that the remaining shape in the middle of Figure 2.12(b) is a square with sides of length $c - b$. The area of the big square can either be written as the square of its sides or as the sum of the areas of all the shapes it contains.

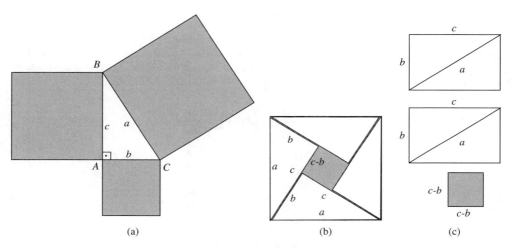

Figure 2.12: (a) Graphical interpretation of Pythagoras's theorem. The area of the larger, shaded square is equal to the sum of the areas of the other two. (b) Construction used to prove the theorem. Four identical triangles are drawn inside the largest of the three squares. (c) Rearrangement of the five shapes into two rectangles and a small square.

In Figure 2.12(c), the four right-angled triangles can be rearranged to give two rectangles. So, the area of the large square (a^2) is equal to the areas of two rectangles, each with area cb, and a square, with area $(c - b)^2$. This gives us the equation

$$a^2 = 2cb + (c - b)^2$$
$$a^2 = 2cb + c^2 - 2cb + b^2 \tag{2.10}$$

Which results in the algebraic form of Pythagoras's theorem:

$$a^2 = c^2 + b^2 \tag{2.11}$$

Ancient geometers kept their distance from numerical measurement. As we saw, they created new facts by logical argument, not the accumulation of numerical evidence. This was because they found the imprecision of measurement both unavoidable and unsatisfactory, and because they wanted their statements to be general, not tied to particular numerical examples. However, there are some deceptively simple and important questions that cannot be approached by this method. For example, it is not possible to trisect any given angle with the ancient methods. Inelegant though it may have seemed to our ancestors, today's physical and natural scientists make as much use of experimental measurement as they do of deductive logic.

2.4.   *Distance between two points*

Example 2.4: Spatial autocorrelation in ecological variables

Atlantic cod, Gadus morhua

Most geological and biological variables show spatial patterns that are not entirely random. Conditions in nearby spatial locations are more similar than those farther apart, a phenomenon known as **positive spatial autocorrelation**. Soil types follow spatial gradients. Plants of the same species are clustered because of limitations in dispersal ability. Even highly mobile species such as predatory fish (Figure 2.13) may aggregate at food-rich patches. Hence, spatial autocorrelation is propagated through different parts of the food web. To describe spatial patterns in ecological variables, we must be able to measure the similarity between the values of the variables at different locations and the geographical closeness of these locations.

Figure 2.13: The estimated distribution of cod around UK waters (Matthiopoulos *et al.* 2008).

A **distance** is a nondirectional (i.e. unsigned) measure of the closeness of two points. In one-dimensional space, the distance between two points with coordinates x_1 and x_2 is simply

$$d = |x_1 - x_2| \tag{2.12}$$

In two-dimensional space, the distance between two points, with Cartesian coordinates (x_1, y_1) and (x_2, y_2) can be calculated from Pythogoras's theorem by treating the one-dimensional distances along each of the two axes as the smaller sides of a right-angled triangle (Figure 2.14).

Applying Equation (2.11) to this triangle gives what is known as **Euclidean distance in two dimensions**

$$d = \sqrt{(x_1 - x_2)^2 + (y_1 - y_2)^2} \tag{2.13}$$

This expression can be extended to as many dimensions as required. Although physical space only has three (height, width, depth), there are several nongeographical applications (see Chapter 12) that are considered in n dimensions.

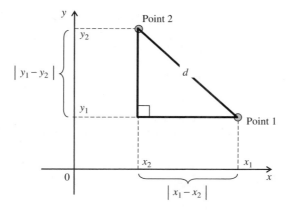

Figure 2.14: As the crow flies. The Euclidean distance (d) between two points is the hypotenuse of the triangle formed by their horizontal and vertical distances.

Example 2.5

Euclidean distances in three-dimensional space ($n = 3$) can be written in terms of width (x), depth (y) and height (z)

$$d = \sqrt{(x_1 - x_2)^2 + (y_1 - y_2)^2 + (z_1 - z_2)^2} \tag{2.14}$$

Distances in one-dimensional space ($n = 1$) can also be seen as Euclidean

$$d = \sqrt{(x_1 - x_2)^2} = |x_1 - x_2| \tag{2.15}$$

Euclidean distances are not the most suitable measure for some applications. For example, in studies of population distribution and individual movement, we must account for the fact that organisms may need to circumnavigate obstacles. In these cases, the distance between two points is defined as the shortest biologically plausible path. Other types of distance measures are needed when the units of measurement in different dimensions are not the same. For example, in phylogenetics, qualitatively different aspects of an organism's phenotype and genotype need to be combined into a measure of similarity. Each of these characteristics is a dimension in the n-space in which the comparison takes place. Although the Euclidean distance could be used, it is clearly not appropriate, because the scale of measurement of these characteristics can artificially inflate or deflate the importance of one over another. Alternative measures (such as the **Mahalanobis distance**) have been devised to deal with this problem.

2.1: Correlogram

The correlogram is a graph of the similarity between two points in space (or time) as a function of the distance (or time lag) between them. For example, given the spatial distribution of a species within a 50 × 50 m rectangular region (Figure 2.15(a)) the similarity in density between neighbouring points appears as patchiness in the overall distribution. The command correlogram() within the package spatial will generate an autocorrelation plot (Figure 2.15(b)) indicating how rapidly the similarity between points drops with distance between them. The implementation details are beyond the scope of this book but it is useful to know how to interpret the output: the autocorrelation function takes values between 1 (implying total similarity between two points) and −1 (implying total dissimilarity). The correlogram always starts from the value 1 because when the distance between two points is almost zero, they are effectively the same point (and, hence, identical). In most spatial data sets (such as the example in Figure 2.15) autocorrelation decays with distance and eventually settles to values around zero. The distance at which this happens is informative about the spatial scale of similarity. In this example, two points that are approximately 10 m apart are expected to be no more similar than two points selected at distances of 20, 30, 40 or 50 m.

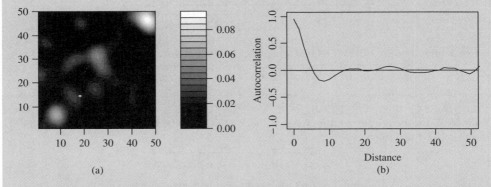

(a)

(b)

Figure 2.15: The spatial distribution of a population (a) shown as a density plot. The correlogram of this distribution (b) indicates how quickly autocorrelation (the similarity between points in space) decays with the distance between them.

2.5. Areas and volumes

The formulae used for calculating the areas and volumes of Euclidean shapes are listed in Section A2.2 in the appendix. The regularity, symmetry, even simplicity of these shapes may often make classical geometry seem irrelevant to biological phenomena. However, as we have already seen (Example 2.3), it is a handy tool in survey design, and, occasionally (see Example 2.6), real-life processes can be approximated by regular shapes.

Example 2.6: Hexagonal territories

Black tilapia, Oreochromis mossambica

Territorial animals living in low-density populations can be thought of as exploiting roughly circular regions around a central point of interest (e.g. a nest). The boundaries of these territories are unlikely to be well defined because the animals have little need to defend them. Such circular boundaries are wasteful because regions between the territories are left unclaimed. In dense populations animals need to divide space as efficiently as possible (mathematicians call such an efficient partition a **tessellation**). The space between territories is taken up and their boundaries become straightened out. If the environment is homogeneous and the territory defenders are equally aggressive, then territories should be regular hexagons. Interestingly, when the necessary conditions are borne out, hexagonal territories are observed in the wild. Barlow (1974) reported that the breeding pits dug by males of an introduced population of Black Tilapia in a large outdoor pond at Berkeley, California, were polygonal. The majority were pentagons, although a shift towards hexagons occurred at higher population densities. In the idealised example of Figure 2.16, the area of the circles at low population density (Figure 2.16(a)) represents preferred territory size. The area of the hexagons formed under high density (Figure 2.16(b)) represents the minimum required territory size. An index of territory size flexibility can therefore be devised using simple geometrical formulae from Section A2.2.

$$\text{territory flexibility} = \frac{\text{Preferred territory size}}{\text{Minimum territory size}} = \frac{2\pi r^2}{3\sqrt{3}a^2} \tag{2.16}$$

where r is the radius of the circular territory and a is the length of the side of the hexagonal territory.

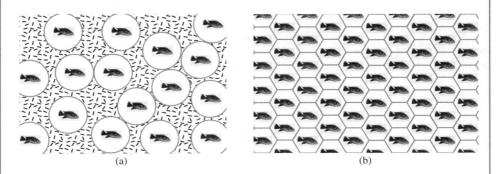

(a)   (b)

Figure 2.16: Theoretical territory arrangements at low and high population density.

2.2: Voronoi tessellations

The tool most frequently used for mapping animal territories or plant exclusion zones is the **Voronoi** (or **Dirichlet**) tessellation. The idea is simple, although the internal implementation by R is quite complicated. Given a set of points in space, straight boundaries are drawn between nearest neighbours in such a way that each boundary bisects perpendicularly the line joining the points. Depending on the study species, the data points may represent relatively fixed locations of territorial displays, the position of an animal's nest or the position of the individual plant. The data need to be in two lists corresponding to the x and y coordinates of the focal points of each territory. For example,

```
x<-c(38.63, 16.34, 22.38, 78.08, 39.93, 82.12, 45.31)
y<-c(30.13, 18.90, 34.51, 92.25, 25.60, 23.08, 93.28)
```

The following code can be used to generate the tessellation in Figure 2.17

```
require(tripack)  # loads package with tessellation functions
mo<- voronoi.mosaic(x,y) # calculates the tessellation
plot.voronoi(mo, main=NULL, sub=NULL, do.points=FALSE) # plotting
points(x,y) # adds the focal points on the plot
```

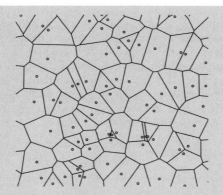

Figure 2.17: Voronoi tesselation generated for a set of focal points.

It is important to remember that the assumptions underlying Voronoi maps are purely geometrical and only under certain circumstances (e.g. a crowded population) are the boundaries in a Voronoi map also biologically meaningful.

2.6.   Measuring angles

Example 2.7: The bearing of a moving animal

Calanus finmarchicus, *a swimming copepod*

The **bearing** of a point B from a point A is the inclination formed between the line segment joining them and the north–south line through A (Figure 2.18(a)). Given six positional observations of an animal as it moves through space, we are interested in calculating the corresponding five bearings (Figure 2.18(b)). We therefore need some system for measuring angles.

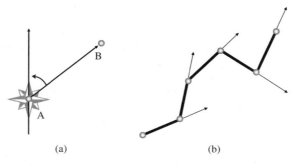

(a)   (b)

Figure 2.18: (a) The bearing is the inclination of the segment AB with the SN axis through A; (b) successive bearings of a moving animal.

There are two standard units for measuring angles: **degrees** and **radians**. One full rotation is divided into 360 equal angles called degrees. Each degree is divided into 60 minutes and each minute is divided into 60 seconds. A degree is denoted $1°$, a minute of a degree is $1'$ and a second is $1''$. For example, $30.28° = 30°16'48''$. The system of degrees stems from ancient Babylon when the numeral system was based on 60 and astronomers used 360 to roughly describe the number of days in a year.

Radians do not have the nice property of dividing the circle into an integer number of units. However, they are better for doing mathematics because they relate the angles in a circle to the circle's radius. If an angle equal to one radian is placed at the centre of a circle of radius r, then it forms an arc whose length is r (Figure 2.19).

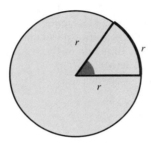

Figure 2.19: One radian is the angle that forms an arc equal in length to the circle's radius.

The circumference of a circle with radius r is $2\pi r$ (Section A2.2). This implies that the total number of radians in a circle is $\frac{2\pi r}{r} = 2\pi$. Therefore, to convert any angle d in degrees to an angle ϕ in radians, we may use the fact that $360° = 2\pi$ rad to write the following scaling relationship (see Section 1.12)

$$\frac{360}{d} = \frac{2\pi}{\phi} \tag{2.17}$$

Some cornerstone conversions can be found in Section A3.2 of the appendix.

Ⓡ 2.3: Angle conversions

If a particular number is to be interpreted by R as an angle, then it is essential to give this number in radians, otherwise the results will be incorrect. For example, as far as R is concerned, $1.571 (= \pi/2)$ is a right angle but 90 is not. If you have a data set that is in degrees, here is a little function that will do the conversion.

```
rad<-function(angle.in.degrees) {return(pi*angle.in.degrees/180)}
```

Note that R recognises the special constant π as pi. Once you have put this function into R, you can use it for your conversions. For example, the command rad(1:360) will convert into radians all 360 of the angles that can be measured on a circle using integer degrees (i.e. 1,2,3,4,...).

2.7. *The trigonometric circle*

Example 2.8: The position of a seed following dispersal

Dandelion seeds (genus Taraxacum*)*

Consider a single movement by a seed of a wind-dispersing plant resulting from a brief gust of wind. We are interested in describing the final position of the seed in relation to the plant. Cartesian coordinates (Section 1.16) will enable us to do this by placing a frame of reference (the origin of the axes) at the position of the plant (Figure 2.20(a)). We can then describe the final position as x steps to the east of the mother plant and y steps to the north. However, this is not a very natural way to describe the movement. If we were asked to describe it verbally, we would probably prefer to say that it travelled a certain distance in a particular direction. The distance from the parent plant defines a circle, and the direction of the seed can be measured as the angle (θ) from an arbitrary baseline (Figure 2.20(b)).

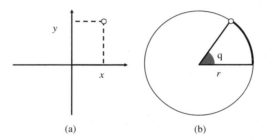

(a)   (b)

Figure 2.20: The position of the seed in relation to the parent plant given in (a) Cartesian coordinates (aligned with longitude and latitude) and (b) in terms of direction and distance from the parent plant. Direction is measured counter-clockwise from the longitude axis.

Irrespective of their relative biological appeal, both of the diagrams in Figure 2.20 define the position of the seed, so they must be equivalent. Trigonometry allows us to combine and switch between these two definitions of relative position. We must first unify the two diagrams into one, called the **trigonometric circle**, which is made up of a Cartesian coordinate system and a circle of radius 1 centred at the axes' origin. The four regions of the plane created by the Cartesian axes are called **quadrants** (Figure 2.21). In Section 2.9 we will see how this simple graphical device can be extended to radii other than 1, but first we need to define some new functions.

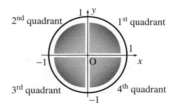

Figure 2.21: The trigonometric circle divides the plane into four quadrants that are conventionally numbered as shown.

2.8. Trigonometric functions

The radius joining the point (1,0) to the centre of the trigonometric circle forms an angle $\theta = 0$ with the positive half of the x-axis. If we start moving this point counter-clockwise along the circle (Figure 2.22), its distance from the origin remains the same but the angle θ increases. Now, consider vertical and horizontal projections cast by the point on the two Cartesian axes (the dashed segments in Figure 2.22).

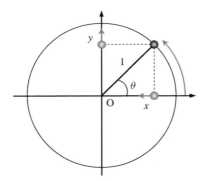

Figure 2.22: A point rotating along the trigonometric circle. As the angle θ changes, so do the projections of the point on the two Cartesian axes.

These projections have the special names **cosine** and **sine** and are written as

$$\begin{aligned}\cos(\theta) &= x \\ \sin(\theta) &= y\end{aligned} \qquad \text{(On the trigonometric circle)} \qquad (2.18)$$

The sine and cosine are specified to a particular value of the angle θ because as θ changes, so do the projections x and y. For example, as θ increases from 0 to $\pi/2$ ($= 90°$) the distance (x) along the horizontal axis decreases to 0 while the distance y along the vertical axis increases to 1. For brevity, these functions are often written without the brackets, e.g. $\sin(\theta) = \sin\theta$.

These definitions are specific to the trigonometric circle (radius 1) but can be easily generalised to a circle of any radius by a scaling operation. To retrieve Figure 2.22 from a diagram drawn on a general circle of radius r, we need to divide the aspects of the diagram by r. The sine and cosine of the angle will remain the same because this operation does not affect the angle. However, the projections x and y of the segment r are divided by the scalar r (Figure 2.23(a)).

This gives us the general expression for sine and cosine in a circle of radius r

$$\begin{aligned}\cos\theta &= \frac{x}{r} \\ \sin\theta &= \frac{y}{r}\end{aligned} \qquad \text{(On any circle)} \qquad (2.19)$$

This scaling also gives us a definition of the sine and cosine independent of circles: For the angle θ of any right-angled triangle (Figure 2.23(b)), the sine and cosine are

$$\begin{aligned}\cos\theta &= \frac{\text{adjacent side}}{\text{hypotenuse}} = \frac{x}{r} \\ \sin\theta &= \frac{\text{opposite side}}{\text{hypotenuse}} = \frac{y}{r}\end{aligned} \qquad (2.20)$$

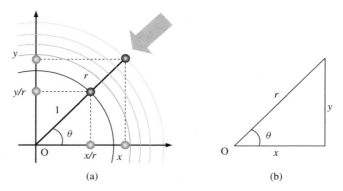

(a)   (b)

Figure 2.23: The sine and cosine for any circle of radius r can be defined by converting it to a trigonometric circle. This is done by dividing all lengths by r. (a) In the example shown here (r > 1), division corresponds to a shrinking; (b) in general, the sine and cosine of an angle in a right-angled triangle can be defined in terms of the triangle's sides.

The ratios of the two smaller sides of the triangle are often useful for calculations. They are known as the **tangent** and **cotangent**,

$$\tan\theta = \frac{\text{opposite side}}{\text{adjacent side}} = \frac{y}{x}$$

$$\cot\theta = \frac{\text{adjacent side}}{\text{opposite side}} = \frac{x}{y}$$

(2.21)

How do we calculate the values of trigonometric functions? As is often the case with computational issues, the most efficient method is far removed from the quantity's definition. For example, the values of the sine and cosine are efficiently approximated by the following expressions

$$\cos\theta = 1 - \frac{\theta^2}{2!} + \frac{\theta^4}{4!} - \frac{\theta^6}{6!} + \cdots$$

$$\sin\theta = \theta - \frac{\theta^3}{3!} + \frac{\theta^5}{5!} - \frac{\theta^7}{7!} + \cdots$$

(2.22)

I will explain where these expressions come from in Section 4.12. The numbers with exclamation marks in these expressions are called **factorials**, defined as the product of all integers from 1 up to the number of the factorial (e.g. $6! = 1 \times 2 \times 3 \times 4 \times 5 \times 6$). As you can gather, these numbers can quickly become very big and the calculations in Equation (2.22) can be labour-intensive. Thankfully, these formulae have been efficiently implemented in calculators and scientific software.

℞ 2.4: Trig functions in R

The trigonometric functions available in R are `sin()`, `cos()` and `tan()`. Unlike typeset maths, in R you always need to use the brackets (so, `cos(pi)` will return −1, but `cospi` will give you an error message). A cotangent function for a given angle `theta` can be defined as follows (remember that all the angles need to be in radians – see R2.3).

```
cot<-function(theta) {return(1/tan(theta))}
```

2.9. Polar coordinates

The movement of an organism in Cartesian coordinates from the origin (0,0) to a new point (x, y) can equivalently be described by the length and direction of its displacement (r, θ) (see Example 2.8). Here, this equivalence is derived mathematically. The general definition of trigonometric functions in Equations (2.19) can be rewritten

$$
\begin{aligned}
x &= r \cos \theta \\
y &= r \sin \theta
\end{aligned}
\tag{2.23}
$$

This isolates the Cartesian coordinates on the left-hand side, and each coordinate is written in terms of the distance (r) and direction (θ) from the origin. This alternative way of locating points gives us the system of **polar coordinates**. For some biological applications, such as seed dispersal (Example 2.8) or animal movement (Example 2.9), polar coordinates are a more intuitive mathematical description.

Example 2.9: Random walks

I will illustrate random walks by considering the movement of a swimming copepod on wide, still water. The behavioural simplicity of this organism makes it easier to capture the features of its movement with a simple model and the homogeneity of the movement substrate means that complications such as obstacles and water currents can be ignored.

A good starting point for a model of movement is the **isotropic random walk** (Hughes, 1995) which assumes that an organism takes steps of fixed length r (e.g. a single swimming beat) in a uniformly random direction. A simple isotropic random walk with eight directions is illustrated in Figure 2.24(a). The animal chooses each of the eight directions about once out of every eight steps.

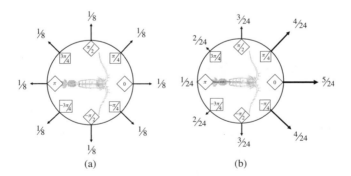

(a)   (b)

Figure 2.24: (a) Isotropic random walk with eight possible directions of movement and (b) a correlated random walk with a tendency to maintain the current direction of movement. The animal's current position is the centre of the circle and the current direction is from west to east. Positive angles measure left-hand turns and negative values turns to the right.

How can we simulate n steps of the animal's movement using this simple set of rules? Here, I simply describe the logical procedure (the **algorithm**). The computer implementation is, as yet, too demanding.

❶ Any point of origin will do. The animal is released at (0,0) with initial direction 0 (i.e. heading east). Its direction of movement is denoted by ϕ.

❷ Consider the next step (let's call it i, so that the current direction of movement is written ϕ_i.)

❸ Select one of the eight changes of direction randomly, each with equal likelihood. The choice made at the ith step is θ_i.

❹ The animal's direction of movement will therefore become $\phi_{i+1} = \phi_i + \theta_i$

❺ The new position of the animal in Cartesian coordinates is $x_{i+1} = x_i + r\cos(\phi_{i+1})$ and $y_{i+1} = y_i + r\sin(\phi_{i+1})$

❻ Go to step 2 above until all n steps have been taken.

Example output of a simulation based on the above algorithm is shown in Figure 2.25(a). Because the animal is allowed to spin randomly, every time it takes a step it regularly crosses its own path. In reality, not even a copepod in a cue-less environment should be expected to move in so simple a fashion. As argued by Bovet and Benhamou (1988) a bilaterally symmetrical animal with even a small degree of cephalisation will tend to maintain its current direction of movement. Modifying the rules of movement to incorporate this tendency gives rise to the **correlated random walk** (Bovet and Benhamou, 1988; Turchin, 1998; Figure 2.24(b)). The simulation algorithm only needs to be changed slightly by ensuring that the forward direction is selected more frequently than the backward direction. These new rules result in somewhat different patterns from the isotropic walk (compare Figures 2.25(a) and 2.25(b)). The animal crosses its own path less frequently and the overall path is less folded.

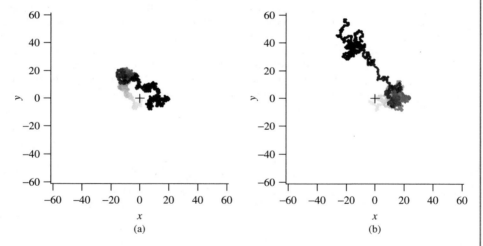

Figure 2.25: Examples of 1000 steps from (a) an isotropic and (b) a correlated random walk. The cross indicates the release point and darker shades represent more recent locations.

2.10. Graphs of trigonometric functions

The trigonometric functions are fundamentally different from the functions of Chapter 1 because they are not one-to-one (i.e. the same sine, cosine or tangent value can be obtained from different angles).

Example 2.10

Consider a point placed at the polar coordinates $(1, \theta)$. If we allow it to come full circle to exactly the same position (Figure 2.26(a)), then its total rotation will be $\theta + 2\pi$ but its polar and Cartesian coordinates will be exactly the same. So, the cosine, sine and tangent of the two angles will be the same. If, instead of originally moving the point in a counter-clockwise direction θ, we moved it clockwise to $-\theta$ (Figure 2.26(b)) then its projection on the x-axis would be the same despite the point being in a different position. Finally, if it was allowed to travel to $\pi - \theta$, its projection on the y-axis would be the same (Figure 2.26(c)).

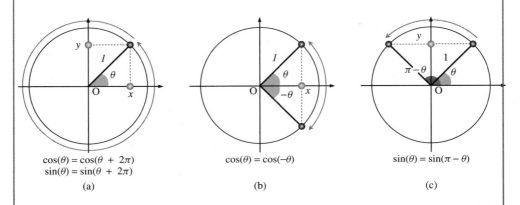

$$\cos(\theta) = \cos(\theta + 2\pi)$$
$$\sin(\theta) = \sin(\theta + 2\pi)$$

(a)

$$\cos(\theta) = \cos(-\theta)$$

(b)

$$\sin(\theta) = \sin(\pi - \theta)$$

(c)

Figure 2.26: Three examples in which different angles produce exactly the same values for the sine, cosine or both.

To visualise the behaviour of sine and cosine, we start at the Cartesian coordinates $(1,0)$. This point's radius forms an angle $\theta = 0$ with the x-axis and its projections are 1 on the x-axis and on 0 on the y-axis. Hence, $\cos(0) = 1, \sin(0) = 0$. Allowing the point to move counter-clockwise, to the coordinates $(0,1)$, forms an angle of $\pi/2 \; (= 90°)$ with the x-axis, meaning that $\cos(\frac{\pi}{2}) = 0$ and $\sin(\frac{\pi}{2}) = 1$. Continuing to $\pi (= 180°)$, gives $\cos(\pi) = -1$ and $\sin(\pi) = 0$. Finally, going under the x-axis to $3\pi/2$ gives $\cos(\frac{3\pi}{2}) = 0$ and $\sin(\frac{3\pi}{2}) = -1$. Using a computer, we can fill in the gaps for intermediate angles (Figure 2.27).

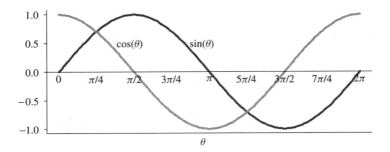

Figure 2.27: Graphs of sine and cosine in the domain 0 to 2π.

These two functions remain constrained on the vertical axis between -1 and 1. In contrast, the graphs of the tan and cot tend to infinity as the values of the x and y projections, respectively, approach zero (see Equations (2.21)). For example, as θ approaches $\pi/2$, the projection on the x-axis tends to zero, sending the ratio y/x to infinity (Figure 2.28). In fact, at exactly $\theta = \frac{\pi}{2}$, the tangent is not defined because it involves division by zero. Once θ exceeds $\pi/2$, even by a little, the tangent is defined again, but with a large, negative value, because now the projection on the x-axis is negative.

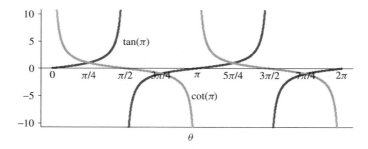

Figure 2.28: Graphs of tangent and cotangent in the domain 0 to 2π.

2.11. Trigonometric identities

You may remember from Chapter 1 that an identity is an equation that is true for all values of its variables. Identities are used as stepping stones in mathematical derivations (e.g. proofs or simplifications). The first few trigonometric identities come directly from the definition of the trigonometric functions. For example, if we divide the definitions of sine and cosine, we obtain a relationship between them and the tangent

$$\frac{\sin\theta}{\cos\theta} = \frac{y/r}{x/r} = \frac{y}{x} = \tan\theta \tag{2.24}$$

Also, if we apply Pythagoras's theorem to the triangle formed by the radius and its projections in the trigonometric circle (see Figure 2.22) we get

$$x^2 + y^2 = 1 \tag{2.25}$$

which immediately yields a really useful relationship between the sine and cosine:

$$\cos^2\theta + \sin^2\theta = 1 \tag{2.26}$$

Note that $(\cos\theta)^2$ is written $\cos^2\theta$, not $\cos\theta^2$, because that would imply that the angle is first squared before taking its cosine.

Other identities can be obtained by the sort of argument illustrated in Figure 2.26. Here are some examples that you should be able to verify by drawing them on the trigonometric circle:

$$\begin{array}{ll} \cos(-\theta) = \cos\theta & \sin(-\theta) = -\sin\theta \\ \cos(\theta + n2\pi) = \cos\theta & \sin(\theta + n2\pi) = \sin\theta \quad (n \in \mathbb{Z}) \\ \cos(\pi - \theta) = -\cos\theta & \sin(\pi - \theta) = \sin\theta \end{array} \tag{2.27}$$

Several other trigonometric identities are presented without derivation in Sections A3.6–A3.11 of the appendix. These deal with trigonometric functions of the sum of two angles (Sections A3.7 and A3.8), the products and sums of trigonometric functions (Sections A3.9 and A3.10) and the relationships between angles and sides on oblique triangles (Section A3.11).

Example 2.11: A two-step random walk

The random walks examined in Example 2.9 can be generalised to allow for more than eight angles of directional change and unequal step lengths. Consider two successive steps by such a random walker (Figure 2.29). We know the step lengths r_1, r_2 and its change of direction (θ_2) following the first step. We would like to calculate a general expression for its distance (r) from the origin after the two steps.

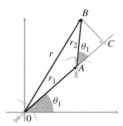

Figure 2.29: Two successive steps in a generalised random walk.

Previously (Section 2.4), we saw how to calculate a distance by treating it as the hypotenuse of a right-angled triangle. Since the animal's steps aren't necessarily perpendicular to each other, we have to construct a suitable triangle ($\triangle\, OBC$ in Figure 2.29). The application of Pythagoras to this triangle gives

$$r^2 = OC^2 + BC^2$$
$$r^2 = (OA + AC)^2 + BC^2$$
$$r^2 = (r_1 + r_2 \cos\theta_2)^2 + (r_2 \sin\theta_2)^2 \tag{2.28}$$
$$r^2 = r_1^2 + 2r_1 r_2 \cos\theta_2 + r_2^2 \cos^2\theta_2 + r_2^2 \sin^2\theta_2$$
$$r^2 = r_1^2 + 2r_1 r_2 \cos\theta_2 + r_2^2(\cos^2\theta_2 + \sin^2\theta_2)$$

Using the identity in Equation (2.26) gives

$$r = \sqrt{r_1^2 + 2r_1 r_2 \cos\theta_2 + r_2^2} \tag{2.29}$$

2.12. *Inverses of trigonometric functions*

Trigonometric functions can be used to convert polar coordinates (Section 2.9) into horizontal and vertical (Cartesian) coordinates. The inverse operation is also possible, but requires the concept of inverse trigonometric functions. Consider the function

$$y = \sin\theta \tag{2.30}$$

Its inverse, called the **arcsine**, is the angle θ whose sine is y

$$\theta = \arcsin y \tag{2.31}$$

The inverses of cos and tan (written **arccos** and **arctan**) have similar interpretations. An alternative notation for inverse trigonometric functions is $\sin^{-1} y, \cos^{-1} y, \tan^{-1} y$.

Inverse trigonometric functions are calculated with the aid of approximations similar to those used for the sine and cosine (see Equation (2.22) and Section 4.12 for more on series expansions) so, in most cases, we need to reach for a calculator or a computer. Two facts are important in performing these inverse operations: the sine and cosine take values between -1 and 1, so y in Equation (2.31) needs to be constrained between these values; and the trigonometric functions are not one-to-one, several angles can have the same sine, cosine and tangent (see Section 2.10). To avoid confusion, we therefore need to define both the domain and the range of inverse trigonometric functions:

$$\begin{aligned} \theta &= \arcsin y & -1 \leqslant y \leqslant 1 & \quad -\tfrac{\pi}{2} \leqslant \theta \leqslant \tfrac{\pi}{2} \\ \theta &= \arccos y & -1 \leqslant y \leqslant 1 & \quad 0 \leqslant \theta \leqslant \pi \\ \theta &= \arctan y & -\infty \leqslant y \leqslant \infty & \quad -\tfrac{\pi}{2} \leqslant \theta \leqslant \tfrac{\pi}{2} \end{aligned} \qquad (2.32)$$

While the inverse functions do not exist for values outside these ranges of y, there are values of θ outside these ranges that will satisfy the equations for a given y. For example, $\arcsin(0.87) \cong \tfrac{\pi}{3}$, but $\theta = \pi - \tfrac{\pi}{3}$ will also give the same sine. If you need help in finding these other values, have a look at Equations (2.27).

Example 2.12: Displacement during a random walk

The term **displacement** formally includes the Euclidean distance and direction of a move. It is therefore a collective word for polar coordinates (if the origin of the axes coincides with the origin of the move). The example in Figure 2.30 shows ten steps of a correlated random walk with a fixed step length and only four possible directions of movement (this is also called a random walk on a square grid).

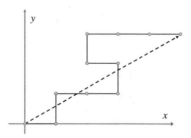

Figure 2.30: Ten steps of a random walk on a grid.

The net movement is five steps to the east and three steps to the north. We would like to find the displacement achieved by the ten steps. The distance covered can be found from Pythagoras's theorem as $\sqrt{5^2 + 3^2} \cong 5.83$. The direction can be found by examining the values $\arcsin(3/5.83) \cong 0.54$ and $\arccos(5/5.83) = 0.54$. So, the angle is 0.54 radians (or 0.17π, about $31°$). It is interesting that despite the discrete nature of this random walk (fixed step length, only four directions per step) neither the angle nor the distance is an integer. Conceivably, such a random walk on a grid could be used to approximate a continuous random walk on a larger scale, by grouping every ten steps together into one.

2.5: Inverse trig functions in R

These are `asin()`, `acos()`, `atan()` and `atan2()`. All return values in radians. The values returned by the first three of these conform to the ranges given in Equation (2.32). Other possible values for the angle need to be calculated by the user. The function `atan2(y,x)` aims to resolve this ambiguity when both the cosine (x) and the sine (y) of an angle are known. For example, consider the type of random walk discussed in Example 2.12. A net move of three steps to the south and six steps to the west would give us a tangent of $\frac{-3}{-6} = \frac{1}{2}$. Notice that the two negative signs vanish through the division, so some information is lost. The command `atan(0.5)` gives the answer 0.46 ($\cong 0.15\pi$). This answer lies in the first quadrant, despite the fact that the move ended in the third. In contrast, the command `atan2(-3,-6)` gives the more appropriate answer -2.60($\cong -0.83\pi$).

2.13. *Trigonometric equations*

The main principles of equation solving (Chapter 1) remain the same, but the last step of successfully solving a trigonometric equation will involve the use of an inverse trigonometric function.

Example 2.13

The equation $3\sin x + 2 = 1$ can be solved by first isolating the trigonometric term, $\sin x = -1/3$ and then applying the arcsine on both sides of the equation to get $x = \arcsin(-1/3) \cong -0.34$.

Not all trigonometric equations have a solution (e.g. any expression that leads to an arcsine or arccosine of a number outside the domain $(-1,1)$). Furthermore, even if a solution exists, it cannot always be found algebraically. There are usually two different reasons for this: either the equation contains several different trigonometric functions or it contains the variable both inside and outside a trig function. In the first case, it is sometimes possible to make progress by recognising (or forcing the equation into) an algebraic expression that appears in a trigonometric identity (see Example 2.14). Failing that, numerical approximation is the last recourse.

Example 2.14

Consider the equation $\sin x + \cos x = 1$. It contains both the sine and cosine of the unknown x. The simplest trigonometric identity relating these two is $\cos^2 x + \sin^2 x = 1$. Notice what happens if we square both sides of the equation (all the necessary identities can be found in Section A3 in the appendix).

$$(\sin x + \cos x)^2 = 1$$
$$\sin^2 x + 2\sin x \cos x + \cos^2 x = 1$$
$$(\sin^2 x + \cos^2 x) + (2\sin x \cos x) = 1$$
$$1 + \sin 2x = 1 \qquad (2.33)$$
$$\sin 2x = 0$$
$$2x = \arcsin 0$$
$$x = 0$$

We only get one answer from this last step because arcsine returns only values in the range $[-\pi/2, \pi/2]$. However, there are other values of x that satisfy the equation $\sin 2x = 0$. For example, π will also give a sine equal to zero, so another solution of the equation is $\pi/2$.

Example 2.15: VHF tracking for terrestrial animals

Our last attempt to develop this field method (Example 2.3) ended with the requirement that the operator equalises the signal at the two receivers by rotating the bracket that links their aerials. This time, we are told that the intensity of the signal can be used to estimate the distance of the animal from each receiver. We are given the following geometric formula for estimating the length of the segment l (called the **median** from the apex A of the triangle to the middle of the opposite side BC)

$$l = \sqrt{\frac{2d_1^2 + 2d_2^2 - d^2}{4}} \tag{2.34}$$

We would like to estimate the animal's direction in relation to the midpoint of the bracket.

To calculate the angle θ (Figure 2.31(a)) formed by the bracket BC and the line segment AD, we first apply the identity known as the cosine law (Section A3.11) twice, for different triangles:

$$(\triangle\ ADC)\ d_1^2 = l^2 + \left(\frac{d}{2}\right)^2 - 2ld\cos\theta \tag{2.35}$$

$$(\triangle\ ADB)\ d_2^2 = l^2 + \left(\frac{d}{2}\right)^2 - 2ld\cos(\pi - \theta) \tag{2.36}$$

Bearing in mind that $\cos(\pi - \theta) = -\cos\theta$, we may now subtract these two equations by parts to get

$$d_1^2 - d_2^2 = -4ld\cos\theta \tag{2.37}$$

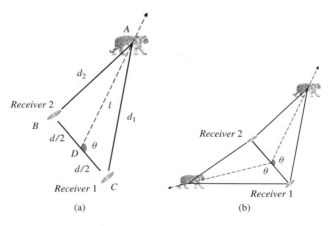

Figure 2.31: (a) Given the distances d_1 and d_2 we want to calculate the angle θ; (b) this problem has two solutions.

This involves the required angle θ in a trigonometric term. Solving for this term

$$\cos\theta = \frac{d_2^2 - d_1^2}{4ld} \tag{2.38}$$

This contains known quantities, apart from l. Substituting it into Equation (2.34) gives

$$\cos\theta = \frac{d_2^2 - d_1^2}{2d\sqrt{2d_2^2 + 2d_1^2 - d^2}} \tag{2.39}$$

The required angle can now be found:

$$\theta = \arccos\left(\frac{d_2^2 - d_1^2}{2d\sqrt{2d_2^2 + 2d_1^2 - d^2}}\right) \tag{2.40}$$

This only returns an answer in the range $[0, \pi]$. An additional answer in the range of negative angles also exists (Figure 2.31(b)).

2.14. Modifying the basic trigonometric graphs

Example 2.16: Nocturnal flowering in dry climates

The flower of a caper plant, Capparis spinosa L.

The timing of flowering in dry climates has evolved as a response to water loss. In caper plants, the white petals open at night, as relative humidity increases and temperature drops (Rhizopoulou *et al.*, 2006). Consequently, the number of visits by pollinating insects fluctuates with the state of flowering and, therefore, the time of day.

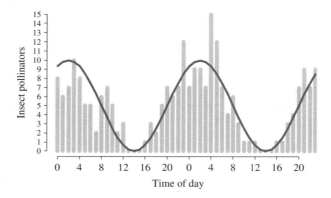

Figure 2.32: Attendance of insect pollinators on the flowers of a caper plant measured at hourly intervals during a two-day period (observations shown as bars, underlying pattern shown as a curve).

We wish to approximate the underlying pattern of the observations in Figure 2.32 by the curve shown in that figure. The pollinator counts follow a simple, regular pattern that resembles the graph of the sine or cosine. The features in the data can be summarised as follows:

❶ The independent variable is time (not angle).
❷ The number of pollinators peaks at about 2 am.
❸ There are hardly any pollinators at 2 pm.
❹ The pattern is repeated every day.
❺ The curve fluctuates between 0 and 10 (although chance events mean that the actual number of pollinators may exceed this maximum).

How can we describe this pattern mathematically by a function $f(t)$, using as a basis a trigonometric function such as the cosine?

Two operations may be used to convert the function cos into the required function $f(t)$: the graph could be shifted to the left/right or it could be stretched/contracted. In addition, these two operations can be applied in the vertical or horizontal direction. Let us take as our starting point the function $f(t) = \cos t$ in the domain (0,48) (Figure 2.33(a)). Operating on the horizontal direction is equivalent to changing the independent variable. The time required to complete a full cycle is called the **period**. We would like a function with a period of 24 instead of 2π hours. This can be done by scaling time as follows (see also Section 1.12):

$$f(t) = \cos\left(\frac{\pi}{12}t\right) \tag{2.41}$$

This multiplication by $\pi/12$ (a number smaller than 1) results in a slowing down of the oscillations (Figure 2.33(b)). Next, we can correct the position of the peaks and troughs of the oscillation. In Figure 2.33(b), the peaks occur at 0:00. We would like to shift the curve so that the peaks occur at 2:00. This is a shift of two hours to the right, achieved by subtracting 2 from the time variable (Figure 2.33(c)).

$$f(t) = \cos\left(\frac{\pi}{12}(t-2)\right) \tag{2.42}$$

Note that the multiplier $\pi/12$ is applied to the whole difference, to scale hours to radians.

Focusing next on the vertical axis, the curve in Figure 2.33c fluctuates between -1 and 1 (an interval of 2). This can be converted to an interval of 10 (the amplitude of the fluctuations in Figure 2.32) by multiplying with a scaling factor of $10/2 = 5$

$$f(t) = 5\cos\left(\frac{\pi}{12}(t-2)\right) \tag{2.43}$$

The resulting curve (Figure 2.33(d)) can now be shifted upwards to ensure that it never becomes negative and that its minimum is zero. Since half of the amplitude is under the x-axis, we can do this by adding 5 to the whole function

$$f(t) = 5 + 5\cos\left(\frac{\pi}{12}(t-2)\right) \tag{2.44}$$

This is the required approximation to the data (Figure 2.33(e)).

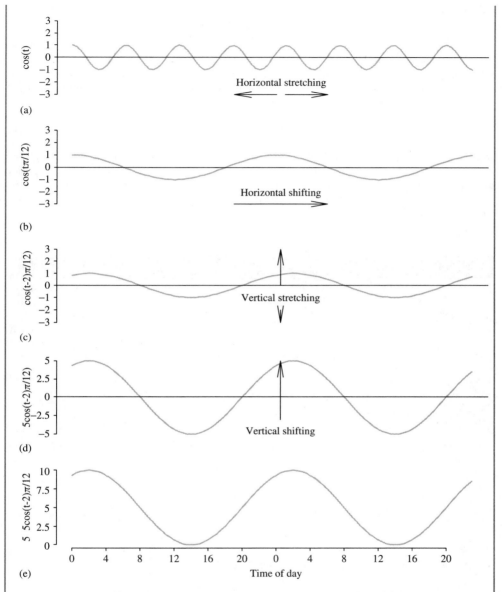

Figure 2.33: Transformation of the function cos *t* into one approximating the pattern of pollinator attendance shown in Figure 2.32. The transformation is achieved in four successive steps by stretching and shifting the original curve in the vertical and horizontal directions.

In general, the transformation of cos into a new periodic function can be written as

$$f(t) = \frac{f_{max} + f_{min}}{2} + \frac{f_{max} - f_{min}}{2} \cos\left(\frac{2\pi}{T}(t - t_0)\right) \qquad (2.45)$$

where f_{min} and f_{max} are the minimum and maximum of the new function (replacing -1 and 1), T is the new period (replacing 2π) and t_0 is the location of the new peak (replacing 0). A simpler version of this expression is

$$f(t) = a + b \cos \omega(t - t_0) \qquad (2.46)$$

In this relationship, a determines the baseline of the fluctuations (in fact, it is the long-term average value of the function), b determines the amplitude, ω is inversely related to the period and t_0 is the position of the peak. The inverse of the period is known as the **frequency**. So, whereas the period is the time it takes to complete one full cycle (e.g. 24 h), the frequency is the number of cycles that can be performed in a unit of time ($1/24 = 0.042$).

ℝ 2.6: Adding arrows to plots

Arrows such as those shown in Figure 2.33 can be drawn by the command `arrows(x0,y0, x1,y1)`, in which `x0, y0` are the coordinates of the point from which the arrow begins and `x1, y1` are the coordinates of the end point. These are measured in the coordinate system of the plot to which the arrows are to be added. If more than one arrow is required, then we can replace `x0,y0,x1,y1` by lists of values using the `c()` command. For example, to pinpoint the maximum and minimum values in the sine graph:

```
th<-seq(0,2*pi, by=pi/20)
plot(th,sin(th), xlim=c(0,2*pi), ylim=c(-2,2), type="l")
arrows(c(pi/2,3*pi/2), c(1.5,-1.5), c(pi/2,3*pi/2), c(1.1,-1.1))
```

The first line creates a sequence of values from 0 to 2π at intervals of $\pi/20$. The second line plots the sine function, leaving enough space in the y-axis for the arrows. The third line adds the arrows.

2.15. Superimposing trigonometric functions

Example 2.17: More realistic model of nocturnal flowering

We now consider a slightly different pattern of pollination attendance for the caper plant example (Example 2.16). It stands to reason that, while the flowers of the plant are closed, no insects will be visiting it. The flowers open at about 8 pm so we would expect a sharp increase of visits around that time, reaching the same peak as before. This pattern (Figure 2.34) is simple and periodic, but it is not horizontally symmetric (the peaks are sharper than the troughs). How are we to approximate such shapes using trigonometric functions?

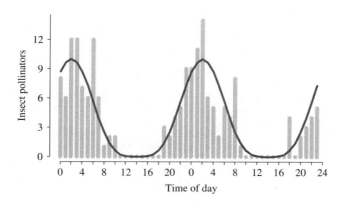

Figure 2.34: Flowers are closed during daytime, so attendance by insects is zero for that period.

It turns out that any periodic function can be emulated by the summation (superposition) of sines and cosines in a special kind of expression called a **trigonometric polynomial**, which can be derived as follows: Equation (2.46) was obtained by stretching and shifting the cosine curve. The cosine term of this function can be expanded with the aid of the second identity in Section A3.7,

$$
\begin{aligned}
f(t) &= a + b \cos \omega(t - t_0) \\
&= a + (b \cos \omega t_0) \cos \omega t + (b \sin \omega t_0) \sin \omega t
\end{aligned} \tag{2.47}
$$

The terms inside the brackets do not depend on t, the time of day. We can simplify the notation by renaming $b \cos \omega t_0 = \beta$ and $b \sin \omega t_0 = \gamma$. Note how similar this is to a conversion from polar to Cartesian coordinates (Equations (2.23), Section 2.9): just like any point on the plane can be uniquely specified by either polar or Cartesian coordinates, so can any combination of values (β, γ) be constructed by an appropriately chosen combination of values $(b, \omega t_0)$. This change of notation gives us

$$
f(t) = a + \beta \cos \omega t + \gamma \sin \omega t \tag{2.48}
$$

Consider several of these functions, each with its own parameters, and each describing a different periodic pattern

$$
\begin{aligned}
f_1(t) &= a_1 + \beta_1 \cos \omega_1 t + \gamma_1 \sin \omega_1 t \\
f_2(t) &= a_2 + \beta_2 \cos \omega_2 t + \gamma_2 \sin \omega_2 t \\
f_3(t) &= a_3 + \beta_3 \cos \omega_3 t + \gamma_3 \sin \omega_3 t \\
&\vdots
\end{aligned} \tag{2.49}
$$

If we add these together, we get

$$
\begin{aligned}
&f_1(t) + f_2(t) + f_3(t) + \cdots = \\
&a_1 + \beta_1 \cos \omega_1 t + \gamma_1 \sin \omega_1 t + a_2 + \beta_2 \cos \omega_2 t + \gamma_2 \sin \omega_2 t + a_3 + \beta_3 \cos \omega_3 t + \gamma_3 \sin \omega_3 t + \cdots
\end{aligned} \tag{2.50}
$$

The notation can be simplified by calling the resulting function $f(t)$, incorporating all the constants $a_1, a_2, a_3, \ldots$ into a single symbol (a) and collecting the sine and cosine terms into two groups

$$f(t) = a + (\beta_1 \cos \omega_1 t + \beta_2 \cos \omega_2 t + \beta_3 \cos \omega_3 t + \cdots) \\ + (\gamma_1 \sin \omega_1 t + \gamma_2 \sin \omega_2 t + \gamma_3 \sin \omega_3 t + \cdots)$$

(2.51)

We assume, in addition, that the terms $\omega_1, \omega_2, \omega_3, \ldots$, relating to frequency, are integer multiples of a particular value, so

$$f(t) = a + (\beta_1 \cos \omega t + \beta_2 \cos 2\omega t + \beta_3 \cos 3\omega t + \cdots) \\ + (\gamma_1 \sin \omega t + \gamma_2 \sin 2\omega t + \gamma_3 \sin 3\omega t + \cdots)$$

(2.52)

This expression is a **trigonometric polynomial** and it is capable of reproducing simple patterns like the one in Example 2.17, as well as more complicated ones (see Figure 2.35 in Example 2.18).

Example 2.18

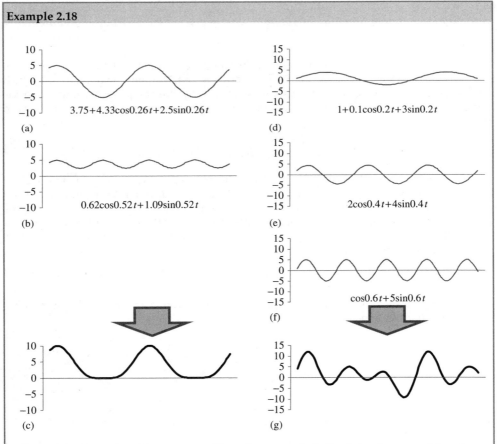

(a) $3.75 + 4.33\cos0.26t + 2.5\sin0.26t$

(b) $0.62\cos0.52t + 1.09\sin0.52t$

(c)

(d) $1 + 0.1\cos0.2t + 3\sin0.2t$

(e) $2\cos0.4t + 4\sin0.4t$

(f) $\cos0.6t + 5\sin0.6t$

(g)

Figure 2.35: Examples of two superpositions. The example in the first column combines two curves (parts (a) and (b)) to generate the curve in part (c). This is the insect attendance curve shown in Example 2.17. The example in the second column has three component curves (parts (d), (e) and (f)) and leads to a more complicated pattern (part (g)).

2.16. Spectral analysis

Different ecological processes may be associated with characteristic periods (diurnal patterns, seasonality, tidal cycles, sunspots, etc.). When observed in isolation, each of these periods may hint at a particular underlying process. Unfortunately, when several of these processes operate together, they may conceal each other's signature patterns. When dealing with real data, particularly population time series, it is useful to be able to **decompose** a complicated graph into its constituent parts. **Spectral analysis** achieves this by identifying the relative contribution of different frequencies to an observed time series. The mathematical mechanics of this are less important than knowing how to apply the method and interpret its output. The plot generated by spectral analysis (known as the **periodogram, spectral diagram, power spectrum** or simply the **spectrum**) represents the contribution (known as the **spectral density**) of each frequency in the time series. So, in a strongly diurnal time series measured in hours, the frequency $1/24$ will have a high spectral density. Similarly, in the spectra of time series affected by seasonal climate patterns, the frequency $1/12$ will have a high spectral density (assuming the unit of time is one month).

Example 2.19: Dominant frequencies in density fluctuations of Norwegian lemming populations

The wide-amplitude fluctuations of lemming populations in Fennoscandia were introduced in Example 2.1. A typical example, after Kausrud *et al.* (2008), is shown in Figure 2.36(a). Even casual observation of the time series reveals some degree of regularity in the occurrence of peak densities. This is confirmed by Figure 2.36(b): the peak of spectral density occurs at a frequency of approximately 0.15 (indicated by an arrow). Since there were two surveys performed each year, the unit of time separating successive counts is 0.5 y. The dominant period of the fluctuations is therefore $0.5 \times \frac{1}{0.15} = 3.3$ years.

Traditionally, the focus of spectral analysis has been on extracting the dominant frequencies of the time series, in the hope that these can be used as the tell-tale sign of a particularly influential ecological mechanism. For example, a mathematical model based on a particular hypothesis (e.g. mustelid predation on lemmings) may predict a regular four-year cycle. The detection of high spectral densities associated with the frequency 0.25 in a real time series of annual population counts would lend support to this hypothesis.

More recently, there has been an effort to provide ecological interpretation of the properties of the entire spectrum rather than just its peak values (Akçakaya *et al.*, 2003). For example, examining where the bulk of spectral density is, along the frequency axis, may suggest whether the fluctuations in abundance are mainly affected by trophic interactions (usually leading to higher frequencies), long-term environmental change (usually giving rise to low frequencies) or both. Such discussions have borrowed terminology from optics (Figure 2.37). Hence, reddened spectra associate high spectral densities with low frequencies while in blue spectra, high frequencies are more influential. Extending the analogy, all frequencies make approximately the same contribution in a white spectrum, in the same way that all colours are equally represented in white light.

Our ability to detect the influence of extremely high or extremely low frequencies in a time series depends on the scale and resolution of observation. The **resolution** of the data is associated with the sampling rate which sets a limit on the maximum observable frequency.

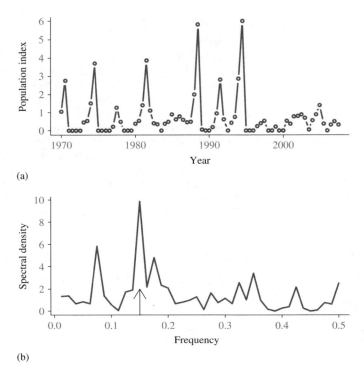

(a)

(b)

Figure 2.36: (a) Lemming population density in Finse, in the south of Norway, represented by index values between 0 and 6. Two counts were made in each year (spring and autumn); (b) the spectral diagram of the time series.

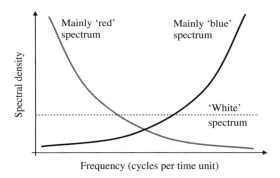

Figure 2.37: The colours of ecological time series are derived from their spectral diagrams.

Changes occurring faster than this cannot be resolved. On the other hand, the **scale** of the data is determined by the duration of the time series. This defines the lowest frequency observable in the data ($=$ duration^{-1}). Changes slower than this cannot easily be detected. The fact that most ecological time series are short implies that long-term (low-frequency) trends in particular will generally go undetected (Inchausti and Halley, 2002). This has prompted the criticism that spectral analysis may not be a useful approach to analysing ecological data (Turchin, 2003). However, this limitation need not apply to the environmental variables that impact ecological processes such as temperature or rainfall for which long and finely-resolved time series are common. Also the power of spectral analysis lies in the perspective it offers, particularly when it comes to the interpretation of variability on different scales of time (Halley, 1996).

Example 2.20: Spectral analysis of oceanographic covariates

Pacific sardine, Sardinops sagax

The Pacific Decadal Oscillation (Figure 2.38(a)) is a temperature pattern with enormous significance for the relative abundance of sardines and anchovies in the Pacific Ocean (Chavez *et al.*, 2003). The power spectrum of this time series shows a number of distinct peaks, indicating strong periodic (sinusoidal) oscillations at certain frequencies. However, this spectrum also offers us knowledge of things beyond sinusoidal patterns. For example, the general roll-off of power (fitted line if Figure 2.38(b)) is a '1/f-noise' pattern, which indicates that random variability is distributed evenly between all timescales (Halley, 1996).

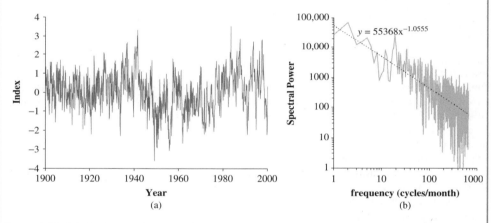

Figure 2.38: (a) The index of the Pacific Decadal Oscillation and (b) its spectral diagram using a linearising transformation on the y-axis.

2.7: Objects

When a particular command generates different types of output, R may package those together into parcels called objects. These always include the main output of the command but they may also include the specifications of the code used to obtain it, some intermediate calculations and potentially useful by-products of the calculations. The output of the command `spectrum()` is such an object. This is how it is obtained from the data of Example 2.19:

```
p<-
c(1.05,2.75,0,0,0,0,0.45,0.54,1.51,3.7,0,0,0,0,0.22,1.27,0.48,0,0,0,0.38,
0.55,1.4,3.85,1.1,0.42,0.35,0,0.37,0.5,0.9,0.61,0.78,0.6,0.47,0.5,2,5.81,
0.05,0,0,0.19,0.94,2.81,0.61,0,0.44,0.75,2.85,6,0,0,0,0.24,0.4,0.55,0,0,
0.23,0,0,0.54,0.38,0.78,0.8,0.88,0.7,0.05,0.58,0.88,1.4,0.39,0,0.33,0.51,
0.32)
```

```
sp<-spectrum(p, plot=FALSE)
```

The command `spectrum()` has an intrinsic plotting function which I have silenced here by setting the option `plot` to FALSE. To see the components of the object `sp` in a table, type `summary(sp)`. The names of the components are in the first column. To see the object in its entirety, simply type its name, `sp`. To extract a particular component, type the name of the object, then the symbol $ and then the name of the component. In the object sp, we are interested in plotting spectral densities (`sp$spec`) against frequencies (`sp$freq`). The following command gives a plot similar to Figure 2.36(b):

```
plot(sp$freq, sp$spec, xlab="Frequency", ylab="Spectral density", type="l")
arrows(0.15,0,0.15,3)
```

2.17. Fractal geometry

Despite its far-reaching applications in human technology and the occasionally convincing approximation of natural phenomena (see Example 2.6 and Figure 2.39(a)), Euclidean geometry is limited. Natural objects are often more complicated than Euclidean shapes: sketching a triangle requires just three lines, but a convincing image of a sand dune, a cloud or a crashing wave requires quite a bit more skill. Furthermore, with natural systems, new layers of structure are revealed as we look closer (forest, tree, leaf, leaf cell, etc.), each as rich in complexity as the layers above it. Such high complexity at multiple spatiotemporal scales has defeated theoreticians at every turn since Euclid's time. **Fractal geometry** represents a new approach to this old problem. Instead of trying to describe complexity, fractal geometry tries to summarise it and find simple mathematical rules that can give rise to very complicated patterns. If, in addition, these summaries and rules are found to apply at several scales simultaneously, then maybe complexity can be made less daunting.

The idea of revealing all-pervading symmetries in nature is not new. Ancient geometers (like Pythagoras), alchemists and renaissance thinkers (such as Da Vinci) strived towards this aim (Figure 2.39(b)). Modern mathematicians have just been more successful because of the discovery that exceedingly complex patterns can be generated by simple mathematical rules. These days, the Internet and poster-art shops are full of pictures of the Mandelbrot snowman or the Julia set, bearing testament to the appeal of these mathematical patterns to the aesthetics of nonmathematicians. The part that most people tend to forget is that these beautiful and

(a)   (b)

Figure 2.39: (a) Euclidean geometry may occasionally succeed in describing natural shapes such as the snail shell; (b) mathematicians like Leonardo Da Vinci have always tried to decompose natural shapes into constituent components and to quantify their underlying symmetries. © Luc Viatour GFDL/CC.

complicated patterns can be generated with a few lines of computer code that any interested layman could understand.

The observation that certain infinitely detailed patterns are simple to describe mathematically gives us hope that we can get a handle on natural complexity. The term 'fractal' (Mandelbrot, 1982) derives from the Latin word for 'broken' because fractal shapes (curves, surfaces, solids) are continuous but not smooth. The Koch arc in Example 2.2 is a good example of such a jagged curve. This aspect of fractals distinguishes them from nonfractal objects and reveals them to be rich in new features under repeated magnification.

Example 2.21: Availability of coastal habitat

Northern fulmar, Fulmarus glacialis

Northern fulmars are large, long-lived sea birds that breed on cliffs in the North Atlantic and North Pacific. A study focusing on the fulmars of Mainland, the largest island in Orkney, Scotland, aims to calculate their available breeding habitat. As a first step, the total length of coastline needs to be calculated. The lengths taken up by human habitation or beaches will then be subtracted to yield an estimate of available rocky shore.

One way to measure the coastline is by using a ruler of given length (l). In Figure 2.40(a), using $l = 5$ km yields an estimate of around 19 lengths (= 95 km). Increasing the resolution to half that length (Figure 2.40(b)) yields an increased estimate of $2.5 \times 45 = 112.5$ km. This happens because a smaller ruler can better describe alcoves and peninsulas. Using an even smaller length would give a higher estimate as more features of the Orkney coastline were more accurately traced. This dependence of total curve length on the unit of measurement is a distinguishing property of fractal curves.

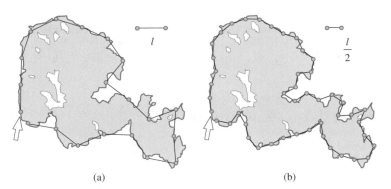

(a)   (b)

Figure 2.40: Mainland of the Orkney island complex in the north of Scotland. The available coastline is measured twice with rulers of length *l* and *l*/2, starting from the point indicated by the arrow.

How fine a resolution would we need to answer the question for an animal the length of a fulmar (about 0.5 m bill-to-tail)? Certainly a lot finer than that implied by a ruler of 5 or even 2.5 km. What about a smaller animal like a limpet? If the coast is a fractal curve, then do smaller organisms 'perceive' more habitat than larger ones (Morse *et al.*, 1985)?

In fractal curves, the relationship between the measured length of the curve (L) and the unit of measurement (l) is expressed by the function

$$L(l) = cl^{1-D} \tag{2.53}$$

where c is a constant of proportionality and D is a constant, characteristic of the curve, called the **fractal dimension**.

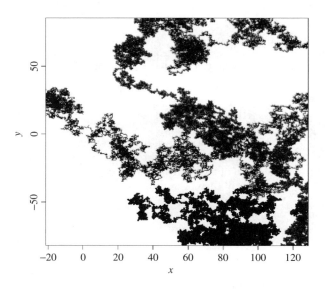

Figure 2.41: The isotropic random walk: a space-filling fractal. This example comprises 10^5 steps.

For curves on the plane, D takes values between 1 and 2 and it quantifies the sinuosity of a curve and its ability to fill 2-dimensional space. Hence, a straight line has a fractal dimension of 1 and a complicated curve such as the trace of a random walker (Figure 2.41) has a fractal dimension of 2.

Example 2.22: Fractal dimension of the Koch curve

The fractal dimension of the Koch curve from Example 2.2 can be calculated as follows: If the curve is measured using a ruler of length l (Figure 2.42(a)), its length is l. The finer resolution of $l/3$ (Figure 2.42(b)) yields the length $4l/3$.

Figure 2.42: Measuring the length of the Koch curve.

The Koch curve is a fractal, so it must satisfy Equation (2.53) that relates measured length to measurement scale. Specifying this function to the two measurement scales l and

$l/3$ produces the following system of two coupled equations:

$$l = cl^{1-D}$$

$$\frac{4l}{3} = c\left(\frac{l}{3}\right)^{1-D} \tag{2.54}$$

Dividing these by parts gets rid of l and c

$$\frac{4}{3} = \left(\frac{1}{3}\right)^{1-D} \tag{2.55}$$

Using the properties of powers, we can simplify this to

$$\frac{4}{3} = \frac{1}{3}\left(\frac{1}{3}\right)^{-D} \tag{2.56}$$

$$4 = 3^D$$

This last equation has the unknown D as an exponent. In Chapter 3, we will show how to use logarithms to solve such equations. In this particular example, it can be shown that $D = 1.26$, implying that the Koch curve is a complicated, nonlinear shape (because $D > 1$) that is, nevertheless, far from being space-filling (because $D < 2$).

The explanatory power of fractals has already had considerable impact on ecology. The online resource by Kenkel and Walker reviews applications relating to organism size and number of individuals, the movement of organisms, ecotonal boundaries, environmental transects, dispersal and the spread of disease and landscape fragmentation. As you can imagine, the promise of quantifying natural patterns using simple summaries such as the fractal dimension triggered a gold rush: biologists started looking again at their data in the hope that they would conform to the properties of fractals. It was easy to get carried away. Although fractal objects look a lot more natural than Euclidean shapes, it is important to note that natural structures do not conform to the strict mathematical construct of a fractal (Turchin, 1996; Halley *et al.*, 2004). In the real world, self-similarity and complexity can only hold over a finite range of scales. More recent developments, using so-called **multifractals**, attempt to capture such deviations from strictly fractal behaviour.

Further reading

A great online resource on Euclid's *Elements* can be found at http://aleph0.clarku.edu/~djoyce /java/elements/elements.html. It has a searchable database of proven results and the Java animations of the diagrams really make all the difference for comprehension. The quintessential geometrical interpretation of biological form and pattern is D'Arcy Thompson's *On Growth and Form*, beautifully illustrated and full of as yet unexplored ideas, despite its age. In the same spirit, the wonderfully illustrated book by Ball (1999) is rich in examples of natural patterning. An in-depth and technical discussion of spectral and time-series analyses is given by Chatfield (2004). The standard introduction to fractals for scientists is Mandelbrot (1982). Popularised discussions on the science of complexity (incorporating fractals) can be found in Gleick (1987) and Liebovitch (1998). Kenkel and Walker provide a regularly updated online document at: http://www.umanitoba.ca/faculties/science/botany/LABS/ ECOLOGY/FRACTALS/fractal.html.

References

Akçakaya, H.R., Halley, J.M. and Inchausti, P. (2003) Population level mechanisms for reddened spectra in ecological time series. *Journal of Animal Ecology*, **72**, 698–702.

Ball, P. (1999) *The Self-made Tapestry: Pattern Formation in Nature*. Oxford University Press. 287pp.

Barlow, G.W. (1974) Hexagonal territories. *Animal Behaviour*, **22**, 876–878.

Batzli, G.O. (1993) Dynamics of small mammal populations: a review. In *Wildlife 2001: Populations* (Eds McCullogh, D.R. and Barrett, R.H.). Elsevier Science, London, pp. 831–850.

Bovet, P. and Benhamou, S. (1988) Spatial analysis of animals' movement using a correlated random walk model. *Journal of Theoretical Biology*, **131**, 419–433.

Chatfield, C. (2004) *The analysis of time series: An introduction*. Chapman & Hall, Boca Raton. 333pp.

Chavez, F.P., Ryan, J., Lluch-Cota, S.E. and Ñiquen, M. (2003) From Anchovies to Sardines and Back: Multidecadal Change in the Pacific Ocean. *Science*, **299**, 217–221.

Elton, C.S. (1924) Periodic fluctuations in the number of animals: their causes and effects. *Journal of Experimental Biology*, **2**, 119–163.

Gleick, J. (1987) *Chaos: Making a New Science*. Cardinal, Suffolk. 352pp.

Halley, J.M. (1996) Ecology, evolution and 1/f noise. *Trends in Ecology and Evolution*, **11**, 33–37.

Halley, J.M., Hartley, S., Kallimanis, A.S., Kunin, W.E., Lennon, J.J. and Sgardelis, S.P. (2004) Uses and Abuses of Fractal Methodology in Ecology. *Ecology Letters*, **7**, 254–271.

Hartshorne, R. (2000) *Geometry: Euclid and Beyond*. Springer.

Hastin, H.M. and Sugihara, G. (1993) *Fractals: A User's Guide for the Natural Sciences*. Oxford University Press, Oxford. 235pp.

Hughes, B.D. (1995) *Random walks and random environments. V1: Random walks*. Oxford University Press, New York. 631pp.

Inchausti, P. and Halley, J. (2002) The long-term temporal variability and spectral colour of animal populations. *Evolutionary Ecology Research*, **4**, 1033–1048.

Kausrud, K.L., Mysterud, A., Steen, H., Vik, J.O., Østbye, E., Cazelles, B., Framstad, E., Eikeset, A.M., Mysterud, I., Solhøy, T. and Stenseth, N.C. (2008) Linking climate change to lemming cycles. *Nature*, **456**, 93–97.

Liebovitch, L.S. (1998) *Fractals and Chaos, Simplified for the Life Sciences*. Oxford University Press, New York. 268pp.

Mandelbrot, B.B. (1982) *The Fractal Geometry of Nature*. Freeman & Co., New York. 468pp.

Matthiopoulos, J., Smout, S.C., Winship, A., Thompson, D., Boyd, I.L. and Harwood, J. (2008) Getting beneath the surface of marine mammal–fisheries competition. *Mammal Review*, **38**, 167–188.

Morse, D.R., Lawton, J.H., Dodson, M.M. and Williamson, M.H. (1985) Fractal dimension of vegetation and the distribution of arthropod body lengths. *Nature*, **314**, 731–733.

Nisbet, R.M. and Gurney, W.S.C. (1982) *Modelling Fluctuating Populations*. John Wiley & Sons, Ltd, Chichester.

Rhizopoulou, S., Ioannidi, E., Alexandredes, N. and Argiropoulos, A. (2006) A study of functional and structural traits of the nocturnal flowers of *Capparis spinosa* L. *Journal of Arid Environments*, **66**, 635–647.

Thompson, D'A. (1961) *On Growth and Form*. Cambridge University Press. (Canto edition, 1990, 346 pp).

Turchin, P. (1996) Fractal analyses of animal movement: A critique. *Ecology*, **77**, 2086–2090.

Turchin, P. (1998) *Quantitative Analysis of Movement: Measuring and Modeling Redistribution in Plants and Animals*. Sinauer Associates, Massachusetts.

Turchin, P. (2003) *Complex Population Dynamics: A Theoretical/Empirical Synthesis*. Princeton University Press, Princeton. 450pp.

3

How to change things, one step at a time
(Sequences, difference equations and logarithms)

'In nature, things move violently to their place, and calmly in their place'

Sir Francis Bacon (1561–1626), English philosopher

Quite aptly, the cause for all science is causality itself. Our understanding of the natural world and our ability to make predictions rely on finding tractable relationships between cause and effect. Some relationships are obvious and can be formulated verbally; others are not and need to be explored analytically. Discovering and quantifying more complex or even counter-intuitive causal relationships is the stuff of exciting science. One way in which complexity arises from simplicity is via **iteration** (applying the same operation repeatedly and using the output of one repetition as the input of the next). In Chapter 1, functions were used to quantify the impact of one variable on another. Here, we focus on sequences, a particular class of function with integer-valued independent variables. Section 3.1 introduces some new notation and two important examples of sequences, known as the **arithmetic** and **geometric progression**. Interpreting the steps of a sequence as units of time leads to **difference equations** (Sections 3.2 and 3.3), models with wide application in systems that evolve in discrete time, such as the populations of species with well-defined, annual life cycles. Given a set of **initial conditions** (Section 3.4), a difference equation can be **solved** (Section 3.5) and **predictions** made about the future state of a system. Special solutions of the system, called **equilibrium**

How to be a Quantitative Ecologist: The 'A to R' of Green Mathematics and Statistics, First Edition. Jason Matthiopoulos.
© 2011 John Wiley & Sons, Ltd. Published 2011 by John Wiley & Sons, Ltd.

solutions (Section 3.6) represent a state of no change. If the study system spontaneously moves towards it, then the equilibrium solution is called **stable** (Section 3.7). The visual and mathematical tools introduced in Section 3.8 to determine the stability of equilibria are then used to explore the rich dynamics of unstable systems (Section 3.9). Examples motivated by population growth in discrete and continuous time give rise to the **exponential function** (Section 3.10) and its inverse, the **logarithmic function** (Section 3.11). The chapter ends with a discussion of how to deal with algebraic equations whose unknown variable appears inside a logarithm (Section 3.12).

3.1. Sequences

Example 3.1: Reproductive output in social wasps

European paper wasp, Polistes dominula

Colonies of European paper wasps are founded by single queens that have been mated in the previous year and spent the winter dormant. During late spring, a queen begins to build the colony's nest and produce her first worker daughters. She then focuses solely on the production of offspring, initially worker-females and then new founder-females and males. If a queen typically produces an initial spring batch of 100 workers and then a constant daily number of about 35 eggs during the summer, her cumulative egg production in the summer is as given in Table 3.1.

Table 3.1

Days	0	1	2	3	...	n
Eggs	100	135	170	205	...	$100 + n\,35$

The number of eggs $100 + n35$ produced at some point is therefore written as a simple function of time n. Two simplifying assumptions are required to use this simple expression for prediction:

❶ There is such a thing as a 'typical' wasp queen, i.e. the number of eggs laid initially and the number of eggs added daily does not differ between queens. If individuals of the species vary considerably, the model refers to all of them but applies to none in particular.
❷ The number of eggs laid each day is constant. This assumption would be violated by random daily variations or if the queen's rate of production during the summer declined due to senescence.

This is an example of a **sequence**, generally defined as an ordered set of terms $a_0, a_1, a_2, \ldots$ indexed by zero and the natural numbers. Another way of thinking of a sequence is as a function $a(n)$ whose independent variable takes values from the natural numbers. The notation a_n is used instead of $a(n)$ to stress the discreteness of a sequence's independent variable.

In the sequence of Example 3.1, the next term (a_{n+1}) is obtained by adding a constant to the current term (a_n). Such a sequence is called an **arithmetic progression** and is defined **iteratively** by the equation

$$a_{n+1} = a_n + m \tag{3.1}$$

where m is a constant, ($= 35$ in Example 3.1). Alternatively, the nth term of an arithmetic progression can be written directly using the very first term (in Example 3.1, $a_0 = 100$) and the constant m

$$a_n = a_0 + nm \tag{3.2}$$

Equations (3.1) and (3.2) are examples of the **iterative** and **general definition** of a sequence. They are mathematical descriptions – models – of a biological process. Equation (3.1) describes egg production iteratively from one day to the next, and Equation (3.2) gives total production after n days in the summer. Since any sequence is also a function, it has a graph (Figure 3.1). This comprises disjointed points along a line because Equation (3.2) is the general expression for a linear function but its domain is not continuous.

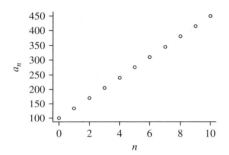

Figure 3.1: The graph of the arithmetic progression with $a_0 = 100$ and $m = 35$.

Example 3.1 focuses on productivity, but could the model be extended to describe colony size? Not quite, because it ignores the possibility that several of the queen's offspring die during the summer. Furthermore, social insects are rather unusual as a general model of population change because not all colony members reproduce. More complex demographies can be captured by another important type of sequence, the **geometric progression**, introduced by the following example:

Example 3.2: Unrestricted population growth

Blue tit (Cyanistes caeruleus), a sexually reproducing animal

A population of birds has a constant annual mortality $d = 0.6$, i.e. 40% of the animals alive at the beginning of the year survive to the end of it. The sex ratio in the population is 1:1. The birds start reproducing in their first year and each adult female produces on average seven hatchlings, of which about half survive to adulthood. We want to model population size (a_n) with passing years (n) assuming that it was originally founded by four breeding pairs.

In a sexually reproducing population, the **birth rate** is the number of new offspring per adult animal (male and female). Given a 1:1 sex ratio, the number of females in year n is $a_n/2$. Each female produces seven hatchlings. Therefore, the birth rate is $7\times$ No of females/Total Population $= (7a_n/2)/a_n = 3.5$. The **recruitment rate** is the number of chicks produced by each adult that make it to adulthood, here half the birth rate, $3.5/2 = 1.75$. The iterative equation for this biological system is

$$a_{n+1} = a_n + \text{recruitment} - \text{adultdeaths} \tag{3.3}$$

Let us now specify the timing of events during a single year. Egg-laying occurs during a narrow time interval in spring, so this can be taken as the beginning of the year. In contrast, mortality (of chicks and adults) occurs continuously. Total adult mortality during year n is $0.6a_n$. The eggs laid in the spring of year n will eventually yield the population's next batch of recruits, so total recruitment is $1.75a_n$. The iterative definition now becomes

$$a_{n+1} = a_n + 1.75a_n - 0.6a_n$$
$$a_{n+1} = (1 + 1.75 - 0.6)a_n \tag{3.4}$$
$$a_{n+1} = 2.15a_n$$

In year 0, the founding population of four pairs had size $a_0 = 8$. To predict population size in year 1 we can use Equation (3.4)

$$a_1 = 2.15a_0 \tag{3.5}$$

The same iterative relationship can be used to calculate population size in year 2,

$$a_2 = 2.15a_1 = 2.15(2.15a_0) = 2.15^2 a_0 \tag{3.6}$$

The population sizes in years 3 and 4 can similarly be found to be $2.15^3 a_0$ and $2.15^4 a_0$. Substituting the initial population size into the latter, $a_4 = 170.9$.

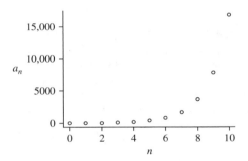

Figure 3.2: The graph of the geometric progression with $m = 2.15$ and $a_0 = 8$.

In just four years, we went from a small integer population to a large, noninteger one. Although 0.9 of a blue tit may appear to make little sense biologically, here we are predicting the average (expected) number of birds from one year to the next (remember, births and deaths were also expressed as noninteger averages). We will revisit these issues in Section 9.21. In the above example, population size after several years was calculated iteratively,

$$a_{n+1} = ma_n \tag{3.7}$$

However, even after the first four iterations, a pattern emerged that hinted at the sequence's general definition:

$$a_n = m^n a_0 \tag{3.8}$$

This function has its independent variable in the exponent and its graph is no longer linear (Figure 3.2).

Ⓡ 3.1: Iterating the general definition of a sequence

To generate the first 21 terms of the arithmetic progression in Example 3.1 using the sequence's general definition, we must first create an index n and then apply to it the formula $a_n = a_0 + nm$ to obtain a sequence a.

```
n<-0:20
a<-100+n*35
```

Here, n is a list of integers and so is a. The same result can be achieved by merging the two lines of code into one

```
a<-100+(0:20)*35
```

To retrieve the number of eggs laid by the tenth day of summer we need to type a[11]. This is because the index n includes the 0th day (incorporating all eggs produced during the spring) but lists in R are indexed from 1 upwards.

3.2. Difference equations

In the examples used to introduce arithmetic and geometric progressions, the index n of a sequence $\{a_n\}$ was time-related: in Example 3.1, n represented the days since the beginning of summer and in Example 3.2, it was the number of years since the blue tit population was founded. Sequences are often used to describe the development of systems in discrete time (t). The iterative definitions of such sequences are called **difference equations**. Many, but not all, take the form

$$a_{t+1} = a_t + f(a_t) \tag{3.9}$$

Equation (3.9) moves the system from its present state (a_t) to its future state (a_{t+1}) by adding an incremental change (f). This increment, or **rate of change**, may itself be a function of a_t. Rewriting Equation (3.9) in the following way explains why these models are called *difference* equations.

$$f(a_t) = a_{t+1} - a_t \tag{3.10}$$

Difference equations are not always additive like Equation (3.9) and, sometimes, the future of the system depends explicitly on time. A more general model, called a **first order** difference equation, is

$$a_{t+1} = f(a_t, t) \tag{3.11}$$

There are many important and more specific versions of Equation (3.11). In particular, a **linear** difference equation is one that has no higher-order terms of the form a_t^2. Linear difference equations take the form $a_{t+1} = c_1 a_t + c_0$. A **linear affine** difference equation is one that only contains terms with a_t (e.g. $a_{t+1} = c_1 a_t$). An example of a **nonlinear affine** difference equation is $a_{t+1} = c_2 a_t^2 + c_1 a_t$. If the coefficients of the terms a_t are constant (c_1, c_2) then the equation is called **autonomous**. In contrast, if any of these coefficients depends on time, then the difference equation is called **nonautonomous** (e.g. $a_{t+1} = c_1(t)a_t + c_0$). The term c_0 of nonaffine equations may be a constant, in which case we have a **homogeneous** difference equation. A **nonhomogeneous** equation has a term $c_0(t)$ which varies with passing time.

All of the above definitions are organised in the table of Section A4.3 in the appendix. I apologise for burdening you with all this terminology. It may seem that no ecologist will ever need to consider something as perverse as a 'nonhomogeneous, nonautonomous, nonlinear difference equation'. However, when these arcane terms are interpreted biologically, they appear rather more understandable and essential.

Example 3.3: More realistic models of population growth

Consider a population which, in year t, comprises P_t adults (both male and female). In Example 3.2 we examined the general model of population growth as a balance between mortality and recruitment

$$P_{t+1} = P_t + \text{recruitment} - \text{adultdeaths} \qquad (3.12)$$

If d is mortality (proportion of adults dying during the year) and b is recruitment (number of offspring per parent that reach adulthood), then

$$\begin{aligned} P_{t+1} &= P_t + bP_t - dP_t \\ &= (1 + b - d)P_t \end{aligned} \qquad (3.13)$$

It is customary to simplify notation by replacing the difference $b - d$ by a single number r called the population's **intrinsic growth rate**.

$$P_{t+1} = (1 + r)P_t \qquad (3.14)$$

If the intrinsic growth rate is zero, then $P_{t+1} = (1 + 0)P_t = P_t$, meaning that each individual in the population this year is replaced by exactly one individual in the next. This may either be the same animal, its own offspring or somebody else's. If r is positive (recruitment exceeds mortality), e.g. 0.1, then the population next year is multiplied by 1.1, so will equal 110% of its current size. If r is negative (mortality exceeds recruitment), e.g. -0.1, then the population declines by 10% each year (have a look also at Example 1.51). The minimum value of the intrinsic growth rate ($r = -1$) is achieved when no adults survive ($d = 1$) and no offspring make it to adulthood ($b = 0$). This lower bound for r is realistic because a value smaller than -1 in Equation (3.14) would lead to negative population sizes.

Equation (3.14) is first order, homogeneous, autonomous and linear affine, i.e. the simplest type of difference equation (look at Section A4.3). Would you feel comfortable using it to describe a population of real animals? Its main problem is that it is a geometric progression and therefore predicts that, if $r > 0$, the population will increase indefinitely. That will not be true of real populations whose size is ultimately restricted by limited resources such as prey, nutrients and refuge sites. If population growth rate changes with population size (for example due to competition for limited resources), we say that it is **density dependent**.

To model this phenomenon, let's assume that the population's environment can support a limited and fixed number of animals called the environment's **carrying capacity** (K). As the population approaches this size, available resources become scarce and individuals start to fail in meeting their life-history priorities. Therefore, when close to, but still beneath, the carrying capacity, the population should grow more slowly. If its size exceeds the environment's carrying capacity, then it should start to decrease back towards it. We therefore need to model the intrinsic growth rate as a decreasing function of population size. The simplest possible formulation is to write growth rate as a linear function of density:

$$r = r_{\max}\left(1 - \frac{P_t}{K}\right) \qquad (3.15)$$

The constant $r_{\max}$ is the growth rate of very small populations, relative to the carrying capacity. To convince yourself that Equation (3.15) does what it is supposed to, examine the following four scenarios: (i) $P_t = 0$, (ii) $P_t < K$, (iii) $P_t = K$, (iv) $P_t > K$.

Placing this back into the population model of Equation (3.14) yields what is known as the **discrete logistic population model**

$$P_{t+1} = P_t + r_{\max}\left(1 - \frac{P_t}{K}\right)P_t \qquad (3.16)$$

This is a homogeneous, autonomous, nonlinear affine difference equation. To see this more clearly, the RHS of Equation (3.16) can be rewritten as a second order polynomial in P_t

$$P_{t+1} = \left(-\frac{r_{\max}}{K}\right)P_t^2 + (1 + r_{\max})P_t \qquad (3.17)$$

So, the addition of the most obvious biological feature (limited resources) has immediately led to a nonlinear difference equation. If we added density-independent terms this would stop being an affine equation. For example, adding a net influx of immigrants (I) into the population from somewhere else gives:

$$P_{t+1} = P_t + r_{\max}\left(1 - \frac{P_t}{K}\right)P_t + I \qquad (3.18)$$

Another process that could give rise to a constant term is harvesting, although it is hard to imagine why anyone would want to have an annual cull of blue tits. It is, however, more plausible to argue that the rates of recruitment, mortality and immigration are time-dependent. Such temporal variations could result from environmental trends or random perturbations. The same processes could affect the availability of resources and hence the carrying capacity. We can declare this by replacing the symbols $r_{\max}, K, I$ by the subscripted symbols r_t, K_t, I_t. Multiplying out the resulting difference equation gives:

$$P_{t+1} = \left(-\frac{r_t}{K_t}\right)P_t^2 + (1 + r_t)P_t + I_t \qquad (3.19)$$

This is nonhomogeneous, nonautonomous and nonlinear, i.e. the most complicated kind of first order difference equation. We rather quickly arrived at this by incorporating three ubiquitous ecological features into the basic logistic model: density dependence, immigration and environmental change.

ⓇR 3.2: Iterations and loops

Sequences are ideal for introducing the concept of iteration, i.e. the repeated application of a set of commands. If the results of one iteration do not depend on preceding iterations, this can be done easily, as we saw in Example 3.1. In that example, I used the general definition of a sequence, so the value of each term depended only on the first, not the previous, term. However, difference equations are iterative. This calls for a different programming device called a **loop**, which comprises two things: the **loop declaration** and its **main body**. The R syntax is as follows:

```
for (counter in min:max)
    {
    # Main body of the loop
    }
```

For example, the following piece of code will produce the same results as R3.1.

```
a<-c() # creates an empty list
for (n in 0:20) # Loop will be repeated 21 times (from 0 to 20)
    {
    a<-c(a, 100+n*35) # adds a new element to the existing list
    }
```

The brackets in lines 3 and 5 enclose the main body of the loop. Several lines of commands can be enclosed in the main body but in this example there is only one, which uses the concatenation command `c()`, to add to the current list a, the new value `100+n*35`. Since this value contains the counter n, it will be recalculated with each iteration of the loop. To see how the loop operates, try inserting the command `print(a)` as a new line in between lines 4 and 5. Also, note that the brackets and main body of the loop are right-indented. This is a convenient bookkeeping device that makes it easy to identify where the loop begins and where it ends on the screen or printed page.

For the above task, where each iteration could have been calculated independently of the previous one, the structure presented in R3.1 is preferable to a loop, not only because it is shorter (one line of code instead of five) but it is also considerably (hundreds of times!) faster. However, when the input of an iteration depends on the output of the previous iteration, then a loop is unavoidable. For example, to predict the size of a population with logistic growth, we could use Equation (3.16). We need some numerical values for the intrinsic growth rate and carrying capacity. We also need a value for the number of founding individuals, say, $r_{max} = 0.3, K = 100$ and $P_0 = 8$. The following will generate the first 21 years of the population time series and plot them.

```
p<-c(8) # initialises list
for (t in 1:20) # declares loop.
  {
  p[t+1]<-p[t]+0.3*(1-p[t]/100)*p[t] # logistic model
  }
plot(p)
```

The loop only goes up to 20 because the expression `p[t+1]` in the main body will automatically generate the 21st element of the time series. An inelegant and potentially time-consuming feature of the code in both of the above examples is that the list of population values p is made to grow with each iteration (its first element is set to 8 and further elements are created in the main body of the loop). Ideally, an empty list of length 21 should be created first, and its values simply filled in with each iteration of the loop. If you need to refresh your memory on lists and list-indexing, have a look back at R1.8. There are other types of loops available in R, most notably the `while` structure, which repeats a set of commands while a particular condition holds true.

3.3.   Higher order difference equations

The notation used for first order equations (i.e. $a_{t+1} = f(a_t, t)$) can be extended to **higher order difference equations**, used to describe systems whose state doesn't only depend on the immediately preceding time, but also times before that.

$$a_{t+1} = f(a_t, a_{t-1}, a_{t-2}, \ldots, t) \tag{3.20}$$

Because these equations contain time lags greater than 1, they are also known as **delay-** or **lagged difference equations**. In ecology, they arise in systems that have some sort of **structure**, for example, a population composed of individuals in multiple life-history stages (e.g. juveniles and adults), or whose survival and fecundity is mediated by other species (e.g. predators or prey). The dynamic interactions between these different components build delays into the system.

Example 3.4: Delay-difference equations in a biennial plant

Hemlock, Conium maculatum

Hemlock is a biennial plant that is only capable of producing a few viable seeds in its first year. Assume that a typical plant only produces one seedling during its first year and four seedlings in the second. Only one half of all seedlings survive to be one-year old, and all plants die at the end of their second year. Given that the population consists of two components, we need two symbols to describe its state, S_t and Y_t for seedlings and yearlings respectively. We can write update rules for these two components using the available information:

$$
\begin{aligned}
S_{t+1} &= 4Y_t + S_t \\
Y_{t+1} &= 0.5S_t
\end{aligned}
\tag{3.21}
$$

This is an example of a system of **coupled difference equations**, so called because the variables S_t and Y_t depend on each other (yearlings contribute to the number of seedlings and they, in turn, grow into yearlings). Now, assume that we are only interested in the population of yearlings and we want to derive an equation for the number of yearlings in the next year in terms of the number of yearlings in previous years. We can start by shifting the whole system back by one year:

$$
\begin{aligned}
S_t &= 4Y_{t-1} + S_{t-1} \\
Y_t &= 0.5S_{t-1}
\end{aligned}
\tag{3.22}
$$

Solving the second of these equations for S_{t-1} gives $S_{t-1} = Y_t/0.5$. Substituting this into the first equation gives $S_t = 4Y_{t-1} + Y_t/0.5$. Placing this back into the second of Equations (3.21) and tidying up, gives

$$
Y_{t+1} = Y_t + 2Y_{t-1}
\tag{3.23}
$$

So, we have managed to uncouple the equation for yearlings from the seedlings equation. The price is that we now have to deal with a second order difference equation for one of the two components of the population.

3.4.   Initial conditions and parameters

So far, I have mostly focused on the functional form (the structure) of difference equations. However, in every example I have used constants, such as the inherent growth rate of a population or its carrying capacity. These constants are collectively called the model's **parameters**. In addition, if the terms of a time series are seen as stepping stones, and the difference equation as the rule that describes how to get from one stone to the next, we still need to know where the first stone lies (e.g. the initial population size from which to predict future population sizes). Numbers referring to the initial state of a system are collectively called **initial conditions**. Predicting the future outcome of a difference equation on the basis of parameters and initial conditions is known as an **initial value problem**.

Ⓡ 3.3: Setting out initial conditions and parameters

When programming, it is good practice to represent parameters by symbols whose numerical assignments are collected together at the beginning of the code. This gives an overview of the numerical information required by the model. It also makes expressions in the main part of the code resemble their symbolic description, making it easier to find typing errors. Most importantly, in many models the same parameter is repeated several times. Rather than having to trawl through the entire code and change numerical values one at a time, having the parameters collected at the start means that it is only necessary to change the parameter's numerical assignment once. Here is a tidier (if a bit longer) version of the logistic growth model from R3.2.

```
P0<-8          # Initial population size
MAX.Y<-21      # Number of years for which predictions are wanted
R<-0.3         # Intrinsic growth rate
K<-100         # Carrying capacity

p<-rep(P0, MAX.Y)
for (t in 1:(MAX.Y-1)) # The entire upper limit is in brackets
  {
  p[t+1]<-p[t]+R*(1-p[t]/K)*p[t]
  }
plot(p)
```

Note a further difference between this and the code in R3.2. Here, a list p of length MAX.Y is first created outside the loop and then merely updated inside it.

3.5. Solutions of a difference equation

The examples of the arithmetic and geometric progressions show that the general definition is a more powerful tool for calculating a specific future term of a sequence (Figure 3.3).

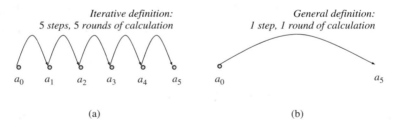

Iterative definition:	*General definition:*
5 steps, 5 rounds of calculation	*1 step, 1 round of calculation*

a_0 a_1 a_2 a_3 a_4 a_5 a_0 a_5

(a)   (b)

Figure 3.3: The general definition of a sequence (b) is computationally more efficient than the iterative definition (a) for calculating the sequence's future terms.

However, it is generally easier and more intuitive to describe a process iteratively and considerable work is required to derive a general definition from the iterative definition of the process. In many cases, the task is impossible. In Section 3.2 I introduced difference equations as the iterative definitions of time-indexed sequences but made no mention of the corresponding general definitions. From now on, I will refer to those as the **solutions** of the corresponding difference equations. In general, the solution of a first order difference equation $a_{t+1} = f(a_t, t)$ is a function $a_t = g(t)$ that satisfies the difference equation when substituted in for a_{t+1} and a_t.

Example 3.5

Consider the difference equation corresponding to an arithmetic progression

$$a_{t+1} = a_t + m \qquad (3.24)$$

for m, a constant. The general definition of an arithmetic progression is

$$a_t = a_0 + tm \qquad (3.25)$$

This expression applies to all values of t including the value $t + 1$. Hence,

$$a_{t+1} = a_0 + (t + 1)m \qquad (3.26)$$

Substituting a_t and a_{t+1} from Equations (3.25) and (3.26) back into the iterative definition in Equation (3.24) gives

$$a_0 + (t + 1)m = a_0 + tm + m \qquad (3.27)$$

Multiplying out the term in brackets verifies that this expression is true.

The solution in the above example holds irrespective of the values of the parameter (m) and initial condition (a_0), so it is called a **general solution** of the difference equation. A **specific solution** is obtained by setting the parameters and initial conditions to particular numerical values.

Example 3.6

The specific solution of $a_{t+1} = a_t + m$ with $a_0 = 2$ and $m = 3$ is $a_t = 2 + 3t$.

Use of the term 'solution' here is consistent with its use in Chapter 1, in the sense that a solution of an equation satisfies the equation. However, in the present context, the solution is a function, not a number.

This section has only shown how to check that a particular candidate expression is indeed the solution of a difference equation, but not how to find it. It is important to remember that not all difference equations can be solved analytically. In such cases, we have to fall back to using a computer to generate specific **numerical solutions** by iteration (see R3.2 and R3.3).

3.6.   Equilibrium solutions

Example 3.7: Harvesting an unconstrained population

Trawler

Consider a population of fish whose growth is currently not constrained by environmental limitations (see Example 3.2), so that its size from year to year is given by

$$P_{t+1} = (1 + r)P_t \qquad (3.28)$$

Here, population size is measured in some units of biomass (e.g. 10^4 tonnes). A new fishery is proposed for this species and the authorities plan to regulate its operation by setting an annual quota c, measured in the same unit as biomass. Further, I make the (unrealistic) assumption that c is the amount taken from the seas, not simply landed legally or discarded due to by-catch regulations. The situation can be modelled as

$$P_{t+1} = (1+r)P_t - c \tag{3.29}$$

The population's intrinsic growth rate is $r = 0.15$ and its current size is estimated as $P_0 = 1320$. For a quota $c = 250$, the predictions of the model for the next few years are $P_1 = 1268, P_2 = 1208.2, P_3 = 1139.43$. The population seems to be declining. If we carry on like this, eventually we will obtain a prediction below zero. At that point, the fish population will be considered extinct. However, if we use $c = 150$ then the future terms of the sequence increase without constraint (albeit at a slower rate than what would occur if no fishing took place).

In short, if too low a quota is used the fishery loses out but using too high a quota causes the fish to become extinct. A better, more sustainable strategy is to keep the population of fish to its current size by an intermediate quota. The answer lies somewhere in the range $150 < c < 250$ but what should the exact value of c be? Keeping the population at its current size means

$$P_0 = P_1 = P_2 = \cdots = P_t = P_{t+1} \tag{3.30}$$

If P^* is the size of this constant population, then Equation (3.29) becomes

$$P^* = (1+r)P^* - c \tag{3.31}$$

This situation is only true for a particular value of c, calculated from Equation (3.31)

$$\begin{aligned} c &= (1+r)P^* - P^* \\ c &= rP^* \\ c &= rP_0 = 198 \end{aligned} \tag{3.32}$$

Therefore, the difference equation $P_{t+1} = 1.15P_t - 198$ with $P_0 = 1320$ has the specific solution $P_t = 1320$ for all future years.

This is an example of an **equilibrium solution** for the difference equation. In this case, $P^* = 1320$ is an **equilibrium** or **fixed point**. More generally, given a difference equation

$$a_{t+1} = f(a_t, a_{t-1}, \ldots) \tag{3.33}$$

a^* is an equilibrium if $a_t = a^*$ for all values of t. Many difference equations have more than one equilibrium and nonhomogeneous difference equations (those containing time explicitly) have none. The way to find an equilibrium is to replace all occurrences of the variable (i.e. $a_{t+1}, a_t, a_{t-1}, \ldots$) by the constant a^*. This yields an algebraic equation that can be solved analytically or numerically (see R3.4).

The equilibria of a first-order, homogeneous difference equation can be visualised in a graph of a_{t+1} against a_t. Any equilibria of the system will lie on a $45°$ line through the origin (see Example 1.49). The difference equation can then be plotted as a continuous function to illustrate all possible transitions of the system from t to $t+1$. Any points of intersection between the line and the curve indicate the equilibrium point(s) of the system.

Example 3.8: Visualising the equilibria

Consider the model of unrestricted growth with harvesting (Example 3.7). Any equilibria of the system will be found on the graph of $f(x) = x$. The difference equation $P_{t+1} = (1+r)P_t - c$ can be plotted as the continuous function $g(x) = (1+r)x - c$ (a first order polynomial, so we expect the graph to be a line). The intersection between these two lines (Figure 3.4) shows the location of the single equilibrium. Its value ($x = c/r$) is found by solving the equation $x = (1 + r)x - c$ for x.

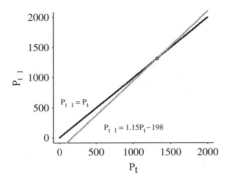

Figure 3.4: Unrestricted growth with harvesting for the parameters $r = 0.15$ and $c = 198$. The equilibrium ($P^* = c/r = 1320$) lies at the intersection of $f(x) = x$ and $g(x) = (1 + r)x - c$.

 3.4: Finding equilibria numerically
The procedure for finding the equilibria of any homogeneous difference equation $a_{t+1} = f(a_t, a_{t-1}, \ldots)$ is to rewrite it as an algebraic equation $a^* = f(a^*)$. Some of these equations will be solvable analytically, others will not. For example, if the equations are first or second order polynomials (e.g. Example 3.8) then they can be solved directly (see Sections 1.11 and 1.13). Higher order polynomials may require numerical approximation (see R1.9).

3.7.   Stable and unstable equilibria

Example 3.9: Parameter sensitivity and ineffective fishing quotas

Example 3.7 examined the fisheries management model $P_{t+1} = (1 + r)P_t - c$ with $P_0 = 1320$. The fundamental assumption of unrestricted population growth determined the model's structure, but there are some other crucial assumptions that relate to its parameters. Specifically, that:

❶ The population's intrinsic growth r is known precisely.
❷ The current population size P_0 is estimated precisely.
❸ The quota c is adhered to precisely.
❹ Most importantly, there are no temporal variations in these values.

Any of these assumptions could easily be wrong. For example, the lower trajectory in Figure 3.5 shows what would happen if, through some external effect, the current population dropped to 1290. The recommended quota $c = 198$ would lead to a reduction in population size and eventual extinction within 30 years. Similarly, if the true population was slightly greater than its estimated size (e.g. 1350), it would more than double within 30 years despite the effect of fishing.

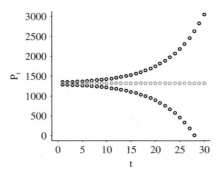

Figure 3.5: Sensitivity of solutions to disturbance. The system at equilibrium is represented by the specific solution plotted in grey. Perturbations placing the population slightly above or below the equilibrium size lead to rapid divergence.

Following these relatively small perturbations, the specific solutions of this system moved away from the equilibrium point. This is an example of an **unstable equilibrium**. Conversely, an equilibrium that attracts the system is called **stable**.

Example 3.10: Stable and unstable equilibria in a density-dependent population

Consider the model of population growth introduced in Example 3.3:

$$P_{t+1} = P_t + r_{max}\left(1 - \frac{P_t}{K}\right)P_t \qquad (3.34)$$

The equilibria of this system are found by setting $P_{t+1} = P_t = P^*$

$$P^* = P^* + r_{max}\left(1 - \frac{P^*}{K}\right)P^* \qquad (3.35)$$

Rearranging Equation (3.35) leads to a quadratic equation with solutions $P^* = 0$ and $P^* = K$. Therefore, the system has two equilibria, corresponding to populations that are either extinct or at their carrying capacity.

Now consider two parameterisations of this model. First, one with $r_{max} = 0.11$ and $K = 1000$. The stability of the two equilibria under these parameter values can be explored by simulating the model for different initial conditions. Each initial condition gives rise to a specific solution (Figure 3.6(a)).

All of these solutions (with the exception of $P_t = 0$) are attracted to the carrying capacity, approaching but never actually attaining it (an **asymptotic approach**). $P_t = 1000$ is an example of a stable solution but $P_t = 0$ is unstable because all specific solutions that start close to zero diverge away from it.

For the second parameterisation of the model, set $r_{max} = -0.11$ and $K = 1000$. As might have been expected from the fact that the growth rate is negative, the extinction point has now become a stable equilibrium, attracting solutions that start from nonzero population sizes (Figure 3.6(b)). Unfortunately, this is where realism ends: in order for the solutions to be attracted to the extinction point, the carrying capacity has now become unstable and repels solutions. This works fine when the population starts below K, but if the initial population is above the environment's carrying capacity, then the solutions move to infinity. Mathematically, this happens because a negative value for r_{max} multiplied by a negative value for $1 - P_t/K$ results in a positive number, and the population increases. It appears that this model has reached its limits.

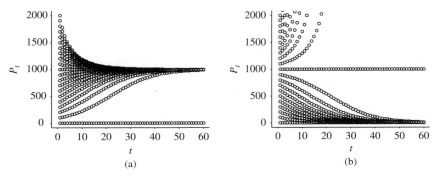

Figure 3.6: The behaviour of any system depends on the values taken by its parameters. In this example, the carrying capacity is stable when the intrinsic rate is positive (a) and unstable when it is negative (b).

Apart from serving as a reminder that models merely mimic, not replicate, reality, this example illustrates two further points. First, an equilibrium is stable if some (but not necessarily all) specific solutions approach it with passing time. Second, the stability of an equilibrium may depend on the specific parameters used.

3.5: Nested loops are (occasionally) useful

Each of the panels of Figure 3.6 comprises the graphs of several specific solutions. Such a composite plot may be constructed by first initiating the graphics device using a command like plot() and then adding to it by using commands such as points() or lines(). If many such repetitions are required, then a loop can be used to achieve them. For example, the following piece of code will produce 21 parabolas with intercept ranging from -20 to 20, at steps of 1.

```
x<- seq(-20,20)
plot(x,x^2, type="l", ylab="f(x)", xlim<-c(-8,8), ylim=c(-20,50))
for (i in -20:20)
    {
    lines(x,x^2+i)
    }
```

However, in the case of Figure 3.6, the curves are not parabolas, they are numerical solutions of a differential equation and *each* one of them needs to be generated by a loop (see R3.2). To generate Figure 3.6 it is necessary to **nest** one loop inside another so that the inner

loop will be repeated as many times as instructed by the outer loop. Here is some code that
will generate something similar to Figure 3.6(a).

```
# Model parameters
T.MAX<-60
R<- 0.11
K<-1000
tt<-seq(1,T.MAX)

# Initialises plotting device with desired labels and dimensions
plot(tt,rep(0,T.MAX),xlab="t",ylab="P",type="p",ylim=c(0,2000))

# Outer loop begins. This examines different initial conditions.
for (i in 1:20)
  {
 po<-i*100 # Generates a new initial condition
 p<-rep(po,T.MAX) # Creates vector to store specific solution

#Inner loop to generate a specific solution for initial value po
 for (t in 1:(T.MAX-1))
    {
    p[t+1]<-p[t]+p[t]*R*(1-p[t]/K) # Population model
    } # Inner loop ends here
   points(tt,p) # Adds new solution to the plot
  } # Outer loop ends here
```

The parameters at the start of the code do not include initial conditions because 20
different ones will be examined by the outer loop. The variable i is just a counter that ranges
from 1 to 20. Initial conditions for each solution are generated by multiplying the counter (i)
of the outer loop by 100. Hence, the first solution will start from 100 and the last from 2000.
The inner loop is indexed by t and is similar to the one presented in R3.2.

3.8.   Investigating stability

The stability of an equilibrium can be investigated graphically and analytically. As is often
the case, graphical methods help to build intuition whereas analytical methods prove more
general results. We have already seen one graphical method, the time series plot of a difference
equation (see Figures 3.5 and 3.6). A different perspective is offered by the **cobweb** plot, in
which time is implicit and, instead, the axes represent the current and future states of the
system (see also Example 3.8).

Example 3.11: Cobweb plot for an unconstrained, harvested population

The difference equation $P_{t+1} = (1 + r)P_t - c$ can be plotted on the
axes P_t and P_{t+1} (Figure 3.7 shows an example). The graph is a line
with slope $(1 + r)$ and intercept $-c$ representing all the possible
single-step transitions between P_t and P_{t+1}. At equilibrium, $P_t =
P_{t+1}$. This defines an identity line with slope 1 and intercept zero.
Any intersection between the first line (all possible transitions) and the identity line (all
conceivable equilibria) marks this system's single equilibrium point. Now, assume that we
started the system at $P_0 = 1200$ (so that the present population size, plotted on the x-axis of

the cobweb plot is $P_t = 1200$, as shown in Figure 3.7(a)). To find the population size (P_1) in the next year, we simply need to move up vertically in the plot until encountering the graph of the difference equation. In the next time step, what was the future has become the present, so the starting value on the x-axis is $P_t = P_1$. To shift the x-coordinate to this new value, we simply need to move horizontally (in the case of Figure 3.7(a) to the right) until meeting the identity line ($P_t = P_{t+1}$). This process can be repeated several times to obtain a visual representation of the system's behaviour. If the system is initialised at two different points (Figure 3.7(b)), one below and one above the equilibrium, the solutions diverge so that the system either goes extinct (if the initial population succumbs to the harvesting) or grows without check (if the initial population is large enough to tolerate a constant annual harvest). This, therefore, is an unstable equilibrium. The behaviour illustrated in Figure 3.7(b) is the equivalent of that shown in Figure 3.5 as a time series plot.

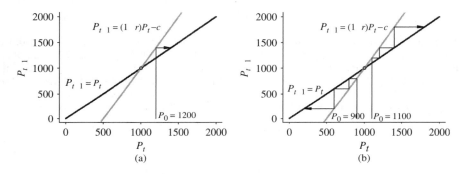

Figure 3.7: (a) First iteration of the construction of a cobweb plot for the model $P_{t+1} = (1+r)P_t - c$ with $r = 1, c = 1000$ and $P_0 = 1200$; (b) three iterations of the model for the two initial conditions $P_0 = 900$ and $P_0 = 1100$.

This example demonstrates that if the parameters and initial conditions of the model have been specified, then stability can be investigated simply by plotting the specific solution (either as a time series or cobweb plot). Similarly, if only the parameters have been specified, then we might try several specific solutions starting from different initial conditions (e.g. Figures 3.6 and 3.7(b)). But what happens if neither the parameters nor the initial conditions have been given? What sort of question can we ask about the stability of the difference equation on its own? An interesting question is: for what parameter values is each equilibrium stable? We therefore seek **analytical conditions for stability**. I will illustrate this with an even simpler example than the previous one.

Example 3.12: Conditions for stability under unrestricted growth

Going back to the simple model of unrestricted growth,

$$P_{t+1} = (1+r)P_t \tag{3.36}$$

In general (i.e. for any value of r), the only way in which present and future populations will be the same in this model is if the population has become extinct $P_t = P_{t+1} = 0$. This **extinction equilibrium** may either be stable (attracting solutions to zero) or unstable (sending them away to infinity) depending on the value of the intrinsic growth rate r. With $r > 0$, the slope of the graph of the difference equation is greater than 1 and the population increases unchecked (Figure 3.8(a)). With $r < 0$, the slope of the graph is smaller than 1 and the population becomes extinct (Figure 3.8(b)).

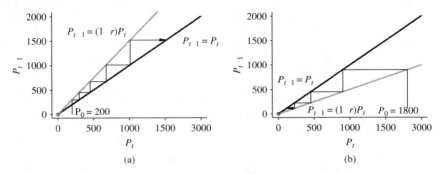

Figure 3.8: Cobweb plots for two different parameterisations of the unconstrained population model. (a) When the intrinsic growth rate is greater than zero (here, $r = 0.5$), the slope of the graph of the difference equation is greater than one (the grey line is steeper than the black one) and the population increases indefinitely. (b) In contrast, when the intrinsic growth rate is less than zero (here, $r = -0.5$), the slope of the graph of the difference equation is less than one (black line is steeper than the grey one) and the population becomes extinct.

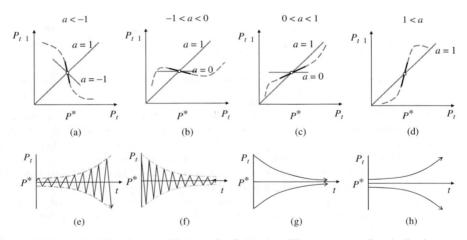

Figure 3.9: Local stability of an equilibrium of a first-order difference equation. In the top row of diagrams, (a)–(d), the dashed curve indicates the idealised graph of a difference equation and the small dark line segments represent the lines with slope a that approximate the slope of the curve in the vicinity of the equilibrium. For comparison, the solid grey lines and line segments show characteristic slopes as indicated alongside them. The diagrams in the bottom row, (e)–(h), depict typical solutions (plotted as time series) around the equilibrium P^* that correspond to the diagrams immediately above them. The four columns of this schematic distinguish four cases: when $a < -1$ ((a) and (e)), solutions in the vicinity of the equilibrium show **divergent oscillations**. When $-1 < a < 0$ ((b) and (f)), the solutions are damped (or convergent) oscillations. When $0 < a < 1$ ((c) and (g)), the solutions are **monotonic convergent** (they approach the equilibrium from one direction only). Finally, when $a > 1$ ((d) and (h)), the solutions **diverge monotonically**.

This relationship between the slope of the graph of the difference equation and the stability of equilibria is far-reaching. It can be shown that the slope of the graph of any difference equation *around* an equilibrium determines the stability of that equilibrium. The next chapter (see Section 4.4) will formally define the slope of a curve as the **derivative** of a function but, in the meantime, here is a useful quantitative result: if the graph of the difference equation near an equilibrium can be approximated by a line of the form (we will call this the **tangent line** to the curve)

$$P_{t+1} = aP_t + b \tag{3.37}$$

then the equilibrium is stable if the slope of this line satisfies the condition

$$|a| < 1 \tag{3.38}$$

Hence, all values of a between -1 and 1 will make solutions near the equilibrium move closer towards it. It is furthermore known that, if the slope is negative, the solutions oscillate around the equilibrium. The schematic in Figure 3.9 provides an intuitive overview of these results, to be applied more fully in the next section. Because of the approximation involved in Equation (3.37), the slope of the tangent line can only deal with stability in the close vicinity of the equilibrium. We therefore say that the above results deal with **local stability** properties. Stability analysis will be formalised in Section 4.11 using the concept of derivatives.

® 3.6: Creating a cobweb plot

The code needs to generate a numerical solution of the difference equation for a given number of time steps (here, specified by the parameter TMAX). The main loop of the program, which achieves this, is therefore not too different from the one in R3.3. Below is some code for creating a cobweb plot for the unconstrained population model in Example 3.12. It is divided into three parts. The first initialises the program and the second creates the main plot containing the graphs of the difference equation and the identity function. The command expression(), used to label this plot, enables printing of pretty mathematical expressions. In this particular example, square brackets are used to indicate subscripts, so expression(P[t]) prints out P_t. The third part of the code carries out the main task of simulating the population growth and plotting the cobweb's arrows. In the particular version shown, there are two separate arrow commands: the first draws the vertical arrow and the second the horizontal one. It is a good exercise to check if the coordinates used to plot the arrows make sense to you. The final important job, just before closing the loop, is to set the current coordinates x1 and y1 to the new population size y2. This allows the next iteration of the loop to recalculate population size by starting from this new value.

```
# General cobweb code

# Part 1: Model parameters & initial conditions
R<-0.5     # Intrinsic growth rate
XMAX<-2000# Maximum population to be plotted
TMAX<-5    # Total number of time steps to be shown
x1<-200    # Sets x1 coordinate to initial population size
y1<-0      # Sets initial y1 coordinate

# Part 2: Main plot
x<-seq(0,XMAX)
plot(x,x*(1+R), xlab=expression(P[t]), ylab=expression(P[t+1]),
   type="l",xlim=c(0,XMAX), ylim=c(0,XMAX))
abline(0,1)
```

```
# Part 3: Main loop used to calculate the specific solution
for (t in 1:(TMAX))
    {
    y2<-x1*(1+R)        # Unrestricted growth model
    arrows(x1,y1,x1,y2, length=0.1, angle=10)
    arrows(x1,y2,y2,y2, length=0.1, angle=10)
    x1<-y2
    y1<-y2
    }
```

The options length and angle regulate the shape of the arrow heads. I chose these particular values by trial and error to make the plot less cluttered. To make the code tidier, the two `arrows` commands can be combined (see R2.6).

```
arrows(c(x1,x1),c(y1,y2),c(x1,y2),c(y2,y2), length=0.1, angle=10)
```

This code can be used for various tasks. To examine the effect of using a different initial condition, experiment with different values of `x1`. To examine the stability properties of a different growth rate, just change the value of `R`. You may also customise this for use with a completely different population model.

3.9. Chaos

We are now ready to take a look at one of the most highly publicised discoveries of modern theoretical science. In common language, **chaos** is a byword for lack of order, often associated with phenomena that defy prediction. As I will argue below, this is true in practical terms. However, it is first necessary to introduce the more abstract concept of **deterministic chaos**; that is, the behaviour of systems that merely give the *impression* of randomness.

Example 3.13: Chaos in a model with density dependence

The discrete logistic model is probably the simplest nonlinear difference equation in ecology

$$P_{t+1} = P_t + r_{max}\left(1 - \frac{P_t}{K}\right)P_t \qquad (3.39)$$

Yet, as Robert May discovered in 1976, this deceptively simple model is capable of some rather complicated behaviour. Specifically, if different populations, each characterised by ever-larger values of r_{max}, are modelled by Equation (3.39), the predicted dynamics become increasingly volatile and ultimately erratic. To replicate May's result we will mainly use time series plots and cobwebs. Consider first a population with the relatively low value $r_{max} = 0.8$ (Figure 3.10). This model is a second order polynomial $(-r_{max}/K)P_t^2 + (1 + r_{max})P_t$ in the variable P_t. This, and the fact the coefficient $(-r_{max}/K)$ of the second order term is negative, means that the graph of the difference equation is a parabola pointing down (Figure 3.10(a)). The intersections between this parabola and the graph of the identity function give two possible equilibria: $P^* = 0$ and $P^* = K$ (see Example 3.10). The local stability of the two equilibria can be evaluated using the methods of Section 3.8. As we will see in Chapter 4, the slopes of the parabola at the two equilibria of this model are:

$$\begin{aligned} slope &= 1 - r_{max} \quad \text{at } P^* = K \\ slope &= 1 + r_{max} \quad \text{at } P^* = 0 \end{aligned} \qquad (3.40)$$

Remember that solutions around the equilibrium are stable if $|slope| < 1$. So, for $r_{max} = 0.8$, we would expect the carrying capacity to be stable and the extinction equilibrium to be unstable. This is borne out by Figure 3.10(b) which shows the specific solution to increase smoothly from 50 individuals to 200 (the carrying capacity).

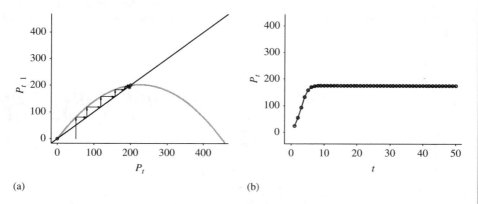

(a)   (b)

Figure 3.10: (a) Cobweb and (b) time series plots of a system with $K = 200, P_0 = 50$ and $r_{max} = 0.8$.

If the intrinsic rate is now increased to a value greater than 1 but still smaller than 2, say $r_{max} = 1.95$, then Equation (3.40) implies a slope of -0.95 around the carrying capacity. The theory in Figure 3.9 suggests, and the simulation in Figure 3.11 confirms, that solutions will oscillate towards the carrying capacity.

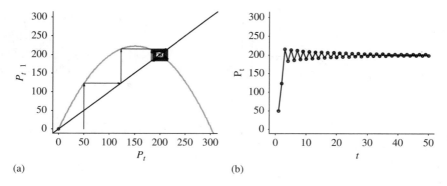

(a)   (b)

Figure 3.11: (a) Cobweb and (b) time series plots of a system with $K = 200, P_0 = 50$ and $r_{max} = 1.95$. Notice how the cobweb of the system around K now looks more recognisably like a spider's web.

If the growth rate is increased further, to $r_{max} = 2.1$, the slope around the carrying capacity becomes -1.1. Based on the summary results of Figure 3.9, this implies divergent oscillations. Compare the schematic in Figure 3.9(a) with what actually happens to solutions (Figure 3.12): instead of diverging, the oscillating solutions seem to be constrained between two values. This apparent discrepancy is due to the fact that the conditions in Figure 3.9 refer to *local* stability: remember that we approximated the slope of the curve close to the equilibrium by a tangent line. So, the fact that, when $r_{max} = 2.1$, the slope around K is large and negative tells us that solutions close to K will oscillate away from it. In contrast, the simulation in Figure 3.12 follows a solution that starts far below K, approaches it and then settles to oscillations away from it. If you are intrigued by the fact that the solutions initially approach the unstable equilibrium K, then remember that the local stability of the extinction equilibrium must also play a role. Since extinction is also unstable (check Equation (3.40)), it repels solutions towards the carrying capacity. We can therefore interpret the system's **global behaviour** (in this case, stable oscillations) as the outcome of the competition between the system's two equilibria.

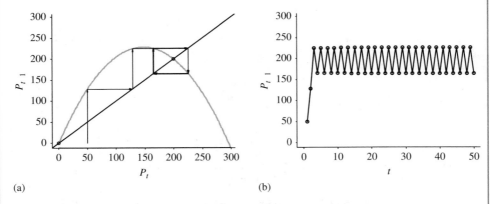

(a)   (b)

Figure 3.12: (a) Cobweb and (b) time series plots for $r_{max} = 2.1$.

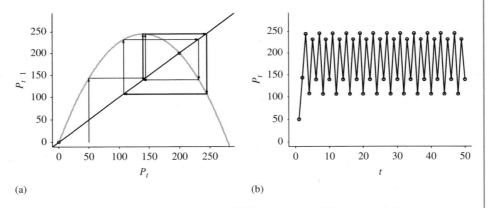

(a)   (b)

Figure 3.13: (a) Cobweb and (b) time series plots for $r_{max} = 2.5$.

Increasing the value of r_{max} even further leaves the local stability of the equilibria unaltered (they both repel solutions) but makes a big difference for the global behaviour of solutions. For $r_{max} = 2.5$ (Figure 3.13), the oscillations repeat themselves exactly every four points. This is twice as long as the two-point oscillations of Figure 3.12, so the phenomenon is called **period-doubling**. With a further small increase to $r_{max} = 2.57$ (Figure 3.14) the solutions undergo period-doubling again, now displaying eight-point oscillations. Further increases lead to rapid period-doubling until, eventually, any periodicity is lost completely (Figure 3.15). At that point, we say that we have **chaotic behaviour**. Note that, although the time series in Figure 3.15(b) looks random, the model generating it is the same, simple model in Equation (3.39).

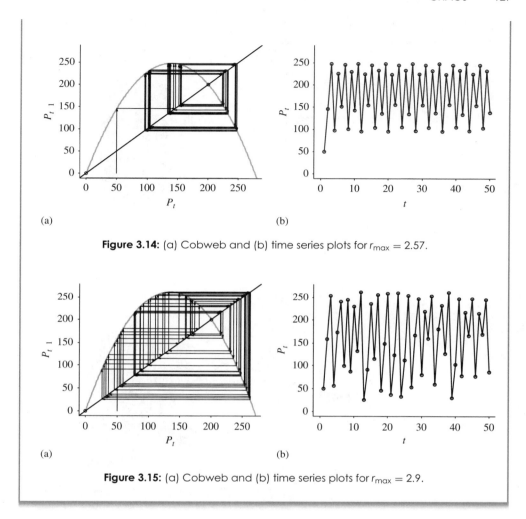

Figure 3.14: (a) Cobweb and (b) time series plots for $r_{max} = 2.57$.

Figure 3.15: (a) Cobweb and (b) time series plots for $r_{max} = 2.9$.

In the above example, I only tried out six carefully selected values of r_{max} to illustrate the onset of chaotic dynamics through period-doubling. A more synoptic graphical device known as a **bifurcation plot** shows population sizes for an entire range of values of a critical parameter such as the growth rate r_{max}. The example in Figure 3.16 is generated by simulating the system for a particular value of r_{max} and storing all population sizes occurring after a certain settling period (say, 50 time units). Repeating the process with several hundreds of different values for r_{max} and generating a scatter plot of population sizes versus r_{max} finally gives the bifurcation plot.

Bifurcation plots show how the system transits from a state of stability to chaos, via period-doubling. Interestingly, the plot may then re-enter parameter regions of relatively regular dynamics. Upon magnification, these regions look exactly like the original bifurcation plot (compare the magnified section in Figure 3.16(b) with the original in Figure 3.16(a)). We have already encountered patterns of infinite self-similarity in Chapter 2. They are called fractals (Section 2.17).

Although, in theory, a chaotic system can be predictable despite the erratic look of its solutions, this is of little practical use. In nature, a system that is predisposed to chaotic behaviour will also be subject to random influences. No matter how small these influences may be, the sensitivity of a chaotic system means that, in practice, its behaviour is unpredictable.

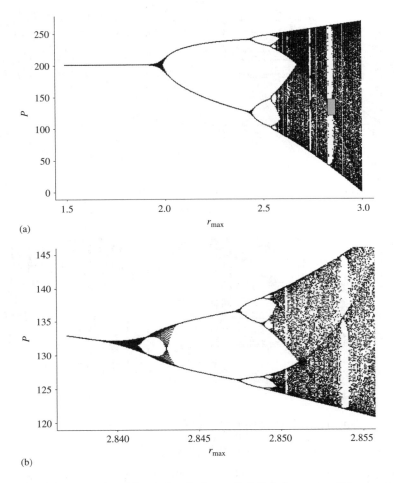

(a)

(b)

Figure 3.16: (a) Bifurcation plot of the system in Example 3.13 in the range $1.5 < r_{max} < 3.0$ and (b) an enlargement of the small region shown as a rectangle in (a).

3.10. Exponential function

Example 3.14: Modelling bacterial loads in continuous time

A lab study has shown that the number of bacteria of a particular species living in humans doubles every day. A biopsy of a patient has shown a load of 1000 bacteria per 100 cm^3 ($= 1 \times 10$ cm^{-3}). Samples taken after $1,2,3,\ldots,t$ days are therefore expected to show loads of

$$2, 2^2, 2^3, \ldots, 2^t \quad \times 10 \text{ cm}^{-3} \tag{3.41}$$

These are recognisable as the terms of the geometric progression $a_t = 2^t$. Similar logic can be used to **hindcast** bacterial loads. Had samples been taken 1,2,3,...,t days before the day of the original biopsy (day 0) they would have shown loads of

$$(\tfrac{1}{2}), (\tfrac{1}{2})^2, (\tfrac{1}{2})^3, \ldots, (\tfrac{1}{2})^t \times 10 \text{ cm}^{-3} \tag{3.42}$$

These are the terms of the geometric progression $a_t = (1/2)^t$. We can combine the two sequences by recalling that, for any positive number c,

$$c^{-x} = \frac{1}{c^x} \tag{3.43}$$

So, the unified sequence is

$$a_t = 2^t \quad (t = \ldots, -2, -1, 0, 1, 2, \ldots) \tag{3.44}$$

Although the bacterial load was estimated on a daily basis (in discrete time) the bacteria multiply continuously. This model can therefore be expanded by redefining the integer number of days t as a continuous variable measuring time in units smaller than days. Hence, bacterial load 36 hours after day 0 will be $2^{1.5} \cong 2.828 \times 10 \text{ cm}^{-3}$.

More generally, continuous processes of unrestricted growth can be modelled by the function

$$f(x) = c^x \tag{3.45}$$

for a given constant $c \geqslant 0$ (the base) and a real variable x (the exponent). Note that I have reverted to the conventional notation $f(x)$ for functions of continuous variables. Equation (3.45) gives the general form of the **exponential function**. Its graph can have one of three forms (Figure 3.17). If $c = 1$, it is the graph of the constant function $f(x) = 1$. For $c > 1$, it is an increasing curve and for $0 \leq c < 1$ it is a decreasing curve that asymptotically approaches zero. The properties of the exponential function are the same as the properties of powers (Chapter 1 and Section A1.1). Most importantly, the exponential function cannot give negative values.

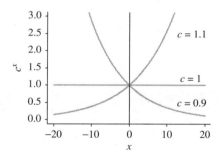

Figure 3.17: Plot of exponential functions for three different bases c.

The most frequently used exponential functions are **base-ten** ($c = 10$) and **base-e** ($c = e \cong 2.718281828$). The significance of the number e (**Euler's number**) will be discussed in Chapters 4 and 5. The exponential function with base e can be written in two equivalent ways

$$e^x = \exp(x) \tag{3.46}$$

So, is any modelling flexibility lost by limiting exponential functions to just two bases? For example, does the fact that both 10 and e are greater than one mean that we are constrained to only model exponential increases? Not really. Exponential declines can also be modelled by introducing a signed coefficient to the exponent. Taking the base-e exponential with positive values of x as an example, the following model can create all three shapes shown in Figure 3.17:

$$e^{ax} : \begin{cases} \text{exponential increase if } a > 0 \\ \text{exponential decline if } a < 0 \\ \text{constant function if } a = 0 \end{cases} \tag{3.47}$$

Equation (3.47) is exactly equivalent to the general definition of the exponential function in Equation (3.45) in the sense that any base c can be emulated by an appropriate choice of the coefficient a. For example, $a = -1$ gives $e^{-x} = (1/e)^x \cong 0.368^x$, which is an exponential function similar to Equation (3.45) with $c < 1$.

Example 3.15: A negative blue tit? Using exponential functions to constrain models

One problem with the discrete logistic equation

$$P_{t+1} = P_t \left(1 + r_{max} \left(1 - \frac{P_t}{K} \right) \right) \tag{3.48}$$

is that the entire population can crash-dive into negative values if it ever becomes sufficiently larger than the carrying capacity. In fact, for a given carrying capacity K and intrinsic growth rate r_{max}, the population at time $t + 1$ will become negative if $P_t > K(1 + 1/r_{max})$. There are several biological variables (population size, anatomical dimensions, proportions and probabilities) for which negative values simply make no sense. When formulating models that involve such quantities, exponentials offer a natural way to constrain them to positive values. For example, a modification of the logistic model called the **Ricker logistic model** is incapable of giving negative populations.

$$P_{t+1} = P_t \exp \left(r_{max} \left(1 - \frac{P_t}{K} \right) \right) \tag{3.49}$$

Both can be seen as specific versions of the basic model $P_{t+1} = P_t r(P_t)$, in which the per capita growth rate is a density-dependent function and they have similar dynamical properties. For example, when $P_t = 0$ or $P_t = K$, the population is at equilibrium. The two models also share their propensity for unstable and chaotic dynamics for large values of the intrinsic growth rate r_{max}.

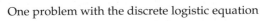

3.7: The exponential function in R

The base-ten exponential function is written simply as `10^x`. The base-e function is written `exp(x)`, so Euler's number can be obtained by typing `exp(1)`.

3.11.   Logarithmic function

For each value x the exponential function $f(x) = c^x$ produces a value y by raising the constant c to the power x. Suppose, instead, that we had the output y and wanted to calculate the input x. We would need to find the inverse $(f^{-1}(y))$ of the exponential function by asking: to what power x do we need to raise the constant c in order to get y as the result? This value is called the **logarithm with base c** (written, $\log_c$). Mathematically,

$$\begin{aligned} f(x) &= c^x = y \\ f^{-1}(y) &= \log_c y = x \end{aligned} \tag{3.50}$$

The exponential function can also be treated as the inverse of the logarithmic function (see Section 1.24 for a refresher on inverse functions).

$$f(x) = \log_c x = y$$
$$f^{-1}(y) = c^y = x$$
(3.51)

The domain of the exponential function is $\mathbb{R}$ and its range is $\mathbb{R}_+$ (i.e. all positive reals). Consequently, any positive input to the logarithmic function will yield a real output. Like trigonometric functions, logarithms and exponentials are hard to calculate by hand, so a calculator or computer will do this for you, using series approximations (Chapter 4). The properties of the logarithmic function (listed in Section A5.1) are a direct consequence of the properties of the exponential function. In most applications, the base of a logarithm will have one of two values. If $c = 10$, then we talk about **common logarithms** denoted '$\log_{10}$' or, simply, '$\log$'. If $c = e$ we talk about **natural logarithms** denoted '$\log_e$' or '$\ln$'. The graph of the logarithmic function (Figure 3.18(a)) is a mirror image of the exponential function (their symmetry axis is the identity line).

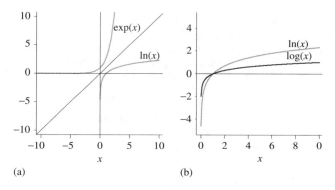

(a)   (b)

Figure 3.18: (a) Comparison of the graphs of the base-e exponential and natural logarithm functions; (b) comparison of the graphs of the common and natural logarithms.

Given that most scientific applications focus on common and natural logarithms, it is useful to know how they relate to each other. Starting with the definition of the base-ten exponential function

$$10^x = y$$
(3.52)

and taking the natural logarithm on both sides, with the aid of property (vi) in Section A5.1, this gives

$$x \ln 10 = \ln y$$
(3.53)

The logarithmic expression corresponding to Equation (3.52) is $x = \log y$. Replacing this in Equation (3.53) gives

$$\log y \ln 10 = \ln y$$
(3.54)

This provides a relationship between the natural and common logarithms of the same number (y). Notice that $\ln 10$ is a constant value, which implies that the two logarithmic functions are proportional to each other. Since $\ln 10 \cong 2.3026$ (i.e. greater than 1), we would expect the graph of $\ln y$ to be steeper than the graph of $\log y$, as confirmed by Figure 3.18(b).

Example 3.16: Log-transforming population time series

Turchin (2003) argues that use of the exponential function in population models has a deeper significance than the mere convenience of avoiding negative population values. By extending the rationale of the Ricker logistic model (Example 3.15), he suggests the following general model for population dynamics

$$N_{t+1} = N_t \exp(f(\cdot)) \tag{3.55}$$

For the present discussion, the function $f(\cdot)$ and its arguments need not be specified. Depending on the model, they could be past population densities, environmental influences or random perturbations. In the case of the Ricker logistic model, it is a simple decreasing function of current population size

$$f(N_t) = r_{max} \left(1 - \frac{N_t}{K}\right) \tag{3.56}$$

This is a first order polynomial of the form $f(N_t) = a_0 + a_1 N_t$ (for $a_0 = r_{max}, a_1 = r_{max}/K$). Turchin further argues that all of the important factors affecting a population can be strung together in an additive fashion inside the function $f(\cdot)$. For example, if we assume that a population's growth rate during a year is affected by its current size (N_t), average rainfall (X_t) and a harvest term (H_t), then the function $f(\cdot)$ could be written as

$$f(N_t, X_t, H_t) = a_0 + a_1 N_t + a_2 X_t + a_3 H_t \tag{3.57}$$

As we will see in Chapter 11, this formulation is appealing because it facilitates the estimation of the coefficients $a_0, a_1, a_2, \ldots$ from real data. Therefore, a useful starting point in exploring real population time series may be to look at the following transformation, which is derived directly from Equation (3.55),

$$f(\cdot) = \ln \left(\frac{N_{t+1}}{N_t}\right) \tag{3.58}$$

One would hope that this would be a linear (or, at least, additive) function of the factors acting on the population (see artificial example in Figure. 3.19)

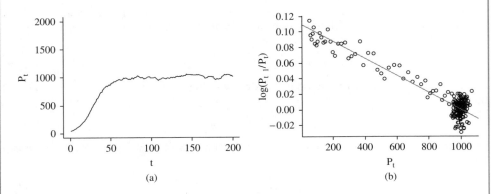

Figure 3.19: (a) A population growing according to the Ricker logistic model with parameter values $K = 1000$ and $r_{max} = 0.11$ under some environmental noise; (b) the transformation suggested in Equation (3.58). The points arrange themselves in a linear pattern. The intercept of the line is approximately r_{max} and the slope is approximately r_{max}/K.

There is a general message in this example. Biologists and statisticians have always been more comfortable dealing with linear relationships. Some nonlinear relationships can be linearised by transformations such as the one in the previous example. Although we are increasingly able to deal with nonlinear models (see Sections 11.7, 11.10 and 11.11), linearising transformations are often used during initial, exploratory data analysis. Most readily, when plotting relationships between variables, it may be useful to log-transform the x or y axes.

3.8: Logs and log-transformations in plots

The main command for calculating all logarithms in R is `log()`. The default is the natural logarithm but other bases can be specified through the option `base`. Hence, `log(10)` will give the value of $\ln(10) = 2.302585$ but `log(10, base=10)` will give the value of $\log(10) = 1$.

Log-plots can be generated by transforming either the data or the axes. A list of values a can be transformed by typing `log(a)`. Specifying log-transformed axes on a plot is just as easy, using the option `log`. Hence, `plot(data, log="y")` gives a plot with a log-transformed y-axis. You can use `log="x"` for the x-axis or `log="xy"` to transform both. Whichever method you choose, be prepared to get warning messages if any of your data are negative or zero.

3.12. Logarithmic equations

A **logarithmic identity** is an equation, involving logarithms, that is valid for all values of the variable (such as the properties of logarithms listed in Section A5.1). A **conditional logarithmic equation** is only valid for certain values of the variable. To solve the equation we need to find these values, but how? The simplest logarithmic equation (using natural logarithms as an example) takes the form

$$\ln(x) = c \tag{3.59}$$

for some known constant c. The required value of x is trapped inside the ln() and, in order to extract it, the ln() function must be inverted. This is done by applying the exponential function to the LHS of Equation (3.59): $e(\ln(x)) = x$. To maintain the balance of the equation, this must be applied to both sides, to obtain:

$$x = \exp(c) \tag{3.60}$$

Given a particular value for c, a calculator can do the rest. More complicated equations involving several logs can be solved in the same way. First, use the properties of the logarithms (Section A5.1) to try and get the equation in one of the following two forms:

$$\ln(f(x)) = c$$
$$\ln(f(x)) = \ln(g(x)) \tag{3.61}$$

where $f(x)$ and $g(x)$ are simple functions (e.g. polynomials), and then take exponentials on both sides. The result should be easier to solve than the original equation.

Example 3.17

Consider the following logarithmic equation

$$\log(x - 6) + \log(x - 7) = 1 - \log 5 \tag{3.62}$$

There are several ways to solve this using the identities in Section A5.1. Here is one:

$$\log((x - 6)(x - 7)) = \log 10 - \log 5$$

$$\log((x-6)(x-7)) = \log 2$$
$$10^{\log((x-6)(x-7))} = 10^{\log 2} \tag{3.63}$$
$$(x-6)(x-7) = 2$$
$$x^2 - 13x + 40 = 0$$

This quadratic equation can be solved to give $x_1 = 8, x_2 = 5$. However, only the first of these is a solution of the original equation because $x = 5$ results in negative numbers inside the logs of Equation (3.62).

Further reading

A great, nontechnical treatment of all the main topics in this chapter is Sandefur (1990) which motivates the theory of difference equations from simple numerical examples and illustrates it with real-world problems, some of them ecological. There are a multitude of good books on discrete time models in ecology. Some excellent ones are Gurney and Nisbet (1998), Case (2000), Kot (2001) and Otto and Day (2007). All of them have sections on instability, and chaos is discussed to variable degrees depending on how relevant it is deemed by the authors to real-world ecology. Less mathematical treatments of the same issues are given by Begon, Townsend and Harper (2005), Turchin (2003) and Coulson and Godfray (2007). The classic layman's reference to chaos is Gleick (1987). A good treatment of logarithmic and exponential functions in biology is given in Chapters 6 and 10 of Batschelet (1979).

References

Batschelet, E. (1979) *Introduction to Mathematics for Life Scientists*. Springer Verlag, Berlin. 643pp.

Begon, M., Townsend, C.A. and Harper, J.L. (2005) *Ecology: From Individuals to Ecosystems*, 4th edition. Wiley Blackwell. 752pp.

Case, T.J. (2000) *An Illustrated Guide to Theoretical Ecology*. Oxford University Press, New York. 449pp.

Coulson, T. and Godfray, H.C.J. (2007) Single-species dynamics. In *Theoretical Ecology: Principles and Applications*. (Eds R.M. May and A. McLean). pp 17–34.

Gleick, J. (1987) *Chaos: Making a New Science*. Cardinal, Suffolk. 352pp.

Gurney, W.S.C. and Nisbet, R.M. (1998) *Ecological Dynamics*. Oxford University Press, New York. 335pp.

Kot, M. (2001) *Elements of Mathematical Biology*. Cambridge University Press, Cambridge. 453pp.

May, R.M. (1976) Simple mathematical models with very complicated dynamics. *Nature*, **261**, 459–467.

Otto, S.P. and Day, T. (2007) *A Biologist's Guide to Mathematical Modelling in Ecology and Evolution*. Princeton University Press, New Jersey. 732pp.

Sandefur, J.T. (1990) *Discrete Dynamical Systems: Theory and Applications*. Oxford University Press, New York. 445pp.

Turchin, P. (2003) *Complex Population Dynamics: A Theoretical/Empirical Synthesis*. Princeton University Press, New Jersey.

4

How to change things, continuously
(Derivatives and their applications)

'And what are these [derivatives]? . . . They are neither finite quantities nor quantities infinitely small, nor yet nothing. May we not call them the ghosts of departed quantities?'
Bishop George Berkeley (1685–1753), Irish philosopher

In the 5th century BC, Heraclitus said that nature is always in a state of flux. As natural scientists and citizens of the 21st century, we are accustomed to the idea of rapid change: modern science has disputed constancy at every level, from the incorruptibility of atoms to the invariance of species. So much so, that Heraclitus's notion may seem to us commonplace – a classic example of a classical philosopher overstating the obvious. Nevertheless, the mathematical tools for modelling change, collectively known as **calculus**, are relatively recent. They were developed by Isaac Newton and Gottfried Leibniz, 22 centuries after Heraclitus's time. Even then, these developments were seen as counterintuitive and flawed (see quote above from George Berkeley who publicly referred to the inventors of calculus as 'infidel mathematicians'). Since then, calculus has motivated most of the developments in physics, astronomy, economics, engineering and ecology. This chapter begins with the concept of **average rate of change** (Section 4.1) and uses it to generate the idea of **instantaneous change** (Section 4.2). It then makes a brief detour to present the notation and meaning of **limits** (Section 4.3) that are then used to formalise the notion of **a function's derivative** (Section 4.4). Differentiation is discussed in Sections 4.5 and 4.6, leading to a set of algebraic shortcuts for calculating derivatives, in particular, the **chain rule** (Section 4.7) which can be used to simplify complicated

How to be a Quantitative Ecologist: The 'A to R' of Green Mathematics and Statistics, First Edition. Jason Matthiopoulos.
© 2011 John Wiley & Sons, Ltd. Published 2011 by John Wiley & Sons, Ltd.

differentiation problems. **Higher order derivatives**, the result of repeatedly differentiating the same function, are discussed in Section 4.8. The concept of differentiation is then extended to functions of many variables in order to introduce **partial derivatives** (Section 4.9). Three important applications of derivatives are left for the end of the chapter: the retrieval of function **minima** and **maxima** (Section 4.10), the analysis of **local stability for difference equations** (Section 4.11) and the approximation of complicated functions with the use of **series expansions** (Section 4.12).

4.1.   Average rate of change

Example 4.1: Seasonal tree growth

A botanist interested in the productivity of a particular tree species measures plant mass repeatedly, at quarterly intervals, rather than just once at maturity. After three years, a total of 12 measurements are available per specimen (Figure 4.1(a)). These discrete observations are successive snapshots from a continuous process (tree growth).

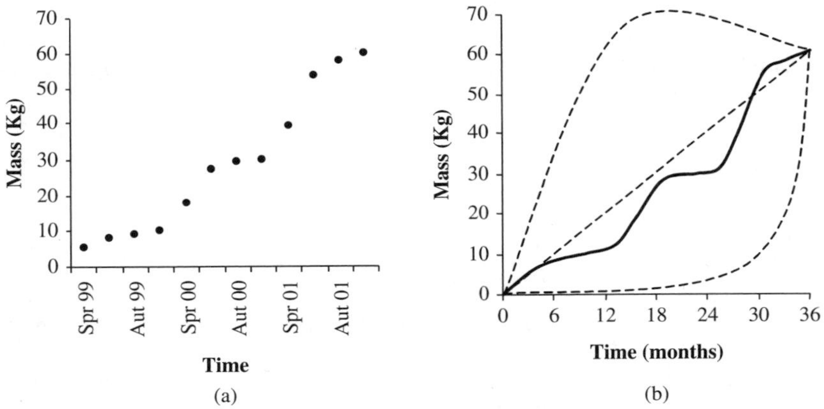

Figure 4.1: (a) Outcome of the botanist's measurements over three years; (b) different possible pathways to achieving the same final mass.

The final biomass of 60 kg could have been attained in many ways (Figure 4.1(b)), but these particular data suggest that: (1) annual growth increases with age, (2) there is more growth in the spring and summer months and (3) individual plants never seem to lose weight.

Chapters 1 and 3 freely moved from polynomial functions to sequences and from there to the exponential function without paying too much attention to issues of continuity. Unlike the polynomial, exponential and logarithmic functions, sequences are discontinuous functions

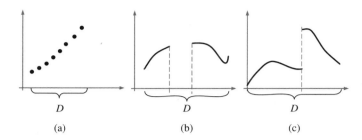

D	D	D
(a)	(b)	(c)

Figure 4.2: Graphs of functions with discontinuities in their domain D showing (a) a discrete function, (b) a gap discontinuity and (c) a jump discontinuity.

(their domain is the set of non-negative integer numbers). This type of discontinuous function is called **discrete**. Other discontinuous functions might include gaps in their (otherwise continuous) domains or jump discontinuities (Figure 4.2).

A **continuous function** $f(x)$ must, first of all, be defined on an uninterrupted domain D. The examples in Figures 4.2(a) and 4.2(b) fail to satisfy this condition. Secondly, a function $f(x)$ is continuous if any two values x_1 and x_2, selected arbitrarily close to each other, yield values $f(x_2)$ and $f(x_1)$ that are also very close. The example in Figure 4.2(c) fails to satisfy this requirement at one point of its domain.

The mass of an undisturbed plant must be a continuous function of time because, even when it is not growing, it still maintains its biomass and cannot instantaneously jump from one weight to another without first going through intermediate values. However, by recording the plant's mass at two different points in time, the botanist measures the average rate of weight change in the time interval between observations. More formally, the **average rate of change** of a continuous function $f(x)$ over an interval $[x_1, x_2]$ is defined as the ratio

$$\frac{\Delta f(x)}{\Delta x} = \frac{f(x_2) - f(x_1)}{x_2 - x_1} \tag{4.1}$$

The Greek letter delta (Δ) is used to denote changes (or differences, or increments) in the values of the function and its independent variable (see Example 1.6).

Example 4.2: Tree growth

$g(t)$ is the function giving plant biomass at time t (measured in months since the beginning of the botanist's experiment). During the second year of observation (from $t_1 = 12$ months to $t_2 = 24$ months) the plant grew from $g(t_1) = g(12) = 10$ kg to $g(t_2) = g(24) = 30$ kg. The average rate of growth over that period was

$$\frac{\Delta g(t)}{\Delta t} = \frac{g(t_2) - g(t_1)}{t_2 - t_1} = \frac{(30 - 10)\ \text{kg}}{(24 - 12)\ \text{month}} \cong 1.67\ \text{kg month}^{-1} \tag{4.2}$$

The unit of time in this example is the month. This does not imply that the average value was calculated over a monthly period, rather, that if the plant grew at the same monthly rate during its entire second year, it would gain 1.67 kg each month.

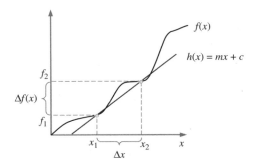

Figure 4.3: A line, called the secant, can be drawn through any two points of the graph of a function. The slope of the secant can be seen as the average rate of change of the function between these two points.

Another way to interpret the average rate of change of a function $f(x)$ is to consider any two points $(x_1, f(x_1))$ and $(x_2, f(x_2))$ on its graph and draw a straight line through them (Figure 4.3). This is called a **secant line** and is described by a function of the form

$$h(x) = mx + c \tag{4.3}$$

By construction, the curve $f(x)$ and the line $h(x)$ coincide at two points, $f(x_1) = h(x_1)$ and $f(x_2) = h(x_2)$. Let us call these values f_1 and f_2 and construct the following two equations from Equation (4.3)

$$f_1 = mx_1 + c \tag{4.4}$$

$$f_2 = mx_2 + c \tag{4.5}$$

Subtracting Equation (4.4) from Equation (4.5) gives

$$f_2 - f_1 = m(x_2 - x_1) \tag{4.6}$$

which can be written as

$$m = \frac{f_2 - f_1}{x_2 - x_1} = \frac{\Delta f(x)}{\Delta x} \tag{4.7}$$

Therefore, the average rate of change of a function between two values of the function's independent variable is equal to the slope of the secant going through the function's graph at these two values.

4.1: Calculating the rate of change from a sequence of regular data

Given a set of n observations (say, the records of tree growth in Example 4.1), collected over regular time intervals (in this case, approximately every 91 days),

```
w<-c(5,8,9,10,20,27,30,31,39,55,59,60) # The weight data
```

we may ask what the average rate of change was in each of these periods. To do this, we need to subtract every observation recorded at time t from every observation recorded at time $t + 1$,

```
n<-length(w)   # Sample size
w1<-w[1:(n-1)] # Weights at time t, going up to n-1
```

```
w2<-w[2:n]      # Weights at time t+1, starting from 2nd measurement
dw<-w2-w1       # List of differences between pairs of weights
```

The result may be converted to an estimate of daily growth rate over each of the 91-day periods,

```
dt<-91
rate<-dw/dt
```

What if we wanted to calculate the growth rate at six-month intervals? This corresponds to the difference between observations at t and observations at $t + 2$. A more general piece of code can be obtained by defining a new parameter ($\texttt{lag}$) as the timescale over which the rate of change is to be calculated. This parameter is an integer multiple of the timescale at which the observations were made (so a lag of 2 is equivalent to six months). It can take values between 0 and n:

```
lag<-2              # Scale for calculation of rate (between 0 and n)
dt<-91              # No of time units between successive observations
n<-length(w)        # Sample size
w1<-w[1:(n-lag)]    # Weights at time t
w2<-w[(lag+1):n]    # Weights at time t+lag
dw<-w2-w1           # List of differences
rate<-dw/dt
```

4.2.   Instantaneous rate of change

The average rate of change of a function is calculated over an arbitrary interval of its independent variable. So, knowing that a plant grows by 60 kg over three years tells us nothing about how it achieves this increase or how quickly it grows at any particular time. These questions relate to the **instantaneous rate of change** of a continuous function $f(x)$.

To calculate the instantaneous rate of change, consider a fixed point $(x_1, f(x_1))$ and a movable point $(x, f(x))$ on the graph of the function. The slope of the secant through these two points can be calculated from Equation (4.7). If x is allowed to approach x_1 (Figure 4.4), then $f(x)$ should also approach $f(x_1)$.

When the value of x becomes almost identical to x_1, the secant is called the **tangent** to the function's graph at the point $(x_1, f(x_1))$ and the slope of the tangent is the **slope of the graph of**

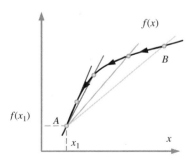

Figure 4.4: As the point B approaches the point A by travelling along the graph of $f(x)$, the slope of the secant gives the rate of change of the function over ever-smaller intervals.

the function at that point; the instantaneous rate of change of $f(x)$ at x_1. Mathematically, this is written using the new notation of **limits** (examined more thoroughly in the following section),

$$\lim_{x \to x_1} \frac{f(x) - f(x_1)}{x - x_1} \text{ or, alternatively, } \lim_{\Delta x \to 0} \frac{f(x_1 + \Delta x) - f(x_1)}{\Delta x} \qquad (4.8)$$

ⓡ 4.2: Visualising the instantaneous rate of change

The following code will generate a plot similar to Figure 4.4. You can look at different functions by changing the original specification (here, $f(x) = x^3$). You can look at different points of the function's graph by changing the value of x (here, $x = 3$). You can look at different magnifications of the plot by modifying the values of Xmin and Xmax. If you tighten the plotting range (e.g. setting x<-3, Xmin<-2.9 and Xmax<-3.1), the secant lines become closer to each other and more aligned with the graph of the function (i.e. they become better approximations of the tangent line).

```
# Definition of the function to be examined
f<-function(x) { return(x^3) }

# Parameters
x<-3          # Point around which to estimate rate of change
y<-f(x)       # y-value of that point
Xmin<- -2     # minimum x-axis value
Xmax<-8       # maximum x-axis value
N<-10         # number of secants to be drawn
Inc<-0.1      # Plotting increment

# Plotting
xs<-seq(Xmin, Xmax, by=Inc)          # list of x values for plotting
plot(xs,f(xs), type="l", col="red")  # graph of the function
i<-seq(N,1,by=-1)                    # sequence of decreasing numbers
dx<-i*(Xmax-x)/N                     # increments on x-axis ahead of x
m<-(f(x+dx)-f(x))/dx                 # secant slopes from Equation (4.7)
c<-f(x)-x*m                          # secant intercepts
for(i in 1:N) abline(c[i],m[i])      # graphs of secants
```

4.3. Limits

The notation $\lim_{x \to a}$ is read 'the limit, as x tends to a'. So, $\lim_{x \to a} f(x) = c$ means that c is the limit of the function $f(x)$ as x tends to a. This expression does not assume that x attains the value a or, indeed, that $f(x)$ becomes equal to c, but merely that as x gets closer to the value a, $f(x)$ approaches c. Why make this distinction? Because sometimes, 'close enough' is the best we can do. A discontinuous function such as the one plotted in Figure 4.5(a) may not be defined at the point a, but it may be defined at all nearby points. So, although in this case, $f(a)$ is nonsense, $\lim_{x \to a} f(x)$ is not: it is equal to c. Even more perverse situations can be handled with limits: the function plotted in Figure 4.5(b) has a jump discontinuity at a, so it is certainly not defined at that point. The limit of the function exists, but it is different depending on whether we approach a from the left or right. The left-hand limit is written $\lim_{x \to a^-} f(x) = c_1$. The right-hand limit is $\lim_{x \to a^+} f(x) = c_2$.

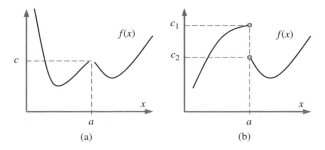

Figure 4.5: Limits are useful for handling discontinuous functions. Neither of the two functions shown above is defined at a. The function in (a) has a breakpoint and the one in (b) has a jump discontinuity. However, apart from these abnormalities, the functions are well behaved elsewhere. This enables us to say what happens to the function f(x) as its independent variable x approaches the value a.

Limits can be manipulated algebraically. For example, given the limits of two functions at the same point, $\lim_{x \to a} f(x) = c_1$ and $\lim_{x \to a} g(x) = c_2$ the limit of their sum is the sum of their limits, $\lim_{x \to a} (f(x) + g(x)) = c_1 + c_2$. More such properties are listed in Section A6.1 and put to use in the following example.

Example 4.3

Consider the two polynomial functions $f(x) = 3x + 2$ and $g(x) = x^2 + 2x - 1$. We would like to calculate $\lim_{x \to 2} (f(x)g(x) - g(x)^{-1})$. The functions are defined at $x = 2$, so that $\lim_{x \to 2} f(x) = \lim_{x \to 2} (3x + 2) = 8$ and $\lim_{x \to 2} g(x) = \lim_{x \to 2} (x^2 + 2x - 1) = 7$. Therefore, the required limit can be found as follows using the properties of Section A6.1,

$$\lim_{x \to 2} (f(x)g(x) - g(x)^{-1})$$

$$= \lim_{x \to 2} \left((f(x)g(x)) - \frac{1}{g(x)} \right)$$

$$= \lim_{x \to 2} (f(x)g(x)) - \lim_{x \to 2} \left(\frac{1}{g(x)} \right) \quad \text{(property i)} \tag{4.9}$$

$$= \lim_{x \to 2} (f(x)) \lim_{x \to 2} (g(x)) - \frac{\lim_{x \to 2} (1)}{\lim_{x \to 2} (g(x))} \quad \text{(properties ii, iii)}$$

$$= (8 \times 7) - 1/7 = 55\tfrac{6}{7}$$

Since all three functions $f(x) = 3x + 2$, $g(x) = x^2 + 2x - 1$ and $(f(x)g(x) - g(x)^{-1})$ exist at 2, we could just as easily have obtained the result without limits. So, this example understates the usefulness of limits, which really come into their own at points where the function does not exist, as seen in the following example.

Example 4.4

Consider the function

$$f(x) = \frac{2x^2 - 12x + 18}{x - 3} \cdot x \in \mathbb{R}, x \neq 3 \tag{4.10}$$

which is not defined at $x = 3$. Using the concept of the limit, we can examine what happens as x approaches 3

$$\lim_{x \to 3} f(x) = \lim_{x \to 3} \frac{2x^2 - 12x + 18}{x - 3} \qquad (4.11)$$

The numerator of this fraction can be factored as follows,

$$2x^2 - 12x + 18 = 2(x^2 - 6x + 9) = 2(x - 3)^2 \qquad (4.12)$$

Substituting this back into Equation (4.11), gives

$$\lim_{x \to 3} f(x) = \lim_{x \to 3} \frac{2(x - 3)^2}{x - 3} = \lim_{x \to 3} 2 \lim_{x \to 3} (x - 3) = 2 \cdot 0 = 0 \qquad (4.13)$$

The original function, Equation (4.10) is not defined at 3. In contrast, a little algebra led to the expression $2(x - 3)$, which is defined. If these functions are equivalent, does this not lead to a mathematical paradox? Yes, it probably would, if the value $x = 3$ hadn't been cunningly side-stepped by use of limits.

The reason that Equation (4.10) ran into trouble and needed to be handled using limits was that, at $x = 3$, it attempted a division by zero. Although the operation $1/0$ is a big mathematical no-no, we know from basic arithmetic that a fraction with a very small denominator will tend to be large. Conversely, a fraction with a very large denominator will tend to be very small. Limits now enable us to formalise these intuitive notions as follows,

$$\lim_{x \to 0} \frac{1}{x} = \infty \qquad (4.14)$$

$$\lim_{x \to \infty} \frac{1}{x} = 0 \qquad (4.15)$$

Infinity is not a number, it can be approached but never reached. Therefore, no variable can attain it and no algebraic operation can involve infinity explicitly. Hence, in Equation (4.14) $1/x$ *tends* to infinity as x *tends* to zero and in Equation (4.15), $1/x$ *tends* to zero as x *tends* to infinity.

Example 4.5: Methane concentration around termite mounds

Termite mounds

In the absence of wind, the concentration of methane emitted from the top of a termite mound decays with distance (d) from the source according to the following (fictitious) function

$$f(d) = \frac{0.4d + 0.7}{d + 1} \qquad (4.16)$$

Methane exists in background concentrations in the environment, independently of any given mound. To calculate this ambient concentration we need to ask what the value of the function $f(d)$ is very far from the mound or, mathematically,

$$\lim_{d \to \infty} f(d) = \lim_{d \to \infty} \frac{0.4d + 0.7}{d + 1} \qquad (4.17)$$

Note that $\lim\limits_{d\to\infty} (d + 1) = \infty$ so the above limit could be tackled with Equation (4.15) if it weren't for the fact that d also exists in the numerator. There is no obvious way to determine what happens to the fraction as d increases, because both the numerator and the denominator will increase. To overcome this, first note that

$$\frac{0.4d + 0.7}{d + 1} = \frac{0.4d + (0.4 + 0.3)}{d + 1} = \frac{0.4(d + 1) + 0.3}{d + 1} = 0.4 + \frac{0.3}{d + 1} \tag{4.18}$$

Substituting Equation (4.18) back into Equation (4.17) gives

$$\lim_{d\to\infty} f(d) = \lim_{d\to\infty} \left(0.4 + \frac{0.3}{d + 1}\right) = \lim_{d\to\infty} (0.4) + 0.3 \lim_{d\to\infty} \left(\frac{1}{d + 1}\right) = 0.4 \tag{4.19}$$

The concentration of the substance at the entrance of the mound can be found by simply setting x to zero, $f(0) = 0.7$ (no need to use limits in this case).

ⓡ 4.3: Infinity

R is sensible in its treatment of infinity, compared to most computer languages. For example, typing `1/0` or `-1/0` will return `Inf` and `-Inf`, indicating that R has interpreted these two fractions as limits. However, typing something like Equation (4.10),

```
> x<-3
> (2*x^2-12*x+18)/(x-3)
[1] NaN
```

which is numerically equivalent to 0/0, gives the response `NaN` ('Not a Number'), even though the limit at 3 can be calculated, as we saw in Example 4.4. R can also use infinity as an approximate value when machine precision is exceeded. For example, typing `1/10^-1000`, will return `Inf`. It also tries to be sensible in its plotting of functions that have discontinuities involving infinity. For example, the following plot will have a break at $x = 3$

```
x<-seq(1, 10, by=.1)
plot(1/(3-x), type="l")
```

4.4. The derivative of a function

In Section 4.2 the instantaneous rate of change of a function $f(x)$ at a point x was defined as

$$\lim_{\Delta x\to 0} \frac{f(x + \Delta x) - f(x)}{\Delta x} \tag{4.20}$$

If Equation (4.20) is calculated for all values of x in the function's domain, we obtain a complete mapping between the function and its rate of change.

Example 4.6: Plotting change in tree biomass

A rough sketch of the rate of change of biomass is drawn next to a plot of biomass in Figure 4.6. The tree underwent three periods of growth, so we should expect the rate of change to present three distinct peaks. During the winter months the tree does not grow, so the rate of change comes down to zero. Also, since the tree never loses weight, its rate of change is always positive.

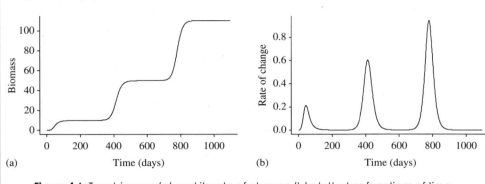

(a)   Time (days)   (b)   Time (days)

Figure 4.6: Tree biomass (a) and its rate of change (b) plotted as functions of time.

This new function is called the **derivative** of $f(x)$, it is denoted by $f'(x)$ and its value at x is given by

$$f'(x) = \lim_{\Delta x \to 0} \frac{f(x + \Delta x) - f(x)}{\Delta x} \tag{4.21}$$

A function $f(x)$ is called **differentiable at x** if the limit in Equation (4.21) exists at x. The function is called simply **differentiable** if the limit in Equation (4.21) exists for all values of x in the function's domain. There are five different notations for derivatives:

$$f'(x) = \frac{df}{dx} = \frac{df(x)}{dx} = \frac{d}{dx}f(x) = \dot{f}(x) \tag{4.22}$$

The dot-notation $\dot{f}(x)$ is reserved for functions of time and is rarely seen in the ecological literature. I will use both the f' and df/dx notations throughout this book. The ds in $df(x)/dx$ are used to indicate infinitely small changes in $f(x)$ and x. The d and Δ notations are linked by the relationship between instantaneous and average rate of change

$$\frac{df}{dx} = \lim_{\Delta x \to 0} \frac{\Delta f(x)}{\Delta x} \tag{4.23}$$

For a particular function $f(x)$, it is possible to use the rules of limits (see Section A6 in the appendix) to find an expression for its derivative, $f'(x)$.

Example 4.7: Linear tree growth

If, in between the times $t_1 = 0$ and t_2, a plant grows approximately linearly, its mass may be described by a function of the form

$$g(t) = mt + c \tag{4.24}$$

where c is the mass of the tree at t_1. From the definition of the derivative,

$$g'(t) = \lim_{\Delta t \to 0} \frac{g(t + \Delta t) - g(t)}{\Delta t} \qquad (4.25)$$

Equation (4.24) can help specify the expressions $g(t + \Delta t)$ and $g(t)$ in Equation (4.25)

$$g'(t) = \lim_{\Delta t \to 0} \frac{m(t + \Delta t) + c - mt - c}{\Delta t} = \lim_{\Delta t \to 0} \frac{m\Delta t}{\Delta t} = \lim_{\Delta t \to 0} m = m \qquad (4.26)$$

So, the rate of change of mass is constant and equal to the line's slope.

Hence, the derivative of a linear function is constant. Geometrically, the derivative can be interpreted as the slope of a function $f(x)$ at every point of its domain. So, increasing functions have positive derivatives and vice versa. Further, when a function is constant, or just as it goes from increasing to decreasing, its derivative (rate of change) will be zero.

There are two situations in which the derivative cannot be defined. The first, sensibly enough, is when the function itself is not defined. For example, the function $1/(1 - x)$ has no derivative at $x = 1$. The second is when the graph of the function does not have a unique slope.

Example 4.8

The derivative of a function involving absolute values cannot be defined wherever $f(x) = 0$. The function $f(x) = |x^2 - 9|$ becomes zero at $x = \pm 3$ (Figure 4.7) so, rather ambiguously, its slope is both negative and positive, depending on whether these two points are approached from their left or right.

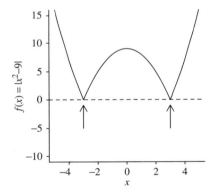

Figure 4.7: The derivative of this function is defined everywhere apart from the values $x = \pm 3$.

Visually, a function is differentiable if its graph has no 'kinks' such as the ones appearing in Figure 4.7. Functions that can only be differentiated for parts of their domains are called **piece-wise differentiable**. The function in Example 4.8 is piece-wise differentiable over the intervals $(-\infty, -3), (-3, 3)$ and $(3, +\infty)$. The fractal shapes that we examined in Section 2.17 are examples of functions that are continuous everywhere but differentiable nowhere.

4.5. *Differentiating polynomials*

Example 4.9: Spatial gradients

Cuvier's beaked whale, Ziphius cavirostris

Cuvier's beaked whales, like many marine animals, show a strong association with areas of steep sea-bed. To model the spatial distribution of these animals it is often necessary to generate maps of slope from sparse and irregular depth data. To simplify this illustration, consider a data set of depth measurements collected along a single, linear transect (the points in Figure 4.8). To describe these bathymetry data mathematically, I will use the following sixth order polynomial,

$$d = -3.991 \times 10^{-17}l^6 + 5.754 \times 10^{-13}l^5 - 3.025 \times 10^{-9}l^4$$
$$+ 7.006 \times 10^{-6}l^3 - 6.706 \times 10^{-3}l^2 + 1.250l - 109 \tag{4.27}$$

This gives depth (the sea-bed **relief**) as a function of transect length for the entire transect. Its coefficients are estimated from the data (see R4.4, below). The graph of this function (the curve in Figure 4.8) closely approximates the 13 data points. So, now the question is, how can we calculate the slope of the sea-bed?

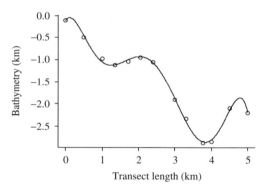

Figure 4.8: Sea-bed profile along a bathymetry transect. The measurements are shown as circles together with the graph of the approximating polynomial. Generating this graph requires that the coefficients of the polynomial are specified with greater precision than they are shown in Equation (4.27) – so don't worry if you can't replicate this graph from Equation (4.27).

The process of finding the derivative (slope) of a function is called **differentiation**. We saw in Example 4.7 how to differentiate a linear function using limits. This section shows you how to obtain the derivative of any polynomial. To build up to this, start with the derivative of a constant function $f(x) = c$ for some number c. Intuitively, we would expect that the rate of change of this unresponsive function should be zero. Indeed, for any two values of the independent variable, say x and $x + \Delta x$, we have $f(x) = c$ and $f(x + \Delta x) = c$. Placing these facts into the definition of the derivative gives

$$\frac{df(x)}{dx} = \lim_{\Delta x \to 0} \frac{f(x + \Delta x) - f(x)}{\Delta x} = \lim_{\Delta x \to 0} (c - c)\frac{1}{\Delta x} = 0 \qquad (4.28)$$

Or, more simply,

$$(c)' = 0 \qquad (4.29)$$

In Section A7.2 of the appendix, this result is listed as rule 1. Now, consider the function $f(x) = cg(x)$ where c is a constant and $g(x)$ is another function. To find the derivative of $f(x)$ we write

$$\begin{aligned}
\frac{df(x)}{dx} &= \lim_{\Delta x \to 0} \frac{cg(x + \Delta x) - cg(x)}{\Delta x} \\
&= \lim_{\Delta x \to 0} c\frac{g(x + \Delta x) - g(x)}{\Delta x} \\
&= c \lim_{\Delta x \to 0} \frac{g(x + \Delta x) - g(x)}{\Delta x} \qquad (4.30) \\
&= c\frac{dg(x)}{dx}
\end{aligned}$$

Or, more simply,

$$(cg(x))' = cg'(x) \qquad (4.31)$$

This is listed as rule 3 in Section A7.2. Finally, let n be a positive integer and consider the function $f(x) = x^n$. Once again, we write out the definition of the derivative for this function to get

$$\frac{df(x)}{dx} = \lim_{\Delta x \to 0} \frac{(x + \Delta x)^n - x^n}{\Delta x} \qquad (4.32)$$

To expand the numerator in Equation (4.32), we need the following identity from Section A1.3,

$$a^n - b^n = (a - b)(a^{n-1} + a^{n-2}b + \cdots + ab^{n-2} + b^{n-1}) \qquad (4.33)$$

Applying this to the numerator in Equation (4.32) gives

$$\begin{aligned}
&\frac{df(x)}{dx} \\
&= \lim_{\Delta x \to 0} \frac{((x + \Delta x) - x)((x + \Delta x)^{n-1} + (x + \Delta x)^{n-2}x + \cdots + (x + \Delta x)x^{n-2} + x^{n-1})}{\Delta x} \\
&= \lim_{\Delta x \to 0} \frac{\Delta x((x + \Delta x)^{n-1} + (x + \Delta x)^{n-2}x + \cdots + (x + \Delta x)x^{n-2} + x^{n-1})}{\Delta x} \qquad (4.34) \\
&= \lim_{\Delta x \to 0} ((x + \Delta x)^{n-1} + (x + \Delta x)^{n-2}x + \cdots + (x + \Delta x)x^{n-2} + x^{n-1})
\end{aligned}$$

Since Δx has vanished from the denominator, we can now allow Δx to reach zero, to obtain

$$\begin{aligned}
\frac{df(x)}{dx} &= x^{n-1} + x^{n-2}x + \cdots + xx^{n-2} + x^{n-1} \\
&= x^{n-1} + x^{n-1} + \cdots + x^{n-1} + x^{n-1} \qquad (4.35) \\
&= nx^{n-1}
\end{aligned}$$

Or, more simply,

$$(x^n)' = nx^{n-1} \qquad (4.36)$$

This is listed as rule 2 in Section A7.2. Now, let us put the last two results together to try and differentiate a typical polynomial term,

$$(cx^n)' = c(x^n)' = cnx^{n-1} \tag{4.37}$$

Example 4.10

The derivative of $f(x) = 0.456x^3$ can now be found directly from Equation (4.37),

$$\frac{df}{dx} = (0.456x^3)' = 0.456 \times 3x^2 = 1.368x^2 \tag{4.38}$$

The plot is shown in Figure 4.9.

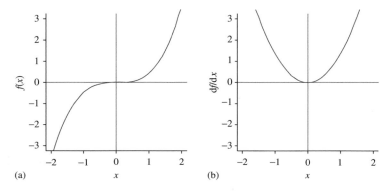

Figure 4.9: (a) The graph of the function $f(x) = 0.456x^3$ and (b) its derivative. This function's derivative is positive throughout, because the function is always increasing. On the left of the y-axis, the slope declines to zero as $f(x)$ levels off. The slope then increases again as $f(x)$ becomes steeper. When the function is flat (at the axes' origin), its derivative is zero.

The moral of this example is that general results lead to mathematical shortcuts. In the case of derivatives, such shortcuts mean that we no longer need to use limits explicitly to differentiate a function. The downside is that we need to refer to the sizeable table of derivatives in Section A7.2. All of the remaining rules in Section A7.2 can be generated from the definition of the derivative, although I will not torture you with these proofs here. You just need to be able to combine these rules sequentially to obtain the derivatives of ever-more complicated functions.

Example 4.11

The derivative of the function $f(x) = 3x^3 + 4x^2 - 5x + 8$ is the sum of four derivatives (see rule 4 in Section A7.2)

$$\frac{df}{dx} = \frac{d(3x^3)}{dx} + \frac{d(4x^2)}{dx} + \frac{d(-5x)}{dx} + \frac{d(8)}{dx} \tag{4.39}$$

We can use rule 1 to deal with the last term in the sum and rule 3 to pull out the constants in the first three terms

$$\frac{df}{dx} = 3\frac{d(x^3)}{dx} + 4\frac{d(x^2)}{dx} - 5\frac{d(x)}{dx} \tag{4.40}$$

Finally, applying rule 2 three times gives

$$\frac{df}{dx} = 9x^2 + 8x - 5 \tag{4.41}$$

This approach can now be extended to the general polynomial function,

$$f(x) = a_n x^n + a_{n-1} x^{n-1} + \cdots + a_2 x^2 + a_1 x + a_0 \tag{4.42}$$

which can be differentiated as follows

$$\frac{df}{dx} = n a_n x^{n-1} + (n-1) a_{n-1} x^{n-2} + \cdots + 2 a_2 x + a_1 \tag{4.43}$$

Note that the derivative of a polynomial function of order n is (also) a polynomial function, of order $n - 1$.

Example 4.12: Spatial gradients

We now return to the calculation of sea bottom slope from the approximate polynomial function for sea bottom depth

$$d = -3.991 \times 10^{-17} l^6 + 5.754 \times 10^{-13} l^5 - 3.025 \times 10^{-9} l^4$$
$$+ 7.006 \times 10^{-6} l^3 - 6.706 \times 10^{-3} l^2 + 1.250 l - 109 \tag{4.44}$$

Although the arithmetic is rather unpleasant, you can check that the function for the slope is

$$d' = -2.395 \times 10^{-16} l^5 + 2.877 \times 10^{-12} l^4 - 1.210 \times 10^{-8} l^3$$
$$+ 2.102 \times 10^{-5} l^2 - 1.341 \times 10^{-2} l + 1.250 \tag{4.45}$$

Plots are shown in Figure 4.10.

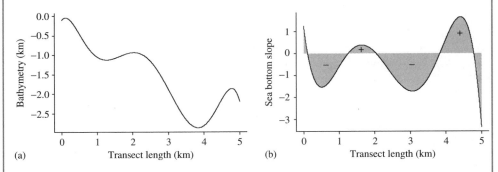

Figure 4.10: (a) The approximating function for sea depth and (b) its derivative, indicating regions of positive and negative slope (increasing and decreasing sea depth). The shape of these graphs is sensitive to the precise numerical values of the polynomial's coefficients. These plots are based on more precise coefficients than those shown in Equations (4.44) and (4.45).

4.4: Fitting polynomial functions to data

Most of the discussion on model-fitting will be left for Chapter 11, but here I illustrate just how easy it is to obtain a continuous polynomial function from an irregular set of points. The data used for Examples 4.9 and 4.12 are:

```
#Length in m
l<-c(0,500,1000,1350,1720,2050,2400,3000,3300,3770,4000,4500,5000)
#Depth in m
d<-c(100,500,980,1120,1040,950,1050,1900,2330,2880,2850,2100,2200)
```

The command used to estimate the coefficients of the polynomial is `lm()`, shorthand for 'linear model'. This command requires us to specify a model (in this case, a sixth order polynomial) and the data to be used, as a data frame.

```
dat<- data.frame(l,d) # Creates a data frame with l and d
depth.model<-lm(d~l+I(l^2)+I(l^3)+I(l^4)+I(l^5)+I(l^6), data=dat)
```

A few comments about the declaration of the model within the command `lm()`. The names of the variables (here, `l` and `d`) are taken from the data frame. The response variable (`d`) appears on the left of the symbol $\sim$, which is used instead of the equation sign. The terms participating in the model are declared on the right of $\sim$. Terms that appear by name in the data frame (here, `l`) are simply written by name in the model declaration. When these terms are raised to a power, they are enclosed in the command `I()`. This syntax is exclusive to model-fitting commands such as `lm()`.

The command `lm()` returns an object of type `model` which, in the above code, is named `depth.model`. This object contains different bits of information about the fitted model, among them the estimates of its coefficients. To see them, simply type `depth.model`. The plot in Figure 4.8 was generated by predicting from this model, using the following code

```
plot(l/1000,-d/1000,xlab="Trasect length(km)",ylab="Bathymetry(km)")
fl<-seq(0,5000)# Regular transect lengths to be used for prediction
lines(fl/1000,-predict(depth.model, newdata=data.frame(l=fl))/1000)
```

4.6.   *Differentiating other functions*

Not all functions are polynomials. Chapters 1, 2 and 3 introduced power, trigonometric and exponential functions. If the derivative of a function is defined at a given point x, then it can be found using the rules in Section A7.2. The best way to perfect this skill is by practising on examples.

Example 4.13

There are two ways to find the derivative of the function $f(x) = (x + 2)(x - 4)$: we can either multiply the expression out and differentiate as a polynomial,

$$(f(x))' = ((x + 2)(x - 4))' = (x^2 - 2x - 8)' = 2x - 2 \qquad (4.46)$$

or we can differentiate using the **product rule** (Section A7.2, rule 5)

$$f'(x) = ((x + 2)(x - 4))' = (x + 2)'(x - 4) + (x + 2)(x - 4)'$$
$$= (x - 4) + (x + 2) = 2x - 2 \qquad (4.47)$$

Example 4.14: Consumption rates of specialist predators

Termite assassin bug (Salyavata variegata) – a specialist predator

Termite assassin bugs are a group of species in the *Reduviidae* family. Nymphs of the assassin bug in Costa Rica are specialist predators with an elaborate hunting strategy (McMahan 1982): the nymph waits, camouflaged by the entrance hole of a termite mound, impales an emerging worker termite with its beak and sucks it dry. Then, using its forelegs, it grasps the empty carcass and pushes it down the entrance hole. After some time, it withdraws the dead termite shell. Attached to it is often another termite, a live worker that had been attempting to clean up the nest by disposing of the dead body. The assassin then kills this termite and uses its body to attract more.

We may be interested in quantifying the number of termites that can be consumed in an hour by the assassin bug, as a function of the number of workers in the vicinity of the entrance hole. In ecological models the **functional response** of a consumer is defined as the amount of food consumed per time unit, per capita as a function of the amount of food available. Most textbooks in ecology (e.g. Begon, Townsend and Harper 2006) distinguish between three types of functional response. Each type is characterised by different biological assumptions. Specialist predators are usually modelled by functional responses of Type I or II.

A Type I response assumes that the amount consumed is proportional to the amount of available food (x),

$$f(x) = \alpha x \tag{4.48}$$

Note the absence of an intercept in this expression indicating that no food is eaten when none is available. The proportionality constant α is greater than zero, because consumption is assumed to increase with increasing food availability. In addition, Equation (4.48) says that if we offer an additional amount (Δx) of food to the predator in each unit of time, it will consume an amount $\alpha \Delta x$ additional to what it was taking. Therefore, consumption increases at a constant rate with increasing food availability.

$$\frac{df}{dx} = \alpha \tag{4.49}$$

The problem with the Type I functional response is that it ignores the time it takes for a predator to process a prey (in this example, this involves pulling out the bait with the victim attached, impaling it and sucking it dry). Expecting that the uptake can be increased simply by swamping the consumer with food is unrealistic, so a Type II functional response takes account of saturation,

$$f(x) = \frac{\alpha x}{1 + \beta x} \tag{4.50}$$

This is of a slightly different form to the general saturation function introduced in Example 1.58. You may remember that, there, I had used an expression of the form

$$f(x) = \frac{ax}{b + x} \tag{4.51}$$

You can verify that these two expressions are the same by setting $a = \alpha/\beta$ and $b = 1/\beta$ in Equation (4.51).

The behaviour of the function in Equation (4.50) is the ecological equivalent of what economists call the 'law of diminishing returns', usually phrased in terms of benefit attained by additional increments of resource. In the terminology of this chapter, the derivative of a Type II functional response is a decreasing function of food availability. To find the derivative of Equation (4.51) we need to use the quotient rule from Section A7.2:

$$\frac{df}{dx} = \left(\frac{\alpha x}{1 + \beta x}\right)'$$

$$(\text{Rule 3}) \quad = \alpha \left(\frac{x}{1 + \beta x}\right)'$$

$$(\text{Rule 7}) \quad = \alpha \frac{(x)'(1 + \beta x) - (1 + \beta x)'x}{(1 + \beta x)^2} \quad (4.52)$$

$$= \alpha \frac{(1 + \beta x) - \beta x}{(1 + \beta x)^2}$$

$$= \frac{\alpha}{(1 + \beta x)^2}$$

The graphs of the two derivatives (Figure 4.11) give additional insight into the assumptions of the two functional responses. Specifically, in a Type II functional response, for every additional unit of food available, the consumer increases its intake by an ever-decreasing amount.

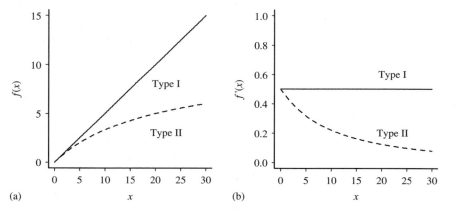

(a)

(b)

Figure 4.11: (a) Graphs of Type I and Type II functional responses and (b) their derivatives. The parameter values used for these plots were $\alpha = 0.5$ and $\beta = 0.05$.

The maximum amount of food that can be eaten by the consumer per unit time is the asymptote of Equation (4.50),

$$\lim_{x \to \infty} \frac{\alpha x}{1 + \beta x} = \lim_{x \to \infty} \frac{\frac{\alpha x}{\alpha x}}{\frac{1 + \beta x}{\alpha x}} = \lim_{x \to \infty} \frac{1}{1/\alpha x + \beta/\alpha} = \frac{\alpha}{\beta} \quad (4.53)$$

Note that the functional response is an *average* rate. So, although it tells us how consumption changes with changing prey availability, its parameters and shape are specific to the timescale at which consumption is measured. Clearly, the maximum consumption (α/β) of the assassin bug during a 1 hr interval is lower than its maximum lifetime consumption.

⟨R⟩ 4.5: Working out derivatives by computer

In R1.7 I mentioned that the symbolic capabilities of R are limited. The good news is that R can use the rules of differentiation in Section A7.2 to find the derivative of any function, symbolically. The command for this is simply `D()`. You need to inform R that the argument of `D()` is a mathematical expression with the command `expression()`, and then specify the independent variable. You can try this with a specific polynomial like the following,

```
> D(expression(3*x^3+0.5*x^2-2*x+1),"x")
3 * (3 * x^2) + 0.5 * (2 * x) - 2
```

But you can also try it with a purely symbolic expression, like the Type II functional response in Example 4.14.

```
> D(expression(a*x/(1+b*x)),"x")
a/(1 + b * x) - a * x * b/(1 + b * x)^2
```

This last expression is not as neat as Equation (4.52), but it is essentially the same result,

$$\underbrace{\frac{\alpha}{1+\beta x} - \frac{\alpha\beta x}{(1+\beta x)^2}}_{\text{R output}} = \frac{\alpha(1+\beta x) - \alpha\beta x}{(1+\beta x)^2} = \underbrace{\frac{\alpha}{(1+\beta x)^2}}_{\text{eq. (4.52)}} \qquad (4.54)$$

Often, you may need to tidy things up after R has finished. Possibly the best way to work with complicated derivatives is to figure them out first by pen and paper and then check to see if R can come up with an expression that agrees with your result.

To calculate the derivative of a function at a particular value of its variable and for particular values of its parameters, you need to use the command `eval()`. For example, the derivative for the values $\alpha = 3$, $\beta = 0.1$ and $x = 100$, is

```
> a<-3; b<-0.1; x<-100
> eval(D(expression(a*x/(1+b*x)),"x"))
0.02479339
```

4.7. The chain rule

Mathematical and scientific thinking deals with complicated problems by breaking them up into components that can be handled more easily with existing methods. This analytical path is fruitful, assuming that the results can then be synthesised into a solution for the original problem. The **chain rule** is a tool for differentiating complicated functions by treating them as composites of other, simpler, ones.

Example 4.15

The function $f(x) = (x^3 + 2x - 1)^{40}$ can be thought of as the composite of $f(g) = g^{40}$ and $g(x) = x^3 + 2x - 1$.

Example 4.16

The function

$$f(x) = \frac{\sqrt{2x-1}}{2+\sqrt{2x-1}} \qquad x \geq 1/2 \qquad (4.55)$$

Can be broken up into $f(g) = g/(2+g)$, $g(h) = \sqrt{h}$ and $h(x) = 2x - 1$.

Note how, in these examples, one component function is nested within the other. In Example 4.15, f is a function of g which is a function of x. By implication, f is a function of x. We can write this as $f(x) = f(g(x))$. Similarly, in Example 3.16, $f(x) = f(g(h(x)))$.

The skill of dealing with such complicated functions is to break them up into components that can be differentiated. The derivatives of these simpler functions can then be combined by the chain rule. For a composite function $f(x)$ that consists of two differentiable functions $f(g)$ and $g(x)$, the chain rule takes the form

$$\frac{df}{dx} = \frac{df}{dg}\frac{dg}{dx} \qquad (4.56)$$

This expression contains three derivatives of two types: two of the derivatives are with respect to the variable x and one is with respect to the variable g. In such situations the notation df/dx is preferred to the notation f', because the latter is ambiguous about what the independent variable is.

Example 4.17

The function $f(x) = (x^3 + 2x - 1)^{40}$ can be written in terms of the simpler functions $f(g) = g^{40}$ and $g(x) = x^3 + 2x - 1$. Their derivatives are $df/dg = 40g^{39}$ and $dg/dx = 3x^2 + 2$. The derivative df/dx is given from the chain rule

$$\frac{df}{dx} = 40g^{39}(3x^2 + 2) \qquad (4.57)$$

To write it in terms of only one independent variable, we use the fact that $g(x) = x^3 + 2x - 1$.

$$\frac{df}{dx} = 40(x^3 + 2x - 1)^{39}(3x^2 + 2) \qquad (4.58)$$

Use of the chain rule is not restricted to functions of only two components. For a function $f(x)$ that can be written as a combination of three nested functions $f(g(h(x)))$, the chain rule can be applied twice, as follows

$$\frac{df}{dx} = \frac{df}{dg}\frac{dg}{dx} = \frac{df}{dg}\left(\frac{dg}{dh}\frac{dh}{dx}\right) = \frac{df}{dg}\frac{dg}{dh}\frac{dh}{dx} \qquad (4.59)$$

Example 4.18

The function $f(g(h(x)))$ from Example 4.16 has $f(g) = g/(2+g)$, $g(h) = \sqrt{h}$ and $h(x) = 2x - 1$. Its derivative is found as follows:

$$\frac{df}{dx} = \left[\frac{d}{dg}\left(\frac{g}{2+g}\right)\right] \times \left[\frac{d}{dh}(h^{\frac{1}{2}})\right] \times \left[\frac{d}{dx}(2x-1)\right]$$

$$= \left[\frac{(2+g)-g}{(2+g)^2}\right] \times \left[\frac{1}{2}h^{-\frac{1}{2}}\right] \times [2] \tag{4.60}$$

$$= \frac{2}{\sqrt{h}(2+g)^2}$$

$$= \frac{2}{\sqrt{2x-1}(2+\sqrt{2x-1})^2}$$

Trigonometric and exponential/logarithmic functions can be dealt with equally well by the chain rule.

Example 4.19: Diurnal rate of change in the attendance of insect pollinators

In Example 2.16 the following relationship was suggested to describe the attendance of insect pollinators at the flower of a caper plant

$$f(t) = 5 + 5\cos\left(\frac{\pi}{12}(t-2)\right) \tag{4.61}$$

The derivative of this function describes when, during the day, the number of insect pollinators is increasing or decreasing most sharply (Figure 4.12). We can write Equation (4.61) as $f(g(t))$ with $f(g) = 5 + 5\cos(g)$ and $g(t) = \pi(t-2)/12$. Hence,

$$\frac{df}{dt} = \frac{d}{dg}(5 + 5\cos(g))\frac{d}{dt}\left(\frac{\pi}{12}(t-2)\right)$$

$$= -5\sin(g)\frac{\pi}{12} = -\frac{5\pi}{12}\sin\left(\frac{\pi}{12}(t-2)\right) \tag{4.62}$$

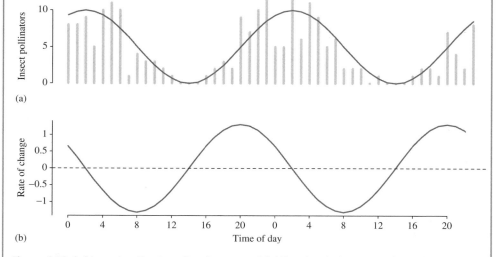

Figure 4.12: (a) Insect pollinator attendance and (b) its rate of change during a 24 h period.

4.6: Implementing the chain rule

R knows how to use the chain rule when you call the command `D()`. Here is an illustration using the insect attendance function of Example 4.19.

```
> D(expression(5+5*cos(pi/12*(t-2))), "t")
-(5 * (sin(pi/12 * (t - 2)) * (pi/12)))
```

4.8.   Higher order derivatives

Example 4.20: Spatial gradients

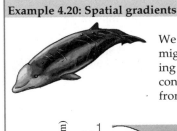

We now return to the question of how Cuvier's beaked whales might respond to different aspects of depth (Figure 4.13(a)) during their foraging for cephalopods. Mammalian physiology sets constraints on dive depth and duration. However, quite apart from their ability to reach their prey, they also need to make

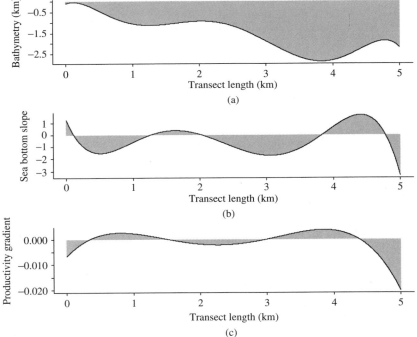

Figure 4.13: (a) Sea depth determines whether a whale is able to dive to the bottom or not; (b) if seabed slope is related to ocean productivity it may determine whether the whale finds food there; (c) if the whale uses prey density as a cue, then the slope of the slope of depth (ocean productivity or prey density gradient) may cause it to move more directionally towards spatial hotspots in the distribution of food.

sure that prey exists at that location. We may postulate that regions of steep (positive or negative) sea-bottom slope have higher primary productivity and are therefore likely to be rich in prey (Figure 4.13(b)). So, whales might be expected to show a preference for particular combinations of depths and slopes for two different biological reasons (physiology and prey availability). But how would a whale navigate to food? Highly developed top predators are unlikely to perform isotropic random walks (see Example 2.9), they are more likely to be sampling the environment to determine in which direction the density of prey seems to be increasing. If sea-bottom slope (Figure 4.13(c)) has a simple and strong relationship with cephalopod density, and whales swim up gradients of prey density, we may observe a response to the slope of the slope of sea depth. For example, we may notice more directional movement, as the whale is better able to detect the gradient in prey density. In regions where prey density changes little, the whales might be expected to move less for two different reasons: when density is high because they are feeding, and when density is low because they have no clear cues as to which way to move.

If the derivative of a function is, itself, a function, can it also be differentiated? In other words, for a differentiable function $f(x)$ can we calculate the quantity

$$\frac{d}{dx}\left(\frac{df}{dx}\right) \text{ or, equivalently, } (f'(x))' \tag{4.63}$$

The answer is yes, assuming that this **second derivative** exists under the provisos of Section 4.4. This repeated application of differentiation on the same function is written

$$\frac{d^2f}{dx^2} \text{ or, } f''(x) \tag{4.64}$$

We could continue this process, to find a function's nth derivative.

$$\frac{d^nf}{dx^n} \text{ or } f^{(n)}(x) \tag{4.65}$$

in the second of these expressions I have put n in brackets, to help distinguish between the nth power ($f^n(x)$) and the nth derivative ($f^{(n)}(x)$) of the function.

Example 4.21
The first and second derivatives of $f(x) = -3x^2 + 5x - 2$ are $f'(x) = -6x + 5$ and $f''(x) = -6$. Derivatives of order higher than two will be zero.

As this example illustrates, polynomials are infinitely differentiable (i.e. have no discontinuities or kinks anywhere in their domain). Such well-behaved functions are called **smooth**.

Ⓡ 4.7: Working out higher order derivatives

The output of the command D() (see R4.6) is an expression. To find the second derivative of a function, the following nested structure can be used

```
> D(D(expression(-3*x^2+5*x-2),"x"),"x")
-(3 * 2)
```

4.9.   Derivatives of functions of many variables

Example 4.22: The slope of the sea-floor

The examples involving sea bed slope were, until now, kept artificially simple by considering depth measurements along a single transect (plotted as a two-dimensional graph of sea depth against transect length). However, the sea bed is a surface in 3D (Figure 4.14), and some parts of it are steeper than others. But how can we extend the concept of slope from curves to surfaces? To make the discussion somewhat easier, let us consider a simpler surface (Figure 4.14(b)) that captures the main features of 4.14(a) but is easier to describe mathematically,

$$f(x,y) = (0.00002x^3 - 0.0039x^2 + 0.2x - 6.7) \times \\ (-0.000012y^3 + 0.0027y^2 - 0.16y - 7.4) - 67 \tag{4.66}$$

This gives depth as a function of two variables, latitude (x) and longitude (y) (refresh functions of many variables by looking back at Section 1.25).

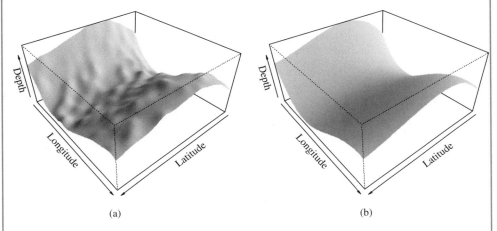

(a)   (b)

Figure 4.14: (a) A detailed depiction of a section of the seabed and (b) a plot of its approximate mathematical description.

So, how can we have a concept of 'steepness' (or slope, or gradient) for any point on a surface like the one in Figure 4.14? In Section 4.4, the steepness of a curve was defined with the aid of a tangent line, so a **tangent plane** must be used for surfaces. To visualise a tangent plane at a point $P_1 = (x_1, y_1, z_1)$ of a surface S, consider the following process (Figure 4.15): the surface S is the graph of some function $z = f(x, y)$. We take a plane that is parallel to the x- and z-axes and intersects the y-axis at y_1. This slices the surface S, creating a curve C_x which can be thought of as the graph of some function $z = f(x)$. Since $z = f(x)$ is a function of one variable, we can calculate its slope and draw the tangent line to C_x at the point P_1. We now repeat this process, this time slicing the surface in parallel to the y- and z-axes, to obtain another curve C_y and its tangent line at P_1. The tangent plane is defined by the two tangent lines to the curves C_x and C_y at the point P_1.

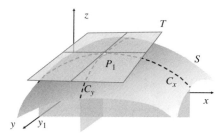

Figure 4.15: To find the tangent plane T of a given surface S at the point $P_1 = (x_1, y_1, z_1)$, we can cut two perpendicular slices intersecting at the point (x_1, y_1). These form two curves (C_1, C_2) along the surface S that meet at the point P_1. For each of these two curves, we can draw the tangent lines at P_1. These two intersecting lines define the tangent plane.

So, the concept of slope for surfaces requires two values characterising the surface's slope in the x and y directions. To calculate these we need to fix the second independent variable and differentiate with respect to the first and then repeat the process by fixing the first variable and differentiating with respect to the second. These are called **partial derivatives** and they are written using slightly modified notation. For a function $f(x, y)$, the partial derivative with respect to x, is written

$$\frac{\partial f}{\partial x} \quad \text{or} \quad \frac{\partial f(x, y)}{\partial x} \quad \text{or} \quad \frac{\partial}{\partial x} f(x, y) \tag{4.67}$$

The notation $f'(x)$ for ordinary derivatives would be too ambiguous for partial derivatives because the prime symbol (') does not indicate which independent variable is being used. Therefore, an alternative notation for $\partial f / \partial x$ is $f_x(x, y)$.

Example 4.23: The slope of the sea-floor

Let us now find the partial derivatives of the surface in Example 4.22.

$$f(x, y) = (0.00002x^3 - 0.0039x^2 + 0.2x - 6.7) \times$$
$$(-0.000012y^3 + 0.0027y^2 - 0.16y - 7.4) - 67 \tag{4.68}$$

The partial derivative with respect to x is found by treating y as a constant. Equation (4.68) consists of two polynomials (one in x and one in y) multiplied together, and a constant (-67) at the end. If we are treating y as a constant, then, to all intents, the second polynomial is just a number (let's call it a, to emphasise this)

$$f(x, y) = (0.00002x^3 - 0.0039x^2 + 0.2x - 6.7)a - 67 \tag{4.69}$$

This can be written as a typical polynomial in x, involving a in the coefficients

$$f(x, y) = a0.00002x^3 - a0.0039x^2 + a0.2x - (6.7a + 67) \tag{4.70}$$

We are now ready to differentiate,

$$\frac{\partial f}{\partial x} = a0.00006x^2 - a0.0078x + a0.2$$

$$= (0.00006x^2 - 0.0078x + 0.2)a \tag{4.71}$$

$$= (0.00006x^2 - 0.0078x + 0.2)(-0.000012y^3 + 0.0027y^2 - 0.16y - 7.4)$$

The partial derivative with respect to y is found in a similar way,

$$\frac{\partial f}{\partial y} = (0.00002x^3 - 0.0039x^2 + 0.2x - 6.7)(-0.000036y^2 + 0.0054y - 0.16) \tag{4.72}$$

Some applications may require the use of higher order partial derivatives. The idea is simple, combining the concepts from this and the previous section. The second order partial derivative with respect to, say, x can be written

$$\frac{\partial^2 f}{\partial x^2} \quad \text{or} \quad \frac{\partial^2 f(x,y)}{\partial x^2} \quad \text{or} \quad \frac{\partial^2}{\partial x^2} f(x,y) \quad \text{or} \quad f_{xx}(x,y) \tag{4.73}$$

With second order partial derivatives, however, we have the option of differentiating with both variables, one after the other. This is written

$$\frac{\partial^2 f}{\partial x \partial y} \quad \text{or} \quad \frac{\partial^2 f(x,y)}{\partial x \partial y} \quad \text{or} \quad \frac{\partial^2}{\partial x \partial y} f(x,y) \quad \text{or} \quad f_{xy}(x,y) \tag{4.74}$$

It is important to remember that the result is the same no matter what order the partial derivatives are taken in

$$\frac{\partial^2 f}{\partial x \partial y} = \frac{\partial^2 f}{\partial y \partial x} \tag{4.75}$$

Example 4.24

Here, we calculate the partial derivatives of $f(x,y) = x/(x+y)$ up to the second order. Using the quotient rule (Section A7.2), the first order partial derivatives are

$$f_x = \frac{(x+y) - x}{(x+y)^2} = \frac{y}{(x+y)^2} \tag{4.76}$$

$$f_y = \frac{0 - x}{(x+y)^2} = -\frac{x}{(x+y)^2} \tag{4.77}$$

Differentiating Equations (4.76) and (4.77) with respect to x and y gives the second order derivatives

$$f_{xx} = \frac{0 - y[2(x+y)]}{(x+y)^4} = -\frac{2y}{(x+y)^3} \tag{4.78}$$

$$f_{yy} = -\frac{0 - x[2(x+y)]}{(x+y)^4} = \frac{2x}{(x+y)^3} \tag{4.79}$$

$$f_{xy} = \frac{(x+y)^2 - y[2(x+y)]}{(x+y)^4} = \frac{x-y}{(x+y)^3} \tag{4.80}$$

$$f_{yx} = -\frac{(x+y)^2 - x[2(x+y)]}{(x+y)^4} = \frac{x-y}{(x+y)^3} \tag{4.81}$$

As expected from Equation (4.75), the last two give exactly the same result.

(R) 4.8: Working out partial derivatives

The command `D()` requires us to specify the independent variable of differentiation. So, calculating a partial derivative in R is really quite simple. The partial derivative f_x of the function from Example 4.24 is

```
> D(expression(x/(x+y)),"x")
1/(x + y) - x/(x + y)^2
```

Second order partial derivatives are just as easy but may need some algebraic simplification by hand

```
> D(D(expression(x/(x+y)),"x"),"x")
-(1/(x + y)^2 + (1/(x + y)^2 - x * (2 * (x + y))/((x + y)^2)^2))
```

4.10. Optimisation

Example 4.25: Maximum rate of disease transmission

In a perfectly mixed population (see Example 1.21 for a reminder of the law of mass action), the rate of transmission of a nonlethal disease can be modelled as

$$\gamma(N) = cbN(K - N) \qquad (4.82)$$

where K is population size, N is the number of carriers and c, b are constants. The number of disease carriers can vary in the interval $0 \leq N \leq K$. We would like to know at what stage of the spread of the disease its rate of transmission will be the highest. Assuming that carriers never recover, we would expect that transmission will be low in the early and late stages because, at those times, there are either few carriers or susceptible individuals. The rate of transmission will be highest at some intermediate value, but which one? Plotting Equation (4.82) (Figure 4.16(a)) reveals that there is a clear maximum halfway between 0 and K. Before it reaches the maximum, the graph increases (positive slope) and after the maximum it declines. Equation (4.82) is a second order polynomial (see Example 1.21) and all polynomials are smooth functions (see Section 4.8). Therefore, for the slope of the function to go from positive to negative, it must pass through zero. At that point, the rate of transmission is maximised. Indeed, the graph of the derivative of this function (Figure 4.16(b)) is a straight line that hits the x-axis at about $K/2$. To show this mathematically, we can calculate the derivative of the function, set it equal to zero and solve the resulting equation for N

$$\frac{d\gamma}{dN} = 0$$
$$[cbN(K - N)]' = 0$$
$$cb(K - 2N) = 0 \qquad (4.83)$$
$$N = \frac{K}{2}$$

So, the maximum transmission rate occurs when half of the total population has been infected. We may also ask what the value of that rate will be. To find this, we simply put $N = K/2$ back into Equation (4.82) to find $\gamma_{\max} = cbK^2/4$.

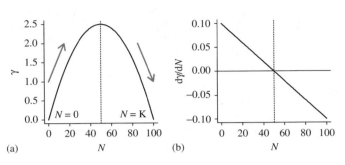

Figure 4.16: (a) Graph of the function $\gamma(N) = cbN(K - N)$ for the specific values $K = 100$, $cb = 0.001$; (b) graph of the slope of the function.

To find the maximum value of the function in the previous example, I first defined the domain of the function and then checked that the function was smooth. I then reasoned that transmission would be low at the extremes of the function's domain and so looked for a point in between. In order for an increasing, smooth function to start decreasing, it has to go through a point of zero slope. That point, the maximum, was therefore found by setting the function's derivative to zero and solving the resulting algebraic equation to find the number of carriers at which the disease attained its highest transition rate.

The process of finding a function's maximum or minimum is called **optimisation**. The idea is straightforward, but there are just a few points to remember: First, optimisation is not easy when dealing with discontinuous or nondifferentiable functions. For example, functions involving absolute values do not have a slope at the points where they touch the x-axis, yet these are the very points where their minima occur (see Example 4.8). Second, a function may have several **local optima** (Figure 4.17). Of those, the most extreme are called **global optima**. Third, it is possible that a function has optima at the extremes of its domain (Figure 4.17). These points may even be global optima but the slope of the function at these points will not necessarily be zero.

All optimisation problems have three fundamental ingredients: the **objective function**, the **optimisation method** and the **optimisation constraints**. The objective function is the mathematical function whose optima are being sought. It may contain one or more independent variables and, in many applications, it represents some concept of profitability or net cost.

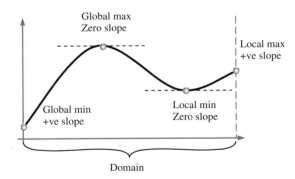

Figure 4.17: In a smooth function, the optima will either have slope zero or lie at the extremes of the function's domain.

The optimisation method may either be analytical (as in Example 4.25) or numerical (see R4.9 below). The constraints represent the domain of the objective function.

There are three main areas in ecological research that use optimisation extensively: The first is **statistical estimation** and **inference**, to be visited in Chapters 10 and 11. In this context, the objective function represents the likelihood of a model and its parameters being true, given a particular set of observations. The objective of the exercise is to find such models and parameter values that are deemed highly likely under the data. The second area is **resource management**. In this context, the objective function contains a mixture of wildlife and economic priorities and we seek an optimum compromise between these, subject to market, legal or natural constraints. The third area is **optimality theory**. In brief, this asserts that natural selection has yielded optimal organisms within the constraints of their environments. Optimality theorists construct objective functions involving behavioural or life-history trade-offs and then attempt to interpret the biological meaning of the resulting optimal strategies.

Example 4.26: The marginal value theorem

Common starling, Sturnus vulgaris

The following is a typical foraging trade-off taken from pages 48–51 of Krebs and Davies (1993): starlings feed their young mainly on cranefly larvae. Parents perform frequent foraging trips (up to 400 day^{-1}) to provision their chicks. The number of larvae that each adult bird can hold in its beak is limited, and the ability of adults to collect additional larvae decreases with load. Therefore, the efficiency of provisioning is a balance between the number of trips performed and the average number of larvae carried in each trip. The starlings have a decision to make: when should they stop foraging and return to their nest to unload? The efficiency of foraging can be modelled as a process of saturation (see Example 1.58)

$$f(t) = \frac{at}{b+t} \tag{4.84}$$

where t is the time the bird spends feeding during any single trip, a is the maximum number of larvae that can be caught in one trip and b represents the feeding time at which half of that maximum is achieved. For the parameter values $a = 10$ larvae and $b = 1.3$ min this has the shape shown in Figure 4.18(a).

Optimal foraging theory does not explain how the bird will make its decision, but merely assumes that animals have evolved to make the best possible decision. The optimal decision, in this example, is the one that maximises the average provisioning rate (r), on the timescale of a single trip

$$r = \frac{\text{Larvae brought back per trip}}{\text{Trip duration}} \tag{4.85}$$

For simplicity, we assume that there is only one feeding patch that is not depleted by the activities of the starlings. If T is the total time that a starling needs to go back to the nest, feed the chicks and return to the foraging patch, then the provisioning rate can be written

$$r = \frac{f(t)}{T+t} = a\frac{t}{(b+t)(T+t)} \tag{4.86}$$

An example of a graph of this function for particular parameter values is shown in Figure 4.18(b). This graph has a peak, indicating that an optimal strategy exists within the

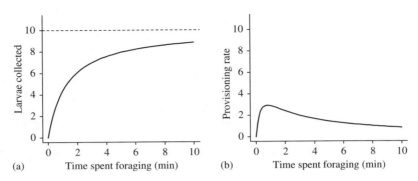

Figure 4.18: (a) The foraging efficiency of starlings asymptotes to a maximum as they spend more time in a foraging patch; (b) for a given commuting time between the nest and the foraging patch, there is an optimum rate of provisioning.

function's domain. To find it, we must first differentiate with respect to t

$$\left(a\frac{t}{(b+t)(T+t)}\right)' = a\frac{(t)'(b+t)(T+t) - t[(b+t)(T+t)]'}{[(b+t)(T+t)]^2}$$

$$= a\frac{(b+t)(T+t) - t[(b+t) + (T+t)]}{[(b+t)(T+t)]^2} = a\frac{bT - t^2}{[(b+t)(T+t)]^2} \quad (4.87)$$

This result may now be set to zero and solved for the particular value of t (say, t^*) at which the provisioning rate is maximised.

$$a\frac{bT - t^{*2}}{[(b+t^*)(T+t^*)]^2} = 0 \quad (4.88)$$

Equation (4.88) is satisfied when $bT - t^{*2} = 0$, implying, $t^* = \pm\sqrt{bT}$. Only the positive of these two solutions makes biological sense (feeding time cannot be negative). Therefore, the optimal feeding time $\sqrt{bT}$ increases if either the half-saturation time (b), or the commute to the patch (T) increases. The maximum rate of provisioning is found by replacing the $\sqrt{bT}$ answer in Equation (4.86)

$$r_{max} = a\frac{\sqrt{bT}}{(b + \sqrt{bT})(T + \sqrt{bT})} \quad (4.89)$$

which can be simplified by dividing both top and bottom by $\sqrt{bT}$

$$r_{max} = a\frac{1}{\dfrac{(b + \sqrt{bT})(T + \sqrt{bT})}{\sqrt{bT}}} = a\frac{1}{\dfrac{(b + \sqrt{bT})}{\sqrt{b}}\dfrac{(T + \sqrt{bT})}{\sqrt{T}}} = a\frac{1}{(\sqrt{b} + \sqrt{T})^2} \quad (4.90)$$

Using the values $a = 10$, $b = 1.3$ and $T = 0.5$, we find $r_{max} = 2.93$ larvae.min^{-1}.

Is this result specific to the particular saturation function Equation (4.84)? If we didn't have the particular curve in Figure 4.18(a), but only assumed that foraging efficiency $f(t)$ declines with load, what would this theory predict about the birds' behaviour? The rate of provisioning is $r = f(t)/(T + t)$ with derivative

$$\left(\frac{f(t)}{T + t}\right)' = \frac{f(t)'(T + t) - f(t)}{(T + t)^2} \quad (4.91)$$

Once again, the position (t^*) of the maximum (r_{max}) may be found by setting the numerator of this expression to zero. Because we do not have any specifics for the function $f(t)$ we cannot actually solve for t^*, but we can at least obtain the following condition

$$f(t^*)' = \frac{f(t^*)}{T + t^*} \qquad (4.92)$$

What does this tell us about behaviour? The ratio on the right is, in fact, the maximum rate of provisioning r_{max}. In plain language, therefore, this condition says that to achieve the optimum provisioning rate, the birds should stop foraging when their instantaneous foraging rate equals the maximum provisioning rate. This result, proposed by Charnov in 1976, is known as the **marginal value theorem**. Although it appears to describe a rule of thumb that an individual could easily follow, it assumes that birds know (through trial and error) what the maximum provisioning rate is. How this information is collected and evaluated by the birds is not clear and this is one of the weak points of optimality theory. Krebs and Davies (1993) themselves call these 'economic decisions'. The theory's critics have pointed out that (1) animals are not necessarily so efficient, (2) environments don't stay constant long enough for such efficiency to be achieved and (3) optimising one aspect of behaviour (e.g. provisioning rate) ignores possible conflicts with other life-history priorities. These opposing views have ranged from the mildly critical (Janetos and Cole, 1981) to the outright polemic (Pierce and Ollason, 1987). Modern optimality theory has addressed some of these concerns (Clark and Mangel, 2000) but retains its 'black box' aura regarding the cognitive mechanism of decision-making.

Optimisation is not restricted to functions of one variable. To find the optimum of a surface (rather than a curve) we need to look for peaks/valleys and also examine the surface's boundaries (Figure 4.19). Optima in the interior of the domain of the function will be points where both the partial derivatives (see Section 4.9) are zero. Optima at the boundaries may be points with only one partial derivative at zero (if the optimum lies along a boundary), or even none (if the optimum lies at a corner of the domain where two smooth boundaries meet).

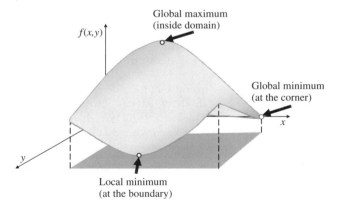

Figure 4.19: Functions of two variables can have optima inside their domain, along the boundaries of the domain or at its corners.

4.9: Numerical optimisation

Numerical optimisation in R comes in several flavours (here illustrated with the problem from Example 4.26). For simpler functions, one possibility is to first work out the derivative by hand and then ask R to solve the algebraic equation $f'(x) = 0$. If this is a polynomial equation, then the command `polyroot()` can be used (see R1.9). If it is not, then the command `uniroot()` should be used. For example, the following two lines will solve the equation $bT - t^2 = 0$ with parameters $b = 1.3, T = 0.5$, in the domain $0 \leq t \leq 10$,

```
f<-function(t,T,b) {b*T-t^2}              # Defines the function
g<-uniroot(f, lower=0, upper=10,T=0.5,b=1.3) # Requests the solution
```

Here, the domain to be searched is specified by the values of `lower` and `upper`. Within `uniroot()` the function f is declared by name only and the parameters T and b are specified separately. The result is an object that I named g. To see the position of the maximum, simply type `g$root`.

There is, however, also a selection of R commands that will carry out the entire optimisation process without requiring the user to calculate derivatives. For example, the command `optimize()` will find the maximum or minimum of a given function in a particular domain. The following two lines retrieve the location and value of the maximum provisioning rate (Equation (4.86) in Example 4.26).

```
r<-function(t,T,b,a) {a*t/((b+t)*(T+t))}
g<-optimize(r, lower=0, upper=10, a=10, b=1.3, T=0.5, maximum=TRUE)
```

The domain (0 to 10) was again specified using the constraints `lower` and `upper`. Other inputs included the parameter values of the function (a,b and T) as well as the fact that we required the maximum instead of the minimum. The output is an object containing the position of the maximum (to see it, type `g$maximum`) and the value of the function at the maximum (type `g$objective` to see it).

Still more flexibility in specifying the optimisation process is offered by the commands `nlm()` and `optim()`, which I do not cover until Chapter 10. These extend the basic optimisation concept to objective functions with multiple independent variables.

4.11. Local stability for difference equations

With our considerably extended mathematical toolkit now containing limits and derivatives, we can revisit the important issue of stability in discrete, dynamical systems of the form $N_{t+1} = f(N_t)$. An equilibrium of this system is a value N^* such that, if the system is placed exactly on it, it remains there forever (i.e. $N_t = N_{t+1} = N^*$, see Section 3.6). The simple examples of Section 3.8 led to the conclusion that the slope of the graph of the function f at the equilibrium point determines the stability of that equilibrium. Specifically, if the absolute value of the slope is smaller than 1, the equilibrium will be stable. Since we can now calculate the slope as the derivative of the function, we can investigate the stability of more complicated dynamical systems. The approach can be broken down into seven steps:

❶ Define the model.
❷ Find the equilibria.
❸ Write the update rule f as a function of a continuous variable x.
❹ Find the derivative of the function with respect to x.
❺ Calculate the derivative at the equilibria.
❻ Derive local stability conditions.
❼ Interpret these conditions biologically.

I illustrate this long recipe with two examples below.

Example 4.27: Unconstrained population growth

The analysis is structured around the seven steps mentioned above.

❶ *Define the model:*
The population model of unrestricted growth in blue tits is (see Example 3.3),

$$P_{t+1} = (1 + r)P_t \tag{4.93}$$

where r is the intrinsic growth rate ($r > -1$). Intuitively, if this year's population size is being multiplied by a number smaller than 1 (i.e. $r < 0$), we would expect the population to eventually become extinct. Let us see if this expectation can be verified mathematically.

❷ *Find the equilibria:*
By setting $P_t = P_{t+1} = P^*$ in Equation (4.93) we find $P^* = (1 + r)P^*$. We require the value of P^* that verifies this equation for any value of r (i.e. not necessarily $r = 0$). Therefore, the system's only equilibrium is $P^* = 0$.

❸ *Write the update rule f as a function of a continuous variable x:*
This can be obtained directly from Equation (4.93)

$$f(x) = (1 + r)x \tag{4.94}$$

❹ *Find the derivative of f with respect to x:*

$$f'(x) = (1 + r) \tag{4.95}$$

❺ *Calculate the derivative at the equilibria:*
Since the slope is a constant function, specifying it at the equilibrium $P^* = 0$ simply returns the constant,

$$f'(0) = (1 + r) \tag{4.96}$$

❻ *Derive local stability conditions:*
The general condition for the stability of an equilibrium is $|f'(P^*)| < 1$ (see Section 3.8). Using Equation (4.96), this condition can be specified as $|1 + r| < 1$, which can be manipulated to get a simpler result

$$\begin{aligned} |1 + r| &< 1 \\ -1 < 1 + r &< 1 \\ -2 < r &< 0 \end{aligned} \tag{4.97}$$

Since, by definition, the intrinsic growth rate is always bigger than -1, the part $r > -2$ in this inequality is always satisfied, leaving the simpler condition $r < 0$.

❼ *Interpret stability conditions biologically:*
The above have shown that the single equilibrium of the system is stable if $r < 0$. A stable equilibrium attracts solutions that happen to be in its neighbourhood. Since this is the only equilibrium of the system, it will either attract solutions or let them move away to infinity. This confirms the intuitive notion that, if the per capita growth rate is negative, the population will ultimately become extinct, if not, the population will increase indefinitely.

Local stability analysis is powerful because *its conclusions do not depend on particular parameter values*. On the other hand, the interpretation of how local stability translates to the global behaviour of solutions is somewhat ad hoc: if the equilibrium is locally stable, we don't know for sure that solutions will be attracted to it from any starting value. Conversely, if the equilibrium is unstable, we cannot be certain where solutions will go once repelled. Here is a less trivial example that underlines both the power and limitations of local stability analysis.

Example 4.28: Density dependence and proportional harvesting

Many exploited populations are subject to density dependence and proportional harvesting. Which harvesting regimes can deplete such populations to the point of extinction?

❶ *Define the model:*
The discrete logistic model, incorporating a proportional harvest rate, is

$$P_{t+1} = P_t + r_{max}\left(1 - \frac{P_t}{K}\right)P_t - hP_t \qquad (4.98)$$

Here, h is the proportion ($0 \le h \le 1$) of the current population taken by the end of the year. This formulation assumes that the harvesting quota is set on the basis of an accurate estimate for P_t and it is adhered to precisely. Assume that $r_{max} > 0$, so that in the absence of exploitation the population does not become extinct.

❷ *Find the equilibria:*
Setting $P_t = P_{t+1} = P^*$, in Equation (4.98) gives

$$\left[r_{max}\left(1 - \frac{P^*}{K}\right) - h\right]P^* = 0 \qquad (4.99)$$

This will be true for the following two values (giving us two possible equilibria)

$$P_1^* = 0 \text{ and } P_2^* = K\left(1 - \frac{h}{r_{max}}\right) \qquad (4.100)$$

The first equilibrium corresponds to extinction through overexploitation and the second is the size of the resource population under harvesting. We now need to determine the conditions under which these will be stable.

❸ *Write the update rule f as a function of a continuous variable x:*

$$f(x) = x\left(1 + r_{max}\left(1 - \frac{x}{K}\right) - h\right) \qquad (4.101)$$

❹ *Find the derivative of f with respect to x:*

$$f'(x) = x'\left(1 + r_{max}\left(1 - \frac{x}{K}\right) - h\right) + x\left(1 + r_{max}\left(1 - \frac{x}{K}\right) - h\right)'$$
$$= \left(1 + r_{max}\left(1 - \frac{x}{K}\right) - h\right) - x\frac{r_{max}}{K} \qquad (4.102)$$
$$= 1 + r_{max} - h - 2x\frac{r_{max}}{K}$$

❺ *Calculate the derivative at the equilibria:*
For the extinction equilibrium,

$$f'(P_1^*) = 1 + r_{max} - h \qquad (4.103)$$

For the sustainable harvesting equilibrium,

$$f'(P_2^*) = 1 - r_{max} + h \qquad (4.104)$$

❻ *Derive local stability conditions:*
The condition for the stability of P_1^* is $|1 + r_{max} - h| < 1$, which is expanded as

$$|1 + r_{max} - h| < 1$$
$$-1 < 1 + r_{max} - h < 1 \qquad (4.105)$$
$$-2 < r_{max} - h < 0$$

Since $r_{max} > 0$ and $h < 1$, the smallest value that can ever be taken by $r_{max} - h$ is -1. Hence, the part $r_{max} > -2$ in Equation (4.105) will always be satisfied and the stability condition becomes $h > r_{max}$.

For stability of P_2^* we require that $|1 - r_{max} + h| < 1$. This simplifies to $-2 < h - r_{max} < 0$. So, if $h > r_{max} - 2$ and $h < r_{max}$ it is possible to exploit the population sustainably. The condition $h < r_{max} - 2$ implies $r_{max} > 2$, but in such a case, both equilibria are unstable.

❼ *Interpret stability conditions biologically:*
The above stability results are summarised in Figure 4.20. Extinction occurs when the harvesting rate (h) exceeds the population's intrinsic growth rate (r_{max}). If the harvesting rate is below this threshold value, then the population is maintained, but its dynamics depend on the value of the quantity $r_{max} - h$. If it is less than 2, then the population settles to the value $P_2^* = K(1 - h/r_{max})$. This equilibrium population size is somewhat smaller than the carrying capacity K of an unexploited population. However, if $r_{max} - h$ is greater than 2, then both equilibria are unstable. What does this mean? Local stability analysis cannot tell us exactly, but three things could happen: the population could go to infinity, it could fluctuate periodically or it could fluctuate chaotically (see Section 3.9). To find out exactly which, we need to simulate the model for the particular parameters of interest.

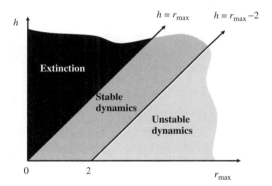

Figure 4.20: Summary of stability analysis in terms of the parameters h and r_{max}.

4.12. Series expansions

Those of us who went to school after calculators became commonplace were misled into thinking that expressions like $\sqrt{4.2}$, $\sin(3.12)$ or $\ln(2.72)$ are easy to deal with. However, high school maths harbours a dirty secret: although it is easy to define the concept of the square root, the sine and the log, it is rarely possible to calculate their numerical value from these definitions. Without a calculator, how would you go about finding the number which, when squared, gives you 4.2? Perhaps a good first guess is that it would be slightly larger than 2, since $2^2 = 4$? Trial and error quickly comes up with 2.05 ($2.05^2 = 4.2025$), but this is still not perfect, it's an approximation. In many cases, such approximations are the best we can do, and believe it or not, calculators don't do it any differently, just more systematically and precisely (if you punch $\sqrt{4.2}$ in a calculator, you will get 2.049390153). This section covers the basic concepts and tools of function approximation. I do this, not out of a misplaced devotion to mathematical completism, but because many results in theoretical ecology are obtained by approximation (see Section 5.11).

I focus on smooth functions, i.e. those that can be differentiated as many times as desired. The essential breakthrough in approximation is the realisation that the derivatives of functions can be calculated analytically, even if their numerical values cannot. Hence, in general it can be proved that $(\sqrt{x})' = 0.5/\sqrt{x}$, $(\sin x)' = \cos x$ and $(\ln x)' = 1/x$ even though it is not possible to calculate $\sqrt{4.2}$, $\sin(3.12)$ or $\ln(2.72)$, exactly. How can this fact be put to good use? Let's take a closer look at the definition of the derivative from Equation (4.8)

$$f'(x_1) = \lim_{x \to x_1} \frac{f(x_1) - f(x)}{x_1 - x} \tag{4.106}$$

If Δx is very small (i.e. if x is very close to x_1), this limit can be approximated by

$$f'(x_1) \cong \frac{f(x_1) - f(x)}{x_1 - x} \tag{4.107}$$

If $f(x_1)$ and $f'(x_1)$ can be calculated for some reference value x_1 but not for a nearby value x, then the value of $f(x)$ can be approximated by rearranging Equation (4.107) as follows:

$$f(x) \cong f(x_1) + f'(x_1)(x - x_1) \tag{4.108}$$

This is a **linear approximation** of $f(x)$ because it can also be written $f(x) \cong f'(x_1)x + [f(x_1) - f'(x_1)x_1]$, which is a linear function of x.

Example 4.29

To calculate the approximate value of $\sqrt{4.2}$ the key is to notice that we know the nearby value $\sqrt{4} = 2$. In the notation of Equation (4.108), $x_1 = 4$, $f(x_1) = \sqrt{4}$, $f(x) = \sqrt{4.2}$, $x = 4.2$ and $f'(x_1) = 0.5/\sqrt{x_1} = 1/4$. Substituting in these values,

$$\sqrt{4.2} \cong \sqrt{4} + \frac{1}{4}(4.2 - 4) = 2.05 \tag{4.109}$$

This answer moves in the right direction (the correct answer is indeed slightly greater than 2) and it's not too poor an approximation compared to the one given by a calculator (2.049390153) or our own guesswork (2.05). If the same reference point is used to approximate

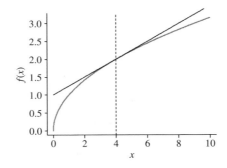

Figure 4.21: Linear approximation of the function $f(x) = \sqrt{x}$ around the value $x = 4$. The function is shown in grey and the approximation in black.

values of the function farther away, the approximation deteriorates (Figure 4.21). For example, approximating $\sqrt{9.1}$ by using 4 as the reference point would yield 3.275. In contrast, using the more proximate number 9 as a reference point gives the more accurate 3.017.

In general, the quality of the approximation at ever-greater distances from the reference point can be improved by using more flexible curves. For example, the **quadratic approximation** of a function around a reference value x is given by the relationship

$$f(x) \cong f(x_1) + f'(x_1)(x - x_1) + \frac{1}{2}f''(x_1)(x - x_1)^2 \tag{4.110}$$

nth order approximations can also be obtained. They are known as **Taylor expansions** or **Taylor series**.

$$f(x) \cong f(x_1) + f'(x_1)(x - x_1) + \frac{1}{2}f''(x_1)(x - x_1)^2 + \frac{1}{3!}f'''(x_1)(x - x_1)^3 + \cdots$$
$$\cdots + \frac{1}{n!}f^{(n)}(x_1)(x - x_1)^n \tag{4.111}$$

where $n!$ is the factorial of the number n, defined as $n! = 1 \times 2 \times \cdots \times n$ and $f^{(n)}$ denotes the nth derivative of the function f. These results are not proved here but I demonstrate their application with the following example.

Example 4.30

To approximate the function $f(x) = e^x$, first note that

$$f(x) = f'(x) = \cdots = f^{(n)}(x) = e^x \tag{4.112}$$

The only reference point that will readily give numerical values for all of these is $x = 0$ (Taylor expansions using zero as their reference point are also known as **Maclaurin expansions**). Specifying these expressions to zero gives

$$f(0) = 1, \, f'(0) = 1, \ldots, f^{(n)}(0) = 1 \tag{4.113}$$

Using Equation (4.111), the required approximation for any (non-negative) value of x is

$$e^x \cong 1 + x + \frac{x^2}{2} + \frac{x^3}{6} + \cdots + \frac{x^n}{n!} \tag{4.114}$$

Figure 4.22 shows the quality of the approximation improving with Taylor series of increasing order n.

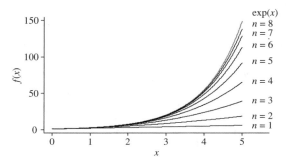

Figure 4.22: Taylor approximations of increasing accuracy (increasing order n) of the function $f(x) = e^x$. The actual graph of the function is shown at the top, in grey.

As the number of terms in the Taylor expansion tends to infinity, the approximation converges to the truth. When this happens, the distance of the point being approximated from the reference point becomes irrelevant. This is a tremendously powerful result: *Taylor series use information about the exact behaviour of a function at just one of its points to infer the value of the function at every other point.*

Further reading

The classic textbook on calculus is Thomas and Finney (2000). Michael Spivak has written some excellent books on calculus, ranging from the informal (Spivak, 1995) to the rigorous (Spivak, 2006). Great introductions to calculus from a biologist's perspective are in Batschelet (1979) and Adler (2005). Most theoretical ecology textbooks provide brief refreshers on calculus either as introductory chapters or appendices. Of the above books, only Thomas and Finney (2000) have good coverage of multivariable calculus. A specific (advanced) reference for calculus of many variables is Marsden and Tromba (2003). You will find more about the use of optimality theory in behaviour and life history in Stephens and Krebs (1987), Krebs and Davies (1993) and Begon, Townsend and Harper (2006). McNamara and Houston (1999) and Clark and Mangel (2000) are the definitive references on the implementation of modern optimality models. Local stability analysis for difference equations is well covered in several excellent textbooks, e.g. Gurney and Nisbet (1998), Case (2000), Kot (2001), Adler (2005) and Otto and Day (2007). If you are particularly interested in models of sustainable and optimal harvesting, you will find some good theoretical sections in Kot (2001).

References

Adler, F.R. (2005) *Modelling the Dynamics of Life: Calculus and Probability for the Life Scientists.* Thompson Brooks/Cole. Belmont, California. 876pp.

Batschelet, E. (1979) *Introduction to Mathematics for Life Scientists.* Springer Verlag, Berlin. 643pp.

Begon, M., Townsend, C.A. and Harper, J.L. (2006) *Ecology: From Individuals to Ecosystems*, 4th edition. Wiley Blackwell. 752pp.

Case, T.J. (2000) *An Illustrated Guide to Theoretical Ecology.* Oxford University Press, New York. 449pp.

Charnov, E.L. (1976) Optimal foraging: The marginal value theorem. *Theoretical Population Biology*, **9**, 129–136.

Clark, C.W. and Mangel, M. (2000) *Dynamic State Variable Models in Ecology.* Oxford University Press, New York. 304pp.

Gurney, W.S.C. and Nisbet, R.M. (1998) *Ecological Dynamics.* Oxford University Press, New York. 335pp.

Janetos, A. and Cole, B. (1981) Imperfectly optimal animals. *Behavioral Ecology and Sociobiology*, **9**, 203–210.

Kot, M. (2001) *Elements of Mathematical Biology.* Cambridge University Press, Cambridge. 453pp.

Krebs, J.R. and Davies, N.B. (1993) *An Introduction to Behavioural Ecology.* Blackwell, Malden, Massachusetts. 420pp.

Marsden, J.E. and Tromba, A.J. (2003) *Vector Calculus.* Freeman & Co, New York. 704pp.

McMahan, E.A. (1982) Bait-and-capture strategy of a termite assassin bug. *Insectes Sociaux*, **29**, 346–351.

McNamara, J.M. and Houston, A.I. (1999) *Models of Adaptive Behaviour: An Approach Based on State.* Cambridge University Press. 388pp.

Otto, S.P. and Day, T. (2007) *A Biologist's Guide to Mathematical Modelling in Ecology and Evolution*. Princeton University Press, New Jersey. 732pp.

Pierce, G.J. and Ollason, J.G. (1987) Eight reasons why optimal foraging theory is a complete waste of time. *Oikos*, **49**, 111–118.

Spivak, M. (1995) *Hitchhiker's Guide to Calculus*. Mathematical Association of America.

Spivak, M. (2006) *Calculus*. Cambridge University Press.

Stephens, D.W. and Krebs, J.R. (1987) *Foraging Theory*. Princeton University Press, New Jersey.

Thomas, G.B. and Finney, R.L. (2000) *Calculus and Analytic Geometry*. Addison Wesley.

5

How to work with accumulated change
(Integrals and their applications)

'Lastly, looking not to any one time, but to all time, if my theory be true, numberless intermediate varieties, linking closely together all the species of the same group, must assuredly have existed.'
Charles Robert Darwin (1809−1882), English naturalist

Prediction is the holy grail of theoretical science. It requires us to wield our scientific understanding with sufficient mathematical technique to be able to say something about the future of a system by looking at its past. The basic idea is simple (although not always feasible): if we know how a system changes and where it's been, we can calculate where it's going to go. This idea was already encountered in Chapter 3, in the context of discrete dynamical systems, it was called an **initial value problem**. Continuous processes are not too different, but they are expressed in terms of infinitesimal, instantaneous change, as formalised by the derivative. The derivative is the mathematical rule that describes how a system changes at any point in time, so the function that gives rise to this derivative can be seen as a complete record of the system's history and future. Therefore, this chapter begins by explaining how the operation of differentiation can be reversed (Section 5.1) to yield functions called **antiderivatives**. In Section 5.2, this process is simplified by introducing the concept and notation of the **integral** and the fundamental rules of **integration** are presented in Section 5.3. A brief detour, in Section 5.4, reviews the **summation** operator and uses it to explain the graphical interpretation of integrals as the **area under the graphs** of functions (Section 5.5). Specifying a starting and ending point to the development of a system (e.g. Genesis to Armageddon) also imposes these limits to the corresponding integral (Section 5.6). These are called **definite integrals** and Section 5.7 examines three important properties that allow us to manipulate them. Section 5.8 focuses on

improper integrals, i.e. definite integrals that are in some way unbounded. Attention then shifts to the powerful tool of **differential equations** (Section 5.9), models of the natural world involving continuous rates of change. Some types of differential equations can be **solved analytically** (Section 5.10 illustrates how this is done) but, for those that cannot, it is essential to be able to use methods of **numerical solution** and **local stability analysis** (Section 5.11).

5.1. Antiderivatives

Example 5.1: Invasion fronts

Rhododendron ponticum, *an invasive species*

Aggressive colonisers have the potential to eliminate native competitors and disturb local ecosystems. Such **invasive species** tend to be either newly introduced, exotic organisms or species that have somehow been released from the control of their predators, or resource limitations. *Rhododendron ponticum* was introduced to the UK from Spain in the eighteenth century as an ornamental plant. Its tolerance to the northern climate was enhanced by selective breeding and hybridisation. It was first recognised as a problem in the 1950s and its spread has since become the UK's highest alien plant control priority.

An **invasion front** can be visualised as the expansion of the range of an invasive species around its perimeter. This may give different spatial patterns depending on the mechanism of dispersal. Short-range seed dispersal is likely to yield invasion fronts that follow the features of the terrain (Stephenson *et al.*, 2007). Long-range dispersal (such as that achieved by the garden merchant distribution network that would have originally traded *R. ponticum*) would give more complicated spatial patterns (Figure 5.1(a)). At cross-section, following a line perpendicular to its perimeter, the invasion front resembles an advancing wave: more recently colonised locations have low densities ahead of the invasion while locations behind the invasion front have reached carrying capacity. For the purposes of this example, I assume that the front travels at a constant speed, say a m·y^{-1} (Figure 5.1(b)). In general, it is important to know what distance $l(t)$ the invasion will have travelled t years after first introduction. You can probably do this calculation in your head, to find the answer at. Granted, this is an easy problem but it can be used to illustrate some harder concepts.

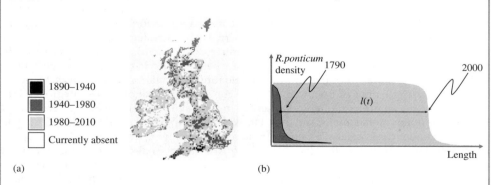

(a)   (b)

Figure 5.1: (a) Spread of *R. ponticum* in the UK since its introduction; (b) cross-section of an idealised invasion front in one-dimensional space showing its progress over 50 years.

We want to formulate a function $l(t)$ of the length travelled by the invasion over a time interval t. Speed is the rate of change of distance with time. Therefore, the instantaneous rate of change of $l(t)$ is

$$\frac{dl}{dt} = a \tag{5.1}$$

The derivative of a linear function is a constant, equal to the slope of its graph (Section 4.4). Therefore, the function we are looking for is of the form

$$l(t) = at + b \tag{5.2}$$

Indeed, differentiating Equation (5.2) gives Equation (5.1). There is, however, an apparent discrepancy. The intuitive answer of at is similar, but not identical, to Equation (5.2). What is the physical interpretation of the constant b? If Equation (5.2) is specified to $t = 0$, when the species was first introduced, this gives $l(0) = a \times 0 + b = b$. More generally,

$$l(t) = at + l(0) \tag{5.3}$$

The interpretation is now easier. After time t the invasion front will have travelled a distance of at metres plus the distance it had already travelled when records began. In our example, records began with the species's first introduction, so the length $l(0)$ is zero, making the formal answer identical to the intuitive guess. In the terminology of Chapter 3, the constant a is a **parameter** for this model and $l(0)$ is the **initial condition**.

In Example 5.1, the instantaneous rate of change gives rise to a model of distance as a function of time. This simple model can be used to *predict* how far the front will be at any given time in the future. To get to this model, differentiation was turned on its head, leading to functions, such as Equation (5.3), that are called **antiderivatives**.

Example 5.2: Diving in seals

A male grey seal (Halichoerus grypus)

Grey seals in the North Sea have a varied diet which, among other prey, includes gadoids, sand eels and salmon. When foraging for sand-burrowing sand eels (family *Ammodytidae*), the seals first dive to the bottom and then spend as much time as possible disturbing the sandy sediment with their flippers to uncover the fish. As part of a study on grey seal diving, an animal is fitted with a tag which consists of a turbine speedometer, a pressure sensor, a memory bank and a satellite transmitter. The transmitter can only send the stored information when the seal's head is above the water. The tag begins to record information as soon as the pressure sensor registers a value above a certain threshold, so there is an unknown lag between the true start of the dive and the start of the record. During a particular dive, the data suggest that the seal's speed through the water followed the relationship

$$s(t) = 0.78t^2 - 15.6t + 92 \tag{5.4}$$

Here, time is measured in min and speed in m·min^{-1}. The record of the dive lasted 22 min. We want to calculate what distance the seal travelled through the water and what its acceleration was at any point of its dive.

The second question is easier than the first. Acceleration is defined as the rate of change of speed. Therefore, taking the derivative of speed with respect to time gives a constant acceleration of $1.56t - 15.6$. The length of the path $l(t)$ can be found as the antiderivative of speed. Equation (5.4) is a second order polynomial, so, according to Section 4.5, it must be the derivative of a third order polynomial

$$l(t) = a_3 t^3 + a_2 t^2 + a_1 t + a_0 \qquad (5.5)$$

Differentiating this with respect to time,

$$\frac{dl}{dt} = 3a_3 t^2 + 2a_2 t + a_1 \qquad (5.6)$$

Speed is the rate of change of length, $s(t) = dl/dt$. Therefore, the coefficients of Equations (5.4) and (5.6) must be equal: $3a_3 = 0.78, 2a_2 = -15.6$ and $a_1 = 92$, which implies $a_3 = 0.26, a_2 = -7.8$ and $a_1 = 92$. Placing these values back into Equation (5.5) gives

$$l(t) = 0.26t^3 - 7.8t^2 + 92t + a_0 \qquad (5.7)$$

Once again, an unknown, additive constant appears at the end of this antiderivative. If t is set to zero, then $l(0) = a_0$ but, in this case, there is no reason to believe that a_0 is zero because of the lag between the start of the dive and the start of the recording. There is just no way of knowing how much length the seal had covered before $t = 0$. Generally, *informative models cannot exist in a vacuum of data*. Therefore, the final answer is

$$l(t) = 0.26t^3 - 7.8t^2 + 92t + l(0) \qquad (5.8)$$

These calculations give biologically realistic results: the acceleration plot (Figure 5.2(a)) starts with negative and ends with positive values. This is because the seal slows down during its descent and speeds up as it resurfaces. As the seal reaches the middle of its dive its speed reaches a minimum (Figure 5.2(b)), possibly because it is engaging with its prey. The distance travelled through water (Figure 5.2(c)) increases sharply at the beginning and end of the dive but only slowly in the middle. Since we don't know how much distance the seal had covered when the recording began, any one of the curves shown in Figure 5.2(c) (as well as an infinite number of similarly shifted curves) could represent the truth.

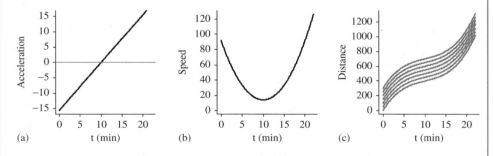

Figure 5.2: (a) Acceleration, (b) speed and (c) possible distances covered by the seal under water (five examples of antiderivatives are shown, each with a different initial condition).

In general, if $F(x)$ is the derivative of a function, an infinite number of antiderivatives $f(x) + c$ can be constructed for different values of the constant c.

5.2.   Indefinite integrals

The set of all antiderivatives $f(x) + c$ of a function $F(x)$ is called the **indefinite integral** of $F(x)$ with respect to x and is written

$$\int F(x)dx = f(x) + c \tag{5.9}$$

There is nothing technically new in this expression; it is simply shorthand for the operation outlined in the previous section. The function $F(x)$ is called the **integrand**. Equation (5.9) contains the new symbol $\int$ (an elongated letter 'S') and a new use for the infinitesimal increment dx. The reason for this notation will become clearer in Section 5.4, but first some examples of its use.

Example 5.3

The indefinite integral of the function $F(x) = 2x$ is

$$\int F(x)dx = x^2 + c \tag{5.10}$$

The indefinite integral of the function $g(x) = 3x^3 + 2x + 1$ is

$$\int g(t)dt = \frac{3}{4}x^4 + x^2 + x + c \tag{5.11}$$

Both of these examples give the integral of a polynomial function by reversing the first few rules of differentiation in Section 4.5. Similar reasoning can be used to reverse some of the other rules. This leads to integration rules 1–8 in Section A8.1 of the appendix. Although these look a bit overwhelming, there is no need to remember them by heart. You can familiarise yourself with them by firstly taking the derivatives of the functions in the right-hand column and making sure that the results are, indeed, the corresponding entries in the left-hand column. Since the integral of a function is also a function, you can try to find its derivative. That derivative will be the function that was integrated in the first place

$$\frac{d}{dx}\int F(x)dx = F(x) \tag{5.12}$$

This last expression, called the **fundamental theorem of calculus**, is the point of convergence between differentiation and integration.

Example 5.4: Allometry

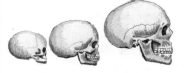

Allometric growth in human skulls

Allometry is the study of relationships between body size and shape. Organisms do not grow **isometrically**. Rather, proportions change with size (e.g. juvenile mammals have comparatively large heads, the limb proportions of arthropods alter in successive moults, etc.).

The rate of growth of the length (l) of the intestinal tract of a mammalian carnivore is defined as the length added to the intestine with every gram gained, as a function of total mass (m). The specific relationship is

$$\frac{dl}{dm} = 0.243m^{-0.25} \tag{5.13}$$

We seek a relation that gives intestinal length as a function of mass ($l(m)$).

$$l(m) = \int 0.243m^{-0.25}dm + c \tag{5.14}$$

Notice that the exponent in the integrand is a rational number,

$$l(m) = \int 0.243m^{-\frac{1}{4}}dm + c \tag{5.15}$$

Using rules 3 and 6 from the table of integrals in Section A8.1 gives

$$l(m) = 0.243 \int m^{-\frac{1}{4}}dm + c$$

$$= 0.243 \left(\frac{4}{3}m^{\frac{3}{4}} \right) + c \tag{5.16}$$

$$= 0.324m^{0.75} + c$$

This would have to be the final answer if there was no additional information about the biological first principles of the process modelled here. The additive constant (c) resulting from integration is (as in Examples 5.1 and 5.2) the initial condition of this model. It corresponds to the intestinal length of an animal that has no mass. It is impossible for this to be anything other than zero, otherwise the animal would consist exclusively of a mass-less gut! Therefore,

$$l(m) = 0.324m^{0.75} \tag{5.17}$$

The exponent of Equation (5.17) is positive and the exponent of Equation (5.13) is negative, so the intestinal tract is growing at a decreasing rate.

Example 5.5

The indefinite integral of the function $f(x) = \frac{4}{x} + 2x - 3$ is found as follows

$$\int f(x)dx = \int \left(\frac{4}{x} + 2x - 3 \right) dx + c$$

$$= \int \frac{4}{x}dx + \int 2xdx + \int (-3)dx + c \tag{5.18}$$

$$= 4\ln x + x^2 - 3x + c$$

It is crucial to mention that *these simple rules cannot find the integrals of all functions*. Many functions require more elaborate techniques that we will examine in Section 5.3 and most functions simply cannot be integrated analytically.

5.3. *Three analytical methods of integration*

This section introduces three techniques for finding the indefinite integrals of more complicated functions than those listed in Section A8.1. These are: **integration by substitution**, **integration by parts** and **integration by partial fractions**.

5.3.1.   Integration by substitution

This first method is essentially the chain rule (Section 4.7) applied to integration. It is used when the integrand $f(x)$ is a composite of two other functions $f(x) = f(g(x))$

$$\int f(g(x))dx \tag{5.19}$$

The functions may be separated by writing them as $f(u)$ and $u = g(x)$ where u is a **dummy variable** used to facilitate the intermediate stages of the calculation. The method is particularly useful if it is known how to find the following integral

$$\int f(u)du \tag{5.20}$$

To express Equation (5.19) in terms of u, the following substitution must be used

$$u = g(x) \text{ and } dx = du/g' \tag{5.21}$$

This last expression is shorthand that looks like algebra with infinitesimal amounts. I will not justify it here further, other than by citing the approximate argument

$$g' = \frac{du}{dx} \cong \frac{\Delta u}{\Delta x} \tag{5.22}$$

Hence, solving for $\Delta x \cong \Delta u/g'$. Placing Equation (5.21) back into the original integral,

$$\int f(g(x))dx \underset{\substack{g(x)=u \\ dx=du/g'}}{=} \int f(u)\frac{1}{g'}du \tag{5.23}$$

I have put the substitution under the equals sign for easy reference. It is often the case that the derivative g' is a constant, allowing Equation (5.23) to be integrated.

Example 5.6

$f(x) = (x + 5)^3$ is made up of the functions $g(x) = x + 5$ and $f(u) = u^3$. The integral of $\int (x + 5)^3 dx$ is not known but the integral $\int u^3 du$ can be calculated. The substitution $u = x + 5$ (note that $(x + 5)' = 1$, so $du = dx$) gives

$$\int (x + 5)^3 dx \underset{\substack{x+5=u \\ dx=du}}{=} \int u^3 du = \frac{u^4}{4} + c \tag{5.24}$$

Having found the integral in terms of u, a back-substitution gives the final answer in terms of the original variable x,

$$\int (x + 5)^3 dx = \frac{(x + 5)^4}{4} + c \tag{5.25}$$

Example 5.7: Stopping invasion fronts

 A pest management programme that involves cutting and cauterising rhododendron plants is being gradually ramped up. This increase in effort has made the speed of movement of the invasion a decreasing function of time,

$$\frac{dl}{dt} = \frac{a}{b+t} \qquad (5.26)$$

where t is the time since the initiation of pest control and a, b are two known parameters. We would like to obtain a function describing the length travelled by the wave a time t after the pest control programme started.

$$l(t) = \int \frac{a}{b+t} dt = a \int \frac{1}{b+t} dt \qquad (5.27)$$

The substitution $u = b + t, dt = du$ gives

$$l(t) = a \int \frac{1}{u} du = a \ln u + c \qquad (5.28)$$

The final answer is $l(t) = a \ln(b + t)$ (since the question asks how much the front has travelled *after* the management programme was started, I have dropped the integration constant).

Fortunately, the method of substitution may work even when g' turns out to contain the variable x. For example, g' may cancel out with some other part of the integrand.

Example 5.8

$$\int x \exp(-x^2) dx \underset{\substack{=\\u=-x^2\\dx=du/-2x}}{} \int x \exp(u) \frac{1}{-2x} du = -\frac{1}{2} \exp(u) = -\frac{1}{2} \exp(-x^2) \qquad (5.29)$$

5.3.2. Integration by parts
This is a useful technique if the integrand is a product of two functions. Again, the objective is to simplify the integral into a known form. The method relies on the product rule for differentiation (see Section A7.2). Consider two differentiable functions $f(x)$ and $g(x)$. The derivative of their product is

$$(fg)' = f'g + g'f \qquad (5.30)$$

Integrating both sides of this gives

$$\int (fg)' dx = \int f'g dx + \int g'f dx \qquad (5.31)$$

Note that $\int (fg)' dx = fg$. Hence, rearranging,

$$\int f'g dx = fg - \int g'f dx \qquad (5.32)$$

Here, each integral involves one function and the derivative of another. How does that help? Well, consider the following example:

Example 5.9

$$\int x \exp x\,dx \tag{5.33}$$

This integral is similar to the one in Example 5.8 but cannot be solved by substitution (try it and see). Nevertheless, it is a product of the identity function x and the exponential function $\exp x$. Have a look at the left-hand side of Equation (5.32). With a great leap of intuition, let's make the following substitutions $g = x, f' = \exp x$. To write Equation (5.33) in the form of Equation (5.32) it is necessary to find the derivative of g and the antiderivative of f'

$$g' = (x)' = 1$$
$$f = \int f'dx = \int \exp x\,dx = \exp x \tag{5.34}$$

Equation (5.32) can now be applied:

$$\int x \exp x\,dx = x \exp x - \int \exp x\,dx$$
$$= (x - 1)\exp x \tag{5.35}$$

With integration by parts, the key is to choose the right substitutions. In this example, choosing the alternative set of substitutions $g = \exp x, f' = x$ would give the rather horrible expression $\int x \exp x\,dx = \frac{1}{2}x^2 \exp x - \frac{1}{2}\int x^2 \exp x\,dx$.

Amazingly, integration by parts can produce results even when the integrand isn't obviously a product of two functions.

Example 5.10

The innocent-looking integral

$$\int \ln x\,dx \tag{5.36}$$

is not in the table of elementary integrals of Section A8.1 and neither is it a prime candidate for integration by parts, since only one function is involved. Rather perversely, it is possible to think of the integrand as the product of the logarithmic function and one: $\int 1 \times \ln x\,dx$. The derivative of $\ln x$ is known, but not its integral. So, the only available substitution is $f' = 1, g = \ln x$, giving the corresponding expressions $f = x$ and $g' = 1/x$. Placing all four of these expressions into Equation (5.32) produces the result

$$\int 1 \times \ln x\,dx = x \ln x - \int x\frac{1}{x}dx$$
$$= x(\ln x - 1) \tag{5.37}$$

5.3.3. Integration by partial fractions

You should consider this method when faced with a function like the following

$$\int \frac{p(x)}{q(x)}dx \tag{5.38}$$

where $p(x)$ and $q(x)$ are polynomials.

I will only examine the case where the denominator is, or can be factored into, a product of *different* linear factors $(ax + b)$. For example, the expression $q(x) = 3x^2 + 5x - 2 = (x +$

2)$(3x - 1)$ qualifies but the expression $q(x) = (x + 2)(3x - 1)^2 = (x + 2)(3x - 1)(3x - 1)$ does not, because the last two terms are identical. Additionally, I will restrict attention to fractions in which the order of the numerator $p(x)$ is lower than the order of the denominator $q(x)$ (although this requirement can be relaxed with some further effort). I illustrate the approach with a second order denominator,

$$\int \frac{a_0 x + b_0}{(a_1 x + b_1)(a_2 x + b_2)} dx \tag{5.39}$$

This expression satisfies the two requirements (the numerator's order is lower than the denominator's and the two factors in the denominator are unique). A useful result from algebra says that any expression of the form

$$\frac{p(x)}{(a_1 x + b_1)(a_2 x + b_2) \cdots (a_k x + b_k)} \tag{5.40}$$

(assuming the order of the numerator is less than that of the denominator) can be broken up into partial fractions

$$\frac{A_1}{(a_1 x + b_1)} + \frac{A_2}{(a_2 x + b_2)} + \cdots + \frac{A_k}{(a_3 x + b_k)} \tag{5.41}$$

where $A_1, A_2, \ldots, A_k$ are constants. Why is this useful? If the original integral can be written in this form, then it can simply be treated as the sum of integrals of the form $\int A/(ax + b)dx$. Each of these can then be calculated by the method of substitution (see Example 5.7). Applying partial fractions to the integrand gives

$$\frac{a_0 x + b_0}{(a_1 x + b_1)(a_2 x + b_2)} = \frac{A_1}{(a_1 x + b_1)} + \frac{A_2}{(a_2 x + b_2)} \tag{5.42}$$

Integrating the right-hand side gives

$$A_1 \int \frac{1}{(a_1 x + b_1)} dx + A_2 \int \frac{1}{(a_2 x + b_2)} dx = \frac{A_1}{a_1} \ln (a_1 x + b_1) + \frac{A_2}{a_2} \ln (a_2 x + b_2) \tag{5.43}$$

This is the general solution for the integral in Equation (5.39), which has a second order polynomial in the denominator, but exactly the same approach can be applied to higher order polynomials. The problem is that the values of A_1 and A_2 are not known. They can be found as follows: first, the right part of Equation (5.42) is converted to the same denominator as its left part:

$$\frac{a_0 x + b_0}{(a_1 x + b_1)(x + b_2)} = \frac{A_1(a_2 x + b_2) + A_2(a_1 x + b_1)}{(a_1 x + b_1)(a_2 x + b_2)} \tag{5.44}$$

This equation will be true if the numerators on both sides are equal

$$a_0 x + b_0 = A_1(a_2 x + b_2) + A_2(a_1 x + b_1) \tag{5.45}$$

Multiplying out the right-hand side and collecting the terms of x gives

$$a_0 x + b_0 = (A_1 a_2 + A_2 a_1)x + (A_1 b_2 + A_2 b_1) \tag{5.46}$$

For the equation to hold, the coefficients of x and the constant terms on both sides must be equal. This observation yields two coupled equations

$$a_0 = A_1 a_2 + A_2 a_1$$
$$b_0 = A_1 b_2 + A_2 b_1 \tag{5.47}$$

Which can be solved (see Section 1.14) to find

$$A_1 = \frac{b_1 a_0 - a_1 b_0}{b_1 a_2 - a_1 b_2}, A_2 = \frac{a_2 b_0 - b_2 a_0}{a_2 b_1 - b_2 a_1} \tag{5.48}$$

Example 5.11

To solve the integral

$$\int \frac{x}{(x+2)(2x-1)} dx \tag{5.49}$$

first note that $a_0 = 1, b_0 = 0, a_1 = 1, b_1 = 2, a_2 = 2, b_2 = -1$. The general solution (see Equation (5.43)) will be

$$A_1 \ln(x+2) + \frac{A_2}{2} \ln(2x-1) \tag{5.50}$$

The values $A_1 = 2/5$ and $A_2 = 1/5$ can then be calculated from Equation (5.48).

Given an indefinite integral that cannot be solved with the methods presented here, it may be a good idea to trawl through specialised books or catalogues of integrals to see if yours is listed. If you are expecting to be doing a lot of analytical work with integrals, you may consider investing in (proprietary) software packages such as MATHEMATICA or MAPLE that have extensive integration libraries.

5.4. Summation

Even the elementary operation of addition can get laborious if it involves a large number of repetitions. This section introduces the **summation notation**, an aid in the task of bookkeeping with large sums. Consider a sequence of n terms

$$a = \{a_1, \ldots, a_n\} \tag{5.51}$$

As in Chapter 3, the members of the sequence are indexed by non-negative integers i, indicating the position of a general member (a_i) in the sequence. In Equation (5.51) the index i takes values between 1 and n. The sum A of all the terms of the sequence is known as a **series** and it is written

$$A = a_1 + \cdots + a_n \tag{5.52}$$

There is an alternative way of writing Equation (5.52) which proves more economical for sequences containing complicated mathematical expressions. To begin with, a suitable symbol is needed to declare the beginning of a sum, one that will be reserved exclusively for this purpose. To avoid forfeiting the use of the Latin letter 'S' (for 'sum'), the upper-case, Greek letter sigma ($\sum$) is used. Whenever you see this symbol followed by a mathematical expression you must interpret it as the sum of terms that are of the form given by that mathematical expression. For example, $\sum a_i$, indicates addition of terms of the sequence a.

The sigma notation is not yet fully specific because it does not indicate how many terms are being added. For instance, we don't know if $\sum a_i$ represents the sum of the first two, three, four or all n terms of the sequence. This is achieved by providing a range of values for the index i. The sum of the first five terms of the sequence would be written as

$$\sum_{i=1}^{5} a_i \tag{5.53}$$

and the sum of all the terms would be written as

$$A = \sum_{i=1}^{n} a_i \tag{5.54}$$

Example 5.12: Metapopulations

Glanville fritillary butterfly (Melitaea cinxia) the iconic species of metapopulation ecology

In the late 1960s, Richard Levins coined the term **metapopulation** to describe a group of populations weakly connected by emigration and immigration. Local populations are thought to occupy patches of suitable habitat that are surrounded by hostile or undesirable habitat. Metapopulation persistence is seen as the dynamic balance between chance extinctions and recolonisations through immigration.

Although metapopulation ecology takes a rather simplistic view of spatial dynamics, it is nevertheless a powerful abstraction. Most obviously, it applies to species living in fragmented environments, but even continuous gradients in environmental variables (Figure 5.3(a)) can be adequately described if the response of the organisms to their environment presents a threshold of suitability (Figure 5.3(b)).

Consider ten patches indexed by $i = 1, \ldots, 10$, each with a corresponding population P_i. If P is the size of the entire metapopulation, then

$$P = \sum_{i=1}^{10} P_i \tag{5.55}$$

Now, consider the number of individuals that, during a given time interval, moved from one patch to another. If n_{ij} denotes the number of animals that were initially observed at the ith patch and were finally found at the jth patch, then the following table can be constructed to fully describe movement between all patches.

		Origin						
		1	2	...	j	...	...	10
	1	$n_{1,1}$	$n_{1,2}$	...	$n_{1,j}$	...	...	$n_{1,10}$
	2	$n_{2,1}$	$n_{2,2}$					$\vdots$
Destination	$\vdots$	$\vdots$						$\vdots$
		$\vdots$			$\vdots$			$\vdots$
	i	$n_{i,1}$		...	$n_{i,j}$	...		$n_{i,10}$
	$\vdots$	$\vdots$			$\vdots$			$\vdots$
	10	$n_{10,1}$	...	...	$n_{10,j}$	...	...	$n_{10,10}$

The diagonal elements of this table refer to animals that either stayed put or decided to return to their patch of origin by the end of the migration interval. The total number of animals in the table should be equal to the size of the metapopulation. Given this table, how would you manually add all the n_{ij} s to verify this? A rather inefficient approach might be to take them in random order. A better approach would be to add together all the row or column totals. Adding together the column totals is written

$$P = \sum_{i=1}^{10} n_{i1} + \sum_{i=1}^{10} n_{i2} + \sum_{i=1}^{10} n_{i3} + \cdots + \sum_{i=1}^{10} n_{i10} \tag{5.56}$$

Each term of this expression refers to a different column of the table, that's why each term has a fixed numerical value in place of the index j. Equation (5.56) can be further simplified by using the summation notation again

$$P = \sum_{j=1}^{10} \sum_{i=1}^{10} n_{ij} \tag{5.57}$$

It is also possible to specify parts of this sum. For example, the number of animals that moved is

$$\sum_{j=1}^{10} \sum_{i=1,i \neq j}^{10} n_{ij} \tag{5.58}$$

and the number of animals that didn't move is

$$\sum_{i=1}^{10} n_{ii} \tag{5.59}$$

I mentioned above that, in a metapopulation, the component subpopulations are weakly connected. This means the migration is not a frequent phenomenon. Mathematically this is written as follows:

$$\sum_{i=1}^{10} \sum_{j=1,j \neq i}^{10} n_{ij} \ll \sum_{i=1}^{10} n_{ii} \tag{5.60}$$

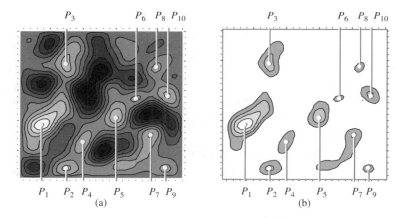

Figure 5.3: (a) Contour plot of habitat suitability (lighter shades indicate more suitable locations; (b) the distribution of habitats naturally forms patches of high suitability for the animals.

A particularly interesting type of sum, known as an **infinite series**, is one that uses infinity as the summation limit. At first sight, it might appear that adding together an infinite number of terms should yield infinity. Indeed, this is often the case. However, rather counter-intuitively, some infinite series have a finite numerical value. In such cases, we say that the infinite series **converges**. This may happen if the sign of the things being added alternates between − and + (so that quantities are also being subtracted from the total), or if the terms being added gradually become infinitely small.

Example 5.13

It can be shown that the infinite series $\sum_{n=1}^{\infty} \frac{1}{n}$ does not converge. However, the series $\sum_{n=1}^{\infty} \frac{1}{n^2}$ does. The reason for the difference is that $1/n^2$ diminishes much faster than $1/n$ as n tends to infinity.

5.1: Summation in R

The command `sum()` will add together the terms of a sequence. For example,

```
> n<-100
> a<-seq(1,n)
> sum(a)
[1] 5050
```

A comparison between the two infinite series of Example 5.13 can be made by setting n to ever-greater values (for very large n, R will run out of memory). As n increases, the value of the first series keeps on changing while the value of the second converges to $\cong 1.64$.

```
> n<-100
> sum(1/seq(1,n))
[1] 5.187378
> sum(1/seq(1,n)^2)
[1] 1.634984

> n<-10000
> sum(1/seq(1,n))
[1] 9.787606
> sum(1/seq(1,n)^2)
[1] 1.644834
```

5.5. Area under a curve

Example 5.14: Swimming speed in seals

Example 5.2 ignored the fact that, although animal movement is continuous, telemetry data are discrete. Consider a grey seal that moves through water at a speed that varies smoothly with time (Figure 5.4(a)). The tag measures and transmits the seal's speed at regular intervals $\Delta t = 5$ min (Figure 5.4(b)).

An observation period of 50 min will result in a sequence of 11 observations.

$$s = \{s_0, \ldots, s_{10}\} \tag{5.61}$$

An approximation of the length of the seal's path during the period of observation can be obtained as follows: firstly, the speed data in Equation (5.61) are multiplied by Δt to give a sequence of incremental lengths

$$l = \{s_0 \Delta t, \ldots, s_{10} \Delta t\} = \{l_0, \ldots, l_{10}\} \tag{5.62}$$

The assumption in doing this is that, during the time interval Δt, the speed of the animal remains approximately constant. An estimate of the total path length is obtained by adding all the terms of the sequence in Equation (5.62)

$$l_{Tot} \cong \sum_{i=0}^{10} l_i = \sum_{i=0}^{10} s_i \Delta t \qquad (5.63)$$

The term $s_i \Delta t$ is the product of a single measurement of speed s_i by the time interval Δt. As it happens, speed and time are also the dependent and independent variables in Figure 5.4, so each such product $s_i \Delta t$ can be treated as the area of a rectangle with height s_i and width Δt (Figure 5.5(a)). These rectangles may be arranged in the order of occurrence of the

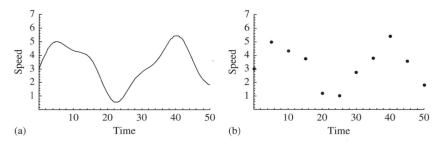

Figure 5.4: (a) Graph of the seal's speed as a function of time and (b) the corresponding data.

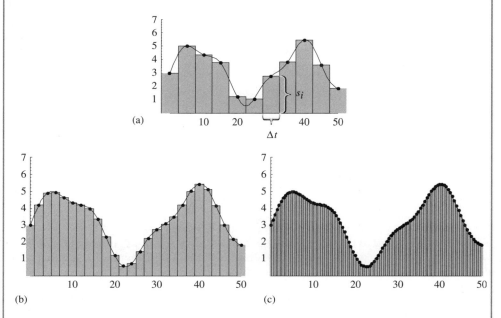

Figure 5.5: (a) Observations recorded at intervals of 5 min. The areas of the bars of the histogram represent the approximate length travelled by the seal during the 5 min interval. Finer approximations can be obtained by getting data at 2 min and 30 sec intervals (parts (b) and (c)).

corresponding observations s_i and superimposed onto the graph of speed (Figure 5.5(a)). The sum in Equation (5.63) was introduced as an approximation of the total length covered by the animal but it can also be interpreted as the total area enclosed by all the rectangles.

The approximation can be improved by reducing the time interval between observations. At $\Delta t = 2$ min the estimated length of the path is

$$l_{Tot} \cong \sum_{i=0}^{25} s_i \Delta t \tag{5.64}$$

and at $\Delta t = 30$ sec,

$$l_{Tot} \cong \sum_{i=0}^{101} s_i \Delta t \tag{5.65}$$

The approximating sums in this example are known as **Riemann sums**. As the time interval Δt becomes smaller, the sequence of points gives an increasingly better approximation of the continuous graph (see Figures 5.5(b) and (c)) and the total area enclosed by the set of rectangles converges to the area under the graph of speed. As Δt becomes very small, the number of data points ($n = 50$ min$/\Delta t$) becomes very large. The plot of the discrete data becomes indistinguishable from the graph of the continuous function and the length of the seal's trajectory through water is calculated exactly by the Riemann sum

$$l_{Tot} = \lim_{\Delta t \to 0} \sum_{i=0}^{n} s_i \Delta t \tag{5.66}$$

In Section 5.1 the exact value of the length travelled was obtained as an antiderivative of speed. So, the integral of speed (the set of all antiderivatives) and the limiting sum in Equation (5.66) must be the same:

$$\int s(t)dt = \lim_{\Delta t \to 0} \sum_{i=0}^{n} s_i \Delta t \tag{5.67}$$

This expression is not quite complete: although the upper limit n of the sum tends to infinity, the overall interval of summation is fixed ($t_{min} = 0$ to $t_{max} = 50$ min, in the seal example). However, the integral has no specified limits. To mirror the same information on both sides of this equation, we write

$$\int_{t_{min}}^{t_{max}} s(t)dt = \lim_{\Delta t \to 0} \sum_{i=0}^{n} s_i \Delta t \tag{5.68}$$

The numbers attached to the integration sign are the lower and upper **limits of the integral**. Use of the word 'limit' in this context differs from Section 4.3. The limits of integration simply define the extremes of the domain of the integrand.

Some points on notation regarding Equation (5.68): As the time interval Δt becomes infinitesimally small, it is replaced by its alter-ego dt, and the Greek sigma denoting the sum is replaced by the elongated Latin 'S'. Also, the customary notation for sequences s_i is replaced by the notation $s(t)$ of continuous functions.

So, this section has achieved two things: first, it provided a geometrical interpretation for the integral as the area under a curve. Such visualisation tools are always helpful for those of us who prefer to think in terms of graphs rather than algebra. The graphical interpretation of the integral as the area under the graph of the integrand complements the interpretation of the derivative as the slope of the function being differentiated. Second, in order to discuss finite areas under curves, it was necessary to define the limits of integration. Hence, the resulting integrals are no longer indefinite, as the next section discusses.

5.6.   Definite integrals

The integral of a function $f(x)$ over a pre-specified interval $[a, b]$ of the independent variable x is called a **definite integral**. It is written

$$\int_a^b f(x)dx \tag{5.69}$$

Example 5.15: Swimming speed in seals

The grey seal diving data in Example 5.2 were described by the function

$$s(t) = 0.78t^2 - 15.6t + 92 \tag{5.70}$$

The corresponding function for path length is, in general, given by the indefinite integral which was calculated (see Example 5.2) as

$$l(t) = 0.26t^3 - 7.8t^2 + 92t + l(0) \tag{5.71}$$

where $l(0)$ is the length already covered by the seal before the start of the data. The tag transmitted data for 22 min. We wish to calculate the length travelled by the seal between the times $t_1 = 0$ and $t_2 = 22$. This is found in the following way:

$$l(t_2) - l(t_1) = [0.26t_2^3 - 7.8t_2^2 + 92t_2 + l(0)] - [0.26t_1^3 - 7.8t_1^2 + 92t_1 + l(0)]$$

$$= 0.26(t_2^3 - t_1^3) - 7.8(t_2^2 - t_1^2) + 92(t_2 - t_1) \tag{5.72}$$

$$= 1017.28m$$

The constant of integration has vanished, leaving a simple numerical answer.

This result is more general that it seems at first. We were given the derivative $f(x)$ of some unknown function $F(x)$

$$\frac{dF}{dx} = f(x) \tag{5.73}$$

The function $F(x)$ was calculated as the indefinite integral of the rate dF/dx, which resulted in an expression of the form

$$F(x) = I(x) + c \tag{5.74}$$

You can think of $I(x)$ as a particular antiderivative of the function $F(x)$; the one that has $c = 0$. The accumulated change over a particular interval $[a, b]$ of x was written as the difference between the overall value (at the end of the interval) and the value that had been achieved at the beginning.

$$F(b) - F(a) = I(b) + c - (I(a) + c)$$
$$= I(b) - I(a) \tag{5.75}$$

In this operation, the constant c of the indefinite integral vanishes. In general, if we can calculate the indefinite integral of a function (i.e. any one of its antiderivatives), its definite integral is

$$\int_a^b f(x)dx = I(b) - I(a) \tag{5.76}$$

Example 5.16

To calculate the definite integral

$$\int_2^4 3xdx \tag{5.77}$$

we first find an antiderivative of the integrand $f(x) = 3x$,

$$F(x) = \frac{3}{2}x^2 + c \tag{5.78}$$

and then specify this to the limits of integration to get $F(2) = 6$ and $F(4) = 24$. Finally, one is subtracted from the other to get to the answer

$$\int_2^4 3xdx = \frac{3}{2}x^2 \Big|_2^4 = F(4) - F(2) = 18 \tag{5.79}$$

The second step in Equation (5.79) uses the vertical bar notation to indicate that the indefinite integral has been found but it has yet to be specified to its limits.

Another example: to calculate the definite integral

$$\int_1^4 f(x)dx = \int_1^4 \left(\frac{2}{x} + 2x + x^{-1/2}\right) dx \tag{5.80}$$

we first find the antiderivative of the integrand $f(x) = 2/x + 2x + x^{-1/2}$. This is equal to $F(x) = 2\ln(x) + x^2 + 2x^{1/2}$. The definite integral is given by

$$\int_1^4 f(x)dx = 2\ln(x) + x^2 + 2x^{1/2} \Big|_1^4 \tag{5.81}$$
$$= (2\ln(4) + 16 + 4) - (2\ln(1) + 1 + 2) \cong 19.77259$$

In all of the above examples the result is a numerical value, not a function. This always happens in definite integrals with numerical limits of integration and numerical coefficients.

Ⓡ 5.2: Numerical integration

Integration is one of the most challenging problems of applied mathematics because the vast majority of functions that are conjured up during research simply cannot be integrated analytically. Numerical integration then becomes a convenient way out. The idea behind it is to approximate the value of a definite integral by a Riemann sum such as the ones in Section 5.5. Thankfully, the algorithms for doing this are nicely packaged within the command `integrate()`. To use it, you need to specify the function of interest and the limits of integration. I illustrate its use by calculating numerically the value of a definite integral that can also be done analytically. Taking Equation (5.80) in Example 5.16, the implementation is

```
> f<-function(x) {2/x+2*x+1/sqrt(x)}
> integrate(f, 1, 4)
19.77259 with absolute error < 7e-11
```

Pleasingly, this result is exactly the same as the one obtained analytically (see Equation (5.81)). It may be tempting to eschew pencil and paper calculations in favour of numerical approximation. There are two good reasons for not doing so: numerical approximation only works when all the parameters and integration limits have been specified – it is therefore of little use for mathematical proofs. Also, numerical approximation is more time consuming. If you are planning to be repeating the calculation several times within a larger piece of code, it is good to check whether you can obtain an analytical solution.

5.7. Some properties of definite integrals

As often happens in maths, the introduction of a new concept (such as the definite integral) is followed by the search for rules for manipulating it. This section presents three such rules.

In Section 5.5 the definite integral was interpreted as the sum of the areas of infinitesimally small bars. This sum is the same, irrespective of how the addition is performed. For example, the interval of integration can be broken into arbitrarily many subintervals. I illustrate this using the two subintervals $[a, c]$ and $[c, b]$ that make up the interval $[a, b]$ (Figure 5.6)

$$\int_a^b f(x)dx = \int_a^c f(x)dx + \int_c^b f(x)dx \qquad (5.82)$$

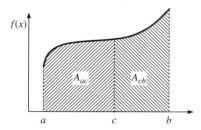

Figure 5.6: Definite integrals are additive. The area under the curve between a and b can be seen as consisting of the sum of areas over two or more subintervals $A_{ab} = A_{ac} + A_{cb}$.

Example 5.17: Total reproductive output in social wasps

Assume that a queen wasp produces 35 eggs per day for the first two months (61 days) of summer and its productivity in the third month decreases linearly until, by the 92nd day, it is zero. Mathematically, these facts are formalised by the following step function

$$f(t) = \begin{cases} 35 & \text{if } 0 < t \le 61 \\ -1.129t + 103.871 & \text{if } 61 < t \le 92 \end{cases} \tag{5.83}$$

The second row of Equation (5.83) is the general function for a line ($f(t) = mt + c$) with slope $m = -1.129$ and intercept $c = 103.871$. These two values have been calculated by solving the system of equations $35 = m61 + c$ and $0 = m92 + c$.

The queen's total summer output, say, P_s, is given by the definite integral

$$P_s = \int_0^{92} f(t)dt \tag{5.84}$$

It is easiest to deal with this integral by splitting it up into two parts

$$P_s = \int_0^{61} f(t)dt + \int_{61}^{92} f(t)dt$$

$$= \int_0^{61} 35dt + \int_{61}^{92} (-1.129t + 103.871)dt \tag{5.85}$$

$$= 35t\Big|_0^{61} + (-0.565t^2 + 103.871t)\Big|_{61}^{92}$$

$$\cong 2678$$

An easier way to calculate this value, which only works because this is a simple function, is to notice that the area under the curve comprises the area of a rectangle (35×61) and the area of a right-angled triangle ($35 \times (92 - 61)/2$).

Another useful fact about definite integrals is that swapping the limits of integration changes the sign of the integral. Here is why:

$$\int_a^b f(x)dx = I(b) - I(a) = -(I(a) - I(b)) = -\int_b^a f(x)dx \tag{5.86}$$

What is the physical interpretation of this? It implies that definite integrals are directional. For example, a calculation involving the reproductive output of the wasp from the end to the beginning of the summer (i.e. taking time in reverse), would imply the *removal* of the queen's output during this period.

In Section 5.5 definite integrals were interpreted as the area *under* a curve, implying that the graph of the function was entirely above the x-axis, but it is also possible that the integrant takes negative values (Figure 5.7). When this happens, the areas below the x-axis will be negative and, thus, subtracted from the total.

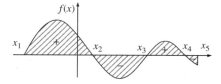

Figure 5.7: Area between the curve and the x-axis.

How we deal with this situation depends on how the result is to be used. If the negative sign is meaningful (e.g. representing some form of attrition, such as mortality in a population) then we need do nothing at all.

Example 5.18: Net change in number of birds at migratory stop-over

Barnacle goose, Branta leucopsis

A particular marshland serves as a stop-gap in the migration route of Barnacle geese. Observations collected during a single 24-hour period suggest a very simple pattern of arrivals and departures.

$$r = 0.5 - 5.56 \times 10^{-4}t \qquad (5.87)$$

where r is the net rate of change of birds per minute and t is time (in min) since the start of observations. Therefore, at the start of the day ($t = 0$), an average of 0.5 birds arrive each minute and at the end of the day ($t = 1440$), an average of 0.3 birds depart. We would like to calculate the net change in the number of birds during that entire day. Here, departures need to be removed from the total so the integral can be calculated directly

$$\int_0^{1440} 0.5 - 5.56 \times 10^{-4}t = 0.5t - 2.78 \times 10^{-4}t^2|_0^{1440} \cong 144 \qquad (5.88)$$

On the other hand, to calculate the total area enclosed between the graph and the x-axis, the sign of the integral needs to be changed for the subintervals with a negative integrand. So, the area enclosed by the graph in Figure 5.7 and the x-axis, over the interval $[x_1, x_5]$ can be calculated as follows

$$A = \int_{x_1}^{x_2} f(x)dx - \int_{x_2}^{x_3} f(x)dx + \int_{x_3}^{x_4} f(x)dx - \int_{x_4}^{x_5} f(x)dx \qquad (5.89)$$

Example 5.19: Total number of arrivals and departures at migratory stop-over

We are interested in calculating the number of geese seen crossing the boundaries of the study site during the 24-hour period of interest, assuming that geese only arrive in the first part of the day and only depart during the second part. The change happens at the time when $r = 0$. This can be calculated from Equation (5.87) as

$t \cong 900\,\text{min}$. The answer to this question is therefore

$$\int_0^{900} 0.5 - 5.56 \times 10^{-4}t - \int_{900}^{1440} 0.5 - 5.56 \times 10^{-4}t \cong 306 \tag{5.90}$$

Example 5.20

To find the total area (A) enclosed between the graph of the function $f(x) = -2x^2 + 4$ and the x-axis in the interval [-2,1], note that the function $f(x)$ takes negative values in the interval $[-2, -\sqrt{2})$ and positive values in the interval $(-\sqrt{2}, 1]$. Therefore,

$$A = -\int_{-2}^{-\sqrt{2}} (-2x^2 + 4)dx + \int_{-\sqrt{2}}^{1} (-2x^2 + 4)dx$$

$$= \int_{-\sqrt{2}}^{-2} (-2x^2 + 4)dx + \int_{-\sqrt{2}}^{1} (-2x^2 + 4)dx$$

$$= \left(-\frac{2}{3}x^3 + 4x\right)_{-\sqrt{2}}^{-2} + \left(-\frac{2}{3}x^3 + 4x\right)_{-\sqrt{2}}^{1} \tag{5.91}$$

$$= \left(\frac{16}{3} - 8 - \frac{4}{3}\sqrt{2} + 4\sqrt{2}\right) + \left(-\frac{2}{3} + 4 - \frac{4}{3}\sqrt{2} + 4\sqrt{2}\right)$$

$$= \frac{2}{3}(1 + 8\sqrt{2}) \cong 8.21$$

Compare this with the value of the definite integral.

$$\int_{-2}^{1} (-2x^2 + 4)dx = \left(-\frac{2}{3}x^3 + 4x\right)_{-2}^{1}$$

$$= \left(-\frac{2}{3} + 4\right) - \left(\frac{16}{3} - 8\right) \tag{5.92}$$

$$= 6$$

5.8.   Improper integrals

What happens to the value of a definite integral if one of its limits, or even the integrand itself, tends to infinity? Astonishingly, getting a numerical value for such **improper integrals** may not be a lost cause. There are examples of infinite functions whose integrals are finite. Such integrals are called **convergent**.

Consider, first, the example of Figure 5.8(a): the upper limit of integration is not set to a numerical value so the integral is unbounded. Since the curve extends to infinity, it may be expected that the area under it will as well. Not necessarily so. If the curve approaches zero fast enough, we end up incrementing our estimate of the total area by infinitesimal (negligible) amounts.

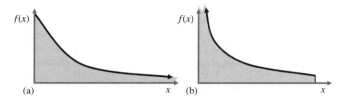

Figure 5.8: Calculating the areas shown in grey in these two graphs involves improper integrals. (a) The upper limit of integration tends to infinity; (b) the integrand itself tends to infinity as the lower limit of integration tends to zero.

To speak formally about this situation, the following piece of notation is needed

$$\int_a^\infty f(x)dx = \lim_{b\to\infty} \int_a^b f(x)dx \tag{5.93}$$

Hence, an improper integral with an infinite limit of integration can be thought of as a definite integral in which the corresponding integration limit tends to infinity. This helps a great deal because, although we haven't got any rules for calculating improper integrals of this form, we most certainly have rules for calculating definite integrals (see Sections 5.6 and 5.7) and limits (see Section 4.3). The idea is simple: first calculate the definite integral in terms of the integration limits a and b and then allow the upper limit to tend to infinity. The same idea applies to explosive functions (e.g. Figure 5.8(b)). I will illustrate both, with the following example.

Example 5.21

Consider the integral

$$\int_a^b \frac{1}{(x-1)^c}dx \tag{5.94}$$

For the moment, the constants a, b, c are left unspecified in value but I assume $a, b, c > 0$ and $c \neq 1$. This integrand approaches zero asymptotically as x increases and it explodes at $x = 1$. The indefinite integral can be calculated by substitution (see Section 5.3), ignoring the integration constant,

$$\int \frac{1}{(x-1)^c}dx \underset{\substack{u=x-1\\dx=du}}{=} \int \frac{1}{u^c}du = \frac{u^{1-c}}{1-c} = \frac{(x-1)^{1-c}}{(1-c)} \tag{5.95}$$

Therefore, the corresponding definite integral is

$$\int_a^b \frac{1}{(x-1)^c}dx = \frac{(x-1)^{1-c}}{(1-c)}\bigg|_a^b = (1-c)^{-1}\left\{(b-1)^{1-c} - (a-1)^{1-c}\right\} \tag{5.96}$$

I now specify the integral to two values of the exponent, $c = 0.5$ and $c = 2$. The first value gives a function that decreases more slowly compared to the second. I will integrate these two functions in the interval $[2, \infty)$ using the general result in Equation (5.96). First, the function that decreases more slowly,

$$\int_2^\infty \frac{1}{(x-1)^{0.5}}dx = \lim_{b\to\infty} \int_2^b \frac{1}{(x-1)^{0.5}}dx \tag{5.97}$$

$$= \lim_{b\to\infty}\left[2\left\{(b-1)^{0.5} - 1\right\}\right] = 2(\lim_{b\to\infty}\sqrt{b-1}) - 1 = \infty$$

Compare this with the faster-decreasing function

$$\int_{2}^{\infty} \frac{1}{(x-1)^2} dx = \lim_{b\to\infty} \int_{2}^{b} \frac{1}{(x-1)^2} dx = 1 - \lim_{b\to\infty} \frac{1}{b-1} = 1 \qquad (5.98)$$

Both of the examples in Equations (5.97) and (5.98) are of the type shown in Figure 5.8(a) (infinite limit of integration). However, the second example encloses a finite area but the first doesn't. To look at a situation of the type shown in Figure 5.8(b) (infinite integrand), I examine the same two functions, this time in the interval (1,2]. So, now I have bounded the area on the right-hand side, but have allowed the lower limit of integration to approach the value at which the function becomes infinite. For $c = 0.5$, the function increases more slowly as its variable tends to 1,

$$\int_{1}^{2} \frac{1}{(x-1)^{0.5}} dx = \lim_{a\to1} \int_{a}^{2} \frac{1}{(x-1)^{0.5}} dx = \lim_{a\to1} \left[2(1 - \sqrt{a-1}) \right] = 2 \qquad (5.99)$$

In contrast, setting $c = 2$ gives a more explosive function around 1,

$$\int_{1}^{2} \frac{1}{(x-1)^2} dx = \lim_{a\to1} \int_{a}^{2} \frac{1}{(x-1)^2} dx = \lim_{a\to1} \frac{1}{a-1} - 1 = \infty \qquad (5.100)$$

This talk of infinitely large or infinitesimally small quantities may seem far removed from the stark biological reality of short lifespans, delimited habitats and small data sets. Nevertheless, infinity is a useful metaphor for 'very large', 'very far away' and 'after a sufficiently long time'. Questions involving such phrases do occasionally pop up in ecology and it is important to know how to formulate them mathematically.

Example 5.22: Failing to stop invasion fronts

Example 5.7 showed how a gradual increase in pest control effort resulted in a decreasing invasion speed:

$$\frac{dl}{dt} = \frac{a}{b+t} \qquad (5.101)$$

However, it was not clear whether this decrease would be enough to stop the invasion. Will there come a time when the length l increases no longer? The distance traversed as a function of time can be obtained by integrating Equation (5.101). If the management strategy is effective, then the total distance (l_∞) will tend to a constant value irrespective of how much time passes. This can be investigated through the definite integral

$$l_\infty = \int_{0}^{\infty} \frac{a}{b+t} dt \qquad (5.102)$$

Calculation by substitution gives

$$l_\infty = \int_{0}^{\infty} \frac{a}{b+t} dt = \lim_{T\to\infty} \int_{0}^{T} \frac{a}{b+t} dt = a \lim_{T\to\infty} \left\{ \ln(b+T) - \ln(b) \right\} \qquad (5.103)$$

This limit equals infinity because $\lim_{T\to\infty} \ln T = \infty$. So, even though the invasion is slowed down, it is never actually stopped.

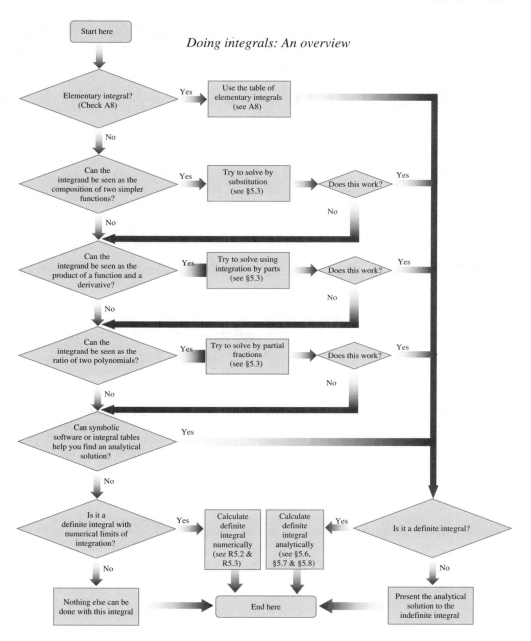

Start here

Doing integrals: An overview

Elementary integral?
(Check A8)

Yes → Use the table of elementary integrals (see A8)

No ↓

Can the integrand be seen as the composition of two simpler functions?

Yes → Try to solve by substitution (see §5.3) → Does this work? → Yes

No

No ↓

Can the integrand be seen as the product of a function and a derivative?

Yes → Try to solve using integration by parts (see §5.3) → Does this work? → Yes

No

No ↓

Can the integrand be seen as the ratio of two polynomials?

Yes → Try to solve by partial fractions (see §5.3) → Does this work? → Yes

No

No ↓

Can symbolic software or integral tables help you find an analytical solution?

Yes

No ↓

Is it a definite integral with numerical limits of integration?

Yes → Calculate definite integral numerically (see R5.2 & R5.3)

No ↓

Nothing else can be done with this integral

Calculate definite integral analytically (see §5.6, §5.7 & §5.8) ← Yes

Is it a definite integral?

No ↓

Present the analytical solution to the indefinite integral

End here

® 5.3: Calculating improper integrals

R can deal with improper integrals. I illustrate using the function of Example 5.21. There, we found that Equation (5.98) integrates to 1. This is also the result produced by R but, since this is a numerical approximation, an estimate of the (small) numerical error is also provided

```
> f<-function(x) {1/(x-1)^2}
> integrate(f,2,Inf)
1 with absolute error < 1.1e-14
```

Not all integrals converge:

```
> f<-function(x) {1/(x-1)^0.5}
> integrate(f,2,Inf)
Error in integrate(f, 2, Inf) : the integral is probably divergent
```

In most cases, R will try to deal with integrals that become infinite at one or the other limit of integration. Here are two examples, one convergent, the other divergent

```
> f<-function(x) {1/(x-1)^0.5}
> integrate(f,1,2)
2 with absolute error < 3.1e-13
```

```
> f<-function(x) {1/(x-1)^2}
> integrate(f,1,2)
Error in integrate(f, 1, 2) : the integral is probably divergent
```

5.9.   Differential equations

This and the next two sections belong to the wider field of **dynamical systems**. Discrete dynamical systems (difference equations) were examined in Chapter 3. Here, I discuss the form and origin of continuous dynamical systems, known as **differential equations**, involving the derivative df/dx of an unknown function $f(x)$. As always, solving the equation means finding a value or expression for the unknown quantity. Integrals are used to retrieve the unknown function $f(x)$ from information involving its derivative df/dx.

Example 5.23: A differential equation for a plant invasion front

Example 5.1 investigated the distance $l(t)$ covered by a biological invasion moving at constant speed (say, a). The mathematical description of this problem is a simple differential equation

$$\frac{dl}{dt} = a \tag{5.104}$$

The antiderivative of this expression

$$l(t) = at + b \tag{5.105}$$

represents the **general solution** of the differential equation (techniques for solving some differential equations will be presented in Section 5.10). In Example 5.1, biological information was used to interpret the integration constant b as the **initial condition** $l(0)$

$$l(t) = at + l(0) \tag{5.106}$$

The combination of a differential equation with parameter values and initial conditions forms an **initial value problem**. Specifying the initial condition to a particular value, say $l(0) = 0$, gives a **specific solution** of the differential equation.

If the terms 'general solution', 'specific solution' and 'initial value problem' seem familiar, it's because they were encountered back in Chapter 3. The similarities with difference equations do not stop there. Differential equations can also have equilibrium solutions, representing a situation of no change. Mathematically, these solutions satisfy the condition:

$$\frac{df}{dx} = 0 \tag{5.107}$$

Specific solutions initiated close to these equilibria may approach them or be repelled by them. Such equilibria are then called **stable** or **unstable** respectively. Specific solutions around equilibria can have as rich a palette of behaviours as those of Chapter 3: monotonic convergence and divergence, damped cycles, divergent oscillations, stable cycles, chaos. Section 5.11 will summarise how to find the equilibria of a continuous dynamical system and analyse their stability.

As with all mathematics, there is no limit to the complexity of a differential equation. Mathematicians have developed particular methods for specific types of equations, so, in most cases, solving a differential equation first requires us to identify its type. If an equation contains functions and their derivatives, all with respect to a single independent variable, it is called **ordinary**. If it contains partial derivatives (Section 4.9) it is a **partial differential equation**. The highest order derivative in an equation determines the equation's order. If the unknown function doesn't itself appear in the differential equation, or if it appears additively, then the equation is linear.

Example 5.24

The following are ordinary, linear, first order differential equations

$$\frac{df}{dx} = cx \qquad\qquad \frac{df}{dx} = f + cx \tag{5.108}$$

Here is an example of a linear second order differential equation

$$\frac{d^2f}{dx^2} = f + cx \tag{5.109}$$

and two examples of second order, nonlinear equations

$$\frac{d^2f}{dx^2} = f^3 + cx \qquad\qquad f\frac{d^2f}{dx^2} = f + cx \tag{5.110}$$

Finally, here is an example of a second order, nonlinear partial differential equation

$$f\frac{\partial^2f}{\partial x^2} + f\frac{\partial^2f}{\partial y^2} = f + cx \tag{5.111}$$

5.10. Solving differential equations

I will focus on a particular class of equations called **separable**. These are ordinary, first order differential equations of the form

$$\frac{dy}{dx} = g(x)h(y) \tag{5.112}$$

Here, $g(x)$ and $h(y)$ are any two functions of the independent and dependent variables and y is shorthand for the unknown function $f(x)$. This type of differential equation is special because the functions $g(x)$ and $h(y)$ each have only one of the two variables as their argument (either x or y) and the two functions $g(x)$ and $h(y)$ participate as a product. This equation is solved by calculating the integrals on either side of the following expression

$$\int \frac{1}{h(y)}dy = \int g(x)dx \tag{5.113}$$

Note that here, the ys and xs are separated on either side of the equality sign (hence the name of this type of differential equation). Would you like to know where this result comes from? If so, read on. Would you just like to learn how to apply it? Then skip to the examples below.

Equation (5.112) can first be written

$$\frac{1}{h(y)}\frac{dy}{dx} = g(x) \tag{5.114}$$

On the RHS of this equation is the function $g(x)$. On the LHS there is the more complicated-looking expression $\frac{1}{h(y)}\frac{dy}{dx}$. Are you satisfied that this expression is a function? If not, consider the following:

❶ A composite of functions $h(y) = h(f(x))$, is also a function.
❷ The inverse of a function $(1/h(y))$, is also a function.
❸ The derivative dy/dx of a function is also a function.
❹ The product of two functions $\left(\frac{1}{h(y)}\right) \times \left(\frac{dy}{dx}\right)$, is also a function.

If the two functions on either-side of Equation (5.114) are equal, their integrals must also be equal

$$\int \frac{1}{h(y)}\frac{dy}{dx}dx = \int g(x)dx \tag{5.115}$$

The RHS of this looks like a typical indefinite integral, the LHS less so but it can be simplified considerably. It can be recast, using the facts $y = f(x)$ and $\frac{dy}{dx} = f'(x)$,

$$\int \frac{1}{h(f(x))}f'(x)dx \tag{5.116}$$

Applying to this the substitution method from Section 5.3,

$$\int \frac{1}{h(f(x))}f'(x)dx \underset{\substack{y=f(x)\\dx=dy/f'(x)}}{=} \int \frac{1}{h(y)}f'(x)\frac{dy}{f'(x)} = \int \frac{1}{h(y)}dy \tag{5.117}$$

Placing this back into Equation (5.115) gives the desired result in Equation (5.113).

Example 5.25: Exponential population growth in continuous time

Differential equations are useful models of population dynamics for species with continuous reproduction (**overlapping generations**) such as bacteria or humans. Consider a population of bacteria which, at time t, has size $P(t)$. The per capita rate of change r is negative if more bacteria die than are being generated. The rate of population change is simply

$$\frac{dP}{dt} = rP \tag{5.118}$$

By analogy to the discrete population models in Chapter 3, r is called the population's **intrinsic growth rate**. Don't be confused by the fact that it has been specified as a daily rate: time units are needed, even in continuous time (if the time unit was an hour, the model would still be the same, but the value of r would be $\frac{1}{24}$ of its daily value). Although t does not appear explicitly on the RHS, this is still a separable equation, so Equation (5.113) can be used,

$$\int \frac{1}{P} dP = \int r dt \tag{5.119}$$

Each of these two integrals can be dealt with

$$\ln(P) + C_1 = rt + C_2 \tag{5.120}$$

Two integration constants make their appearance but they can be combined into one ($C = C_2 - C_1$), with no loss of generality,

$$\ln(P) = rt + C \tag{5.121}$$

This is now a logarithmic equation (Section 3.12). Taking exponentials on both sides,

$$P = e^{rt+C} \tag{5.122}$$

This can be rearranged by introducing the parameter $a = e^C$

$$P = ae^{rt} \tag{5.123}$$

Biologically, this parameter is the initial size of the population (to see why, write Equation (5.123) for $t = 0$). Therefore, the general solution of Equation (5.118) gives population size as a continuous function of time, subject to a parameter (r) and an initial condition ($P(0)$),

$$P(t) = P(0)e^{rt} \tag{5.124}$$

This expression can be used to plot different specific solutions of the differential equation (Figure 5.9).

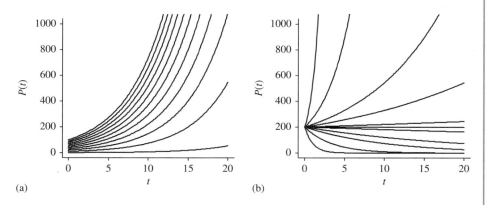

(a)   (b)

Figure 5.9: Specific solutions of the unconstrained growth model using (a) the same intrinsic rate ($r = 0.2$) with different initial conditions and (b) the same initial condition ($P(0) = 200$) with different values of the population's intrinsic growth rate r.

Example 5.26: Constrained growth in continuous time

The logistic model for population growth in continuous time takes the form:

$$\frac{dP}{dt} = r\left(1 - \frac{P}{K}\right)P \tag{5.125}$$

This is a modification of the unconstrained growth model with the introduction of the term $1 - P/K$ which moderates the growth of the population as it moves towards the carrying capacity K (see also Example 3.3). This is an ordinary, first order, nonlinear differential equation. It is also separable,

$$\int \frac{1}{(1 - P/K)P}dP = \int r\,dt \tag{5.126}$$

I first focus on the harder integral on the LHS. The denominator can be seen as the product of two first order polynomials in the variable P (i.e. P and $1 - P/K$). Also, the numerator is of order 0 and the denominator is of order 2. This may therefore be approached by partial fractions (see Section 5.3),

$$\frac{1}{(1 - P/K)P} = \frac{A}{P} + \frac{B}{1 - P/K} \tag{5.127}$$

Rearranging gives

$$1 = P(B - A/K) + A \tag{5.128}$$

which implies $A = 1$ and $B = 1/K$ (see Section 5.3). Therefore, Equation (5.126) can be written

$$\int \left\{\frac{1}{P} + \frac{1}{K(1 - P/K)}\right\} dP = \int r\,dt$$
$$\int \frac{1}{P}dP + \int \frac{1}{K - P}dP = \int r\,dt \tag{5.129}$$

All of these integrals are now doable (the second of the three can be solved with substitution, try $u = K - P$),

$$\ln P - \ln(K - P) = rt + C \tag{5.130}$$

This logarithmic equation may now be solved for P,

$$\ln \frac{P}{K - P} = rt + C$$
$$\frac{P}{K - P} = e^{rt+C} \tag{5.131}$$
$$P = \frac{Ke^{rt+C}}{1 + e^{rt+C}}$$

The notation can be simplified by introducing the parameter $a = e^C$

$$P = \frac{Kae^{rt}}{1 + ae^{rt}} \tag{5.132}$$

At $t = 0$,

$$P(0) = \frac{Ka}{1 + a} \tag{5.133}$$

Solving this for a gives $a = P(0)/(K - P(0))$. Substituting into Equation (5.132) and simplifying gives the general solution of the logistic growth model

$$P(t) = \frac{KP(0)e^{rt}}{K + P(0)(e^{rt} - 1)} \tag{5.134}$$

Equation (5.134) can be plotted for different parameter values and initial conditions (Figure 5.10). If you followed this example, you can start feeling smug because it pulled together almost all the technical strands of this chapter.

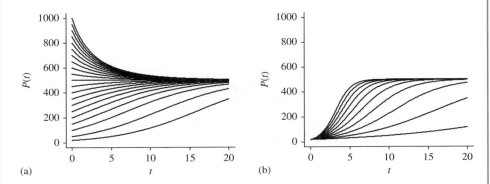

(a)

(b)

Figure 5.10: Specific solutions of the logistic growth model using (a) the same intrinsic rate ($r = 0.2$) with different initial conditions and (b) the same initial condition ($P(0) = 20$) with different intrinsic growth rates. Carrying capacity is the same in both cases ($K = 500$).

Examples 5.25 and 5.26 illustrate **closed-form solutions**, i.e. those solutions that can be expressed in terms of elementary operations (such as division) and functions (such as the exponential). Closed-form solutions are great because they enable us to predict the state of the system at any point in the future from a set of parameters and initial conditions. A method that provides closed-form solutions for a particular type of differential equation is called an **exact method**. There is a wealth of exact methods for several types of differential equations but, unfortunately, not for most. Hence, approximate, or numerical, methods (see R5.4, below) are a valuable, pragmatic tool.

So, which framework is more appropriate for modelling natural phenomena such as population growth? When are continuous models more appropriate than discrete? If change occurs in short bursts (e.g. a well-defined breeding season) then a discrete model may be best. Discrete models may even be able to deal with up to one continuous process (e.g. mortality) in conjunction with a discrete process (e.g. Example 3.2). When multiple continuous processes interact, a discrete model may be a poor approximation. Consider, for example, a system in which prey density depends on predation as well as natural mortality, and predator density depends on prey consumption. An annual model that calculates prey mortality at the end of each year, on the basis of predator density at the beginning of the year, runs the risk of modelling the 'walking-dead' i.e. predators that died of natural causes mid-year, yet still managed to consume prey until the end of the year. Perhaps there is an obvious solution to this: instead of using the year as the unit of time, why not use the day? This is a good way for a discrete system to approximate a continuous process and for a computer to numerically find the solutions of a continuous model.

5.4: Numerical solution of differential equations

Numerical solutions are always specific solutions – they can only be generated for particular parameter values and initial conditions. They may also take time to compute (especially for more complicated differential equations) and they suffer from all the evils of numerical methods (such as round-up errors). They therefore don't have the elegance of closed-form solutions but sometimes they are the only solutions we can get. R offers some functionality for ordinary differential equations through the package `odesolve`. The package must first be installed, and then loaded. It incorporates two different algorithms aimed at achieving the same task. They are called by the commands `lsoda` and `rk4`. Both of these have the same syntax, but `lsoda` is more precise, so I illustrate its use here. The general syntax is `lsoda(y, times, func, parms)` where `y` represents initial conditions, `times` is a list of times for which the algorithm is to calculate the state of the system, `func` is the differential equation and `parms` is a list of parameter values. I will apply it to the logistic growth model in Example 5.26. This model's closed-form solution can be used for comparison with the numerical approximation. Before calling the command, its inputs must be prepared. Three out of the four inputs are easy to specify,

```
init<-c(p=10)              # Initial population size
times<-seq(0,20,1/24)      # Hourly time instances from 0 to 20 days
parms<-c(K=500, r=0.3)     # Vector of two parameters
```

Now for the argument `func`: the `odesolve` library requires the model to be packaged in a new function (here called `logist`), as follows:

```
logist<- function(times, init, parms)
  {
  p<-init[1]               # Reads the initial condition
  with(as.list(parms),{
     dp<-r*p*(1-p/K)       # Put your model here
     res<-dp
     list(res)
  })}
```

This format can be used for any differential equation by specifying the model (here, `dp<-r*p*(1-p/K)`) and making sure that all appearances of the state variable (here, p and dp) are consistent with the symbols used by the model. Now, the numerical solution can be evaluated and stored in a data frame,

```
solution<-as.data.frame(lsoda(init, times, logist, parms))
```

Note that `logist` is a function within another function (`lsoda`). They both use the arguments `init`, `times` and `parms`. The data frame `solution` contains a column called `time` and a column with the corresponding population sizes (it inherits its name from the initial conditions vector `inits`, here containing the only variable p). These numbers can be used for plotting

```
plot(solution$time, solution$p, type="l", xlab="time", ylab="P(t)")
```

Compare this with the analytical solution from Equation (5.134), by typing (see Figure 5.11):

```
tmax<-20; r<-0.3; K<-500; p0<-10; t<-seq(0,tmax, by=0.1)
plot(t, K*p0*exp(r*t)/(K+p0*(exp(r*t)-1)), type="l", xlab="t", ylab="P(t)")
```

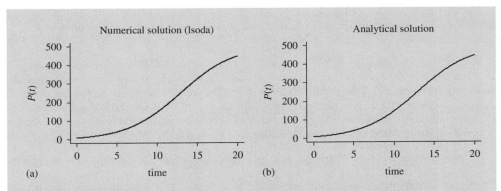

Figure 5.11: Comparison of (a) numerical solutions with (b) analytical solution of the logistic growth model. Parameter values are $r = 0.3, K = 500, P(0) = 10$.

5.11.   Stability analysis for differential equations

Although separable equations comprise only a small part of the group of solvable equations, the bitter truth is that most differential equations have no closed-form solution. Of course, almost every differential equation can be solved numerically but only for specific parameter values and initial conditions. Such specificity is a huge price to pay in science, so an alternative analytical path, the **dynamical systems approach**, has been developed. It has the following rationale: if a system is allowed to exist for a sufficiently long time, it settles into a dynamical pattern that is primarily determined by the interplay between the attractive and repelling forces exerted on specific solutions by the system's equilibria. So, an examination of the position and stability properties of these equilibria may be able to piece together the behaviour of the system even without its general solution.

Example 5.27: Constrained growth in continuous time

The equilibria of the logistic model in continuous time are found by setting

$$\frac{dP}{dt} = r\left(1 - \frac{P^*}{K}\right)P^* = 0 \qquad (5.135)$$

As with discrete systems (Chapter 3), the star notation indicates the equilibria. This is now an algebraic equation that can be solved to find $P^* = 0$ and $P^* = K$.

The question now is how to find out information on the local stability of equilibria without having the full solution for the differential equation. The answer is to use approximation, i.e. replace the model that cannot be solved with a simpler caricature that can. This caricature must be able to mimic the behaviour of the original system sufficiently well around each equilibrium. Previous chapters introduced two methods of approximation. In Chapter 2, complicated periodic functions were decomposed using simpler trigonometric polynomials, and in Chapter 4, Taylor polynomials were used to emulate any function around any one of its points. This latter approach is applicable here. Consider the following general model:

$$\frac{dP}{dt} = f(P) \tag{5.136}$$

This is written in the notation of population models, but it could be any differential equation involving a function $f(y)$ of a state variable y and its derivative dy/dx with respect to an independent variable x. Assuming that the equation $f(P^*) = 0$ can be solved to find the system's equilibria, the behaviour of the differential equation around any given equilibrium P^* can be approximated by a Taylor expansion (see Section 4.12)

$$f(P) = f(P^*) + f'(P^*)(P - P^*) + \frac{1}{2}f''(P^*)(P - P^*)^2 + \frac{1}{3!}f'''(P^*)(P - P^*)^3 + \cdots \tag{5.137}$$

Caution: the derivatives here are with respect to P. Since P^* is a constant number, all the parts of this expression that have P^* as their argument (e.g. $f'(P^*)$) are also constant. They can therefore be replaced by the new symbols $a_1 = f'(P^*), a_2 = \frac{1}{2} f''(P^*), a_3 = \frac{1}{3!} f'''(P^*), \ldots$. Furthermore, note that $f(P) = dP/dt$ and $f(P^*) = 0$. These observations give

$$\frac{dP}{dt} = a_1(P - P^*) + a_2(P - P^*)^2 + a_3(P - P^*)^3 + \cdots \tag{5.138}$$

The variable P appears consistently as part of the expression $P - P^*$, so it can be replaced by the new variable $u = P - P^*$, which simplifies the RHS of Equation (5.138) considerably. The replacement initially seems to complicate the LHS. However, notice that

$$\frac{dP}{dt} = \frac{d(u + P^*)}{dt} = \frac{du}{dt} + \frac{dP^*}{dt} = \frac{du}{dt} \tag{5.139}$$

So, we get the nicer-looking expression,

$$\frac{du}{dt} = a_1 u + a_2 u^2 + a_3 u^3 + \cdots \tag{5.140}$$

This is a precise replacement of the differential equation in Equation (5.136) expressed in terms of the displacement $u = P - P^*$ of the population P from its equilibrium P^*. Equation (5.140) is not (yet) an approximation: given infinite terms in this polynomial, the coefficients $a_1, a_2, a_3, \ldots$ contain all the information that existed in the original differential equation. Remember, though, that we want to approximate the behaviour of solutions *close* to the equilibrium. We may therefore examine what happens if u (the distance of the solution from P^*) is very small. So small, in fact, that the terms that involve its powers ($u^2, u^3, \ldots$) are negligible compared to the first term of Equation (5.140). When this happens, we get

$$\frac{du}{dt} \cong a_1 u \tag{5.141}$$

But how can this tell us whether the equilibrium attracts or repels the solutions that happen to be near it? It can, because Equation (5.141) can be solved analytically – it corresponds to the model of exponential growth that was examined in Example 5.25. Its closed-form solution is $u(t) = u(0)e^{a_1 t}$. If $a_1 < 0$ as time passes, the distance u of the solution from the equilibrium diminishes. In contrast, if $a_1 > 0$, then the distance increases indefinitely (remember, this is an approximation of the real system, so 'indefinitely' in this context means 'away from the neighbourhood of P^*'). Since $a_1 = f(P^*)$, the equilibrium will be locally stable if $f'(P^*) < 0$ and unstable if $f'(P^*) > 0$. What happens when $f'(P^*) = 0$? It is hard to tell. In the exponential model, the population stays where it is initiated (this is called **neutral stability**), but here this is just a local approximation of the original system, so the equilibrium may be either stable or unstable.

Example 5.28: Constrained growth in continuous time

Although the logistic growth model has a closed-form solution, it can be used to illustrate local stability analysis. The first derivative of the model with respect to P is

$$\frac{d}{dP}\left\{r\left(1 - \frac{P}{K}\right)P\right\} = r\left(1 - \frac{2P}{K}\right) \tag{5.142}$$

This may be specified to the two equilibria of the system, to get the following two expressions for $f'(P^*)$.

$$f'(0) = r \quad f'(K) = -r \tag{5.143}$$

The general condition for local stability requires that $f'(P^*) < 0$. When the growth rate r is positive, Equations (5.143) give $f'(0) > 0$ and $f'(K) < 0$, so the carrying capacity is a stable equilibrium. In contrast, when the population's growth rate is negative, we get $f'(0) < 0$ and $f'(K) > 0$, indicating (intuitively enough) that the extinction equilibrium attracts solutions.

The conclusions drawn in the previous example were not tied to any particular value of the parameters r, K but were, instead, expressed as general stability conditions. This is the great advantage of the dynamical systems approach in comparison to numerical solution, made even more apparent by the following example.

Example 5.29: The Levins model for metapopulations

The very first metapopulation model was spatially implicit: it carried no information about the position of habitable patches or the connectivity between them. To address the question of occupancy head-on, the model used a rather unusual state variable, the proportion $p(t)$ of patches that are occupied at any given point in time. The rate of change of this proportion was modelled as the difference between the rate of extinctions and the rate of recolonisations.

$$\frac{dp}{dt} = \text{Colonisation rate} - \text{Extinction rate} \tag{5.144}$$

Levins argued that, if each patch became extinct with equal and constant likelihood, then the number of extinctions occurring at any unit of time would be proportional to occupancy (Extinction rate $= ep$, for some proportionality constant e). The colonisation rate should increase with the number of sites that can send out colonists, but it should also decrease as the empty sites fill up and become unavailable for colonisation. The proportion of donor patches is p and the proportion of available patches is $1 - p$. As a first approach to the problem, it may be assumed that these two effects operate linearly and separately from each other (implying the plausible model Colonisation rate $= cp(1 - p)$ for some constant of proportionality c). The Levins metapopulation model takes the form

$$\frac{dp}{dt} = cp(1 - p) - ep \tag{5.145}$$

This model can also be solved analytically, but here I will look at local stability. The system has two equilibria $p^* = 0$ and $p^* = (c - e)/c$. This second equilibrium is only biologically realistic for values $e \leq c$, because otherwise, it has the nonsensical interpretation of negative occupancy. The general stability condition for this system is

$$(c - e) - 2cp^* < 0 \tag{5.146}$$

Specifying this to $p^* = 0$ indicates that this equilibrium will be stable when $c < e$, meaning that the metapopulation will become extinct when the unitary rate of extinctions exceeds the unitary rate of colonisations. Specifying to $p^* = (c - e)/c$, gives the condition $c > e$, implying the metapopulation's continued existence. As an aside, note that even if the population settles to this equilibrium, the Levins model predicts less than 100% occupancy. For example, even if c is ten times as big as e, we expect equilibrium occupancy of 90% (since, $p^* = (c - 0.1c)/c = 0.9$).

Further reading

A good, comprehensive reference for college calculus is Thomas and Finney (2000). Adler (2005) is an excellent general reference with a biological bias. Accessible sections for differential equations can be found in Batschelet (1979), Krantz (2004) and Adler (2005). If your mind is set on finding a closed-form solution to a particular differential equation, then, if an exact method exists, it is likely to be in Zwillinger (1998). That book also has extensive sections on approximate solutions. The dynamical systems approach has flourished in ecology for almost half a century. The following seminal books all cover the same topics with a different emphasis: Edelstein-Keshet (1988), Gurney and Nisbet (1998), Case (2000), Kot (2001) and Otto and Day (2007). A more mathematical, in-depth treatment of the subject can be found in Verhulst (1985). More advanced ecological models examined using R can be found in Stevens (2009).

References

Adler, F. R. (2005) *Modelling the Dynamics of Life: Calculus and Probability for the Life Scientists.* Thompson Brooks/Cole. Belmont, California. 876pp.

Batschelet, E. (1979) *Introduction to Mathematics for Life Scientists.* Springer Verlag, Berlin. 643pp.

Case, T. J. (2000) *An Illustrated Guide to Theoretical Ecology.* Oxford University Press, New York. 449pp.

Edelstein-Keshet, L. (1988) *Mathematical Models in Biology.* Random House, New York. 586pp.

Gurney, W. S. C. and Nisbet, R. M. (1998) *Ecological Dynamics.* Oxford University Press, New York. 335pp.

Kot, M. (2001) *Elements of Mathematical Biology.* Cambridge University Press, Cambridge. 453pp.

Krantz, S. G. (2004) *Differential Equations Demystified.* McGraw-Hill Professional. 323pp.

Otto, S. P. and Day, T. (2007) *A Biologist's Guide to Mathematical Modelling in Ecology and Evolution.* Princeton University Press, New Jersey. 732pp.

Nisbet, R. M. and Gurney, W. S. C. (1982) *Modelling Fluctuating Populations.* The Blackburn Press. 379pp.

Stephenson, C. M., Kohn, D. D., Park, K. J., Atkinson, R., Edwards, C. and Travis, J. M. (2007) Testing mechanistic models of seed dispersal for the invasive *Rhododendron ponticum* (L.). *Perspectives in Plant Ecology, Evolution and Systematics*, **9**, 15–28.

Stevens, M. H. H. (2009) *A Primer of Ecology with R.* Springer. 388pp.

Thomas, G. B. and Finney, R. L. (2000) *Calculus and Analytic Geometry.* Addison Wesley.

Verhulst, F. (1985) *Nonlinear Differential Equations and Dynamical Systems.* Springer-Verlag, Berlin. 277pp.

Zwillinger, D. (1998) *Handbook of Differential Equations.* Academic Press. 801pp.

6

How to keep stuff organised in tables
(Matrices and their applications)

*'And thou shalt bring in the table, and set in order the things that
are to be set in order upon it'*
King James Bible, Exodus 40:4

Maths is complicated because it deals with hard problems. Particularly when modelling the rich detail of biological systems, things can quickly get unmanageable. That's why good bookkeeping and shorthand notation are essential. This chapter outlines mathematics' way of storing and ordering large amounts of data, in tables called **matrices**. Their definition is motivated by the need to keep track of numbers in systems of linear equations (Section 6.1), but matrices are more than storage devices. They can be manipulated algebraically (Section 6.2), interpreted graphically (Section 6.3) and ultimately used to **solve** systems of linear equations (Section 6.4). Matrices also play a role in the analysis of **multidimensional dynamical systems**, i.e. models that simultaneously track the development of several interacting components of a system through time. Many such systems (e.g. biological populations) comprise units (e.g. individuals) that can occupy only one of several different states but can transit from one state to another with passing time. The models of such systems, called **Markov chains** (Section 6.5), seek to identify the long-term proportion of their time that different units spend in each state. Alternatively, they seek to identify how many units of the system are at any given state in any given time. Such **steady-state distributions** achieved after an initial, **transient period**, are a characteristic of all linear dynamical systems. The matrix concepts of **eigenvalues** and **eigenvectors** (Section 6.6) help deduce the long-term behaviour of linear dynamical systems.

How to be a Quantitative Ecologist: The 'A to R' of Green Mathematics and Statistics, First Edition. Jason Matthiopoulos.
© 2011 John Wiley & Sons, Ltd. Published 2011 by John Wiley & Sons, Ltd.

They are applied to the analysis of mathematical formulations of the life histories of organisms called **Leslie matrix models** (Section 6.7). In Section 6.8, the results from sections 6.5–6.7 are consolidated into a general recipe for analysing the dynamical properties of all linear systems. Finally, Section 6.9 shows how the theory of linear systems can be extended to help analyse more realistic, nonlinear models.

6.1. Matrices

Example 6.1: Plant community composition

Couch grass (Triticum repens), a true grass

A grassland community comprises true grasses (*Poaceae*), sedges (*Cyperaceae*) and rushes (*Juncaceae*). For parts of the year, plants of the three families are difficult to tell apart but we would like to estimate their prevalence (g, s, r) in a particular sample. The typical weights of the plants are 10 g, 12 g and 16 g and their caloric content is 17 MJ·kg^{-1}, 17.8 MJ·kg^{-1} and 15.8 MJ·kg^{-1} respectively. An above-ground sample of 1905 plants weighs 20 kg and has total caloric content 338 MJ. The energetic content contributed to the sample by each plant type will be its total weight in kg multiplied by its energetic value per kg. For example, the energetic content for grass will be $17\text{MJ} \cdot \text{kg}^{-1} \times (0.01g)\text{kg} = (0.17g)\text{MJ}$. These three unknown quantities can be calculated from the following system of equations:

$$\begin{aligned} g + s + r &= 1905 & \text{(Sample abundance)} \\ 0.01g + 0.012s + 0.016r &= 20 & \text{(Sample biomass)} \\ 0.17g + 0.2136s + 0.2528r &= 338 & \text{(Energy content)} \end{aligned} \qquad (6.1)$$

Larger communities, comprising more than three families of plants, could be similarly analysed by constructing additional equations, e.g. referring to different aspects of the chemical composition of the sample.

     The systems of equations solved by replacement in Section 1.14 were constrained to only two variables. With three or more variables, solution by replacement becomes unwieldy (if you have some spare time, try solving the system in Equation (6.1) with the methods of Section 1.14). Yet, there are several ecological applications that involve systems of equations in tens or even hundreds of unknowns.

Example 6.2: Inferring diet from fatty acid analysis

Polar bear, Ursus maritimus

Polar bears are generalists feeding on large prey such as ringed seals, bearded seals, harbour seals, harp seals, beluga whales, narwhals, walruses and even other polar bears. Thiemann *et al.* (2008) applied quantitative fatty acid signature analysis to the body composition of polar bears and their prey in order to infer polar bear diet. Although its statistical implementation is challenging, the basic mathematical idea of fatty acid analysis is enticingly simple. Consider a predator that feeds on n different types of prey. The concentrations of different lipids are identified in the bodies of the predator and its prey (it is assumed that body composition does not vary greatly during the period of specimen collection). If the concentration of the ith lipid in the jth prey type is c_{ij}, and the proportion of that amount that survives metabolism to be detected in the predator is ε_i, then the contribution (a_{ij}) of a single item of this prey type to the concentration of the ith lipid in the predator's body will be

$$a_{ij} = \varepsilon_i c_{ij} \tag{6.2}$$

If the predator eats x_j prey of this type, then the amount contributed by this prey type will be $a_{ij}x_j$. The total concentration (C_i) of this lipid in the predator's body will be the sum of the contributions of all n prey types

$$a_{i1}x_1 + a_{i2}x_2 + \cdots + a_{ij}x_j + \cdots + a_{in}x_n = C_i \tag{6.3}$$

Estimating the diet of the predator, in absolute amounts $\{x_1, \ldots, x_n\}$, requires a total of n equations like Equation (6.3) , each referring to a different fatty acid. The result would be a system of n equations in n unknowns.

This example used subscripts to avoid running out of alphabet letters. Such tidy notation helped to highlight the fact that the system of equations consists of:

❶ coefficients a_{ij} (lipid i contributed from each prey of type j);
❷ variables x_j (consumption of jth prey);
❸ constants C_i (lipid i detected in the predator).

Example 6.3

Consider the system

$$\begin{aligned} x_1 + 2x_2 + 3x_3 &= 2 \\ x_1 - 4x_2 - x_3 &= -3 \\ -2x_1 - x_2 + 5x_3 &= 0 \end{aligned} \tag{6.4}$$

The three types of components can be arranged into three tables:

$$\begin{array}{ccc} \text{❶Coefficients} & \text{❷Variables} & \text{❸Constants} \\ \mathbf{A} = \begin{pmatrix} 1 & 2 & 3 \\ 1 & -4 & -1 \\ -2 & -1 & 5 \end{pmatrix} & \mathbf{x} = \begin{pmatrix} x_1 \\ x_2 \\ x_3 \end{pmatrix} & \mathbf{b} = \begin{pmatrix} 2 \\ -3 \\ 0 \end{pmatrix} \end{array}$$

These rectangular arrays, called **matrices**, may contain numbers, variables, functions or any mixture of the three. The **dimensions** ($r \times c$) of a matrix are determined by the number of its rows (r) and columns (c) (the order matters, number of rows is always quoted first). A **vector** is a one-dimensional matrix. A **row vector** is a $1 \times c$ matrix and a **column vector** is an $r \times 1$ matrix. In the terminology of matrices, a **scalar** is a 1×1 matrix (i.e. just one number or symbol).

Conventionally, on the printed page, shorthand symbols for matrices are written in boldface ($\mathbf{A}$, $\mathbf{x}$ and $\mathbf{b}$ in Example 6.3). Handwritten notes or older texts may indicate matrices by underlining ($\underline{A}, \underline{x}, \underline{b}$) and in more advanced texts, matrices may simply be implied by the context. In some applications it may be useful to use lower case letters for vectors and upper case letters for all other matrices.

In total, Equations (6.4) contain 21 things (nine coefficients, three variables occurring three times each and three constants). This number increases quickly with the number of equations: a system of n equations comprises a total of $2n^2 + n$ numbers or symbols. Although the three matrices $\mathbf{A}, \mathbf{x}, \mathbf{b}$ carry all this information in only three symbols, it is not clear how to write a system of equations using matrices. Leading up to this objective, Section 6.2 will present a set of rules for doing maths with matrices.

ℝ 6.1: Matrices

A matrix can be specified by its dimensions and content within the command `matrix()`. Below, a sequence of numbers from 1 to 12 is used to fill a 4×3 and a 3×4 matrix

```
> x<-1:12
> matrix(x, nrow=4, ncol=3)
     [,1] [,2] [,3]
[1,]    1    5    9
[2,]    2    6   10
[3,]    3    7   11
[4,]    4    8   12
> matrix(x, nrow=3, ncol=4)
     [,1] [,2] [,3] [,4]
[1,]    1    4    7   10
[2,]    2    5    8   11
[3,]    3    6    9   12
```

In these examples all the rows of each column are filled before the next column is reached. This can be changed by using the option `byrow`. The following command creates a 3×4 matrix and fills it with the contents of `x`, row by row.

```
> matrix(x, nrow=3, ncol=4, byrow=TRUE)
     [,1] [,2] [,3] [,4]
[1,]    1    2    3    4
[2,]    5    6    7    8
[3,]    9   10   11   12
```

Specific parts of a given matrix `A` can be accessed by the indexing notation introduced in R1.8 for vectors: the name of the matrix comes first and then, enclosed in square brackets, follow the indexes of the element(s) to be extracted. Elements in rectangular matrices are indexed by both row and column. The following extracts the element in the second row and fourth column of the matrix `A`

```
> A<- matrix(x, nrow=3, ncol=4, byrow=TRUE)
> A[2,4]
[1] 8
```

Always remember that the row index comes first. To extract the entire second row, the column index would need to be left unspecified

```
> A[2,]
[1] 5 6 7 8
```

Similarly, leaving the row reference unspecified would return an entire column

```
> A[,4]
[1]  4  8 12
```

6.2.   Matrix operations

So how do the basic operations of algebra (addition, subtraction, multiplication and division) transfer to matrices? Are there any restrictions to their use and are there any new operations that are unique to matrix algebra? **Matrix addition** and **subtraction** are straightforward, assuming that the two matrices have the same dimensions. For example, the sum of two matrices with dimensions 2×3 is

$$\begin{pmatrix} a_{11} & a_{12} & a_{13} \\ a_{21} & a_{22} & a_{23} \end{pmatrix} + \begin{pmatrix} b_{11} & b_{12} & b_{13} \\ b_{21} & b_{22} & b_{23} \end{pmatrix} = \begin{pmatrix} a_{11}+b_{11} & a_{12}+b_{12} & a_{13}+b_{13} \\ a_{21}+b_{21} & a_{22}+b_{22} & a_{23}+b_{23} \end{pmatrix} \tag{6.5}$$

Multiplication by a scalar is done element-by-element

$$c \begin{pmatrix} a_{11} & a_{12} \\ a_{21} & a_{22} \end{pmatrix} = \begin{pmatrix} ca_{11} & ca_{12} \\ ca_{21} & ca_{22} \end{pmatrix} \tag{6.6}$$

Matrix multiplication is defined rather less intuitively, so I will first explain how it is done, and justify it later. The first rule to remember is that two matrices can only be multiplied if the 'inner' dimensions of the product are the same. This means that the multiplication between two matrices with dimensions $(m \times n)$ and $(n \times p)$ is possible but multiplication between matrices with dimensions $(n \times p)$ and $(m \times n)$ is not. The result of matrix multiplication is a matrix that inherits the outer dimensions of the product. So, multiplying two matrices with dimensions $(m \times n)$ and $(n \times p)$ will give an $(m \times p)$ matrix. Before getting to the more general case, consider multiplication of a row and column vector:

$$\mathbf{a} = (a_1 \quad \cdots \quad a_n) \quad \mathbf{b} = \begin{pmatrix} b_1 \\ \vdots \\ b_n \end{pmatrix} \tag{6.7}$$

This multiplication between a $(1 \times n)$ and an $(n \times 1)$ matrix is possible since the inner dimensions are the same. Its result will be a (1×1) matrix, a scalar. The product (known as the **dot** or **scalar product**) is the sum of the two vectors' element-wise products

$$\mathbf{a} \cdot \mathbf{b} = a_1 b_1 + \cdots + a_i b_i + \cdots + a_n b_n = \sum_{i=1}^{n} a_i b_i \tag{6.8}$$

Example 6.4

$$(1 \quad 2 \quad 3) \begin{pmatrix} 1 \\ 2 \\ 3 \end{pmatrix} = 1 \times 1 + 2 \times 2 + 3 \times 3 = 14 \tag{6.9}$$

Multiplication between any $(m \times n), (n \times p)$ matrices can now be defined with the aid of the dot product. The **matrix product AB** between an $m \times n$ matrix **A** and an $n \times p$ matrix **B**, is an $m \times p$ matrix **C**. The element c_{ij} in row i and column j of **C** is the dot product of the ith row vector of **A** and the jth column vector of **B**.

Example 6.5

Here is the result of multiplication between two general 2×2 matrices **A** and **B**.

$$\begin{pmatrix} a_{11} & a_{12} \\ a_{21} & a_{22} \end{pmatrix} \begin{pmatrix} b_{11} & b_{21} \\ b_{12} & b_{22} \end{pmatrix} = \begin{pmatrix} a_{11}b_{11}+a_{12}b_{21} & a_{11}b_{12}+a_{12}b_{22} \\ a_{21}b_{11}+a_{22}b_{21} & a_{21}b_{12}+a_{22}b_{22} \end{pmatrix}$$

$$= \begin{pmatrix} c_{11} & c_{12} \\ c_{21} & c_{22} \end{pmatrix} \tag{6.10}$$

The resulting matrix **C** comprises the dot products of all possible pairwise combinations of rows from **A** with columns from **B**. For example, the boxed components show the dot product between the first row of **A** and the first column of **B**. The entire operation can be written, simply, $\mathbf{AB} = \mathbf{C}$.

Example 6.6

Here is the result of multiplication between two specific 2×2 matrices **A** and **B**.

$$\mathbf{AB} = \begin{pmatrix} 1 & 2 \\ 3 & 4 \end{pmatrix} \begin{pmatrix} 2 & 3 \\ 4 & 5 \end{pmatrix} = \begin{pmatrix} 2+8 & 3+10 \\ 6+16 & 9+20 \end{pmatrix} = \begin{pmatrix} 10 & 13 \\ 22 & 29 \end{pmatrix} \tag{6.11}$$

Since these are square matrices, the product **BA** is also possible, but the result is different

$$\mathbf{BA} = \begin{pmatrix} 2 & 3 \\ 4 & 5 \end{pmatrix} \begin{pmatrix} 1 & 2 \\ 3 & 4 \end{pmatrix} = \begin{pmatrix} 2+9 & 4+12 \\ 4+15 & 8+20 \end{pmatrix} = \begin{pmatrix} 11 & 16 \\ 19 & 28 \end{pmatrix} \tag{6.12}$$

This last example establishes that matrix multiplication is **noncommutative**: even if **AB** and **BA** are defined, they are not necessarily equal.

So, can matrix multiplication be used to write a system of coupled linear equations (Section 5.1) in shorthand notation? If the coefficients are arranged in a matrix **A**, the variables in a column vector **x** and the constants in a column vector **b**, as was done in Example 6.3, then the system can be written:

$$\mathbf{Ax} = \mathbf{b} \tag{6.13}$$

Example 6.7

According to Equation (6.13), the matrix representation of Equations (6.4) is

$$\begin{pmatrix} 1 & 2 & 3 \\ 1 & -4 & -1 \\ -2 & -5 & 5 \end{pmatrix} \begin{pmatrix} x_1 \\ x_2 \\ x_3 \end{pmatrix} = \begin{pmatrix} 2 \\ -3 \\ 0 \end{pmatrix} \qquad x_1 + 2x + 3x_3 = 2 \tag{6.14}$$

To check this, you need to carry out the matrix multiplication on the LHS of this matrix equation (do this for practice). The multiplication is between a 3×3 and a 3×1 matrix, so it is possible and should yield a 3×1 matrix, which is in agreement with the dimensions of the RHS of Equation (6.14).

In elementary algebra, multiplication by 1 leaves the product unchanged. The equivalent in matrix algebra is the **identity matrix**. The identity matrix $\mathbf{I}_n$ is square ($n \times n$) with zero entries everywhere apart from its diagonal.

$$\mathbf{I}_n = \begin{pmatrix} 1 & 0 & \cdots & 0 \\ 0 & 1 & & 0 \\ \vdots & & \ddots & \vdots \\ 0 & 0 & \cdots & 1 \end{pmatrix} \tag{6.15}$$

Often, when the size of the identity matrix is known, the subscript n is dropped.

Example 6.8

Here is an example of I_3 being multiplied by a column vector

$$\begin{pmatrix} 1 & 0 & 0 \\ 0 & 1 & 0 \\ 0 & 0 & 1 \end{pmatrix} \begin{pmatrix} 1 \\ 2 \\ 3 \end{pmatrix} = \begin{pmatrix} 1 \times 1 + 0 \times 2 + 0 \times 3 \\ 0 \times 1 + 1 \times 2 + 0 \times 3 \\ 0 \times 1 + 0 \times 2 + 1 \times 3 \end{pmatrix} = \begin{pmatrix} 1 \\ 2 \\ 3 \end{pmatrix} \tag{6.16}$$

and one where it is being multiplied by a square 3×3 matrix

$$\mathbf{A}\mathbf{I}_3 = \begin{pmatrix} 1 & 2 & 3 \\ 4 & 5 & 6 \\ 7 & 8 & 9 \end{pmatrix} \begin{pmatrix} 1 & 0 & 0 \\ 0 & 1 & 0 \\ 0 & 0 & 1 \end{pmatrix}$$

$$= \begin{pmatrix} 1 \times 1 + 2 \times 0 + 3 \times 0 & 1 \times 0 + 2 \times 1 + 3 \times 0 & 1 \times 0 + 2 \times 0 + 3 \times 1 \\ 4 \times 1 + 5 \times 0 + 6 \times 0 & 4 \times 0 + 5 \times 1 + 6 \times 0 & 4 \times 0 + 5 \times 0 + 6 \times 1 \\ 7 \times 1 + 8 \times 0 + 9 \times 0 & 7 \times 0 + 8 \times 1 + 9 \times 0 & 7 \times 0 + 8 \times 0 + 9 \times 1 \end{pmatrix} \tag{6.17}$$

$$= \begin{pmatrix} 1 & 2 & 3 \\ 4 & 5 & 6 \\ 7 & 8 & 9 \end{pmatrix}$$

In this second example the operation is commutative, i.e. $\mathbf{A}\mathbf{I}_3 = \mathbf{I}_3\mathbf{A} = \mathbf{A}$.

Sometimes, multiplying a matrix requires its dimensions to be swapped around. **Transposition** rotates a matrix so that its ith row becomes its ith column. The matrix resulting from the transposition of $\mathbf{A}$ is called its **transpose**, $\mathbf{A}^T$. The properties of the transpose are listed in Section A9.2 of the appendix.

Example 6.9

$$\begin{pmatrix} 1 & 3 & 5 \\ 2 & 4 & 6 \end{pmatrix}^T = \begin{pmatrix} 1 & 2 \\ 3 & 4 \\ 5 & 6 \end{pmatrix} \tag{6.18}$$

Some applications call for the sum of the diagonal entries of a square matrix $\mathbf{A}$, a quantity known as the **trace of the matrix**, $\mathrm{Tr}(\mathbf{A})$.

Example 6.10: Movement in metapopulations

In a three-patch metapopulation, the column vector $\mathbf{n}$ represents the number of animals currently in each of the patches and the matrix $\mathbf{M}$ gives the net proportion of animals at a given patch making particular transitions to another during a short period of time (Figure 6.1).

So, if there are currently n_1 animals in patch 1, then the number of animals arriving at patch 2 from patch 1 is $m_{21}n_1$. The number of animals in each of the patches at the end of the time period can be calculated as follows:

$$\mathbf{M}\mathbf{n} = \begin{pmatrix} m_{11} & m_{12} & m_{13} \\ m_{21} & m_{22} & m_{23} \\ m_{31} & m_{32} & m_{33} \end{pmatrix} \begin{pmatrix} n_1 \\ n_2 \\ n_3 \end{pmatrix} = \begin{pmatrix} m_{11}n_1 + m_{12}n_2 + m_{13}n_3 \\ m_{21}n_1 + m_{22}n_2 + m_{23}n_3 \\ m_{31}n_1 + m_{32}n_2 + m_{33}n_3 \end{pmatrix} \tag{6.19}$$

M is called a **transition matrix**. Alternatively, multiplication by a 3×3 matrix **N** containing the patch populations along its diagonal yields a table of the number of animals making different transitions:

$$\mathbf{MN} = \begin{pmatrix} m_{11} & m_{12} & m_{13} \\ m_{21} & m_{22} & m_{23} \\ m_{31} & m_{32} & m_{33} \end{pmatrix} \begin{pmatrix} n_1 & 0 & 0 \\ 0 & n_2 & 0 \\ 0 & 0 & n_3 \end{pmatrix} = \begin{pmatrix} m_{11}n_1 & m_{12}n_2 & m_{13}n_3 \\ m_{21}n_1 & m_{22}n_2 & m_{23}n_3 \\ m_{31}n_1 & m_{32}n_2 & m_{33}n_3 \end{pmatrix} \qquad (6.20)$$

The total size of the metapopulation is $\mathrm{Tr}(\mathbf{N})$. The number of animals that did not move during the time period is $\mathrm{Tr}(\mathbf{MN})$.

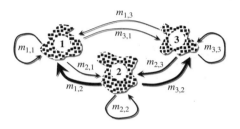

Figure 6.1: A three-patch metapopulation. The direction and thickness of the arrows represent the frequency of movement of migrants between different pairs of patches. In this example, patch 2 acts as a source population and patches 1 and 3 are sinks.

6.2: Matrix operations

Consider the following two matrices:

```
A<-matrix(1:4, 2,2)
B<-matrix(5:8, 2,2)
```

Scalar operations
Multiplication by a scalar (say, 2) is done simply by typing `2*A`. Other element-wise operations, not usually encountered in written mathematics, are also recognised by R. For example, to increment all entries of A by 1, type `1+A`. Division can be handled as well, try `A/2`.

Matrix addition
The matrices A and B have the same dimensions so they may be added,

```
> A+B
     [,1] [,2]
[1,]    6   10
[2,]    8   12
```

Matrix multiplication
This is done with the special operator `%*%`

```
> A%*%B
     [,1] [,2]
[1,]   23   31
[2,]   34   46
```

The identity matrix $\mathbf{I}_n$ for some value of n is created with the command `diag(n)`. To create an $n \times n$ matrix with zero off-diagonal entries and diagonal entries given by a vector v, use the command `diag(v,n,n)`. For example,

```
> diag(1:5, 5,5)
     [,1] [,2] [,3] [,4] [,5]
[1,]    1    0    0    0    0
[2,]    0    2    0    0    0
[3,]    0    0    3    0    0
[4,]    0    0    0    4    0
[5,]    0    0    0    0    5
```

Other operations
The transpose of a matrix A is given by `t(A)` and its trace by `sum(diag(A))`.

6.3. Geometric interpretation of vectors and square matrices

The discussion of coordinate systems in Section 1.16 covered one-dimensional and two-dimensional spaces but higher dimensional spaces are also possible. Certainly, organisms exist in three dimensions and, as we will see in Chapter 12, sample units can be clustered in multidimensional spaces according to several of their characteristics.

A single point in n-dimensional space can be located with the aid of n coordinates. These coordinates can be stored in a vector. Then, the vector can be visualised as a directional line segment, an arrow, joining the axes' origin to the specific point. Why is this a good idea? Because vector algebra can then benefit from the results of geometry and trigonometry. For example, a vector **v** can be summarised by the length of the line segment that it represents. This length is called the vector's **magnitude or norm**, it is written $\|\mathbf{v}\|$ and it can be calculated using the generalised form of Pythagoras's theorem from Section 2.4. Also, the sum of two vectors can be visualised by arranging the arrows one after the other.

Example 6.11: Random walks as sequences of vectors

In a two-dimensional random walk, each step taken by the animal can be thought of as a sequence of increments along the horizontal and vertical directions. The first, second and third steps can be represented by the row vectors

$$\mathbf{x}_1 = (x_1 \quad y_1), \mathbf{x}_2 = (x_2 \quad y_2), \mathbf{x}_3 = (x_3 \quad y_3) \tag{6.21}$$

They can be visualised as three arrows (Figure 6.2(a)) starting at the axes' origin (assuming that the origin represents the animal's position before each step).

After the first two steps, the animal's horizontal position will be $x_1 + x_2$ and its vertical position $y_1 + y_2$ (Figure 6.2(b)). Thus, the final position after two steps is described by the vector $\mathbf{x}_1 + \mathbf{x}_2$. Graphically, this vector joins the origin of the random walk to the tip of the second arrow in Figure 6.2(b). This observation highlights an interesting property of random walks: since addition of vectors is commutative, the outcome of the walk does not depend on the order in which the steps are taken (Figure 6.2(c)).

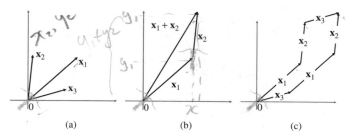

Figure 6.2: (a) Three steps taken by the random walker visualised as vectors from the walker's position before taking the step; (b) graphical interpretation of the sum of two vectors – the origin is the starting point of the random walk; (c) the sums, $x_1 + x_2 + x_3$ and $x_3 + x_1 + x_2$ represent different paths but identical endpoints, because vector addition is commutative.

The cross-fertilisation between geometry and vector algebra can provide insights for both areas. For example, a square $n \times n$ matrix can be thought of as a shape in n-dimensional space.

Example 6.12

Each row in a 2×2 matrix can be interpreted as a vector in 2D space (Figure 6.3(a)),

$$X = \begin{pmatrix} x_1 & y_1 \\ x_2 & y_2 \end{pmatrix} = \begin{pmatrix} x_1 \\ x_2 \end{pmatrix} \tag{6.22}$$

The two vectors define the two sides of a parallelogram (Figure 6.3(b)).

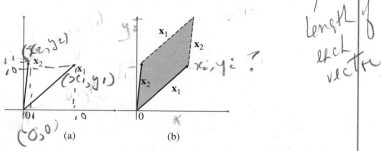

Figure 6.3: (a) The rows of a square 2×2 matrix define two vectors in 2D space; (b) this parallelogram is the graphical interpretation of the matrix **X**.

Just as a 2×2 matrix defines a planar shape (a parallelogram), a 3×3 matrix defines a solid (a parallelepiped). Shapes in higher dimensions are also possible. Geometrical characteristics of these shapes, such as their perimeter, area or volume, can be used to summarise the entire matrix. So, the four numbers in the matrix in Equation (6.22) may be condensed to one value, the area of the corresponding parallelogram. Although this reduction represents loss of information (several parallelograms may have the same area, so this summary does not uniquely specify the matrix), it nevertheless carries vital information about the matrix.

I will illustrate this for a 2×2 matrix (Figure 6.4) but the approach generalises to all square matrices. Let h be the height of the parallelogram from the base x_1. The angles formed between the vectors and the x-axis are denoted ϕ_1 and ϕ_2. The angle formed between the two vectors is $\theta = \phi_2 - \phi_1$. The lengths of the two vectors are written, $\|x_1\|, \|x_2\|$. The area of the shaded region in Figure 6.4(a) is $\|x_1\|h$ (see Section A2.2). The height h can be written as $\|x_2\|\|\sin\theta\| = \|x_2\|\|\sin(\phi_2 - \phi_1)\|$ (I have used the absolute value of the sine to avoid getting a negative height).

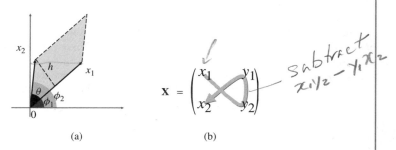

(a)   (b)

Figure 6.4: (a) Notation used to calculate the area of the parallelogram defined by a 2×2 matrix; (b) a mnemonic for calculating the determinant of a 2×2 matrix.

The sine of a difference can be expanded using the trigonometric identity in Section A3.7. Combining all the above facts gives

$$\text{Area} = \|x_1\|\|x_2\|\|\sin\phi_2\cos\phi_1 - \cos\phi_2\sin\phi_1\| \tag{6.23}$$

The definitions of the sines and cosines of the individual angles are

$$\sin\phi_1 = \frac{y_1}{\|x_1\|}, \cos\phi_1 = \frac{x_1}{\|x_1\|}, \sin\phi_2 = \frac{y_2}{\|x_2\|}, \cos\phi_2 = \frac{x_2}{\|x_2\|} \tag{6.24}$$

Putting the expressions in Equations (6.24) into Equation (6.23) gives us a much simpler expression

$$\text{Area} = |x_1y_2 - y_1x_2| \tag{6.25}$$

The expression inside the absolute value is known as the **determinant** of the matrix, written $\det(X)$. Use Figure 6.4(b) to remember the order in which each of the matrix's elements appear in the determinant.

R 6.3: Calculating determinants

It is possible to calculate the determinant of any square matrix using the command `det()`. For example, the determinant of the matrix **A** in Example 6.3 is

```
> A<-matrix(c(1,1,-2,2,-4,-1,3,-1,5), 3, 3)
> det(A)
[1] -54
```

6.4. Solving systems of equations with matrices

Writing a system of linear equations as $\mathbf{Ax} = \mathbf{b}$ (Section 6.2) leads to more compact notation but does not solve the system. In elementary algebra (Section 1.11), the way to solve a linear equation in one unknown, say $ax = b$, is to multiply both sides by a^{-1} to get $x = a^{-1}b$. In the preceding section I carefully avoided discussing the matrix analogue of division, so it is not clear how to do an operation of the type

$$\mathbf{x} = \mathbf{A}^{-1}\mathbf{b} \tag{6.26}$$

Equation (6.26) requires the, as-yet-undefined, matrix $\mathbf{A}^{-1}$, called the **inverse** of $\mathbf{A}$. In perfect analogy with scalars, where $aa^{-1} = 1$, the inverse of a matrix has the property

$$\mathbf{A}\mathbf{A}^{-1} = \mathbf{I} = \mathbf{A}^{-1}\mathbf{A} \qquad (6.27)$$

Given a particular matrix $\mathbf{A}$, two questions arise: does its inverse exist, and how can it be calculated? An inverse matrix can only exist if the original matrix is square (equal number of rows and columns), but not all square matrices have inverses. Those that do are called **nonsingular**, and those that don't are called **singular**. Fortunately, many computer packages, including R (see R6.4), have functions that will tell you if a matrix is nonsingular and calculate its inverse if it is. For the simple case of a 2×2 matrix, the general expression for the inverse is

$$\begin{pmatrix} a & b \\ c & d \end{pmatrix}^{-1} = \frac{1}{(ad - bc)} \begin{pmatrix} d & -b \\ -c & a \end{pmatrix} \qquad (6.28)$$

Do you recognise the expression $ad - bc$ in the denominator as the determinant of the matrix (Section 6.3)? When the values of a, b, c, d are such that $ad - bc = 0$, then the inverse is not defined, due to division by zero. This useful diagnostic generalises to larger matrices: *a square matrix is singular when its determinant is zero*. Consequently, systems of the form $\mathbf{Ax} = \mathbf{b}$, in which $\mathbf{A}$ is nonsingular, have a unique solution (a numerical value for each of the unknowns in the system).

Assuming that $\mathbf{A}$ is invertible (i.e. nonsingular) and that its inverse has been calculated, the next thing to check is whether all the constants in the system are zero ($\mathbf{b} = \mathbf{0}$). Such a system (i.e. $\mathbf{Ax} = \mathbf{0}$) is called **homogeneous**. If a homogeneous system has $\det(\mathbf{A}) \neq 0$ then its only possible solution is $\mathbf{x} = \mathbf{0}$ (all the unknowns are zero). If the vector $\mathbf{b}$ has nonzero elements, then the solution of the system can be calculated as $\mathbf{x} = \mathbf{A}^{-1}\mathbf{b}$.

Example 6.13: Plant community composition

In Example 6.1, g, s, r represented the unknown abundances of grass, sedge and rush in a grassland community. These quantities participated in the linear system of Equations (6.1), which can now be written in matrix form:

$$\begin{pmatrix} 1 & 1 & 1 \\ 0.01 & 0.012 & 0.016 \\ 0.17 & 0.2136 & 0.2528 \end{pmatrix} \begin{pmatrix} g \\ s \\ r \end{pmatrix} = \begin{pmatrix} 1905 \\ 20 \\ 338 \end{pmatrix} \qquad (6.29)$$

The determinant of the matrix of coefficients is -0.000096 (so, not quite zero) and the vector of constants is nonzero. The inverse of the matrix can be found by computer (see R6.4, below) and the solution can be written

$$\begin{pmatrix} g \\ s \\ r \end{pmatrix} = \begin{pmatrix} 4 & 408.33 & -41.67 \\ -2 & -862.50 & 62.5 \\ -1 & 454.17 & -20.83 \end{pmatrix} \begin{pmatrix} 1905 \\ 20 \\ 338 \end{pmatrix} \qquad (6.30)$$

Carrying out this multiplication gives the answers $g \cong 1703, s \cong 65, r \cong 137$.

To understand why a system may have a singular matrix of coefficients, the geometric interpretation of a zero determinant is helpful: remember, the absolute value of the determinant of a 2×2 matrix is the area of the parallelogram defined by the row vectors of that matrix (Section 6.3). The area of a parallelogram is zero only when it is collapsed, i.e. when its sides coincide (Figure 6.5). So, if the segments defined by the row vectors of the square matrix are aligned, the matrix is singular and the corresponding system of equations does not have a unique solution.

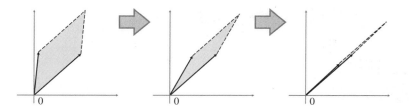

Figure 6.5: As the sides of the parallelogram tend to coincide, its area tends to zero. Therefore, if the row vectors of a square matrix are multiples of each other, the determinant of the matrix tends to zero.

A singular matrix may indicate one of two things: either the system of equations is **inconsistent** (it has no solutions) or it is **underspecified** (it has infinite solutions). To find out which case applies to a particular system, we have to attempt to solve the system by a traditional method such as back-substitution (more efficient methods, such as **row-reduction**, exist for larger systems).

Example 6.14

Consider the system

$$x + 2y = 2$$
$$2x + 4y = 1 \tag{6.31}$$

$2x + 4\left(\dfrac{2-x}{2}\right) = 1$

$4x + 8 - 4x = 2$

This has $\det(\mathbf{A}) = 1 \times 4 - 2 \times 2 = 0$. So, solution by inversion is not possible, but back-substitution may be attempted

$$\left. \begin{matrix} x + 2y = 2 \\ 2x + 4y = 1 \end{matrix} \right\} \quad \left. \begin{matrix} y = (2-x)/2 \\ 2x + 4 - 2x = 1 \end{matrix} \right\} \quad \left. \begin{matrix} y = (2-x)/2 \\ 4 = 1 \end{matrix} \right\} \tag{6.32}$$

We have hit upon an impossibility: 4 is certainly not equal to 1. So, this system is inconsistent. Now, a similar system with different values on the RHS:

$$x + 2y = 1$$
$$2x + 4y = 2 \tag{6.33}$$

Once again, $\det(\mathbf{A}) = 0$. Using back-substitution, $y = \dfrac{1-x}{2} = \dfrac{1}{2}(1-x)$

$$\left. \begin{matrix} x + 2y = 1 \\ 2x + 4y = 2 \end{matrix} \right\} \quad \left. \begin{matrix} y = (1-x)/2 \\ 2x + 2 - 2x = 2 \end{matrix} \right\} \quad \left. \begin{matrix} y = (1-x)/2 \\ 2 = 2 \end{matrix} \right\} \tag{6.34}$$

The equation $2 = 2$ is valid, but it doesn't help in finding a numerical value for the unknown quantity x. So, this system will be satisfied by any one of the infinitely many pairs of values $(x, \frac{1}{2}(1 - x))$ for x and y. This system is underspecified because it just doesn't contain enough information to evaluate both x and y. A careful look at Equation (6.33) reveals why: the second equation is the same as the first, multiplied through by 2.

ⓡ 6.4: Solving a system of linear equations

Typing `det(A)` will test whether a matrix A is invertible. If the determinant is not zero, then the inverse can be found from the command `solve(A)`. The solution of the system can then be calculated as `solve(A) %*% b`. I illustrate the whole process with the grassland community system (Example 6.1):

```
> A<-matrix(c(1,0.01,0.17,1,0.012,0.2136,1,0.016,0.2528),3,3)
> b<-matrix(c(1905,20,338), 3,1)
```

```
> det(A)          #Check that determinant is not zero
[1] -9.6e-05
> solve(A)%*%b    #Find solution
          [,1]
[1,] 1703.3333
[2,]   65.0000
[3,]  136.6667
```

Simply typing `solve(A,b)` will carry out the entire process (and produce an error report if the matrix A is singular).

```
> solve(A,b)
          [,1]
[1,] 1703.3333
[2,]   65.0000
[3,]  136.6667
```

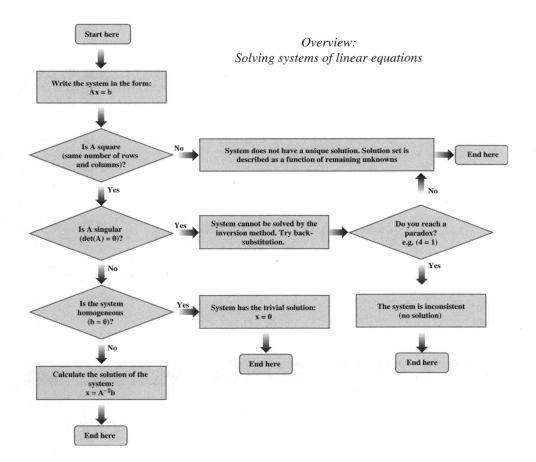

Overview:
Solving systems of linear equations

6.5. Markov chains

Example 6.15: Redistribution between population patches

Consider an isolated three-patch population in which inter-patch movement happens more often than births and deaths, so that the total population remains constant for several time units while local populations change repeatedly through movement. Notation can be expanded to accommodate dynamics, by using $n_{i,t}$ to denote the population size of the ith patch at time t. The column vector $\mathbf{n}_t$ represents the numbers of animals currently in all three of the patches. Further, assume that the transition matrix $\mathbf{M}$ (see Example 6.10) is independent of time and local density. The model for population redistribution can be written in matrix notation:

$$\mathbf{n}_{t+1} = \mathbf{M}\mathbf{n}_t \tag{6.35}$$

This expression may be applied twice to predict the distribution of animals after two time units (using the shorthand notation $\mathbf{M}^2 = \mathbf{MM}$),

$$\mathbf{n}_{t+2} = \mathbf{M}\mathbf{n}_{t+1} = \mathbf{M}(\mathbf{M}\mathbf{n}_t) = (\mathbf{MM})\mathbf{n}_t = \mathbf{M}^2\mathbf{n}_t \tag{6.36}$$

More generally, if the population starts with a distribution $\mathbf{n}_0$, after t time units its distribution will be

$$\mathbf{n}_t = \mathbf{M}^t\mathbf{n}_0 \tag{6.37}$$

This process is an example of a three-state **Markov chain**. In all Markov chains, the transitions are described by an $n \times n$ matrix, with non-negative entries, each of whose columns adds up to 1. In Example 6.15, this unit-sum property of the columns of the transition matrix ensures that total population size is conserved despite the iterative redistribution of animals. This is seen more clearly in the following numerical example.

Example 6.16: Redistribution between population patches

The following transition matrix describes a population that starts with two empty patches. *transition matrix ($\neq 0$)*

$$\mathbf{M} = \begin{pmatrix} 0.8 & 0.3 & 0.1 \\ 0.1 & 0.4 & 0.2 \\ 0.1 & 0.3 & 0.7 \end{pmatrix}, \mathbf{n}_0 = \begin{pmatrix} 100 \\ 0 \\ 0 \end{pmatrix} \tag{6.38}$$

distribution

After the first move, patches 2 and 3 are no longer empty,

$$\mathbf{n}_1 = \mathbf{M}\mathbf{n}_0 = \begin{pmatrix} 0.8 & 0.3 & 0.1 \\ 0.1 & 0.4 & 0.2 \\ 0.1 & 0.3 & 0.7 \end{pmatrix} \begin{pmatrix} 100 \\ 0 \\ 0 \end{pmatrix} = \begin{pmatrix} 80 \\ 10 \\ 10 \end{pmatrix} \tag{6.39}$$

---

but the total population is still 100 animals. The population remains constant even if Equation (6.37) is used to calculate the population distribution after 100 iterations (I did this

$$\left(\begin{bmatrix} 0.8 & 0.3 & 0.1 \\ & & \\ & & \end{bmatrix} - \begin{bmatrix} 1 & 0 & 0 \\ 0 & 1 & 0 \\ 0 & 0 & 1 \end{bmatrix} \right)$$

$-0.2 \quad 0.3 \quad 0.1$

by computer – see R6.5, below)

$$\mathbf{n}_2 = \begin{pmatrix} 68 \\ 14 \\ 18 \end{pmatrix}, \mathbf{n}_3 = \begin{pmatrix} 60.4 \\ 16.0 \\ 23.6 \end{pmatrix}, \mathbf{n}_4 = \begin{pmatrix} 48.8 \\ 18.7 \\ 32.6 \end{pmatrix}, \dots, \mathbf{n}_{10} = \begin{pmatrix} 46.2 \\ 19.2 \\ 34.62 \end{pmatrix}, \dots, \mathbf{n}_{100} = \begin{pmatrix} 46.2 \\ 19.2 \\ 34.62 \end{pmatrix} \quad (6.40)$$

Have a look at the last two terms of Equation (6.40). After ten iterations, the numbers of animals in each patch have settled to constant values. Together, these numbers are known as a **steady-state distribution**. The population has reached a **dynamic equilibrium** (i.e. even though animals are still moving between patches, the local population sizes remain the same). This is an interesting, long-term property of *some* Markov chains. Those that have it are called **regular** and their transition matrix **M** has the property that it loses any zero entries it may originally have had once multiplied by itself a few times. So, to find whether a particular Markov chain has a steady state, first have a look at its transition matrix. If it contains no zero entries, then a steady state is guaranteed. If there are some zero entries, calculate $\mathbf{M}^k$ for increasing values of k. If, after a few iterations, the zeroes disappear, then, once again, a steady state is guaranteed.

Given a regular Markov chain, is it possible to find its steady state directly, without having to calculate successive powers of the transition matrix? Or, to pose the question mathematically, is there a vector **n** that satisfies the matrix equation

$$\mathbf{Mn} = \mathbf{n} \quad (6.41)$$

To solve this equation for the vector **n**, it must first be rewritten as

$$\mathbf{Mn} - \mathbf{n} = \mathbf{0} \quad (6.42)$$

The zero in boldface on the RHS is a column vector of zeroes. I will now factor out the vector **n** using property 7 in Section A9.1. Had this been an algebraic equation, it could be factored this way: $mn - n = (m - 1)n$. Something similar can be done with Equation (6.42), using an identity matrix with the same dimensions as **M**

$$(\mathbf{M} - \mathbf{I})\mathbf{n} = \mathbf{0} \quad (6.43)$$

(but, always remember that in matrix multiplication the order matters, so this couldn't be written n(**M** − **I**)). In the terminology of Section 6.4, Equation (6.43) is a homogeneous system of equations (**M** − **I** is a square matrix, **n** is a vector of unknowns and all the constants on the RHS are zero). Therefore, one of two things is happening: either the matrix **M** − **I** is invertible and the system has the unique solution **n** = **0**, or the matrix is not invertible and therefore the system is either inconsistent or underspecified. An obligatory solution of **n** = **0** would lead to a contradiction. To illustrate this biologically, consider the scenario in Examples 6.15 and 6.16, in which no change occurred in total population size. A population distribution of **n** = **0** would imply that all the animals had somehow vanished.

Since an invertible matrix is now excluded, the only alternative is that **M** − **I** is singular. This is easy to prove for the transition matrix of a two-state Markov chain

$$\mathbf{M} = \begin{pmatrix} x & y \\ 1-x & 1-y \end{pmatrix} \quad (6.44)$$

Here, the numbers x, y are proportions (positive and less than one) and the second row is completed so that the columns add up to 1, as required for Markov chains. The determinant of **M** − **I** is

$$\begin{aligned}
\det(\mathbf{M} - \mathbf{I}) &= \det\left(\begin{pmatrix} x & y \\ 1-x & 1-y \end{pmatrix} - \begin{pmatrix} 1 & 0 \\ 0 & 1 \end{pmatrix} \right) = \det\begin{pmatrix} x-1 & y-1 \\ 1-x & 1-y \end{pmatrix} \\
&= (x-1)(1-y) - (1-x)(y-1) \\
&= -(x-1)(y-1) + (x-1)(y-1) = 0
\end{aligned} \quad (6.45)$$

Example 6.17: Redistribution between population patches

Since the matrix $\mathbf{M} - \mathbf{I}$ is not invertible, a method other than matrix inversion must be used to find the steady state distribution of the system

$$(\mathbf{M} - \mathbf{I})\mathbf{n} = \mathbf{0} \tag{6.46}$$

Using the values from Example 6.16, this expands to

$$\begin{pmatrix} -0.2 & 0.3 & 0.1 \\ 0.1 & -0.6 & 0.2 \\ 0.1 & 0.3 & -0.3 \end{pmatrix} \begin{pmatrix} n_1 \\ n_2 \\ n_2 \end{pmatrix} = \begin{pmatrix} 0 \\ 0 \\ 0 \end{pmatrix} \tag{6.47}$$

which corresponds to the three equations

$$\begin{aligned} -0.2n_1 + 0.3n_2 + 0.1n_3 &= 0 \\ 0.1n_1 - 0.6n_2 + 0.2n_3 &= 0 \\ 0.1n_1 + 0.3n_2 - 0.3n_3 &= 0 \end{aligned} \tag{6.48}$$

Trying to solve this, say by back-substitution, causes one of the equations to drop out (it transforms into the identity $0 = 0$) and leaves a system of two equations and three unknowns. Therefore, two of the unknowns will have to be specified in terms of the third. For example, solving for n_2 and n_3 gives $n_2 = 0.417n_1$ and $n_3 = 0.75n_1$. In vector notation,

$$\mathbf{n} = \begin{pmatrix} n_1 \\ n_2 \\ n_3 \end{pmatrix} = \begin{pmatrix} n_1 \\ 0.417n_1 \\ 0.75n_1 \end{pmatrix} = n_1 \begin{pmatrix} 1 \\ 0.417 \\ 0.75 \end{pmatrix} \tag{6.49}$$

This gives the steady-state sizes of populations 2 and 3 relative to the unknown size of population 1. However, the Markov chain keeps the size of the metapopulation constant to 100 animals, therefore

$$\sum_{i=1}^{3} n_i = 100 \tag{6.50}$$

Using the information from Equation (6.49) to expand Equation (6.50),

$$n_1 + 0.417n_1 + 0.75n_1 = 100 \tag{6.51}$$

leads to the answer $n_1 = 46.15$. Putting this back into Equation (6.49) gives the population's steady-state distribution.

$$\mathbf{n} = \begin{pmatrix} 46.15 \\ 19.24 \\ 34.61 \end{pmatrix} \tag{6.52}$$

which, allowing for round-off errors, is the same as the steady-state distribution estimated numerically in Equation (6.40) .

It is worth taking a step back to consider what the last three examples (6.15, 6.16 and 6.17) have achieved. Starting from a table of connectivities between patches, they produced a prediction of the population's long-term distribution. This approach is applicable to any network, irrespective of the number of patches.

6.5: Matrix powers

For a given matrix M, the expression M^k will not calculate its *k*th power, but, rather, a matrix with the *k* th powers of the entries of M. To calculate repeated matrix multiplications, a loop is needed. I demonstrate, using the matrices in Example 6.16:

```
> M<-matrix(c(0.8,0.1,0.1,0.3,0.4,0.3,0.1,0.2,0.7), 3,3)
> Mk<-M       #This matrix will eventually hold the answer
> k<-5        #The exponent to be used
> for (i in 1:(k-1)) #For kth power, loop must only run k-1 times
+   {
+   Mk<-Mk%*%M
+   }
> Mk
         [,1]     [,2]     [,3]
[1,]  0.52268  0.44679  0.38821
[2,]  0.17884  0.19615  0.20813
[3,]  0.29848  0.35706  0.40366
```

The distribution of animals across the three patches after *k* time units can now be calculated as follows:

```
> n<-matrix(c(100,0,0),3,1)
> Mk%*%n
        [,1]
[1,]  52.268
[2,]  17.884
[3,]  29.848
```

6.6. *Eigenvalues and eigenvectors*

The idea of a steady-state distribution is easier to understand in Markov chains because, after several iterations of the chain, the *absolute* prevalence of each state remains exactly the same (see Example 6.16). However, it is possible for a dynamic process to reach a steady state in which only *relative* prevalence remains constant.

Example 6.18: Growth in patchy populations

Examples 6.15, 6.16 and 6.17 modelled inter-patch movements but assumed that total population size remained fixed at 100 individuals. What would happen if this assumption was relaxed? Consider a two-patch metapopulation which begins with ten animals in patch 1. The annual per capita growth rate (survivors plus new recruits) is 1.1. Animals are allowed to move between the two patches and the population is modelled as follows

$$\mathbf{n}_{t+1} = \mathbf{M}\mathbf{n}_t \tag{6.53}$$

where

$$\mathbf{M} = \underbrace{1.1}_{Growth} \underbrace{\begin{pmatrix} 0.7 & 0.1 \\ 0.3 & 0.9 \end{pmatrix}}_{Movement} = \tag{6.54}$$

Here, the growth term means that the columns of the matrix **M** no longer sum to 1 and, hence, this is no longer a Markov chain. After an initial redistribution phase, both subpopulations increase (Figure 6.6(a)). Despite having identical intrinsic growth rates, the two populations have different relative sizes because patch 2 operates as a sink and patch 1 as a source. Total population size also increases exponentially. However, the relative size of the two populations stabilises to a fixed value within the first ten years (25% of the total population is in patch 1 – Figure 6.6(b)). So, in this case, a steady-state distribution can still be defined, in proportional terms, by the vector (0.25,0.75).

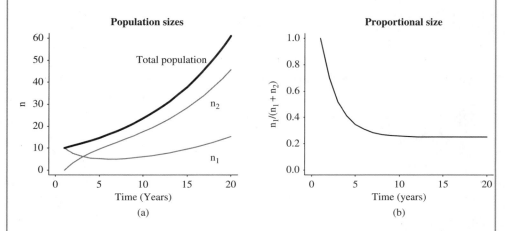

Figure 6.6: (a) In patchy populations with births and deaths as well as movement, relative stability in the size of subpopulations may be achieved without absolute stability in total population size; (b) for example, the size of population 1 relative to population 2 remains constant after approximately ten years, although the total population increases exponentially.

In many situations, given a system of the form,

$$n_{t+1} = Mn_t \tag{6.55}$$

the objective is to find a relative steady-state distribution. To calculate it, the question must first be posed mathematically: if the *relative* sizes of n_t and n_{t+1} remain the same, then they may be related by a simple scaling operation   *scalar = eigenvalue*

$$n_{t+1} = \lambda n_t \tag{6.56}$$

From Equations (6.55) and (6.56),

$$Mn_t = \lambda n_t \tag{6.57}$$

which refers only to the present time. In fact, it refers to any point in time after the establishment of a steady-state distribution, so the subscripts can be dropped

$$Mn = \lambda n \tag{6.58}$$

The only known component in Equation (6.58) is the transition matrix **M**. The unknown scalar λ is called an **eigenvalue** and the unknown vector **n** is called an **eigenvector**. In brief, therefore, given an $m \times m$ matrix **M**, the corresponding eigenvector **n** is a nonzero column vector such that $Mn = \lambda n$, where λ is a scalar known as an eigenvalue.

To determine λ and **n**, Equation (6.58) must be rearranged as follows

$$\begin{aligned} Mn - \lambda n &= 0 \\ Mn - I\lambda n &= 0 \\ (M - I\lambda)n &= 0 \end{aligned} \tag{6.59}$$

This calculation follows the logic of Equations (6.41)–(6.43) in the previous section. Its interpretation is also the same: this is a homogeneous system of linear equations, whose matrix $(\mathbf{M} - \lambda\mathbf{I})$ depends on an unknown parameter λ. If the matrix is nonsingular, there is only one solution $\mathbf{n} = \mathbf{0}$. Depending on the numbers in $\mathbf{M}$, the zero solution is either nonsensical or noninformative. Nonzero solutions for $\mathbf{n}$ can only exist if $\mathbf{M}$ is singular, i.e.

$$\det(\mathbf{M} - \lambda\mathbf{I}) = 0 \tag{6.60}$$

Equation (6.60) is called the **characteristic equation** of M. Solving this algebraic equation will yield the eigenvalues of M. These can then be replaced into Equation (6.59) to calculate the corresponding eigenvectors. First, a simple example.

Example 6.19: Metapopulation growth

The transition matrix of the two-patch metapopulation in Equation (6.54) is

$$\mathbf{M} = \begin{pmatrix} 0.77 & 0.11 \\ 0.33 & 0.99 \end{pmatrix} \tag{6.61}$$

Giving

$$\mathbf{M} - \lambda\mathbf{I} = \begin{pmatrix} 0.77 - \lambda & 0.11 \\ 0.33 & 0.99 - \lambda \end{pmatrix} \tag{6.62}$$

and

$$\begin{aligned} \det(\mathbf{M} - \lambda\mathbf{I}) &= (0.77 - \lambda)(0.99 - \lambda) - 0.11 \times 0.33 \\ &= \lambda^2 - 1.76\lambda + 0.726 \end{aligned} \tag{6.63}$$

So, the 2×2 matrix led to a second order characteristic equation in the eigenvalue λ. The roots of a polynomial (including complex roots – see Chapter 1) can be as many as the polynomial's order, 2 in this case. This observation generalises: an $n \times n$ matrix can have as many as n eigenvalues. Solving Equation (6.63) yields the answers $\lambda_1 = 1.1$ and $\lambda_2 = 0.66$.

Each of these eigenvalues will give a different eigenvector which can be found by solving the equation $(\mathbf{M} - \lambda\mathbf{I})\mathbf{n} = \mathbf{0}$. For the first eigenvalue, this gives

$$\begin{pmatrix} -0.33 & 0.11 \\ 0.33 & -0.11 \end{pmatrix} \begin{pmatrix} n_1 \\ n_2 \end{pmatrix} = \begin{pmatrix} 0 \\ 0 \end{pmatrix} \tag{6.64}$$

The second equation in this system is the same as the first with the signs reversed. So, Equation (6.64) corresponds to the single equation, $0.33n_1 - 0.11n_2 = 0$ with two unknowns. Solving it for one of the two unknowns produces the following relationship between them: $n_2 = 3n_1$. The eigenvector $(1,3)$ is therefore a biologically realistic steady-state distribution, expressing the long-term relative sizes of the two populations. Remember, the eigenvectors can only be interpreted as relative population sizes, so if this eigenvector is converted into a percentage, it gives $(25,75)$, the answer that was found by simulation in Example 6.18. The second eigenvalue gives a biologically unrealistic eigenvector because the system

$$\begin{pmatrix} 0.11 & 0.11 \\ 0.33 & 0.33 \end{pmatrix} \begin{pmatrix} n_1 \\ n_2 \end{pmatrix} = \begin{pmatrix} 0 \\ 0 \end{pmatrix} \tag{6.65}$$

implies $n_1 = -n_2$, so that either one population or the other is negative.

Did you notice how the largest of two positive eigenvalues (the **dominant eigenvalue**) in the previous example was equal to the assumed population growth rate ($= 1.1$) from Equation (6.54)? This is not a coincidence. In general, the dominant eigenvalue of matrix models of exponentially growing populations is equal to the population's growth rate. This fact will be put to good use in the following section.

6.6: Calculating eigenvectors and eigenvalues

The command `eigen(M)` will return an object containing a list of eigenvalues and a matrix with the eigenvectors in its columns. Using the matrix in Equation (6.61),

```
> M<-matrix(c(0.77,0.33,0.11,0.99), 2,2)
> eigen(M)
$values
[1] 1.10 0.66

$vectors
            [,1]        [,2]
[1,] -0.3162278 -0.7071068
[2,] -0.9486833  0.7071068
```

The eigenvalues 1.1 and 0.66 are the same as those calculated in Example 6.19 but the eigenvectors appear to be different. This is because eigenvectors represent relative magnitude. If the eigenvectors produced by R are standardised to have 1 as their first element, they are the same as those calculated in Example 6.19 (i.e. (1,3) and (1,−1)):

```
> n<-eigen(M)$vectors #Extracts eigenvectors from object eigen
> st<-diag(1/n[1,])    #Off-diagonals are zero
> n%*%st
     [,1] [,2]
[1,]    1    1
[2,]    3   -1
```

Can you see how this works? It performs the following multiplication

$$\begin{pmatrix} a & b \\ c & d \end{pmatrix} \begin{pmatrix} 1/a & 0 \\ 0 & 1/b \end{pmatrix} = \begin{pmatrix} 1 & 1 \\ c/a & d/b \end{pmatrix}$$

6.7. Leslie matrix models

Individuals in real populations are unique. They differ in their genetic make-up, their life history stage, the conditions in which they find themselves now and in their past. Depending on the question asked, population models may acknowledge some of these differences by incorporating one or more types of **structure**. Matrices have proven invaluable in formulating and analysing such models.

Example 6.20: Stage-structured seal populations

In Europe, grey seals come ashore on remote islands and coastlines to give birth to their pups in the autumn, to moult in spring and, at other times of the year, to haul out and rest between foraging trips to sea. Females begin to breed at the end of their fifth year. In good conditions, about 90% of adult females will come on shore to breed. Each mature female gives birth to a single white-coated pup, which is nursed for about three weeks before it is weaned and moults into its adult coat. Little is known about the sex ratio in the population but current estimation methods assume (probably incorrectly) that there is

an approximately equal number of males and females. First-year survival is about 70%, but this goes up to approximately 90% for sub-adults and adults. Until recently, there was little evidence of density dependence or large environmental effects in grey seal dynamics, so these numbers can be assumed constant.

Clearly, there are differences between individuals: sub-adults don't breed and pups suffer greater mortality than adults. Thankfully, male seals can be ignored because of the assumptions of density independence and a constant sex ratio. The average number of female offspring produced by each female seal is $0.90/2 = 0.45$.

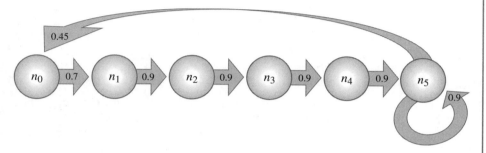

Figure 6.7: The six stages of the life history of a female seal. Stage 0, comprising the year's newborns, is followed by four sub-adult stages. Once animals reach the final, adult stage they become reproductively active and remain so until they die.

It's a good idea to sketch out what is known about the seals' life history (Figure 6.7). To translate this diagram into mathematics, it is essential to define the time at which the year changes. Here, I will take it to be the day before the start of the pupping season. Therefore, the number of sub-adults aged 1 at the end of next year will comprise the surviving pups born at the start: $n_{1,t+1} = 0.7n_{0,t+1}$. These pups are born from the current year's breeding population: $n_{0,t+1} = bn_{5,t}$. The two expressions can be combined into one, hence bypassing the class of newborns

$$n_{1,t+1} = 0.7 \times 0.45n_{5,t} = 0.315n_{5,t} \tag{6.66}$$

The number of 2-year-olds at the end of next year will be the surviving 1-year-olds from this year ($n_{2,t+1} = 0.9n_{1,t}$). Similarly, for ages 3 ($n_{3,t+1} = 0.9n_{2,t}$) and 4 ($n_{4,t+1} = 0.9n_{3,t}$). The adult class receives survivors from ages 4 and 5 ($n_{5,t+1} = 0.9n_{4,t} + 0.9n_{5,t}$). Collectively, these expressions yield a system of coupled difference equations

$$
\begin{aligned}
n_{1,t+1} &= 0.315n_{5,t} \\
n_{2,t+1} &= 0.9n_{1,t} \\
n_{3,t+1} &= 0.9n_{2,t} \\
n_{4,t+1} &= 0.9n_{3,t} \\
n_{5,t+1} &= 0.9n_{4,t} + 0.9n_{5,t}
\end{aligned}
\tag{6.67}
$$

This system can be rewritten in matrix notation as

$$
\underbrace{\begin{pmatrix} n_{1,t+1} \\ n_{2,t+1} \\ n_{3,t+1} \\ n_{4,t+1} \\ n_{5,t+1} \end{pmatrix}}_{\mathbf{n}_{t+1}} = \underbrace{\begin{pmatrix} 0 & 0 & 0 & 0 & 0.315 \\ 0.9 & 0 & 0 & 0 & 0 \\ 0 & 0.9 & 0 & 0 & 0 \\ 0 & 0 & 0.9 & 0 & 0 \\ 0 & 0 & 0 & 0.9 & 0.9 \end{pmatrix}}_{\mathbf{M}} \underbrace{\begin{pmatrix} n_{1,t} \\ n_{2,t} \\ n_{3,t} \\ n_{4,t} \\ n_{5,t} \end{pmatrix}}_{\mathbf{n}_t}
\tag{6.68}
$$

As in all the previous examples of transition matrices, the columns of **M** correspond to the contributing class and the rows correspond to the receiving class. Hence, the value 0.315 of surviving female offspring from each adult female is placed in the fifth column (adult females) and first row (yearlings).

Matrix models for the dynamics of age-structured populations are called **Leslie matrix models**. They take the general form

$$\mathbf{n}_{t+1} = \mathbf{M}\mathbf{n}_t \tag{6.69}$$

There are several questions of interest that we can ask of such models: will the population grow or decline and at what rate? Following a transient phase from some starting conditions, will its composition reach a steady state? If so, how long will it take to achieve this steady state and what will the transient dynamics look like?

Mathematically, the model in Equation (6.69) is a system of coupled difference equations. The ideas concerning the stability of difference equations that we introduced in Chapter 3 can be generalised here, starting with the equilibria of the system that can be found from the equation

$$\mathbf{M}\mathbf{n}^* = \mathbf{n}^*$$
$$(\mathbf{M} - \mathbf{I})\mathbf{n}^* = \mathbf{0} \tag{6.70}$$

For a typical Leslie matrix **M**, containing mostly zeroes along its diagonal, the matrix $(\mathbf{M} - \mathbf{I})$ is rarely singular. Hence, the only solution of Equation (6.70) is $\mathbf{n}^* = \mathbf{0}$.

Example 6.21: Equilibrium of linear Leslie model

Using the Leslie matrix of Example 6.20 we can demonstrate that the only equilibrium corresponds to population extinction ($\mathbf{n}^* = \mathbf{0}$) by calculating the determinant of the matrix

$$\mathbf{M} - \mathbf{I} = \begin{pmatrix} -1 & 0 & 0 & 0 & 0.315 \\ 0.9 & -1 & 0 & 0 & 0 \\ 0 & 0.9 & -1 & 0 & 0 \\ 0 & 0 & 0.9 & -1 & 0 \\ 0 & 0 & 0 & 0.9 & -0.1 \end{pmatrix} \tag{6.71}$$

Using R, or another computer package, gives $\det(\mathbf{M} - \mathbf{I}) \cong -0.11$ which, according to the results of Section 6.4, implies a unique, trivial solution $\mathbf{n}^* = \mathbf{0}$.

More generally, any linear dynamical system of the form $\mathbf{n}_{t+1} = \mathbf{M}\mathbf{n}_t$ will only have the trivial equilibrium $\mathbf{n}^* = \mathbf{0}$ assuming that the entries of **M** are constant, and the matrix itself is nonsingular ($\det(\mathbf{M} - \mathbf{I}) \neq 0$).

The next question is whether the solutions of the system will be attracted to this trivial point or not. In terms of population dynamics, will the system go extinct or will it grow exponentially? In Section 6.6 the dominant eigenvalue of the matrix **M** coincided with the growth rate of the population. In discrete time models, a growth rate greater than 1 means population increase. So, if the dominant eigenvalue is greater than 1, the extinction equilibrium is unstable. In trying to extend this result to all linear systems, two problems arise. First, eigenvalues can be negative. If a Leslie matrix has the two eigenvalues $\lambda_1 = 0.9$ and $\lambda_2 = -1.1$, will it go extinct or not? It turns out that it will not. So, the equilibrium is unstable if any of the eigenvalues (positive or negative) have a distance > 1 from zero. The more general stability condition therefore is $|\lambda| < 1$.

The second problem is that some eigenvalues may be complex numbers of the form $a + bi$ (this may be a good time to brush up on complex numbers by re-reading Section 1.17). Even if this happens, the requirement for stability remains the same: the eigenvalues must be close to zero, within a distance of 1. The Euclidean distance (see Section 2.4) of the eigenvalue from the origin of the complex plane (see Figure 1.16 in Section 1.17) is

$$\sqrt{a^2 + b^2} < 1 \qquad (6.72)$$

So, if a Leslie matrix has the two eigenvalues $\lambda_1 = 0.9$ and $\lambda_2 = 0.9 + 0.5i$, then the population will not go extinct because $\sqrt{0.9^2 + 0.5^2} \cong 1.03$ is greater than 1. The condition in Equation (6.72) is general, applying equally well to real and complex eigenvalues: for a real eigenvalue we have $b = 0$ and hence $\lambda = a$. Therefore, the condition becomes $\sqrt{a^2} < 1$, which is equivalent to $|\lambda| < 1$.

Example 6.22: Stability in a linear Leslie model

A computer may be used to calculate the eigenvalues of the Leslie matrix **M** in Equation (6.68) . Since this is a 5×5 matrix, we expect to find six eigenvalues. They are: 1.06, 0.39+0.60i, 0.39−0.60i, −0.47+0.40i and −0.47−0.40i. Four out of five are complex numbers and two have negative real parts. Their distances from zero are: 1.06, 0.71, 0.71, 0.62 and 0.62. The first is greater than 1, implying an unstable equilibrium. Therefore, this population is predicted to grow exponentially at an annual rate of 1.06 (so, each year it will increase by 6%).

The steady state of the population can be calculated using the theory of Section 6.6. It can be obtained from the eigenvector corresponding to the dominant eigenvalue. In the context of Leslie models, this is interpreted as the relative prevalence of different ages in the population, the population's **stable age structure**.

Example 6.23: Stable age structure in a linear Leslie model

The eigenvector corresponding to the dominant eigenvalue for the grey seal model is $\mathbf{n} = (0.15, 0.13, 0.11, 0.09, 0.52)$. This may be converted to a percentage by multiplying by the constant $100 \left/ \sum_1^5 n_i \right.$, giving $\mathbf{n} = (15, 13, 11, 9, 52)$. So, the yearlings comprise 15% of the population at steady state and the adults 52%. These percentages remain constant, even though the total size of the population increases exponentially.

A final question of interest concerns the transient dynamics of the system as it approaches, or moves away from, the single equilibrium. For example, will the proportion of yearlings fluctuate before it settles to the steady state value of 15% or will it approach it monotonically? Will a population that is heading for extinction do so in a monotonically decreasing fashion, or could it temporarily increase before it collapses? Once again, the dominant eigenvalue provides this information: if it is complex, then the transient dynamics are oscillatory. If it is real (as was the case in Example 6.22), then the approach to steady state is monotonic.

6.7: Analysis of a Leslie model

In practice, analysing the dynamical properties of a linear discrete model is easy. What follows is an interactive session which gradually extracts all the necessary information. First, the Leslie matrix

```
M<-
matrix(c(0,0.9,0,0,0,0,0,0.9,0,0,0,0,0,0.9,0,0,0,0,0,0.9,0.315,0,0,0,
0.9), 5, 5)
```

The determinant indicates whether the system has a unique equilibrium,

```
> det(M-diag(5)) # if non-zero, model has a unique equilibrium
[1] 0.1066715
```

The command `eigen()` always lists the dominant eigenvalue first.

```
> e<-eigen(M)      # calculates eigenvalues, eigenvectors
> de<-e$values[1] # extracts the dominant eigenvalue
> abs(de)          # distance of dominant eigenvalue from zero
[1] 1.062294
```

Conveniently, the command `abs()` works for both real and complex eigenvalues. From the above result, we conclude that the population will grow exponentially at an annual rate of approximately 1.17. The stable age structure of the population (in percentages) can be obtained by

```
> ev<-e$vectors[,1]    # extracts dominant eigenvector
> round(abs(100*ev/sum(ev)),0) # obtains stable age structure
[1] 15 13 11 9 52
```

Finally, to determine whether the approach to equilibrium will be monotonic. If the dominant eigenvalue is complex (i.e. if its imaginary part `Im(de)` is not zero), then the approach to equilibrium will be oscillatory.

```
> Im(de)==0    # if TRUE monotonic, if FALSE oscillatory
[1] TRUE
```

6.8. Analysis of linear dynamical systems

The analytical approach applied to the Leslie matrix model generalises to all linear models whether in discrete or continuous time. Although the details of this process differ between continuous and discrete systems, the overall steps are the following:

❶ Find the equilibria of the system (linear systems will have either a single or an infinite number of equilibria).
❷ Determine if the equilibria are stable.
❸ Determine whether solutions will oscillate or move monotonically towards/away from the equilibria.
❹ If required, find the steady-state structure (i.e. the long-term, relative prevalence of different components of the system).

To help make the comparison with discrete systems, I introduce a continuous example.

Example 6.24: A fragmented population in continuous time

For some species living in fragmented habitats, births and movement may occur all the time, rather than in well-defined breeding seasons. In these cases, it may be more appropriate to model metapopulation dynamics in continuous time. The following is a simple two-patch model with density-independent demography and migration. The population sizes in the two patches are n_1, n_2.

In the absence of migration, the populations grow independently and exponentially (see Example 5.25) at a common per capita rate r. At any given instant, emigration out of patch 1 has the effect of reducing n_1. If this happens at a rate proportional to the size of the local population, the animals are said to be **diffusing** out of the patch. The per capita rate of emigration is a for patch 1 and b for patch 2. There is also some attrition of animals during their transit (e.g. loss to predators while crossing unfavourable inter-patch space). So, if an_1 animals leave patch 1, a total of εan_1 animals arrive at patch 2 (where $0 \leq \varepsilon \leq 1$). The model is written

$$\frac{dn_1}{dt} = rn_1 - an_1 + \varepsilon bn_2$$

$$\frac{dn_2}{dt} = rn_2 - bn_2 + \varepsilon an_1$$

(6.73)

The assumption made implicitly here is that movement between patches is instantaneous. The equilibria n_1^*, n_2^* can be found by setting the rates of change to zero (also see Section 5.11) resulting in a system of linear algebraic equations

$$rn_1^* - an_1^* + \varepsilon bn_2^* = 0$$
$$rn_2^* - bn_2^* + \varepsilon an_1^* = 0$$

(6.74)

This can be written in matrix form

$$\begin{pmatrix} r-a & \varepsilon b \\ \varepsilon a & r-b \end{pmatrix} \begin{pmatrix} n_1^* \\ n_2^* \end{pmatrix} = \begin{pmatrix} 0 \\ 0 \end{pmatrix}$$

(6.75)

We would like to know if this equilibrium will be stable (both local populations become extinct) or unstable (both local populations grow exponentially).

There is a graphical device that will enable us to informally assess the stability properties of any dynamical two-component system known as the **phase-space** (and also as the **state-space** or **phase-plane**). Each of the two axes of a phase-space plot represents a state variable (e.g. the population size of a patch in Example 6.24). Every point in the phase-space uniquely describes a particular state of the system (e.g. the two local population sizes in Example 6.24). If the system is found at a particular state, then in which direction will it tend to move and how quickly? This can be visualised by an arrow that starts from the system's current state, points in the direction in which the system will tend to move and has length proportional to the speed with which the system will move. Placing many of these arrows on a regular grid of points in the phase-space allows us to visualise how the system will move in the long term. If these arrows are pointing towards an equilibrium point, then that equilibrium is likely to be stable. Phase-space plots are particular examples of what mathematicians call **vector fields**.

Example 6.25: Phase-space for a two-patch metapopulation

I consider three scenarios for the parameters in Equation (6.73). In scenario 1 (Figure 6.8(a)), the rate of growth exceeds the overall rate of loss and the vectors of the phase-space point away from the extinction equilibrium. In scenario 2 (Figure 6.8(b)) there is neither growth nor loss, the system has infinite equilibria (arranged along the diagonal line in Figure 6.8(b)). In scenario 3 (Figure 6.8(c)) the rate of population change is negative, so the system eventually moves towards extinction.

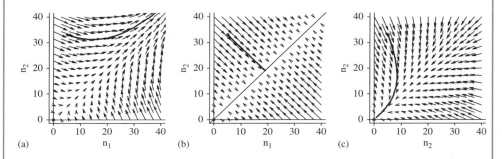

(a) (b) (c)

Figure 6.8: Phase-space plots for three parameterisations of the model in Equations (6.73) . The parameter values are: (a) $r = 0.25, a = b = 0.4, \varepsilon = 0.8$; (b) $r = 0, a = b = 0.4, \varepsilon = 1$; (c) $r = -0.25, a = b = 0.4, \varepsilon = 0.8$. The solid curves represent particular trajectories of the system obtained by numerical approximation (see R6.8) using the starting values $n_1 = 5, n_2 = 33$.

But how can these results be formalised mathematically? Below, I outline the procedure for discrete and continuous systems in parallel. Linear dynamical systems take the general form

$$\underbrace{\frac{d\mathbf{n}}{dt} = \mathbf{Mn} + \mathbf{c}}_{\text{Continuous system}} \qquad \underbrace{\mathbf{n}_{t+1} = \mathbf{Mn}_t + \mathbf{c}}_{\text{Discrete system}} \tag{6.76}$$

where $\mathbf{M}$ is a constant, square matrix and $\mathbf{c}$ a vector of constants. The examples of linear systems examined above had $\mathbf{c} = \mathbf{0}$. Take the system in Example 6.25: it considered two interacting, local populations that were closed to external influences. Had there been a constant influx of immigrants from outside the metapopulation, then these would give a system with a nonzero vector $\mathbf{c}$.

The equilibria of Equations (6.76) are found by setting $d\mathbf{n}/dt = 0$ in the continuous case and $\mathbf{n}_t = \mathbf{n}_{t+1}$ in the discrete case.

$$\underbrace{\mathbf{0} = \mathbf{Mn}^* + \mathbf{c}}_{\text{Continuous system}} \qquad \underbrace{\mathbf{n}^* = \mathbf{Mn}^* + \mathbf{c}}_{\text{Discrete system}} \tag{6.77}$$

Using matrix operations, this yields the following systems of linear equations

$$\underbrace{\mathbf{Mn}^* = -\mathbf{c}}_{\text{Continuous system}} \qquad \underbrace{(\mathbf{M} - \mathbf{I})\mathbf{n}^* = -\mathbf{c}}_{\text{Discrete system}} \tag{6.78}$$

These are of a similar form: a square matrix ($\mathbf{M}$ for the continuous case, and $\mathbf{M} - \mathbf{I}$ for the discrete) multiplied by the vector of equilibrium values ($\mathbf{n}^*$), equated to a vector of constants

(−c). If the square matrix is singular, then the system has infinite equilibria. If it is nonsingular (i.e. invertible), it can be solved to find the unique solution

$$\underbrace{\mathbf{n}^* = -\mathbf{M}^{-1}\mathbf{c}}_{\text{Continuous system}} \qquad \underbrace{\mathbf{n}^* = -(\mathbf{M} - \mathbf{I})^{-1}\mathbf{c}}_{\text{Discrete system}} \qquad (6.79)$$

If the $\mathbf{c}$ is a zero vector (as in Examples 6.20 and 6.24, above), then the system has the unique, trivial solution $\mathbf{n}^* = \mathbf{0}$.

The stability of the equilibria, in both continuous and discrete systems, can be investigated by looking at the eigenvalues λ of the matrix $\mathbf{M}$, given as the solutions of the characteristic equation

$$\det(\mathbf{M} - \mathbf{I}\lambda) = 0 \qquad (6.80)$$

This will always take the form of a polynomial equation of order equal to the rows or columns in the matrix $\mathbf{M}$. The interpretation of the eigenvalues differs between discrete and continuous systems. In discrete systems (Section 6.7), the dominant eigenvalue must be within a distance 1 from the origin of the complex plane. In continuous systems, stability requires that the real parts of all eigenvalues are negative. Complex dominant eigenvalues indicate oscillatory solutions. For quick reference, all of these facts are summarised in Figure 6.9.

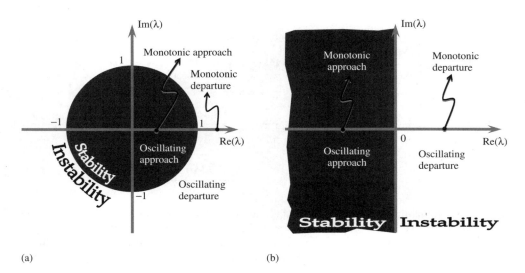

(a)   (b)

Figure 6.9: Overview of the stability of equilibria for (a) discrete and (b) continuous linear systems. The eigenvalues of a dynamical system are plotted in the complex plane. For discrete systems (a), the stability region is a circle of radius 1 around the origin. If the system's dominant eigenvalue lies within this region then the equilibrium is stable. When the dominant eigenvalue is real, solutions will approach the equilibrium monotonically. For continuous systems (b), the equilibrium is stable if the real part of all eigenvalues is negative. If the eigenvalues have no imaginary part, then the system approaches the equilibrium monotonically (Figure modified from Otto and Day, 2007).

Example 6.26: Stability analysis of a two-patch metapopulation

The system in Equation (6.73) is of the form $d\mathbf{n}/dt = \mathbf{Mn}$, so its equilibria are found from the equation $\mathbf{Mn}^* = \mathbf{0}$, where

$$\mathbf{M} = \begin{pmatrix} r - a & \varepsilon b \\ \varepsilon a & r - b \end{pmatrix} \tag{6.81}$$

The system will have infinite equilibria if the determinant of $\mathbf{M}$ is zero. Hence, any combination of the parameters r, a, b, ε that satisfy the equation

$$r^2 - r(a + b) + ab(1 - \varepsilon^2) = 0 \tag{6.82}$$

will result in infinite equilibria. For example, setting $r = 0, \varepsilon = 1$ satisfies this equation irrespective of the values of a, b. What does this mean biologically? A two-patch metapopulation with $r = 0$ is neither producing nor losing individuals within the patches. Also, $\varepsilon = 1$ implies no losses of migrants. So the system should simply redistribute individuals between its two patches, according to the rates a, b. Eventually, at equilibrium, the constant number of individuals (this number depends on the model's initial conditions) will be split between the patches according to a stable distribution. This situation is depicted in Figure 6.8(b) where the system approaches the line of equilibria from its starting position. In that particular example, the line is at $45°$ because the exchange rates a, b are equal.

If, as will most often be the case, Equation (6.82) is not satisfied, then the system will only have the trivial equilibrium $\mathbf{n}^* = \mathbf{0}$. Its stability can be assessed by looking at the system's eigenvalues, found by solving the quadratic equation

$$\det(\mathbf{M} - \mathbf{I}\lambda) = 0 \tag{6.83}$$

which expands to

$$\lambda^2 - \lambda(2r - a - b) + (r - a)(r - b) - \varepsilon^2 ab = 0 \tag{6.84}$$

After some manipulation, the quadratic formula gives the two solutions

$$\lambda_1, \lambda_2 = \frac{2r - a - b \pm \sqrt{(a - b)^2 + 4ab\varepsilon^2}}{2} \tag{6.85}$$

It is not clear if the real parts of these solutions are negative or positive. However, it is certain that there will be no complex part, because the discriminant under the square root is non-negative ($(a - b)^2, a, b, \varepsilon$ are all non-negative). Therefore, whatever the solutions do, they will do it without oscillations.

If we focus on the special case of $\varepsilon = 1$ (no individuals are lost during movement), the eigenvalues become

$$\lambda_1 = r, \quad \lambda_2 = r - a - b \tag{6.86}$$

Stability requires that both of these are negative and, hence, that $r < 0$. This makes biological sense: when no individuals are lost in transit, the metapopulation will only decline if the intrinsic growth rate is negative.

Relaxing the assumption of $\varepsilon = 1$ introduces a new source of mortality and opens the possibility that the population will become extinct even if the local populations have a slight tendency to increase. Extinction will occur if $\lambda_1, \lambda_2 < 0$. The numerator of Equation (6.85) is an expression of the form $\alpha \pm \sqrt{\beta}$ where $\sqrt{\beta}$ is real and positive. Therefore, if $\alpha + \sqrt{\beta}$ is negative, then so is $\alpha - \sqrt{\beta}$. This can be rewritten as a general condition for stability

$$2r - (a + b) + \sqrt{(a - b)^2 + 4ab\varepsilon^2} < 0 \tag{6.87}$$

The expression simplifies considerably when the exchange rates a, b between the two patches are equal (e.g. Figure 6.8). Setting $b = a$ in Equation (6.87) gives

$$r < a(1 - \varepsilon) \tag{6.88}$$

The quantity $1 - \varepsilon$ is the proportion of animals getting lost in transit and, therefore, $a(1 - \varepsilon)$ represents the rate of loss suffered by each patch due to migration. Hence, the entire metapopulation will become extinct if the local growth rate is smaller than the rate of loss during transit. Rather reassuringly, assuming no loss ($\varepsilon = 1$) retrieves the requirement $r < 0$ derived under Equation (6.86) .

 6.8: Two-dimensional vector plots

I provide and explain code that may be used to generate phase-space plots such as those of Figure 6.8. First it is necessary to define the model, i.e. the parameter values (ep $= \varepsilon$, r,a,b) and the two differential equations (specified as the functions dx and dy).

```
#MODEL SPECIFICATION
ep<- 0.8 # Proportion of emigrants that survive
r<- 0.25 # Local population growth rate
a<- 0.4 # Rate of departure from patch 1
b<- 0.4 # Rate of departure from patch 2
# Differential equations for two patches
dx<-function(x,y) {return(r*x-a*x+ep*b*y)}
dy<-function(x,y) {return(r*y+ep*a*x-b*y)}
```

Next, the region of interest can be specified by the user in the form of xrang and yrang. The number of arrows to be used in each direction can also be specified in res. The value res<-15 will create a regular grid of $15 \times 15 = 225$ arrows. To ensure that the features of the phase-space are visible, some trial and error will be required. Too many arrows will swamp the plot and too few will make it hard to visualise the trajectories of possible solutions. The same applies for the length of the arrows. If they are too long, they will mostly lie outside the plotting region and if they are too short, only the arrowheads will be visible, making it hard to differentiate between regions of fast and slow movement. So, the parameter maxLength can be used to specify the length of the biggest arrow in the units of the plot.

```
# PLOTTING PARAMETERS
xrang<-c(0,40)  # Range of x-axis
yrang<-c(0,40)  # Range of y-axis
res<-15          # Number of arrows in each direction
maxLength<-6    # Desired length of biggest arrow
```

Then, the data to be used for plotting the arrows must be prepared. Each arrow must have a start and end point. The x and y coordinates of the start points are generated with the command seq() and stored in the vectors xs and ys. To make sure that the points lie on a grid (rather than on top of each other), the order with which they are repeated differs between xs and ys using the option each. For example, for a grid of size 3, ranging from 1 to 3, the two vectors are xs: (1,2,3,1,2,3,1,2,3) ys: (1,1,1,2,2,2,3,3,3).

```
# ARROW STARTPOINTS
xs<-rep(seq(xrang[1],xrang[2],length.out=res), res)
ys<-rep(seq(yrang[1],yrang[2],length.out=res), each=res)
```

The length and direction of the arrows, as determined by their endpoints, indicate the direction and speed of movement of the system. This is calculated as the increment of movement that the system would perform in the x and y directions over a very brief time interval (i.e. the instantaneous rate of change of the x and y components of the system). Thankfully, these can be obtained directly from the differential equations without solving them, by specifying them to the startpoint of interest. This is done by calling the functions `dx` and `dy` for the vectors of startpoints and storing the results in `dxs` and `dys`. If these increments were plotted directly, they would have lengths s (calculated as a vector of Euclidean distances). However, these lengths may be too small or too big for plotting. Since only the arrows' relative lengths are needed, we may standardise the x and y increments to have a maximum length of maxLength. For example, `maxLength*dxs/max(s)` will do this for the increments along the x-axis. Adding the standardised x and y increments to their corresponding startpoints gives the arrows' endpoints, `xe`, `ye`.

```
#ARROW ENDPOINTS
# Initial increments along x and y axes
dxs<-dx(xs,ys)
dys<-dy(xs,ys)
s<-sqrt(dxs^2+dys^2) # Initial arrow lengths
# x and y coordinates for arrow endpoints
xe<-xs+maxLength*dxs/max(s)
ye<-ys+maxLength*dys/max(s)
```

All that remains now is to plot the results.

```
# PLOTTING
# Creates an empty plot with the required plotting range
plot(0,0,xlab=expression(n[1]),ylab=expression(n[2]),xlim=xrang, ylim=yrang)
abline(h=0) # Draws x-axis
abline(v=0) # Draws y-axis
arrows(xs,ys,xe,ye, length=0.1, angle=12) # Plots the arrows
```

6.9.   Analysis of nonlinear dynamical systems

A dynamical system of difference or differential equations is called **nonlinear** if its variables occur in nonadditive relations, i.e. as part of products with other variables, powers or other nonlinear functions such as the exponential, logarithmic and trigonometric functions.

Example 6.27: The Lotka–Volterra, predator–prey model

Canadian lynx (Lynx canadensis) and snowshoe hare (Lepus americanus): ecology's emblematic predator and prey

This is probably the most famous model in population/community ecology. It takes its name from Alfred Lotka and Vito Volterra, the two theoreticians who proposed it independently in the mid-1920s. A strong motive behind the development of this model was provided by the observation that several populations of animals show cyclic fluctuations in density (also see Example 2.1). Equally intriguing is that the populations of some prey and their predators appear to be cycling in lagged synchrony (i.e. predators reach their peak one or two years after their prey, and, following a decline, they recover only after the prey have been on the increase for one or two years). These observations had been made on historical data sets, such as the century-long Hudson Bay trapping records for Canadian lynx and snowshoe hare.

The scientific approach followed by Lotka and Volterra was to construct a mathematical model based on a set of simple assumptions and investigate whether it would be capable of replicating the observed patterns of cyclicity. They used two variables, prey density (P) and predator density (N) coupled in a dynamical system in continuous time (i.e. two differential equations, one for the predator and one for the prey). They assumed that, in the absence of predators, the prey increased exponentially, according to the equation $dP/dt = rP$ and, upon the extinction of prey, the predators declined exponentially according to the equation $dN/dt = -mN$. If both prey and predators were present, then each predator was assumed to eat according to a Type I functional response (Example 4.14) at a per capita rate aP. Hence, the rate of consumption by the entire predator population was aPN.

Predators convert the energy and matter of their food into maintenance and reproduction. The net production of new predators through both of these processes may be summarised in a function called the **numerical response**. The simplest numerical response is obtained by scaling total consumption by a prey-to-predator conversion efficiency term (ε). Hence, if total consumption is aPN, then the net rate of production of new predators will be εaPN. To simplify notation, the constant product εa can be replaced by a new parameter, b.

The combination of all of the above assumptions gives the model

$$\frac{dP}{dt} = \overbrace{rP}^{\text{Prey growth}} - \overbrace{aPN}^{\text{Prey mortality}}$$

$$\frac{dN}{dt} = \underbrace{bPN}_{\text{Predat. growth}} - \underbrace{mN}_{\text{Predat. mortality}}$$

(6.89)

This model is nonlinear because the terms aPN and bPN involve products of the two state variables.

The questions that may be asked of a nonlinear system such as this are similar to the questions asked about linear systems. However, whereas linear systems either have one or infinite equilibria, nonlinear systems may have any number. Unfortunately, these cannot be found using matrix techniques, but may be calculated by substitution. Graphical techniques such as the phase-space will also work for nonlinear systems, suggesting what the solutions might look like.

Example 6.28: The Lotka–Volterra, predator–prey model

Once the Lotka–Volterra model has approached an equilibrium, the rates of change of the predator and prey populations will no longer change ($dP/dt = 0$, $dN/dt = 0$). This gives two algebraic equations for the equilibria

$$0 = rP^* - aP^*N^*$$
$$0 = bP^*N^* - mN^*$$

(6.90)

By factorising these two equations,

$$0 = P^*(r - aN^*)$$
$$0 = N^*(bP^* - m)$$

(6.91)

it is easier to see that the system has two equilibria. They are $(P_1^*, N_1^*) = (0,0)$ and $(P_2^*, N_2^*) = (m\,/\,b, r\,/a)$. So, the populations are at equilibrium either when they have both become extinct or when their sizes reach the non-negative values $P = m\,/b$ and $N = r\,/a$. The nonzero values are quite sensible, biologically. For example, the equilibrium prey population increases with m (the predators' natural mortality) and decreases with b (the efficiency with which predators convert prey consumed to more predators).

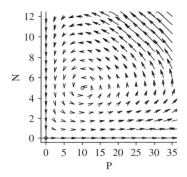

Figure 6.10: An example of the phase-space plot for the Lotka–Volterra model with parameter values $r = 0.2, a = 0.04, b = 0.01, m = 0.1$.

So, this system has two equilibria. Will solutions be attracted to any one of them and, if so, which one? An example of the phase-space plot using specific parameters (Figure 6.10) indicates that when the predators and prey are above their nonzero equilibria, they tend to move back towards them. But this does not necessarily mean that the nonzero equilibrium attracts solutions. Neither is it clear what happens with the extinction equilibrium: if the system is started with zero prey, it follows the y-axis down to predator extinction, but if it is started with zero predators, the solutions follow the x-axis up to infinity.

This example illustrates two of the limitations of purely graphical techniques: they only apply to particular parameterisation of the model and they cannot be conclusive. There is, however, a mathematical result that leads to analytical statements about the stability of the equilibria. It says that *very close* to an equilibrium, a nonlinear system behaves as if it were linear. To determine the form of the linear system in the neighbourhood of any equilibrium some notation is needed. In general, continuous and discrete nonlinear systems involving n variables take the form

$$\frac{dx_1}{dt} = f_1(x_1, \ldots, x_n) \qquad x_{1,t+1} = f_1(x_{1,t}, \ldots, x_{n,t})$$
$$\vdots \qquad\qquad\qquad \vdots$$
$$\underbrace{\frac{dx_n}{dt} = f_n(x_1, \ldots, x_2)}_{\text{Continuous system}} \qquad \underbrace{x_{n,t+1} = f_n(x_{1,t}, \ldots, x_{2,t})}_{\text{Discrete system}}$$

(6.92)

In the continuous case, the function f_i gives the rate of change of the ith variable x_i, in terms of all the variables. In the discrete case, the function f_i gives the update rule that generates $x_{i,t+1}$, the value of the ith variable in the immediate future, as a function of the current values of all the variables.

To determine the stability properties of a particular equilibrium, it may temporarily be defined as the origin of the axes. It can then be shown that the solutions of the nonlinear system behave like the solutions of the following linear ones

$$\underbrace{\frac{d\mathbf{x}}{dt} = \mathbf{M}\mathbf{x}}_{\text{Continuous system}} \qquad \underbrace{\mathbf{x}_{t+1} = \mathbf{M}\mathbf{x}_t}_{\text{Discrete system}} \tag{6.93}$$

where $\mathbf{M}$ is a square matrix containing only constant values. What is great about this result is that, mathematically, we know all about determining the stability of such systems (see Section 6.8). We don't, however, have an analytical expression for the all-important matrix $\mathbf{M}$. This special matrix is called a **Jacobian** and it is defined in terms of the partial derivatives of all the nonlinear functions $f_1, \ldots, f_n$ in the system, each calculated with respect to all the variables $x_1, \ldots, x_n$. Specifically,

$$\mathbf{M} = \begin{pmatrix} \dfrac{\partial f_1}{\partial x_1} & \cdots & \dfrac{\partial f_1}{\partial x_n} \\ \vdots & \ddots & \vdots \\ \dfrac{\partial f_n}{\partial x_1} & \cdots & \dfrac{\partial f_n}{\partial x_n} \end{pmatrix} \tag{6.94}$$

It is easy to get confused with the order in which the derivatives appear in the Jacobian. Keep in mind that the same state variable appears down each column and the same function appears along each row. The next step is to evaluate the partial derivatives in the Jacobian at each one of the system's equilibria (this is equivalent to shifting the origin of the axes to the position of each equilibrium). For a refresher on how to calculate partial derivatives, have a look at Section 4.9.

If you are a bit mystified about where the Jacobian comes from, looking back at the stability analysis of models with a single differential equation (Section 5.11) may help clarify things. In that section, the behaviour of the differential equation around an equilibrium was approximated by the first two terms of a Taylor expansion. This resulted in a model of exponential growth or decline. Simply looking at the sign of the parameter of that exponential model indicated if the solutions would move towards or away from the equilibrium. Similarly, there is a multi-variable version of the Taylor expansion formula that can be used on a system with n differential equations to yield n^2 parameters arranged within the Jacobian.

Although it is hard to tell whether the equilibrium is stable simply by inspecting the signs of the values in the Jacobian, the systems in Equations (6.93) can be subjected to the stability analysis developed in Section 6.8:

❶ Calculate the characteristic equation $\det(\mathbf{M} - \lambda\mathbf{I}) = 0$.
❷ Solve it to find the eigenvalues of the system.
❸ Depending on whether the system is discrete or continuous, determine the equilibrium's stability on the basis of Figures 6.9(a) and 6.9(b).

Example 6.29: Stability analysis of the Lotka–Volerra model

The Lotka–Volterra model is

$$\frac{dP}{dt} = rP - aPN$$
$$\frac{dN}{dt} = bPN - mN \tag{6.95}$$

So, $f_1 = rP - aPN$ and $f_2 = bPN - mN$. The general expression for the Jacobian is

$$\mathbf{M} = \begin{pmatrix} \dfrac{\partial f_1}{\partial P} & \dfrac{\partial f_1}{\partial N} \\ \dfrac{\partial f_2}{\partial P} & \dfrac{\partial f_2}{\partial N} \end{pmatrix} = \begin{pmatrix} r - aN & -aP \\ bN & bP - m \end{pmatrix} \tag{6.96}$$

At the extinction equilibrium $P^* = 0, N^* = 0$, the Jacobian becomes

$$\mathbf{M} = \begin{pmatrix} r & 0 \\ 0 & -m \end{pmatrix} \tag{6.97}$$

The characteristic equation for this is $(r - \lambda)(-m - \lambda) = 0$. Since this is already factorised, the two eigenvalues are $\lambda_1 = r$ and $\lambda_2 = -m$. The second of these is clearly negative because mortality (m) is a positive number. For extinction, both of them are required to be negative (check Figure 6.9(b)), which means that the predator–prey system will only become extinct if $r < 0$ (i.e. when the prey population declines even in the absence of predation).

At the nontrivial equilibrium $P^* = m/b$, $N^* = r/a$ the Jacobian becomes

$$\mathbf{M} = \begin{pmatrix} 0 & -am/b \\ br/a & 0 \end{pmatrix} \tag{6.98}$$

The characteristic equation for this is $\lambda^2 + mr = 0$, which has solutions

$$\lambda_1, \lambda_2 = \pm\sqrt{-mr} \tag{6.99}$$

If the growth rate of the prey population is negative ($r < 0$), then this expression gives a positive eigenvalue ($\sqrt{-mr}$) and the nontrivial equilibrium repels the system towards its eventual extinction. However, in the more interesting case $r > 0$, the eigenvalues are complex numbers (see Section 1.17)

$$\lambda_1, \lambda_2 = \pm i\sqrt{mr} \tag{6.100}$$

What does this imply about the stability of the nontrivial equilibrium? According to Figure 6.9(b), complex eigenvalues imply oscillations, but since the real part of the eigenvalues is zero, the equilibrium will be neither stable nor unstable, it is **neutrally stable**. This means that the solutions oscillate around the nontrivial equilibrium (Figure 6.11(a)) and the two populations perform cycles of constant amplitude (Figure 6.11(b)).

Surely then, the ability of the Lotka–Volterra model to generate cycles is a resounding success for theoretical ecology? The answer is a rather lukewarm 'yes'. Although this model can generate cycles, and thus prove that population cycles could arise out of the interaction between a predator and its prey, its neutral stability is not quite as realistic. Starting the same Lotka–Volterra model at a different set of predators and prey densities gives cycles of a different amplitude and slightly different period (Figures 6.11(c) and 6.11(d)). Such dependence on initial conditions is a problematic feature of this model. It implies that the characteristics of the cycles would change every time the system is perturbed away from its current trajectory. Because in the real world populations are constantly perturbed by environmental events, it is unlikely that the robust, cyclic patterns observed in nature could be completely explained by this model.

Furthermore, the cyclic behaviour of this model is sensitive to its assumptions. For example, introducing density dependence (e.g. as a logistic growth term for the prey) or a more realistic consumption function (e.g. a Type II functional response) turns the neutral cycles into damped oscillations. These alternative models are also nonlinear and can be analysed with the techniques presented in this section.

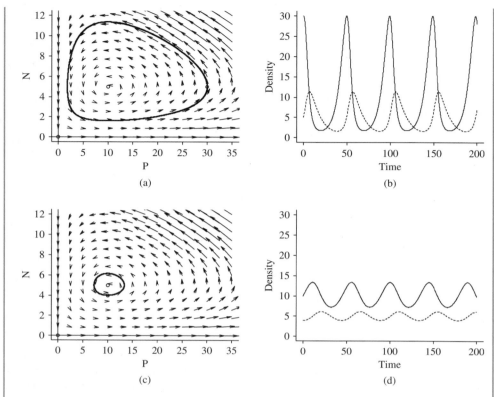

Figure 6.11: Phase-spaces and corresponding time series of the Lotka−Volterra model using two different starting sizes for the predator and prey populations. Parts (a) and (b) show populations starting at $P = 30, N = 5$. The populations in parts (c) and (d) were started at $P = 10, N = 4$. In the time series plots (parts (b) and (d)), the predator population is shown by a dashed curve and the prey population is shown as a solid curve. The parameter values for both models are $r = 0.2, a = 0.04, b = 0.01, m = 0.1$.

⑭ 6.9: Numerical solutions for systems of differential equations

The phase-space trajectories and population time series shown in Figures 6.8 and 6.11 were generated by numerical approximation. The approach to solving systems of differential equations numerically is identical to the one already used for single differential equations, in R5.4. Here, I illustrate with the Lotka−Volterra model.

The library odesolve must be loaded first. Values need to be provided for the parameters (parms) and initial conditions (init) of the model, as well as the time range (times) over which the solution will be calculated.

```
# PARAMETERS
require(odesolve)       # Loads differential equations library
init<-c(p=30, n=5)      # Initial conditions for prey and predators
times<-seq(0,200, 0.1)  # List of time instances
r<- 0.2    # Prey growth rate
a<- 0.04   # Functional response parameter
```

```
b<- 0.01   # Numerical response parameter
m<- 0.1    # Predator mortality
parms<-c(r,a,b,m) # List of all parameters
```

Then, the entire model must be packaged within the function `predprey`.

```
# MAIN FUNCTION DEFINITION
predprey<-function(times, init, parms)
  {
  p<-init[1] # Initialises prey population
  n<-init[2] # Initialises predator population
  with(as.list(parms), {
    dp<-r*p-a*p*n # Prey dif. eq.
    dn<-b*p*n-m*n #Pred. dif. eq.
    return(list(c(dp,dn)))
  })}
```

This function can be called to estimate a particular solution that has the specific characteristics (i.e. parameters, initial conditions and duration) given in the first part of the code, above.

```
# CALL FUNCTION TO ESTIMATE PARTICULAR SOLUTION
solution<-as.data.frame(lsoda(init,times,predprey,parms))
```

To plot the solution as a phase-space trajectory, type

```
plot(solution$p, solution$n)
```

To superimpose to a phase-space plot created as a vector field with the code in R6.8, then, after plotting the vector field, type

```
lines(solution$p, solution$n)
```

To create a time series plot such as the ones shown in Figure 6.11, type

```
plot(solution$t, solution$p, type="l", xlab="Time", ylab="Density",
xlim=c(0,200), ylim=c(0,30))
lines(solution$t, solution$n, lty=2)
```

Further reading

A great introduction to matrix algebra and its applications is Fraleigh and Beauregard (2003). Some textbooks on theoretical ecology (e.g. Batschelet, 1979; Nisbet and Gurney, 1982 and Otto and Day, 2007) have good primers on matrix algebra. Caswell (1989) is the bible for models of structured populations but many other books cover the topic with dedicated chapters (e.g. Nisbet and Gurney, 1982; Case, 2000; Kot 2001; Otto and Day, 2007). Mildly mathematical reviews of many of the ecological topics introduced here can be found in May and McLean (2007) and Begon, Townsend and Harper (2006). The stability analysis of linear and nonlinear dynamical systems is expertly covered in Chapters 7–11 of Otto and Day (2007). Examples of analyses of several ecological models using R can be found in Stevens (2009).

References

Batschelet, E. (1979) *Introduction to Mathematics for Life Scientists*. Springer Verlag, Berlin. 643pp.

Begon, M., Townsend, C.A. and Harper, J.L. (2006) *Ecology: From Individuals to Ecosystems*, 4th edition. Wiley Blackwell. 752pp.

Case, T.J. (2000) *An Illustrated Guide to Theoretical Ecology*. Oxford University Press, New York. 449pp.

Caswell, H. (1989) *Matrix Population Models*. Sinauer Associates Inc., Sunderland, Massachusetts. 722pp.

Fraleigh, R.A. and Beauregard, J.B. (2003) *Linear Algebra*. Addison Wesley. 571pp

Kot, M. (2001) *Elements of Mathematical Biology*. Cambridge University Press, Cambridge. 453pp.

May, R.M. and McLean, A. (2007) *Theoretical Ecology: Principles and Applications*. Oxford University Press. 272pp.

Nisbet, R.M. and Gurney, W.S.C. (1982) *Modelling Fluctuating Populations*. The Blackburn Press. 379pp.

Otto, S.P. and Day, T. (2007) *A Biologist's Guide to Mathematical Modelling in Ecology and Evolution*. Princeton University Press, New Jersey. 732pp.

Stevens, M.H.H. (2009) *A Primer of Ecology with R*. Springer, Dordrecht. 401pp.

Thiemann, G.W., Iverson, S.J. and Stirling, I. (2008). Polar bear diets and Arctic marine food webs: insights from fatty acid analyses. *Ecological Monographs*, **78**, 591–613.

7

How to visualise and summarise data
(Descriptive statistics)

'...this chart lacks schmaltz. Chop off the bottom. Now that's more like it ... The figures are the same and so is the curve ... nothing has been falsified – except the impression that it gives.... The crooks already know these tricks. Honest men must learn them in self-defence.'

How to Lie with Statistics by Darrell Huff (American writer 1913–2001)

It is easy to find disparaging quotes about statistics. Some criticise the malleability of statistical statements, such as the famous ones by Benjamin Disraeli: *'...there are three kinds of lies: lies, damned lies and statistics'* and Gregg Easterbrook: *'Torture numbers, and they'll confess to anything'*. Others focus on the apparently illogical nature of statistical summaries, such as *'the man who drowned crossing a river with an average depth of six inches'*. All of these, however, are condemnations of the misuse and misinterpretation of statistics, not the fundamental need to condense large sets of data into informative images and numerical summaries. Whole data sets are difficult to digest. So, papers published these days rarely contain tables of raw data. However, by their nature, graphical and numerical summaries lead to some information loss. A graph may not be able to represent all aspects of the data and a single summary (such as the average) may fail to represent other salient features (such as the variation between the members of a data set). Most successful analyses use statistical summaries to reveal salient features of the data and generate new ecological hypotheses. By the same token, the choice of summary may also, intentionally or not, conceal vital clues about the nature of the system that generated the data. The purpose of this chapter is to outline the methods that can be used to explore and summarise data sets for final publication but also as an intermediate stage between data

How to be a Quantitative Ecologist: The 'A to R' of Green Mathematics and Statistics, First Edition. Jason Matthiopoulos.
© 2011 John Wiley & Sons, Ltd. Published 2011 by John Wiley & Sons, Ltd.

collection and modelling. Section 7.1 begins with a general overview of the field of statistics and Section 7.2 introduces some of the terminology that will be used to characterise **random variables** in this and the next five chapters. The relationship between **statistical populations** and **samples** is discussed in Sections 7.3–7.4. The data collected for a random variable can be used to construct a **frequency distribution**, describing the commonness of different values of the random variable in the sample (Section 7.5). These distributions are rich in information so, to facilitate comparisons between different samples, they need to be summarised by using descriptive statistics. Interesting properties of frequency distributions pertain to their position (Section 7.6), spread (Section 7.7) and symmetry (Section 7.8). These summaries can be combined into informative graphical representations of the distribution (Section 7.9). All of these ideas can be extended to samples containing more than one random variable (Section 7.10). In such samples, an additional question of interest is whether the variables are related to each other (Sections 7.11 and 7.12) and how to characterise the commonness of combinations of their values (Section 7.13).

7.1.   Overview of statistics

Statistics has emerged out of a need to make sense of an unpredictable, partially observed and imprecisely measured universe. Therefore, the different areas of statistics closely match the activities carried out by natural scientists: observation, description, pattern extraction and prediction.

Doing science is impossible without observations of the natural world. Issues of **data collection**, whether carried out opportunistically or by design, are the first important consideration in statistics. Although the design of experiments and field methodology fall outside the remit of this book, it is always important to consider the data collection process in conjunction with the analysis of the resulting data. Before going out to the field, empirical ecologists spend considerable parts of their lives troubling over how to make the best use of limited funds and time. Experienced field workers are painfully aware that the information they collect must be sufficient and of the appropriate form for the questions they intend to ask. The most elaborate statistical post-processing is no substitute for a well-planned and executed field season. Equally, the strategic decisions that need to be made during planning and data collection are best informed by an understanding of the strengths and limitations of the available statistical tools. Questions of interest at this stage are: 'What attributes of the study system should be observed?', 'How should the data collection effort be distributed over time and space?', 'What is an appropriate unit of observation for the particular study?', 'How can the field techniques and technologies result in spurious results and observation artefacts?'

Once the data have been collected, it is always helpful to inspect and abridge them, preferably in several different ways. This involves numerical and graphical summaries collectively known as **descriptive statistics**. Questions of interest at this stage are 'What is a typical observation?', 'How diverse are the data?', 'Are extreme observations rare?', 'Are there any discernible associations between different attributes of the same system?', 'Do the results present obvious differences between different places or times?'.

Often, at the stage of descriptive statistics the data will suggest patterns that need to be explored further. The identification and quantification of such patterns falls within the area of **inferential statistics** ('inference', means 'conclusion'). Questions of interest at this stage are: 'How likely is it to observe these patterns by chance?', 'What can be said about the entire population from a sample of observations?', 'Do the data confirm theoretical expectations about the system?', 'Which explanatory variables appear to influence the observed responses?'

Many of the questions asked as part of statistical inference require the use of **statistical models**, i.e. mathematical models whose parameters have been estimated from the data as part of a process called **fitting**. A statistical model that has been fit to data can be used to identify possible causal relationships, quantify their strength and even **predict** the state of a system in new circumstances.

7.2.   Statistical variables

All measurable quantities in statistics are treated as variables. So, 'data collection' describes the measurement of values for one or more variables. The physical dimensions and body weights of animals are variables and so is their gender and retinal pigmentation. The distinguishing characteristic of a **statistical** or **random variable** is that its value can change every time it is measured, even if the measurements are done under apparently identical conditions.

Different analytical treatments are reserved for different types of measurements, so it is useful to introduce a taxonomy of statistical variables (Figure 7.1). The first distinction is between qualitative and quantitative variables. **Qualitative** (or **categorical**) variables identify group membership, but not numerical magnitude. Qualitative variables are called **nominal** if their values cannot readily be ordered and **ordinal** if they convey a subjective assessment of magnitude. Gender and colour of eyes are nominal variables but many questionnaire answers are ordinal (e.g. 'What is the public's perception of the importance of global warming? 1. Very important, 2. Moderately important, 3. Not important at all'). **Quantitative** variables are measured on an objective, numerical scale. They can be divided into **continuous variables** (taking values from the set of real numbers) and **discrete variables** (taking values from the set of integer numbers). Sometimes, it is convenient to **discretise** continuous variables. Hence, the exact age of a 27-year-old tree is anywhere in the interval [27,28].

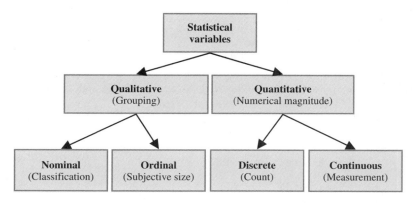

Figure 7.1: Taxonomy of statistical variables.

Example 7.1: Activity budgets in honey bees

European honey bee, Apis mellifera

The behavioural activity (*A*) of worker honey bees can be divided into the following recognisable tasks: resting (Re), patrolling, (Pt), cell-cleaning (Cl), eating pollen (Ea), tending brood (Tb), comb-building and maintenance (Cb) and external activities (Ex) such as guarding, foraging and dance-following. The data on the type of activity performed by any particular bee at any particular time are nominal. Symbols such as Re and Pt are chosen arbitrarily and have no natural order (it makes no sense to say that resting is larger or smaller than patrolling). These symbols are called the **levels of the nominal variable** *A*. However, the energy (*E*) required to carry out any one task is an ordinal and continuous variable. Hence, the energy expended during resting is less than the energy used for patrolling.

 7.1: Factors

In R terminology, qualitative variables are called **factors** and quantitative variables **numeric**. The levels of a factor may be given as text. For example, a data set containing 12 observations of honey bee activities may look like this:

```
A<- c("Re","Cb","Tb","Re","Ea","Re","Pt","Cb","Ex","Cl")
```

Alternatively, each level of a nominal variable may be represented by a serial number. So, if the seven different activities of a bee are allocated an arbitrary number, the data set may look something like this:

```
A<- c(1,6,5,1,4,1,2,6,7,3)
```

However, neither of the above two versions of data entry will be automatically interpreted as a factor. We can check this by asking,

```
> is.factor(A)
[1] FALSE
```

In fact, the first version is seen as a list of labels and the second as a list of numbers. Accidentally misinterpreting a factor as a numeric variable can cause major headaches at the analysis stage, so factors must be declared using the command `factor()`. The command `levels()` can then be used to extract the levels of a factor from a data set. Here are two examples:

```
> A<-factor(c("Re","Cb","Tb","Re","Ea","Re","Pt","Cb","Ex","Cl"))
> levels(A)
[1] "Cb" "Cl" "Ea" "Ex" "Pt" "Re" "Tb"
>
> A<-factor(c(1,6,5,1,4,1,2,6,7,3))
> levels(A)
[1] "1" "2" "3" "4" "5" "6" "7"
```

Data that are already stored in memory can be coerced into factor form using the command `as.factor()`.

7.3. Populations and samples

Data can be collected from any predefined group of **units** (organisms, inanimate objects, locations in space, states of a system at different times, etc.). The total number of observations that can conceivably be made is called the **target population** and a survey that collects all possible observations is called a **census**. Notice the difference between the statistical and ecological uses of the word 'population'. *Statistics are populated by observations, not necessarily individuals.*

For large or potentially infinite populations (e.g. the diameters of all *Asterionella* diatoms in a lake) a census is either impractical or impossible. The next best thing is to examine a subset of the population called a **sample**. One of the most important tasks of statistics is to help draw conclusions about large or infinite populations from finite samples (Chapter 10). Distinguishing between a population and a sample is a matter of definition.

Example 7.2: Production of gannet chicks

Atlantic gannet, Morus bassanus

Depending on the context of the research, the same set of observations can be either a sample or a population. If we are interested in the weights of gannet chicks born this year on Bass Rock, off the east coast of Scotland, then these weights form the statistical population and collecting all of them would constitute a census. If the question relates to all the chicks born this year in Britain, then the measurements on Bass Rock are just a (highly localised) sample from a finite population. If the question relates more generally to the weight of gannet chicks including all past and future years, then the measurements made this year on Bass Rock are a sample from a potentially infinite population.

7.2: Simulating a sample

Once a particular approach of data analysis has been chosen, it is important to validate it by seeing how it would perform if a different sample had been drawn from the same population. Ideally, validation should be done by collecting new data, but this may not be possible. One solution to this problem is to **sub-sample** from a large data set and examine how the method performs with different samples (see Section 10.6 for details). The following example uses the command `sample()` to create a random sample of four observations from a data set of ten

```
> A<-c(1,6,5,1,4,1,2,6,7,3)
> sample(A,4)
[1] 6 1 6 3
```

7.4. Single-variable samples

The clue is in their name: *random* variables are not entirely predictable. Repeated measurements for the same variable may differ. It is, therefore, useful to distinguish between the variable (e.g. weight of gannet chicks on Bass Rock) and a particular measurement of it (e.g. the weight of the 15th chick sampled). Statisticians call single measurements **realisations** of the random variable. Conventionally, upper-case letters are used to symbolise random variables and lower-case letters for their realisations. When considering random variables in isolation, the main objective is to describe just how unpredictable they are. This is examined in Sections 7.5–7.8.

7.5.   Frequency distributions

Example 7.3: Activity budgets in honey bees

The tasks carried out by a ten-day-old honey bee were recorded at hourly intervals for a day, resulting in the following data set:

```
{Cl, Pt, Pt, Re, Tb, Re, Cb, Cb, Re, Pt, Pt, Pt, Re,
 Ea, Ea, Pt, Pt, Pt, Re, Re, Re, Cb, Cb, Pt}
```

From this representation of the data it is rather difficult to see what is the bee's predominant occupation. This can be found by going through the raw data and counting the occurrences of similar tasks (Table 7.1).

Table 7.1

Task	Tally	Count	Proportion
Re	|||||||	7	0.29
Pt	|||||||||	9	0.38
Cl			1	0.04
Ea	||	2	0.08
Tb			1	0.04
Cb	||||	4	0.17
Ex		0	0.00
Total		24	1.00

The number of times that a particular value of a random variable occurs in the data is called an **absolute frequency**. Frequencies can be expressed as proportions of the population (or sample) size (n). They are then called **relative frequencies**.

$$\text{Relative freq.} = \frac{\text{Absolute frequency}}{\text{Population (or sample) size}} \qquad (7.1)$$

If $f(x)$ is the relative frequency of a value x and $\phi(x)$ its absolute frequency, then

$$f(x) = \frac{\phi(x)}{n} \qquad (7.2)$$

Relative frequencies are particularly useful for comparing samples of different sizes.

Example 7.4: Activity budgets from different studies

Another researcher focuses on a 21-day-old honey bee but collects data every two hours for a day. The resulting data set is shown in Table 7.2. This older bee is observed resting as many times as the ten-day-old individual but, of course, the comparison is invalid because the sample sizes differ. The relative frequencies indicate that the older bee rests twice as much as the younger one.

Table 7.2

Task	Tally	Count	Proportion
Re	|||||||	7	0.58
Pt	||	2	0.17
Cl			1	0.08
Ea			1	0.08
Tb		0	0.00
Cb		0	0.00
Ex			1	0.08
Total		12	1.00

The data can be more easily inspected by representing them graphically. For nominal variables, the **bar chart** plots the levels of the nominal variable on one axis and on the other it represents the (absolute or relative) frequencies by the height of a bar. The gaps between the bars indicate that this is a nominal variable. An alternative is the **pie chart**, which splits the 360 degrees of a full circle into slices whose sizes represent the contribution of each level to the total.

Example 7.5: Visualising activity budgets

Whether graphical or numerical, data summaries cause some information loss. Figure 7.2 shows three different plots of a sample of 18 records of honey bee activity.

By grouping the data into frequencies, information is lost on the order in which different activities occurred. Hence, while Figure 7.2 tells us which activity the bee is most likely to be performing at any given time, it can't tell us which activity is most likely to succeed it in the next time instant.

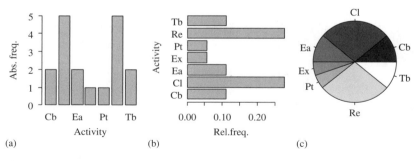

Figure 7.2: Approaches to visualising the frequencies in nominal data: (a) a vertical bar chart of absolute frequencies; (b) a horizontal bar chart with relative frequencies; (c) a pie chart.

In quantitative variables (both discrete and continuous), the equivalent graphical device to the bar chart is called a **histogram**. This is simply a plot of bars, with no gaps between them, representing the absolute or relative frequencies of each value of the variable. Creating histograms of continuous variables requires a **discretisation** of their range into arbitrary classes called **bins**.

Example 7.6: Height of tree ferns

Below are the heights (in m) of 117 tree ferns in a particular area of rainforest, the corresponding table of frequencies (Table 7.3) and histogram (Figure 7.3).

1.40, 1.88, 2.55, 2.18, 1.42, 1.91, 1.96, 1.81, 2.29,
2.43, 1.82, 1.86, 2.06, 2.20, 2.26, 2.33, 1.47, 2.07,
1.90, 1.86, 1.98, 1.88, 2.13, 2.27, 2.02, 2.12, 1.87,
2.06, 1.98, 1.50, 1.66, 1.99, 1.79, 1.61, 1.74, 2.18,
2.25, 2.36, 1.96, 1.92, 1.96, 1.94, 1.82, 1.79, 1.90, 2.06, 1.92, 1.85,
2.03, 1.75, 1.59, 2.28, 1.51, 2.32, 2.08, 2.03, 2.02, 1.86, 1.81, 1.91,
2.06, 1.93, 2.01, 1.75, 1.56, 2.06, 1.64, 2.00, 1.88, 1.73,
1.64, 1.58, 1.90, 1.82, 1.72, 1.55, 1.75, 1.99, 1.89, 1.67,
1.76, 2.08, 1.95, 2.06, 2.16, 2.24, 2.24, 2.28, 1.55, 1.55,
1.68, 1.80, 1.99, 1.64, 1.60, 2.23, 1.544, 1.75, 2.02, 1.89,
1.79, 1.90, 1.67, 1.98, 2.05, 1.78, 1.94, 1.90, 1.81, 1.93,
1.86, 1.92, 2.10, 1.90, 1.92, 1.96, 2.08

Table 7.3

Bin	Frequency	Bin	Frequency
1.30	1	2.00	18
1.40	2	2.10	6
1.50	10	2.20	9
1.60	8	2.30	3
1.70	12	2.40	1
1.80	22	2.50	1
1.90	24	2.60	0

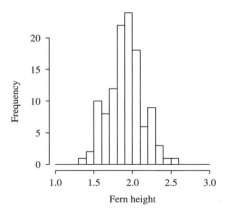

Figure 7.3: Histogram of 117 tree fern heights.

Histograms and bar/pie charts represent graphically the **frequency distributions** of random variables, i.e. the pairing of the possible values of the variable with their observed frequency. Frequency distributions of qualitative variables tell us which is the rarest/commonest level and whether the frequency of different levels is similar or not. Frequency distributions for quantitative variables provide even more information. For example, it is interesting to see whether a histogram is symmetric or **skewed**.

Example 7.7: Gannets on Bass rock

 During a single field season on Bass Rock, researchers visit 1000 nests and collect data on three random variables: the weight of the egg, the age of the parent and the number of neighbouring nests within a 2m radius of each focal nest. The histogram of egg weights is symmetric (Figure 7.4(a)): most of the measurements are concentrated around the centre of the distribution and there are few very heavy or very light eggs. The histogram of parent ages starts from the value 5 and is **positively skewed** (i.e. it has a long tail to the right – Figure 7.4(b)). This is a result of two life history facts, the age at first reproduction (5 y) and the annual mortality of adults that results in smaller numbers of older individuals. The histogram of the number of neighbours is **negatively skewed** (Figure 7.4(c)), possibly because most animals are at the interior of the colony where the density is constant. A few nests that happen to be in sub-standard habitat patches or close to the boundaries of the colony may experience lower densities and have fewer neighbours.

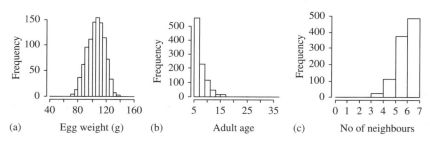

Figure 7.4: Histogram of three random variables relating to 1000 gannet nests in the breeding colony of Bass Rock: (a) the weight of the eggs in the nests; (b) the age of the parent; (c) the number of neighbouring nests.

Since the shape of a histogram can be interpreted biologically, it is important to have objective (i.e. numerical) ways to compare different shape characteristics. Each one of those numerical summaries is called a **statistic**. Four types of statistics are introduced in Sections 7.6–7.8 referring to the centrality, spread, symmetry and compactness of frequency distributions.

⦿ 7.3: Bar charts, pie charts, histograms

To visualise the distributions of qualitative variables, you first need to declare the data as a factor (see R7.1) and then use the command `table()` which tabulates the frequency of occurrence for each factor level. Here is an example:

```
> A<-factor(c("Re","Cb","Tb","Re","Ea","Re","Pt","Cb",
+ "Ex","Cl","Cl","Re","Ea","Cl","Cl","Re","Tb","Cl"))
> table(A)
A
Cb Cl Ea Ex Pt Re Tb
 2  5  2  1  1  5  2
```

The table generated in this way can be used directly in the plotting commands `barplot()` or `pie()`. The following lines will generate plots like those in Figure 7.2:

```
barplot(table(A), xlab="Activity", ylab="Abs. freq.")
n<-length(A)   #Sample size, used in converting abs. to rel. freqs
barplot(table(A)/n, horiz=T, ylab="Activity", xlab="Rel. freq.")
pie(table(A))
```

Generating histograms for quantitative variables is even easier. Given a numerical list of raw data, say `data`, then `hist(data)` will automatically generate a histogram. The option `breaks` within the command `hist()`, allows you to specify the bins of your histogram. The following lines will generate the tree fern histogram in Example 7.6:

```
bins<-seq(1.3, 2.6, by=0.1)
fernHist<-hist(data, breaks=bins, main="", xlab="Fern height")
```

Here, the object generated by `hist()` contains the frequencies associated with each bin. To see them type

```
> fernHist$counts
[1]  1  2 10  8 12 22 24 18  6  9  3  1  1
```

7.6.   Measures of centrality

If asked to describe a distribution, you could make a good start by saying where its 'centre' is located. The most widely known statistic for centrality is the **average**. Some notation will facilitate the discussion. If we are interested in measuring the weight (X) of gannet eggs on Bass Rock, a data set D containing n measurements may be represented symbolically as follows:

$$D = \{x_1, x_2, x_3, \ldots, x_n\} \tag{7.3}$$

The average weight in this sample is defined as the sum of all measurements divided by the sample size. In general, an over-bar is used to denote the average,

$$\bar{x} = \frac{\sum_{i=1}^{n} x_i}{n} \tag{7.4}$$

If the data come from a census of a finite population, then the statistic in Equation (7.4) is called the population **mean**. The mean of an infinite population is also defined, but never observed.

On several occasions, data may not be available in their raw form but, instead, as a table of frequencies. The average can still be calculated by using the frequencies as **weights**.

Example 7.8: Chick rearing in red grouse

Red grouse, Lagopus lagopus scotticus

The following are data on the number (M) of chicks successfully reared by each of 20 red grouse hens in a sample.

$$\{5, 7, 1, 2, 0, 5, 3, 2, 4, 3, 2, 3, 1, 0, 3, 2, 4, 0, 2, 3\} \tag{7.5}$$

Using Equation (7.4), the average of this sample can be calculated as $\bar{m} = 2.6$,

$$\bar{m} = \frac{5+7+1+2+0+5+3+2+4+3+2+3+1+0+3+2+4+0+2+3}{20} \tag{7.6}$$

The result would be exactly the same if the data were sorted in increasing order

$$\bar{m} = \frac{0+0+0+1+1+2+2+2+2+2+3+3+3+3+3+4+4+5+5+7}{20} \tag{7.7}$$

This makes it easier to see how many times each value of the variable occurs in the data. For example, the number 0 occurs three times and the number 1 occurs twice. We can therefore simplify this expression by collecting like terms together:

$$\bar{m} = \frac{3 \cdot (0) + 2 \cdot (1) + 5 \cdot (2) + 5 \cdot (3) + 2 \cdot (4) + 2 \cdot (5) + 0 \cdot (6) + 1 \cdot (7)}{20} \tag{7.8}$$

The numbers in brackets are the values of the variable and the numbers in front of the brackets are the absolute frequencies of each value in the sample.

This result can be written in more general notation to obtain an alternative definition of the average

$$\bar{x} = \frac{\sum\limits_{\text{All}x} \phi(x)x}{n} \tag{7.9}$$

where the sum in the numerator is over all possible values x of the random variable X and $\phi(x)$ is the absolute frequency of each value. Equation (7.2) can be used to obtain a simpler relationship, using relative frequencies

$$\bar{x} = \sum\limits_{\text{All}x} f(x)x \tag{7.10}$$

From Equation (7.10), it is clear that the value of the average will be influenced, not just by the frequency, but also by the particular values of the variable, making it particularly sensitive to extreme values in the data (**outliers**). Therefore, the average may be an unsuitable measure of centrality for skewed distributions (Figure 7.5).

There are two other measures of centrality that are not so sensitive to outliers. The **mode** of a distribution is the value of the variable that occurs with the highest frequency. The mode is particularly useful for describing **multimodal** distributions (i.e. those with more than one peak).

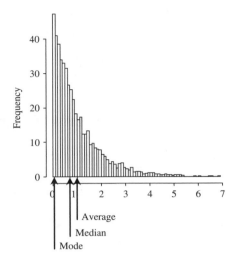

Figure 7.5: Relationship of the three most common measures of centrality illustrated on a heavily positively skewed distribution.

Example 7.9: Swimming speed in grey seals

Multimodality usually indicates that the data are being generated by a **mixture** of different processes. A typical histogram of horizontal speed measurements for an animal foraging in a patchy environment may have two separate modes (Figure 7.6): the animal moves slowly when foraging and quickly when transiting between food patches.

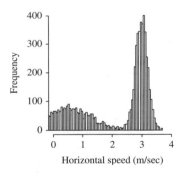

Horizontal speed (m/sec)

Figure 7.6: The mode on the left corresponds to localised movement during foraging. The mode on the right corresponds to directional movement during transit between patches.

The **median** is the middle value of a data set. To find the median, the data first need to be ordered. If the sample size is an odd number, then the median is simply the value in the middle of the sorted data. If the sample size is even, then the median is the average of the middle two observations.

Example 7.10: Median of chicks reared by red grouse

The original chick-rearing data {5,7,1,2,0,5,3,2,4,3,2,3,1,0,3,2,4,0,2,3} can be sorted, to give

$$\{0,0,0,1,1,2,2,2,2,\underbrace{2,3}_{2.5},3,3,3,3,4,4,5,5,7\} \qquad (7.11)$$

Since the sample size is $n = 20$, the median is calculated as the average of the 10th and 11th observations, i.e. 2.5.

There are yet more measures of centrality developed for particular types of data. The **geometric average** is defined as the nth root of the product of all observations and it is a useful statistic if the data refer to percentages or proportions. For example, successive measurements of percentage growth in a biological population may be summarised in this way.

7.4: Average, median, mode

The average and median of a set of data can be obtained directly, with the commands `mean()` and `median()`. For example,

```
> data<-c(14.6, 7.9, 10.2, 15.9, 11.8, 8.7, 12.2, 11.5, 11.0, 10.8)
> mean(data)
[1] 11.46
> median(data)
[1] 11.25
```

The mode is less straightforward. One way to calculate it is from the output of the command `hist()`. This command (see R7.3) produces an object containing the bin labels in the component `$mids` and their corresponding absolute frequencies in the component `$counts`. We can retrieve the list of frequencies and apply to it the command `which.max()`, to find the position of the greatest frequency (1st, 2nd, 3rd ... bin out of the total number of bins). The corresponding bin label can then be read off, at that position. Here it is in action:

```
> h<-hist(data, breaks=seq(5,20,2))  # Generates histogram
> pos<-which.max(h$counts) # Finds the tallest bar
> h$mids[pos] # Finds the value of the variable at the tallest bar
[1] 10
```

7.7. Measures of spread

Measures of centrality represent a 'typical' observation from the data, so it is logical to ask how similar any one datum is to the typical observation, i.e. how variable the data are. Intuitively, variability is the extent of spread of the data, so, one obvious statistic is the **range**, i.e. the difference between the largest and smallest value appearing in the data. However, the range is sensitive to outliers.

Example 7.11

The two data sets in Figure 7.7 have the same range (= 11). They clearly show that variability isn't as easily quantifiable as might be expected: there is more information in the data about variability than is conveyed by the range.

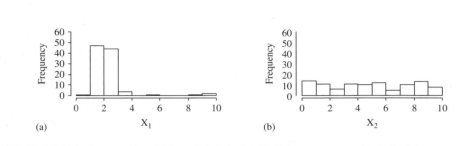

Figure 7.7: Histograms of two different data sets with the same range. Data in (b) are more variable than the data in (a).

The bulk of the data in part Figure 7.7(a) lies between 0 and 4, so it is tempting to discard the outliers from the first data set and simply quote the range of the remaining observations. This is the basis for the **inter-percentile range**, obtained by discarding a fixed percentage of observations on either side of the median. For example, the **inter-quartile range** discards 25% of the smallest and 25% of the largest values, keeping the central 50% of observations. Similarly, the **inter-decile range** keeps the central 80% of observations. These statistics are less likely to be affected by outliers but they are arbitrary because they require us to draw a line in the sand between the main body and the tails of a distribution.

Nevertheless, the range is a sound idea insofar as it measures the spread of a frequency distribution as an interval, a distance between two reference points. Its ambiguity lies in the choice of these two reference points. Rather than forcing a choice, the basic idea of the range may be applied repeatedly to all the data: we may obtain an appropriate measure of distance of every single datum from a suitably selected centre of the distribution and then collapse all these distances into a convenient single summary. The 'appropriateness', 'suitability' and 'convenience' of these decisions will become clearer when more material is covered on expectation (Chapter 9) and inference (Chapter 10). The most useful measure constructed in this way is the **standard deviation**.

To compute the standard deviation we first need to find the distances (also called **residuals**) of observed values from the average. For the ith observation the residual will be

$$(x_i - \bar{x}) \tag{7.12}$$

The residual in Equation (7.12) can be negative, zero or positive. This information is not required because the objective is to summarise the extent of variability around the average. A residual can be converted into an unsigned quantity in two ways: either by taking its absolute value or by raising it to an even power. The latter option, using **squared residuals**, is preferred for good theoretical reasons (see Section 9.4 on expectation).

$$(x_i - \bar{x})^2 \tag{7.13}$$

This operation will give as many squared residuals as the number of original data. A typical squared residual can be obtained as the average of all squared residuals. This is called the **variance** of the data

$$v(x) = \frac{\sum_{i=1}^{n} (x_i - \bar{x})^2}{n} \tag{7.14}$$

For the purpose of measuring the spread of a distribution, the variance is in the wrong units because the residuals have been squared. For example, if the original data referred to length, the variance is measured in units of area. To get back to the original units, we take the square root of the variance, giving the expression for the standard deviation,

$$s(x) = \sqrt{v(x)} = \sqrt{\frac{\sum\limits_{i=1}^{n} (x_i - \bar{x})^2}{n}} \qquad (7.15)$$

Hence, the two data sets in Example 7.11 have exactly the same range but their standard deviations differ considerably: $s(x_1) = 1.44$ and $s(x_2) = 2.94$, indicating that the distribution of values for X_2 is more dispersed.

It is often useful to compare the spread of distributions with different averages. In such cases, the objective may be to determine which of the two distributions is more dispersed *relative* to its average.

Example 7.12: Gannet foraging

The numbers of gannets from the same colony encountered feeding at marine locations with similar environmental characteristics depend on the distance of these locations from the gannet colony. Daily data from an inshore region (area 10^4 m^2), collected over a month, showed an average attendance of 21.4 gannets with a standard deviation of 5.1. Similar data from an offshore region of the same size and habitat type showed average attendance of 4.3 gannets with standard deviation 1.1. Which of the two data sets is more dispersed? The inshore data have the largest standard deviation but in both cases the standard deviation is about 25% of the average. In proportional terms, the appearance of a single extra gannet at the offshore location would cause a greater increase than at the inshore location.

To perform such standardised or proportional comparisons of spread we need to use the **coefficient of variation**, defined as the ratio of the standard deviation over the average

$$c_v(x) = \frac{s(x)}{\bar{x}} \qquad (7.16)$$

7.5: Range and standard deviation

The command `range()` provides the minimum and maximum of a data set (i.e. for any list of values (`data`), it is equivalent to `c(min(data),max(data))`. Any of the following three lines of code will calculate range as an interval:

```
range(data)[2]-range(data)[1]    # Difference between upper and
                                 # lower limit of the range
max(data)-min(data) # Difference between the max and min of the data
diff(range(data))   # The difference between the elements of range()
```

The inter-quartile range of the data can be calculated by `IQR(data)`. Other inter-percentile ranges can be calculated as well. Here is a small function that will calculate the range of the central a% of a list of data.

```
IPR<-function(data, a)
  {
```

```
p<-(100-a) # percentage of data outside the required range
plow<-p/2  # lower percentile
phi<-100-p/2 # upper percentile
dat<-sort(data) # creates ordered data set dat
n<-length(data) # sample size
nlow<-round(n*plow/100) # position, in dat, corresponding to plow
nhi<-round(n*phi/100) # position, in dat, corresponding to phi
return(c(dat[nlow],dat[nhi]))
}
```

The standard deviation of the data can be calculated directly as sd(data). It should be noted that R uses a slightly different formula to Equation (7.15). This involves a correction required when the standard deviation of the sample is to be interpreted as an estimate of the standard deviation of the wider population (see Section 10.4). To calculate standard deviation by Equation (7.15), simply type

```
sqrt(sum((mean(data)-data)^2)/length(data))
```

7.8.   Skewness and kurtosis

To describe the shape of a frequency distribution in more detail, it is necessary to introduce statistics for symmetry and the relative thickness of the distribution's tails. These statistics are called **skewness** and **kurtosis** respectively:

$$skewness = \frac{1}{ns^3} \sum_{i=1}^{n} (x_i - \bar{x})^3 \qquad kurtosis = \frac{1}{ns^4} \sum_{i=1}^{n} (x_i - \bar{x})^4 - 3 \qquad (7.17)$$

For now, it is useful to know how to interpret the values of these statistics in comparison to the characteristics of the baseline distribution (the normal distribution to be discussed in Chapter 9 that has zero skewness and kurtosis – Figure 7.8(a)). A negative value for skewness indicates a longer left-hand tail and a positive value indicates a longer right-hand tail (Figure 7.8(b)). Positive values for kurtosis indicate thicker tails than the baseline distribution (Figure 7.8(c)).

Example 7.13

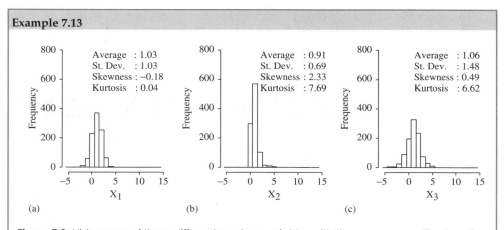

(a)   (b)   (c)

Figure 7.8: Histograms of three different random variables with the same mean. The baseline distribution (a) is symmetric around its mean and has zero kurtosis. In comparison, the distribution in (b) is positively skewed and therefore has a large proportion of observations in its right-hand tail. Finally, the symmetric distribution in (c) has thicker tails than (a) although their peaks are the same.

7.6: Skewness and kurtosis

There are no predefined commands for skewness and kurtosis, so here are two functions that will do the job using Equations (7.17).

```
skew<-function(data)
  {
  n<-length(data)
  mu<-mean(data)
  s<-sd(data)
  return(sum((data-mu)^3)/(s^3*n))
  }

kurt<-function(data)
  {
  n<-length(data)
  mu<-mean(data)
  s<-sd(data)
  return(sum((data-mu)^4)/(n*s^4)-3)
  }
```

7.9. Graphical summaries

The **box** (or **box and whisker**) **plot** communicates information about the centre, spread, skewness and kurtosis of a frequency distribution (Figure 7.9). The median of the distribution is shown as a thick bar. Around this is drawn a rectangle representing the inter-quartile range. Extending beyond this box are two whiskers, generated as multiples of the inter-quartile range. If there are any outliers beyond the whiskers, they are plotted individually. If no point lies outside the whiskers, then the ends of the whiskers are made to coincide with the outermost point.

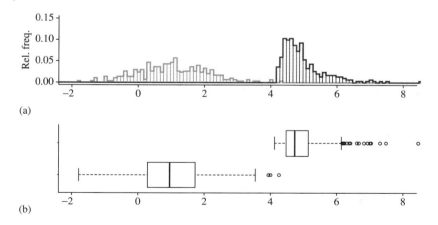

Figure 7.9: The histograms in (a) are based on samples from two different populations. The relative frequency distributions differ in position, spread, skewness and kurtosis. These differences are captured by the corresponding box plots in (b).

The visual information content of a box plot is high: the median represents the centre of the distribution, the box enclosing the bulk of the distribution gives information about its spread and skewness. The same properties are represented by the whiskers, but a comparison between the whiskers and the box also informs about the relative sizes of the main body versus the tails of the distribution.

The box plot is useful for three reasons: firstly, unlike the histogram, it plots the data without the need for binning them into (possibly) arbitrary categories. Secondly, only the salient features of each distribution are represented, the finer, random features of the data are ignored. Finally, it achieves a similar representation for samples of different size and therefore allows easy comparisons.

 7.7: Box and whisker plots

A box and whisker plot for one or more samples of data can be generated with the command `boxplot()`. The option `horizontal` allows the user to swap the axes around. For example, given two samples `d1` and `d2`, a plot similar to Figure 7.9(b) can be generated by typing `boxplot(d1,d2, horizontal = TRUE)`.

7.10.   Data sets with more than one variable

A **sample unit** (e.g. an individual animal or plant, a location in space, a unit of time, etc.) is the subject from which an observation is derived. However, more than one attribute may be measured on any given sample unit (e.g. several aspects of an individual organism such as its physical dimensions, physiological state, genetic traits, etc.). Each of these attributes constitutes a random variable and it may be considered in conjunction with another variable, to examine **relationships** between them. The approach used to do this depends on whether the two variables are qualitative (Section 7.11) or quantitative (Section 7.12).

 7.8: Data frames remembered

The essential storage device used by R, particularly for data sets of many variables, is the data frame (Section 0.10). In simple cases (i.e. when the data come from only one type of sample unit), the rows of the data frame represent sample units and the columns represent different variables.

7.11.   Association between qualitative variables

Example 7.14: Community recovery in abandoned fields

 With the growing worldwide trend of abandonment of agricultural land, it is important to understand the development of plant communities on old fields (Cramer *et al.*, 2008). One of the factors affecting recovery is the type of agriculture to which a field was historically subjected. We may classify this into three broad categories:

❶ Traditional, with limited modification.
❷ Pasture, rather than cropping.
❸ Intensive, industrialised cropping.

The type of recovery may also be classified into three categories:

❶ Full recovery to ancestral community.
❷ Partial recovery (i.e. delayed recovery, or recovery along a novel trajectory).
❸ No recovery (i.e. fields remain in a persistent degraded state).

Table 7.4 shows data on the state of recovery of 108 similar-sized fields, abandoned around the same time, in combination with their historical, agricultural use.

Table 7.4

		Agriculture type			
		Tradit.	*Pasture*	*Intensive*	*Row total*
Recovery type	*Full*	11	2	1	14
	Partial	1	20	13	34
	None	0	12	48	60
	Colunm total	12	34	62	108

Here, the row and column totals give the frequency distributions of the variables 'Recovery type' and 'Agriculture type'. For example, the majority of fields in this sample (62 out of 108) were subjected to intensive farming. The interior entries give the frequency of occurrence of each combination of values of the two variables. For example, the communities of 48 fields that were subjected to intensive farming have not yet recovered.

This is an example of a **contingency table**, more broadly defined as a table of the frequencies of occurrence of the combinations of values of different variables. The table in Example 7.14 is a two-way, 3×3 contingency table because it involves two variables, each with three values. Strictly speaking, the frequency distribution of a single variable may also be seen as a contingency table, so that, for example, the distribution of the variable 'Agriculture type' in the data of Example 7.14 can be thought of as a one-way contingency table with three values.

A logical question to ask concerns association between the qualitative variables in a contingency table; for example, in Example 7.14, is community recovery likely to be dependent upon the type of agriculture that was historically practiced on the field? There are several different statistics for this purpose, but here I mention just two, without explaining how they are calculated: the **phi statistic** (taking values between −1 and 1) can be used for 2×2 contingency tables (i.e. two variables, each with only two values). Absolute values of the phi statistic close to 1 indicate strong association and values close to zero indicate weak association. When dealing with larger contingency tables, the appropriate statistic is **Cramer's V**, which takes values between 0 (weak association) and 1 (strong association).

 7.9: Contingency tables

Given one or more qualitative variables, the command `table()` will calculate the frequencies of occurrence of all combinations of values of these variables and store them in an array. Here is an example with a subset of eight fields from the data of Example 7.14. The numbers 1,2,3 are used as shorthand for 'traditional', 'pasture' and 'intensive' respectively and the letters a,b,c stand for 'full', 'partial' and 'no recovery'.

```
> AgriType<-factor(c(1,3,3,3,1,2,2,3)) # Agricultural treatment
> RecoType<-factor(c("a","c","c","b","a","b","a","c")) # Recovery
> data<-data.frame(cbind(AgriType,RecoType)) # Creates data frame
> table(data)
        RecoType
AgriType 1 2 3
       1 2 0 0
       2 1 1 0
       3 0 1 3
```

The R package vcd has a wealth of commands for visualising and summarising contingency tables. You will first need to install vcd from your nearest CRAN mirror. You may then use the command assocstats() on a contingency table to calculate the phi or Cramer's V association statistics.

7.12.   Association between quantitative variables

Example 7.15: Height and root depth of tree ferns

Tall trees should have deep roots. But how strong is this association for any particular species? The variables in question, 'Tree height' and 'Root depth' are continuous. The sample unit is the individual tree and both variables are measured on every tree to generate a data set of paired values. To visualise their association, the height and root depth of each tree can be treated as the x and y coordinates in a two-dimensional plot (Figure 7.10).

If the association is strong, we would expect that most trees that are taller than average also have deeper-than-average roots. Conversely, shorter-than-average trees should have shallower-than-average roots. Hence, if the association is strong, most trees should be in the unshaded quadrants of Figure 7.10.

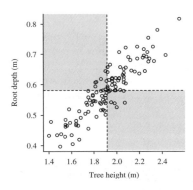

Figure 7.10: A scatter plot of root depth against tree height. Trees taller than average are on the right of the dashed vertical line; trees with roots deeper than average are above the dashed horizontal line.

This graphical representation of two-variable samples is called a **scatter plot**. It is the foremost tool used for the exploratory analysis of association (also see Example 1.52). The pattern seen in Figure 7.10 indicates a possible linear association between the two variables, but can we quantify its strength? The intuitive notions used to describe the strength of association in Example 7.15 can be formalised into the concept of **correlation**, usually denoted by r if the data are a sample and ρ for the entire population.

The most widely used statistic for correlation is **Pearson's correlation coefficient**. The rationale behind it is most easily explained with reference to Figure 7.10. Below, I use the sample notation but the definition for the population is the same. Consider two quantitative random variables X and Y, with corresponding averages $\bar{x}$, $\bar{y}$ and standard deviations s_x, s_y. The differences $\Delta x_i = x_i - \bar{x}$ and $\Delta y_i = y_i - \bar{y}$ indicate whether the pair of measurements (x_i, y_i) lie above or below their respective averages. Multiplying these differences together gives information about their relationship: a positive product indicates that the measurements for both variables were on the same side of their respective means (i.e. in the unshaded regions of Figure 7.10). Conversely, a negative product indicates that the measurements were on the opposite sides of their respective means (i.e. the shaded regions of Figure 7.10). The products thus calculated for all sample units can be averaged together to tell us whether the relationship is predominantly positive or negative. The resulting quantity is called the **covariance** between the two variables,

$$\text{cov}(x, y) = \frac{\sum_{i=1}^{n} (x_i - \bar{x})(y_i - \bar{y})}{n} \tag{7.18}$$

The order in which the two variables are considered does not matter (i.e. $\text{cov}(x, y) = \text{cov}(y, x)$). Extreme observations contribute more to this sum than observations close to the averages. This is a convenient feature because it effectively discounts small variations affecting central points (in Figure 7.10, small random variations can easily shift points that are close to the averages from one quadrant to another, so the closer you look to the centre of the cross, the more random the pattern appears).

Therefore, the covariance will be large (positive or negative) for a strong association, but the scale on which it measures strength depends on the units of the two variables and the spread of their respective distributions. We can force it to remain within the values -1 and 1 by dividing it by the product of the two standard deviations (I won't prove this here) to get the Pearson's correlation coefficient

$$r = \frac{\text{cov}(x, y)}{s_x s_y} \tag{7.19}$$

This expression can be expanded according to the definitions of the standard deviation (Equation (7.15)) and covariance (Equation (7.18)) to get an alternative (and scarier) expression for correlation:

$$r = \frac{\sum_{i=1}^{n} (x_i - \bar{x})(y_i - \bar{y})}{\sqrt{\sum_{i=1}^{n} (x_i - \bar{x})^2} \sqrt{\sum_{i=1}^{n} (y_i - \bar{y})^2}} \tag{7.20}$$

From the derivation of correlation, it should be clear that it can only describe the strength of linear association. Nonlinear patterns that are clearly visible to the eye (Figure 7.11) are not as easily detected by the correlation coefficient.

Example 7.16

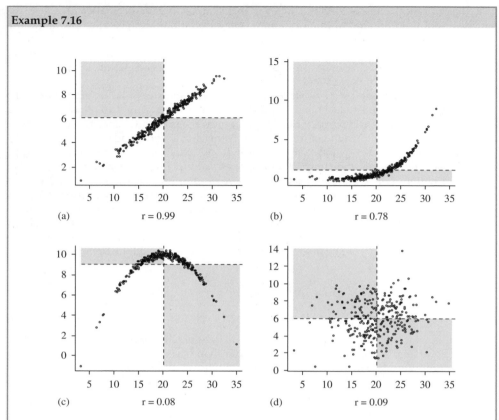

Figure 7.11: Correlation may only be used to quantify the strength of linear association. A strong linear pattern (a) gives a correlation close to 1 but a strong nonlinear pattern (b) gives a much smaller value. Some strong nonlinear patterns (c) may even give correlation values corresponding to randomness (d).

7.10: Scatter plots and correlation

To create a scatter plot of two paired lists of data x, y, simply use plot(x,y) (also see R1.13). For the correlation and covariance, type cor(x,y) and cov(x,y), respectively. If you have a data frame dat containing two or more variables, then pairs(dat) will plot all the pair-wise scatter plots between your variables. Figure 7.12 was generated from a data frame with three variables, x, y and z, using the command pairs().

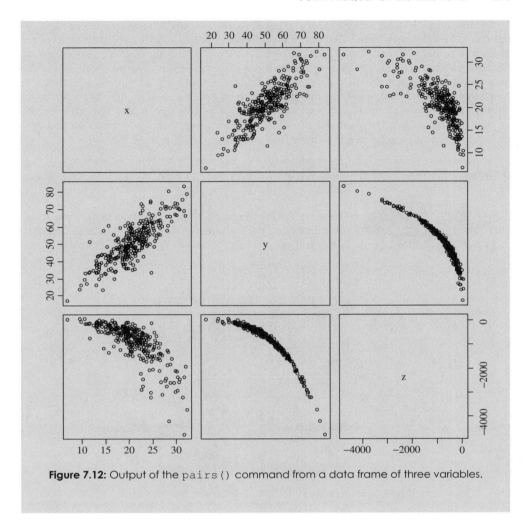

Figure 7.12: Output of the `pairs()` command from a data frame of three variables.

7.13. Joint frequency distributions

Frequencies (absolute or relative) can be calculated for samples of two variables. We may, for example, calculate the number of fern trees whose height is between 1 and 1.2 m *and* whose roots are between 0.9 and 1 m deep. This is called a **joint frequency**, because it refers to values of both variables together. Obtaining joint frequencies for all possible combinations of values of the variables gives their **joint frequency distribution**. In this section, I extend the visual and numerical summaries introduced for the distributions of single variables to the joint frequency distributions of pairs of variables.

The main tools used in this chapter to visualise distributions of single variables were the bar chart (for qualitative variables) and the histogram (for quantitative variables). There are analogues of both of these for visualising joint frequency distributions in three dimensions, but they can get rather confusing (tall bars in the foreground can obscure short bars at the back and the 3D perspective may give a misleading impression of the frequencies in different bins). Instead, it is possible to keep the graphical representation in two dimensions by using shape, size or colour intensity to represent relative frequencies. For qualitative variables, the method of choice is called a **mosaic plot**. Each combination of factor levels is plotted as a rectangle whose area is the product of the frequency of each factor level.

Example 7.17: Mosaics of abandoned fields

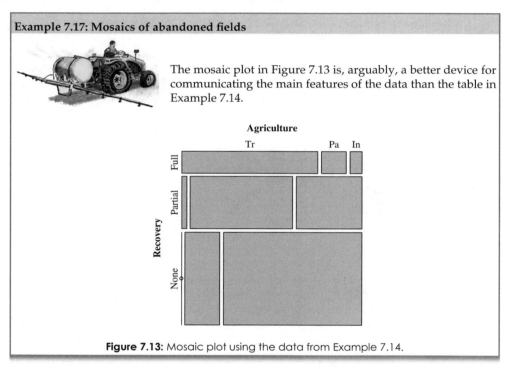

The mosaic plot in Figure 7.13 is, arguably, a better device for communicating the main features of the data than the table in Example 7.14.

Figure 7.13: Mosaic plot using the data from Example 7.14.

An alternative way to visualise joint distributions is to use colours, or shades of grey, to represent frequencies. This leads to the equivalent of the histogram for joint distributions.

Example 7.18: Joint distribution of tree height and root depth

The scatter plot is a quick diagnostic of association between two variables but it can give a misleading impression of the centre and spread of a joint distribution of large samples. Figure 7.14(a) shows a scatter plot of data from 6000 tree ferns. It is rather hard to identify the position of the average tree in this plot. Also, a few outliers tend to dominate our impression of the overall spread of the cloud of points. Alternatively, the plot may be divided up, using a grid, and the number of trees that fall into every grid cell (or bin) can be represented by shading or colour. In the resulting density plot (Figure 7.14(b)), the centre and spread of the joint distribution are clearly discernible.

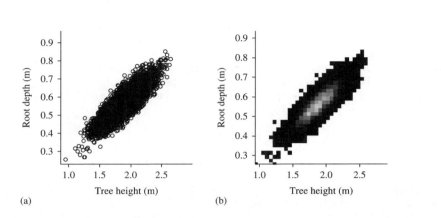

Figure 7.14: (a) Scatter plot of 6000 observations of root depth vs. tree height and (b) density plot of the frequencies of the same data, binned into a 30×30 grid. Light shades of grey indicate high frequencies and darker shades low frequencies. Zero frequencies are shown in white.

The frequency distribution of any one of the variables participating in a joint distribution is called the **marginal distribution** of that variable. Marginal distributions can be visualised as one-dimensional histograms aligned with the corresponding axes of a two-dimensional histogram (Figure 7.15).

Example 7.19: Joint and marginal distributions of tree size

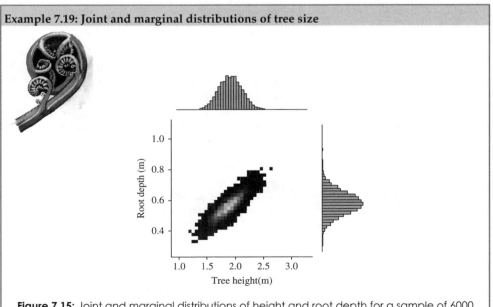

Figure 7.15: Joint and marginal distributions of height and root depth for a sample of 6000 ferns.

A numerical summary of the properties of a joint distribution can be obtained from the summaries (i.e. centre and spread) of the marginal distributions and their association. Usually, these are organised into two matrices: the first is the **multivariate average** ($\bar{x}$), a column vector that collects together the averages of the individual random variables. The second is the **variance–covariance matrix** (**C**), a square matrix that has the variances of the individual variables on the diagonal and the covariances between them in the off-diagonal elements. Therefore, given a paired sample for two random variables X and Y, their joint frequency distribution is summarised as follows:

$$\bar{x} = \left(\begin{array}{c} \bar{x} \\ \bar{y} \end{array} \right) \mathbf{C} = \left(\begin{array}{cc} s_x^2 & \mathrm{cov}(x,y) \\ \mathrm{cov}(y,x) & s_y^2 \end{array} \right) \tag{7.21}$$

Since $\mathrm{cov}(x,y) = \mathrm{cov}(y,x)$, the matrix **C** is symmetric around its diagonal.

 7.11: Summarising joint distributions

The command `mosaic()` in the package `vcd` can be used to generate a mosaic plot from a contingency table. I have used a curtailed version of the data in Figure 7.13,

```
require(vcd)
AgriType<-factor(c(1,3,3,3,1,2,2,3)) # Agricultural treatment
RecoType<-factor(c("a","c","c","b","a","b","a","c")) # Recovery
data<-data.frame(cbind(AgriType,RecoType)) # Creates data frame
mosaic(table(data))
```

Facilities for plotting joint frequency distributions can be found in the package `gregmisc`. Given paired observations in the variables `x` and `y`, a 2D histogram can be obtained by typing `hist2d(x,y)`. By default, this uses 200 bins in each direction and represents density by colours. You may want to change these to suit your needs. For example, a plot using 30 bins and greyscale can be specified as follows:

```
hist2d(x,y), nbins=30, col=c("white",gray(seq(0,0.9,0.1)))
```

Further reading

Introductory books for statistics come in two guises: cookbooks and textbooks. Cookbooks, such as Sokal and Rohlf (1994) and Harris *et al.* (2005), offer self-contained recipes that achieve a particular task. They are great for practising scientists that already have (or, do not particularly care to obtain) a more global understanding of statistics. Textbooks, such as Wild and Seber (2000) and Whitlock and Schluter (2009) build the theory and practice of statistics from the foundations. They require more consistent reading, but offer greater long-term rewards. Almost all of the dedicated textbooks in statistics will expand upon the introductory material presented in this chapter, although most of them tend to treat univariate and multivariate samples in different chapters. Graham (2006) takes a more philosophical (and less mathematical) approach to the basic principles of statistics. Almost all introductory books to R will contain sections for summarising and visualising data. Good points of reference are Crawley (2005, 2007), Cohen and Cohen (2008) and Dalgaard (2008). Finally, if you want to go a little bit deeper into ecological statistics, an excellent reference is Gotelli and Ellison (2004).

References

Cohen, Y. and Cohen, J.Y. (2008) *Statistics and Data with R: An Applied Approach through Examples*. Wiley Blackwell. 618pp.

Cramer, V.A., Hobbs, R.J. and Standish, R.J. (2008) What's new about old fields? Land abandonment and ecosystem assembly. *Trends in Ecology and Evolution*, **23**, 104–112.

Crawley, M.J. (2005) *Statistics: An Introduction using R*. Wiley Blackwell. 342pp.

Crawley, M.J. (2007) *The R Book*. John Wiley & Sons, Ltd, Chichester. 942pp.

Dalgaard, P. (2008) *Introductory Statistics with R*. Springer. 364pp.

Gotelli, N.J. and Ellison, A.M. (2004) *A Primer of Ecological Statistics*. Sinauer Associates, Massachusetts. 510pp.

Graham, A. (2006) *Developing Thinking in Statistics*. Sage. 288pp.

Harris, M., Taylor, G. and Taylor, J. (2005) *Catch-up Maths and Stats for the Life and Medical Sciences*. Scion Publishing. 180pp.

Huff, D. (1954) *How to Lie with Statistics*. Norton, New York. 128pp.

Sokal, R. R. and Rohlf, F.J. (1994) *Biometry: Principles and Practice of Statistics in Biological Research*, 3rd edition. Freeman & Co. 880pp.

Whitlock, M.C. and Schluter, D. (2009) *The Analysis of Biological Data*. Roberts & Co., Colorado. 700pp.

Wild, C.J. and Seber, G.A.F. (2000) *Chance Encounters: A First Course in Data Analysis and Inference*. John Wiley & Sons, Inc., New York. 611pp.

8

How to put a value on uncertainty
(Probability)

'That is probable which for the most part usually comes to pass, or which is a part of the ordinary beliefs of mankind.'
Marcus Tullius Cicero (106–43 BC), Roman philosopher

Some natural phenomena, such as the time of sunrise on a given date or the boiling point of a known liquid under a certain pressure, obey unambiguous laws that make them predictable. Observing such **deterministic** phenomena at some point in the past allows us to predict their future outcome. Much of science up until the 19th century had built its reputation for prediction by selectively focusing on deterministic phenomena. Nondeterministic phenomena were pushed aside, beyond the realm of physical science and the remit of textbooks. As a result, the majority of high-school and university graduates in science and engineering were allowed to believe that most of the natural world is deterministic, when it patently is not. A living person's lifespan, tomorrow's weather or the number of animals in a population three years from now cannot be predicted exactly, even though we may have observed similar processes in the past. These are called **random** or **stochastic** phenomena. As scientific endeavour and the need for prediction went beyond the narrow context of Newtonian mechanics and overflowed into biology and environmental science, the study of stochastic phenomena became unavoidable. So, natural scientists grasped the nettle of stochasticity and reacquainted themselves with the theory of probability, an arcane area of mathematics and logic that had, until then, been relegated to the account books of gamblers, bankers and insurers. Dealing with stochasticity requires a mentality shift: an intuition developed through years of deterministic schooling may not be relied upon to understand uncertainty.

How to be a Quantitative Ecologist: The 'A to R' of Green Mathematics and Statistics, First Edition. Jason Matthiopoulos.
© 2011 John Wiley & Sons, Ltd. Published 2011 by John Wiley & Sons, Ltd.

Probability tries to quantify uncertainty. Although there is still some debate about how it should be defined, the consensus is that probability should be derived from, and confronted with, data. A vital scientific tool used for collecting data is the **experiment**, so, in Section 8.1 the term is unambiguously defined. Section 8.2 discusses the outcomes of an experiment, called **events**, and introduces **Venn diagrams**, an invaluable graphical device for thinking about probability. One definition of probability, as a relative frequency, is presented in Section 8.3 together with the mathematical rules relating to probabilistic calculations (Sections 8.4–8.8). **Bayesian probability**, an alternative definition that folds together experimental data and observer beliefs, concludes this chapter in Section 8.9.

8.1.   Random experiments and event spaces

The popular perception of experiments involves white coats, frothing test tubes or guinea pigs in laboratories. More informed definitions invoke single-variable manipulations under controlled conditions. Surprisingly, many practising scientists struggle when asked to define exactly what an experiment is. In the case of ecology, where much of the data collection is done outside the lab under conditions that cannot be controlled, a broader definition is needed: a procedure or data collection opportunity is called an **experiment** if:

❶ it can be repeated many times under similar conditions (each such repetition is called a **trial**);

❷ it has a well-defined set of possible outcomes (known as the experiment's **event space** and usually denoted by Ω); and

❸ the outcome of any particular trial is unknown before the trial is performed.

Example 8.1: Assumptions of random experiments

The leaning tower of Pisa: reputedly the site of gravitational experiments before the advent of health and safety regulations

Consider the following instances of data collection in the light of the three defining properties of experiments:

- Recording the sex of the next calf born to a baleen whale.
- Recording the sexes of the whale's next two calves.
- Measuring the height of an apple tree in an orchard.
- Counting the number of melanomas on the skin of a wild boar of known age.
- Recording the total cod landings, by weight, in the fishing port of Peterhead on Monday.

Property ❸ is certainly satisfied by all five of these examples. The value of the random variables 'Calf sex', 'Sexes of next two calves', 'Tree height', 'Number of melanomas' and 'Total landings by weight' are not known before the measurements are made. Property ❷ is also satisfied. Event spaces can be written down for all five examples:

- $\Omega = \{male, female\} = \{\male, \female\}$
- $\Omega = \{\male\male, \male\female, \female\male, \female\female\}$
- $\Omega = \mathbf{R}^+$ (any positive real number)
- $\Omega = \{0, 1, 2, 3, \ldots\} = \{0, \mathbf{N}\}$
- $\Omega = \{0, \mathbf{R}^+\}$

With respect to property ❶, none of these experiments can be repeated under *exactly* the same conditions. In examples 1 and 2, the whale may be impregnated by different males and the condition of her reproductive system will differ each time, although these differences may not affect the gender of a whale's offspring. In example 3, tiny variations in sunlight, soil composition and water availability will influence the growth of different apple trees. In example 4, the genetic make-up of different boars will result in variable sensitivity to sunlight, and, as for the total cod landings in example 5, they will be influenced by so many variable factors that no-one can claim much similarity in conditions between this Monday and the next simply because they are the same day of the week in the same place in the country.

Therefore, property ❶ is a matter of interpretation: the influences acting on an experiment are split into three categories, those that are fixed (**conditions**), those that change but are known and measured (**covariates**) and those that are unknown (**stochastic influences**). Covariates will be examined in depth in Chapter 11. At that point, property ❶ will be relaxed even further by merely requiring that each trial happens under known (rather than similar) conditions. For now, keep in mind that *stochasticity or randomness is what remains after the fixed and known influences have been accounted for*.

The event spaces of the last three experiments in Example 8.1 are infinite sets that include unfeasibly large numbers. But how can feasibility be defined? A 100 m-tall apple tree cannot possibly exist, but 10 m-tall trees do. In counting from 10 up to 100 m, do we cross an unknown maximum value? If that maximum is set to some value, say 15 m, does this mean that 14.99 m is possible but 15.01 m is impossible? This chapter shows you how to avoid drawing such hard lines, by attaching to each member of the event space a probability. To do that, the event space must encompass all theoretically possible outcomes, even if their occurrence is practically impossible.

8.2.   Events

Events are collections of experimental outcomes. As such, the notation and language of sets is used to describe events.

Example 8.2: Plant occurrence in survey quadrats

A 0.25 m^2 quadrat is thrown in a field and the occurrence of a particular plant species is recorded. Each throw constitutes a trial. The event space of this experiment comprises the possibilities of absence (A) and presence (P),

$$\Omega = \{P, A\} \tag{8.1}$$

For a two-species experiment, the event space would be

$$\Omega = \{P_1 P_2, P_1 A_2, A_1 P_2, A_1 A_2\} \tag{8.2}$$

Each element of these event spaces represents a different outcome for the corresponding experiment. A single outcome from the event space is called a **simple event**. So, in the two-species experiment, the statement 'both species are present' is the simple event $P_1 P_2$. For brevity, events can be assigned symbols. For example, the presence of both species can be written:

$$E_1 = \{P_1 P_2\} \tag{8.3}$$

Compound events comprise more than one simple event. For example, the absence of at least one plant species from any given quadrat is the compound event E_2,

$$E_2 = \{P_1 A_2, A_1 P_2, A_1 A_2\} \tag{8.4}$$

An event may encompass all the possible outcomes of an experiment: The event E_3, 'the two plant species will be either present or absent' is the same as the event space for the two-species experiment

$$E_3 = \Omega \tag{8.5}$$

Spoken language also allows the construction of impossible events: 'of the two plant species, neither will be present or absent'. Such events, containing no outcomes, are represented by the **empty set**, $\emptyset$.

Relationships between two sets depend on whether they overlap or not. For example, the sets E_1, E_2 above are subsets of E_3,

$$E_1 \subset E_3, \ E_2 \subset E_3 \tag{8.6}$$

Such relationships are visualised using **Venn diagrams** (e.g. Figure 8.1(a)).

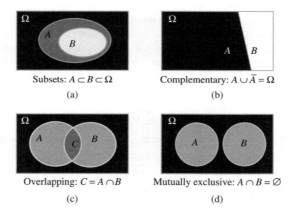

Subsets: $A \subset B \subset \Omega$

(a)

Complementary: $A \cup \bar{A} = \Omega$

(b)

Overlapping: $C = A \cap B$

(c)

Mutually exclusive: $A \cap B = \emptyset$

(d)

Figure 8.1: Venn diagrams representing four different arrangements for events A and B. In (a) and (c) the union of the two events is nonempty, but whereas in (a) one set is contained in the other, in (c) the overlap is partial. In (b) and (d) the events are mutually exclusive but in (b) they are also complementary.

In any given trial of the single-species experiment the plant will either be present (P) or absent (A). Combining these events gives the entire event space. Such events are called **complementary** (so, event P is the **complement** of A – see Figure 8.1(b)). The overbar notation is often used for the complements of events, so that $\bar{P}$ means 'the complement of P' (and is therefore the same as A). Alternatively, the overbar notation can be interpreted as **negation**. Hence, $\bar{P}$ means '*not* present'. For brevity, $\bar{P}$ can be read as '*not* P'. The overbar notation can be used twice to communicate double-negation. The following expression says that if a plant is not absent, then it must be present,

$$\bar{A} = \bar{\bar{P}} = P \tag{8.7}$$

The act of combining two events A and B into one gives their **union**, written $A \cup B$. The union of two complementary events is the entire event space,

$$P \cup \bar{P} = \Omega \tag{8.8}$$

In general, two or more events that, together, take up the entire event space are known as **collectively exhaustive**. Complementary events are exhaustive but exhaustive events need not be complementary, as the following example illustrates.

Example 8.3: Overlapping events

In the two-species experiment, consider the events 'At least one species is present': $E_1 = \{P_1P_2, \bar{P}_1P_2, P_1\bar{P}_2\}$ and 'At least one species is absent': $E_2 = \{\bar{P}_1P_2, P_1\bar{P}_2, \bar{P}_1\bar{P}_2\}$. These events are collectively exhaustive ($E_1 \cup E_2 = \Omega$) but not complementary, because they overlap: they share the event 'One of the two plants is present': $E_3 = \{\bar{P}_1P_2, P_1\bar{P}_2\}$.

The overlap C between two events A and B is called their **intersection** (Figure 8.1(c)) and it is written

$$C = A \cap B \tag{8.9}$$

Two nonoverlapping events have an empty intersection.

$$A \cap B = \emptyset \tag{8.10}$$

Such events are called **mutually exclusive** (Figure 8.1(d)). Complementary events are mutually exclusive, but mutually exclusive events need not be complementary.

Example 8.4: Mutually exclusive events

In the two-quadrat experiment, events $\{PP\}$ and $\{\bar{P}\bar{P}\}$ are mutually exclusive but they are not complementary. In the single-quadrat experiment, events $\{P\}$ and $\{\bar{P}\}$ are complementary.

---

⑧ 8.1: Set operations

Two sets are equal if they have the same unique membership. This can be checked with the command `setequal()`. For example,

```
> setequal(c(1,1,2),c(1,2))
[1] TRUE
```

The union and intersection of two (or more) sets can be calculated with the commands `union()` and `intersect()`,

```
> a<-c(1,3,4,5,6,8,9)
> b<-c(1,3,4,10,12,11)
> union(a,b)
 [1]  1  3  4  5  6  8  9 10 12 11
> intersect(a,b)
 [1] 1 3 4
```

8.3. *Frequentist probability*

Section 8.1 concluded that drawing a hard boundary between 'possible' and 'impossible' is not always a good idea. Instead, it may be more appropriate to think of a smooth transition between them, during which the likelihood of observing increasingly extreme values diminishes. Such a measure of likelihood for different events can be obtained empirically.

Example 8.5: Fluctuating frequency of newborn male wildebeest

Blue wildebeest, Connochaetes taurinus

The adult sex ratio in wildebeest is six females to four males (Mills and Shenk, 1992). To determine if this is the result of sex ratio at birth or higher male mortality, births were observed as they occurred and the sex of each new calf was recorded. The resulting database lists the total number (n) of males born up to and including the time of the Nth birth. Each birth is therefore treated as a trial of the experiment and the relative frequency of males born is the ratio of the total number of males over the total number of births. Results from the first ten births are shown in Table 8.1.

Table 8.1

N	1	2	3	4	5	6	7	8	9	10
n	1	1	1	1	1	2	3	4	4	4
n/N	1.00	0.50	0.33	0.25	0.20	0.33	0.43	0.50	0.44	0.40

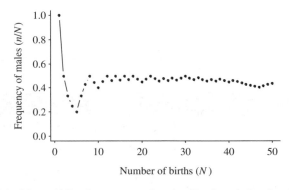

Figure 8.2: Plot of the relative frequency of male offspring during the first 50 births.

A plot of this relative frequency against the total number of births (Figure 8.2) shows that, in the long term, the frequency settles to a particular value about half-way between 0 and 1. Is this different from the observed adult frequency of 0.4? Continuing the observations beyond the first 2000 births, the frequency of male offspring settles to the value 0.5, indicating that the adult sex ratio is the result of differential mortality between males and females.

This 'stabilisation' of the relative frequency as the experiment is repeated an increasing number of times prompts the **law of large numbers**, which states that: every event is associated with a special number called its **probability** (p) such that if the random experiment is repeated a large number of times (N), then the relative frequency (n/N) of the event will tend to this special number. Hence,

$$p = \lim_{N \to \infty} \frac{n}{N} \tag{8.11}$$

A relative frequency is the ratio of two counts (the number of occurrences of an event, n, and the number of trials, N). Since neither of these numbers can be negative, their ratio will also be non-negative. Also, since $n \le N$, the relative frequency will never exceed one. In general therefore, probabilities take values between 0 and 1. A probability near zero indicates a rare (unlikely) event, a probability near 1 indicates a nearly certain event and a probability near 0.5 indicates an event that is as likely to occur as it is not to.

8.4.   Equally likely events

Consider a generic experiment with event space

$$\Omega = \{E_1, E_2, E_3, \ldots, E_n\} \tag{8.12}$$

where $E_1, E_2, E_3, \ldots, E_n$ are simple, mutually exclusive events. When a trial of the experiment is performed, one of these outcomes is certain to occur. Each of these outcomes may have a different probability ($P(E_i)$), but they must all add up to 1:

$$\sum_{i=1}^{n} P(E_i) = 1 \tag{8.13}$$

Example 8.6: Something is certain to occur (even if it is nothing)

If the experiment refers to the sex of a single offspring, then we can be certain that a pregnant wildebeest carries either a male or a female calf, i.e. $P(\sigma) + P(\varphi) = 1$.

If the quantity of interest is the total number (n) of offspring produced by the current population of $N = 103$ female wildebeest, then Equation (8.13) states

$$\sum_{n=0}^{N} P(n) = 1 \tag{8.14}$$

Although it is not known if the probability, $P(20)$, of observing 20 offspring is greater or smaller than the probability, $P(10)$, of observing 10, it is certain that a number between 0 and 103 will be observed. Note that 'zero offspring' is a distinct outcome in this event space and there is no reason to believe that $P(0) = 0$.

Equation (8.13) can be extended to infinite event spaces. The total number of offspring produced in next year's breeding season will depend on next year's breeding population. Since this is not yet known (allowing for recruitment, death and immigration), the event space of offspring numbers includes all non-negative integers,

$$\sum_{n=0}^{\infty} P(n) = 1 \tag{8.15}$$

This is a finite sum with an infinite number of terms, i.e. a convergent, infinite series (Section 5.4).

In some situations, all the possible outcomes of an experiment are assumed to be equally likely, with probability p. This assumption may be imposed by biological fact or it may serve as a convenient null hypothesis. Equation (8.13) gives

$$\sum_{i=1}^{n} P(E_i) = 1 \ \therefore \ \sum_{i=1}^{n} p = 1 \ \therefore \ np = 1 \tag{8.16}$$

and, therefore

$$p = \frac{1}{n} \tag{8.17}$$

Example 8.7: Undirected movement

The bearing of an organism performing undirected movement may be assigned to one of eight possible directions $\Omega = \{N,NE,E,SE,S,SW,W,NW\}$. During any given move, each direction can occur with probability $\frac{1}{8}$.

8.5. The union of events

Statements about events are essentially statements in formal logic. Just as the overbar notation represents the logical operator 'NOT', so do the concepts of union and intersection of events provide representations of the logical operators 'OR' and 'AND'. Hence, $E_1 \cup E_2$ can be interpreted as 'the occurrence of event 1 *or* event 2' and $E_1 \cap E_2$ can be interpreted as 'the occurrence of event 1 *and* event 2'. The probabilities of such compound events can be calculated using simple rules. The rules relating to the union of events are examined here. Those relating to the intersection of events are discussed in the following section.

The probability of the union of two mutually exclusive events (E_1 and E_2) is

$$P(E_1 \cup E_2) = P(E_1) + P(E_2) \tag{8.18}$$

In other words, the probability that either one or the other event occurs as a result of a single trial of the experiment is equal to the sum of their probabilities.

Example 8.8: Seed germination on a gridded landscape

Consider a field (used here as a metaphor for the event space) divided into 192 identical squares by a regular grid drawn with very thin lines (Figure 8.3). A seed is thrown from above, in such a way that it can land in any square with equal probability $p = 1/192$. Outcomes in this event space are assumed mutually exclusive because the seed cannot germinate in two squares simultaneously. Within the landscape, only certain areas are fertile and will allow germination.

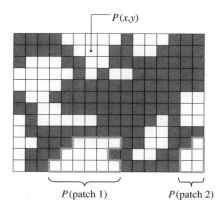

Figure 8.3: An idealised landscape mosaic. Fertile ground is indicated by white.

Here are some probabilities that can be calculated using the addition rule:

$$P(\text{patch 2}) = P(E_2) = 5 \times p \cong 0.026$$

$$P(\text{patch 1}) = P(E_1) = 14 \times p \cong 0.073$$

$$P(\text{patch 1 or patch 2}) = P(E_1 \cup E_2) = 5/192 + 14/192 \cong 0.099$$

$$P(\text{seed grows}) = P(G) = 77/192 \cong 0.401$$

the probability that a seed does not germinate can now be quickly calculated from its complement

$$P(\bar{G}) = 1 - P(G) \tag{8.19}$$

If the two events are not mutually exclusive, the probability of their union is

$$P(E_1 \cup E_2) = P(E_1) + P(E_2) - P(E_1 \cap E_2) \tag{8.20}$$

The Venn diagram in Figure 8.4 is possibly the best way to remember this relationship: Simply adding the probabilities of two overlapping events leads to 'double-counting' of their intersection. This may be corrected by subtracting the probability of the intersection (once) from the total.

$$P(A \cup B) \qquad P(A) + P(B) \qquad P(A \cap B)$$

Figure 8.4: Rationale behind the addition rule for non-mutually exclusive events. The probability of their union can be found by discounting the probability of their intersection from the sum of their respective probabilities.

Example 8.9: Coexisting sparrows

Tree sparrows (Passer montanus) *and house sparrows* (P. domesticus)

Tree sparrows and house sparrows can coexist in hawthorn (*Crataegus monogyna*) bushes in Britain. It is estimated that the probability of encountering *P. montanus* in any given hawthorn bush is 0.1, the probability of encountering *P. domesticus* is 0.4 and the probability of both occurring together is 0.05. The probability that sparrows are found in a given hawthorn bush is, therefore, $0.1 + 0.4 - 0.05 = 0.45$.

8.6.   Conditional probability

Often, the probability that an event B occurs is affected by the prior occurrence of another event A.

Example 8.10: Territoriality and survival in red grouse

Red grouse are territorial, ground-nesting birds. The cocks establish territories in the autumn and birds without territories have considerably higher over-winter mortality than territory-holders (Watson, 1985). Consider the events E_1, 'cock gets a territory in autumn', and E_2, 'cock survives through winter'. These two events are not mutually exclusive. There are, therefore, four possibilities:

❶ Male gets a territory and survives through winter ($E_1 \cap E_2$)
❷ Male gets a territory and does not survive through winter ($E_1 \cap \bar{E}_2$)
❸ Male does not get a territory and survives through winter ($\bar{E}_1 \cap E_2$)
❹ Male does not get a territory and does not survive through winter ($\bar{E}_1 \cap \bar{E}_2$)

A Venn diagram (Figure 8.5) of these four events shows them to be mutually exclusive and, also, collectively exhaustive. To make the transition from events to probabilities, the area taken up by each event in the Venn diagram may be loosely interpreted as the probability associated with that event. Now, assume that a particular territory-holder is identified at the end of autumn. What is the probability that this animal will make it through the winter? If the animal survives, then both events E_1 and E_2 will have occurred. However, using $P(E_1 \cap E_2)$ as the required probability is incorrect because, at the end of autumn, E_1 had already occurred; the bird had already obtained a territory.

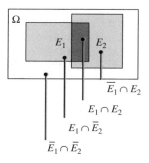

Figure 8.5: Four possible events for territory acquisition and over-winter survival.

To calculate the probability of both events occurring when one of them already has, the event E_1 needs to be 'inflated' so that it occupies the entire event space (Figure 8.6).

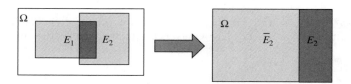

Figure 8.6: If E_1 has already occurred, then, by definition, E_1 is the event space.

Since event E_1 has already occurred, the new event space will inherit only the events $E_1 \cap E_2$ and $E_1 \cap \bar{E}_2$ from the original space, but the probabilities of these events will be inflated by a constant (say, α) so that they add up to 1,

$$\alpha P(E_1 \cap E_2) + \alpha P(E_1 \cap \bar{E}_2) = 1 \tag{8.21}$$

This can be solved for α to find

$$\alpha = \frac{1}{P(E_1 \cap E_2) + P(E_1 \cap \bar{E}_2)} = \frac{1}{P(E_1)} \tag{8.22}$$

Therefore, mathematically, the required inflation of the event space is achieved by dividing the probability $P(E_1 \cap E_2)$ by the probability that E_1 occurs. The result is called the **conditional probability** of E_2 *given* E_1. The word 'given' is represented by a vertical bar '|' between the names of the two events. In summary,

$$P(E_2 \mid E_1) = \frac{P(E_1 \cap E_2)}{P(E_1)} \tag{8.23}$$

Rearranging this gives the definition for the **joint probability** (the intersection) of two events:

$$P(E_1 \cap E_2) = P(E_2 \mid E_1)P(E_1) \tag{8.24}$$

Example 8.11: Territoriality and survival in red grouse

In a particular population of red grouse, 65% of all mature males obtained a territory this autumn. After the end of winter, 30% of the initial population had obtained a territory in autumn and, also, remained alive through the winter. The probability of survival of territory-holders for this year can be found from Equation (8.23)

$$P(E_2 \mid E_1) = \frac{0.30}{0.65} = 0.46 \qquad (8.25)$$

The opposite question could also have been asked (i.e. what is the probability that a cock had established a territory, given that it survived winter?). This would lead to the formula

$$P(E_1 \mid E_2) = \frac{P(E_1 \cap E_2)}{P(E_2)} \qquad (8.26)$$

Equations (8.23) and (8.26) can be rearranged as follows:

$$P(E_1 \cap E_2) = P(E_2 \mid E_1)P(E_1)$$
$$P(E_1 \cap E_2) = P(E_1 \mid E_2)P(E_2) \qquad (8.27)$$

The left-hand sides of both these equations are the same. Therefore,

$$P(E_2 \mid E_1)P(E_1) = P(E_1 \mid E_2)P(E_2) \qquad (8.28)$$

Equation (8.28) can be solved for either of the two conditional probabilities to give an important result known as **Bayes's law**, which is used to reverse the direction of conditional probabilities.

$$P(E_2 \mid E_1) = \frac{P(E_1 \mid E_2)P(E_2)}{P(E_1)} \qquad (8.29)$$

The following example should make it clear that the biological interpretation of $P(E_2 \mid E_1)$ is different to that of $P(E_1 \mid E_2)$

Example 8.12: Territoriality and survival in red grouse

In the grouse population the probability that any cock gains a territory is $P(E_1) = 0.65$ and the probability that any cock survives through the winter is $P(E_2) = 0.35$. Equation (8.25) says that territory-holders survive with a probability $P(E_2 \mid E_1) = 0.46$. This information leads to the probability that an animal found alive at the end of winter had managed to establish a territory by the end of autumn:

$$P(E_1 \mid E_2) = \frac{P(E_2 \mid E_1)P(E_1)}{P(E_2)} = \frac{0.46 \times 0.65}{0.35} \cong 0.85 \qquad (8.30)$$

So, although $P(E_2 \mid E_1)$ and $P(E_1 \mid E_2)$ express the same biological effect (having a territory improves your chances of survival, or fitter birds are better at competing and surviving), they have different values and refer to different things: $P(E_2 \mid E_1)$ is a survival probability and $P(E_1 \mid E_2)$ describes territorial recruitment.

8.7. Independent events

Some events are **independent**, meaning that the probability of the occurrence of one is not affected by the occurrence of the other. In these cases, the conditional probability $P(E_2 \mid E_1)$ is equal to $P(E_2)$ and Equation (8.24) becomes

$$P(E_1 \cap E_2) = P(E_1)P(E_2) \tag{8.31}$$

Example 8.13: Sex of successive calves

The sexes of two successive calves born to the same wildebeest are independent events. Assuming a 1:1 sex ratio at birth, below are the probabilities associated with four different events in this experiment's event space:

$$P(\text{both male}) = P(\male\male) = \frac{1}{2}\frac{1}{2} = \frac{1}{4} \qquad P(\text{both female}) = P(\female\female) = \frac{1}{2}\frac{1}{2} = \frac{1}{4}$$

$$P(\text{different sex}) = P(\male\female) + P(\female\male) = \frac{1}{2}\frac{1}{2} + \frac{1}{2}\frac{1}{2} = \frac{1}{2}$$

The relationship between mutually exclusive, dependent and independent events can be illustrated with Venn diagrams (Figure 8.7). Consider a Venn diagram with two nonoverlapping shapes representing two mutually exclusive events E_1 and E_2 (Figure 8.7(a)). The areas occupied by the events are interpreted as their associated probabilities, so, for example, $P(E_1) < P(E_2)$ and $P(E_2) \cong 0.5$. Since the two events are mutually exclusive, if one of them occurs, the other cannot (so, $P(E_2 \mid E_1) = 0$). Now, imagine moving the rectangle for E_1 towards the right, as shown in Figures 8.7(b)–(e). When there is little overlap between the two events (Figure 8.7(b)), the occurrence of E_1 decreases the likelihood of E_2 (so, $P(E_2 \mid E_1) < P(E_2)$). This happens because, following the occurrence of E_1, the new event space is E_1, and in this new event space, the area taken up by E_2 is proportionately smaller than it was to begin with. When there is a lot of overlap between the two events (Figure 8.7(d)), the opposite happens: the occurrence of E_1 increases the probability of E_2 (so, $P(E_2 \mid E_1) > P(E_2)$). In between these two situations, there is an arrangement (Figure 8.7(c)) where the relative sizes of E_1 and $E_1 \cap E_2$ are precisely the same as the relative sizes of Ω and E_2. In this special case, the two events are independent (so, $P(E_2 \mid E_1) = P(E_2)$). Finally, when E_1 is moved to be completely inside E_2, it is certain that the occurrence of E_1 implies the occurrence of E_2 (so, $P(E_2 \mid E_1) = 1$).

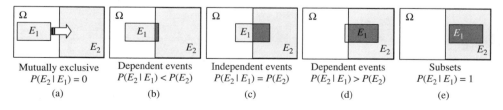

Figure 8.7: An illustration of the continuum between mutually exclusive events and subsets, encompassing the cases of dependent and independent events.

8.8.   Total probability

Example 8.14: Seed germination in a heterogeneous environment

Assume that a particular study area comprises five habitat types: arable, woodland, heathland, open water and grassland. The proportions $p_1, \ldots, p_5$ of the study area taken up by each habitat type differ between them but they are measurable and, hence, known. A dispersing seed can land at any point in the landscape with equal probability. Two types of events are of interest: first, which habitat the seed lands in (written $E_1, \ldots, E_5$) and second whether it germinates or not (written G and $\bar{G}$). Since the seed can land anywhere uniformly randomly, the probability that it lands on the ith habitat type is equal to the availability of that habitat, $P(E_i) = p_i$. Further experiments have been conducted to measure the habitat-specific probabilities of germination, so for the ith habitat type, the value of $P(G \mid E_i)$ is known. The effect of the environment's habitat composition on the total probability of germination $P(G)$ can be calculated as the sum of all the joint probabilities $P(E_i \cap G)$ (see Figure 8.8),

$$P(G) = \sum_{i=1}^{5} P(E_i \cap G) \tag{8.32}$$

or, equivalently, using Equation (8.24),

$$P(G) = \sum_{i=1}^{5} P(G \mid E_i) P(E_i) = \sum_{i=1}^{5} P(G \mid E_i) p_i \tag{8.33}$$

Changing the makeup of the environment, for example by increasing arable land at the expense of other habitat types, may affect the total probability of germination. This new probability can be calculated by adjusting the values of $p_1, \ldots, p_5$ in Equation (8.33).

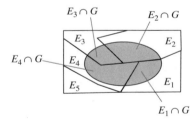

Figure 8.8: The total probability of germination $P(G)$, shown here as the shaded ellipse, is made up of the joint probabilities between G and five mutually exclusive events $E_1, \ldots, E_5$. In this particular example, habitat 5 corresponds to open water and thus there is no overlap between events G and E_5.

In general, given the probabilities $P(E_1), \ldots, P(E_n)$ of n mutually exclusive and collectively exhaustive events, and the conditional probabilities $P(G \mid E_1), \ldots, P(G \mid E_n)$ referring to some other event G, then the probability of G is

$$P(G) = \sum_{i=1}^{n} P(G \mid E_i) P(E_i) \qquad (8.34)$$

This is known as the **rule of total probability**.

8.9. Bayesian probability

The frequentist definition of probability (Section 8.3) is based on the idea that relative frequencies obtained from long-running experiments tend to converge to particular numerical values. However, for experiments that cannot be repeated many times, this has the practical disadvantage that there may simply not be sufficient trials for the relative frequencies to converge.

Example 8.15: Does the sex ratio at birth deviate from 1:1?

 Example 8.2 enquired whether the observed adult sex ratio of 6:4 resulted from asymmetric sex at birth. The answer was unclear after 10 or even 50 births. It was necessary to observe hundreds of births before the relative frequency settled to 0.5. But was it really necessary to pretend that no other information was available to answer this question? In the majority of mammal species, the sex ratio does not, on average, deviate from 1:1. Hence, prior to observing any new data for the study population of wildebeest, a proportion of 0.5 of male births might be considered a good candidate answer. On the other hand, there are good evolutionary arguments (Trivers and Willard, 1973) and experimental evidence (Larson *et al.*, 2001) to suggest that mammalian bodies should (and can) control the sex of their offspring in response to environmental quality. So, it may not be wise to exclude the possibility of a 6:4 birth ratio without the collection of new data. The question is how to combine independent beliefs about the outcome of the experiment with whatever new data the experiment produces?

Scientific hypotheses are statements about how the world works. As scientists, we formulate hypotheses on the basis of existing evidence. Ideally, such evidence includes historical data or proven biological principles. Alternatively, poorer quality evidence might be influential in our thinking. Among them (in decreasing order of reliability) are scientific observations from related but different study systems, anecdotal or lay-person reports and even personal hunches. These hypotheses motivate the collection of new data. In turn, the new data trigger a re-evaluation of the original hypotheses. An alternative interpretation of probability known as **subjective probability** defines it as the *degree of belief* in a particular hypothesis. Under this interpretation, the event space comprises different competing hypotheses and the data are used merely to re-evaluate, not completely define, the likelihood of each. In the simplest case, only two hypotheses are considered: the **null hypothesis** (H_0) represents the starting point for the scientific investigation and conventionally stands for the simplest, most intuitive (and scientifically less interesting) scenario. The null hypothesis is compared and contrasted with an **alternative hypothesis** (H_1 or $\bar{H}_0$).

Example 8.16: Null and alternative hypotheses for wildebeest sex ratio

 In Example 8.14 two different hypotheses were formulated: the null hypothesis stated that the sex ratio at birth is symmetric (so that the proportion p of males in the sample of newborns is 0.5). The alternative hypothesis said that the sex ratio at birth is the same as the adult sex ratio (so that $p = 0.4$). More generally, the truth may be a combination of these two hypotheses. For example, sex ratio may be asymmetric to begin with, and become even more or less so through the effect of differential mortality between the sexes. To keep things simple here, I will focus on the two hypotheses, namely $H_0 : p = 0.5$ and $H_1 : p = 0.4$. We may have an initial idea of the relative plausibility of these two hypotheses. For example, if the environment of the study population was not extreme (neither too lean nor too rich), it may be more plausible to assume that no sex regulation would occur. This initial belief may be formulated somewhat arbitrarily by saying that H_0 is twice as likely as H_1. The event space is restricted to just two hypotheses so, $P(H_0) + P(H_1) = 1$. This, in combination with the initial evaluation of the hypotheses, gives their subjective probabilities: $P(H_0) = \frac{2}{3}, P(H_1) = \frac{1}{3}$.

The initial values of belief associated with each hypothesis are called **prior probabilities**. Example 8.16 makes two important points: first, subjective probabilities can be arbitrary. It is all too easy to incorporate poor quality evidence into the prior probabilities and such cavalier use of subjective probabilities has attracted ferocious criticism from frequentist statisticians. Second, subjective probabilities can (rather perversely) be assigned to anything, even frequentist probabilities: in Example 8.16, a probability of $\frac{2}{3}$ was associated with the probability p of a male offspring being 0.5.

Crucially, subjective probabilities can be updated using new data using Bayes's law, which explains why the area of subjective probabilities is also known as **Bayesian statistics**. The procedure can be illustrated with the simplest case of only two competing hypotheses (H_0 and H_1), but it generalises well to many (even infinite) hypotheses. Bayesian statisticians would write Equation (8.29) in terms of any one of the two hypotheses and the data as follows:

$$P(H \mid data) = \frac{P(data \mid H)P(H)}{P(data)} \tag{8.35}$$

The components of this expression need some explanation. On the left is $P(H \mid data)$, the probability that the hypothesis is true, *given* the new information provided by the data. This **posterior probability** is the desired re-evaluation of the initial (or prior) probability. On the right, the prior probability appears as $P(H)$. This describes the subjective belief in the hypothesis, before the data are collected. The component $P(data \mid H)$ is the conditional probability of the data occurring if the hypothesis is true. This is often called the **likelihood** and is usually directly obtainable from the specifics of the hypothesis (see Example 8.17). Finally, $P(data)$ is the probability of obtaining the data irrespective of which hypothesis is true. This can be calculated as follows: the probability that H_0 is true *and* that it generates the observed data is:

$$P(H_0 \cap data) = P(data \mid H_0)P(H_0) \tag{8.36}$$

Similarly, the probability that H_1 is true *and* that it generates the observed data is

$$P(H_1 \cap data) = P(data \mid H_1)P(H_1) \tag{8.37}$$

The total probability of obtaining the observed data is the sum of all the joint probabilities of the data, under any one of the (mutually exclusive) hypotheses,

$$
\begin{aligned}
P(data) &= P(H_0 \cap data) + P(H_1 \cap data) \\
&= P(data \,|\, H_0)P(H_0) + P(data \,|\, H_1)P(H_1)
\end{aligned}
\tag{8.38}
$$

This is a simple case of the rule of total probability in Equation (8.34).

Example 8.17: Bayesian updating for wildebeest sex ratio

The prior probabilities expressing subjective belief in the two hypotheses of Example 8.16 are $P(H_0) = \frac{2}{3}$ and $P(H_1) = \frac{1}{3}$. After just one birth, the data can either take the value ♂ or ♀. The probability of a male birth under the H_0 is 0.5. The probability of a male birth under the H_1 is 0.4. Therefore, the four likelihoods can be obtained directly from the definitions of the two hypotheses:

$$
\begin{aligned}
P(\text{♂} \,|\, H_0) &= 0.5 & P(\text{♂} \,|\, H_1) &= 0.4 \\
P(\text{♀} \,|\, H_0) &= 0.5 & P(\text{♀} \,|\, H_1) &= 0.6
\end{aligned}
\tag{8.39}
$$

According to Equation (8.38), the total probability of a male offspring is

$$
\begin{aligned}
P(\text{♂}) &= P(\text{♂} \,|\, H_0)P(H_0) + P(\text{♂} \,|\, H_1)P(H_1) \\
&= 0.5 \times \frac{2}{3} + 0.4 \times \frac{1}{3} \\
&\cong 0.467
\end{aligned}
\tag{8.40}
$$

Using similar reasoning, the total probability of a female offspring is $P(\text{♀}) = 0.533$ (the complement of 0.467). How do the data modify the two prior probabilities? Assume that the offspring was a male. H_0 should be strengthened by this single observation, because it predicts a higher proportion of male offspring than H_1. Indeed, applying the version of Bayes's rule in Equation (8.35) gives

$$
P(H_0 \,|\, \text{♂}) = \frac{P(\text{♂} \,|\, H_0)P(H_0)}{P(\text{♂})} = \frac{0.5 \times \frac{2}{3}}{0.467} \cong 0.71
\tag{8.41}
$$

this posterior probability is greater than the prior of $\frac{2}{3} \cong 0.67$ on the basis of the evidence. Bayes's rule can also estimate the posterior probability for H_1,

$$
P(H_1 \,|\, \text{♂}) = \frac{P(\text{♂} \,|\, H_1)P(H_1)}{P(\text{♂})} = \frac{0.4 \times \frac{1}{3}}{0.467} \cong 0.29
\tag{8.42}
$$

Example 8.17 is artificially simple, because it only examines two hypotheses and only a single observation. In fact, the true probability of giving birth to a male calf could take any one of an infinity of values between 0 and 1. To talk about infinite competing hypotheses it is necessary to introduce probability distributions (see Chapter 9). However, extending the basic framework to multiple observations may be done more simply. The appeal of Bayesian statistics is that, *once calculated, the posterior probabilities can be used as prior probabilities for new data.*

Example 8.18: Bayesian updating for wildebeest sex ratio

The posterior probabilities calculated for the two hypotheses in Example 8.17 can now be used as the priors for the second observation. Hence, before the second birth, the priors $P(H_0) = 0.71$ and $P(H_1) = 0.29$. If the second birth was a female, then using the same likelihoods and Bayesian equations as in Equations 8.39–8.41 yields $P(H_0 \mid ♀) = 0.67$ and $P(H_1 \mid ♀) = 0.33$. As the number of trials increases, the probability of H_0 tends to 1, implying a 1:1 sex ratio at birth; however, the number of data required for the answer to converge to the truth depends on the quality of the original subjective priors (Figure 8.9).

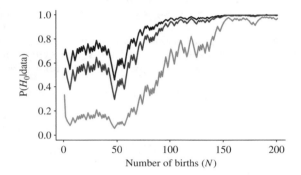

Figure 8.9: Successively updating the posterior probability of the null hypothesis, in this example, eventually leads to complete support for a symmetric sex ratio at birth. This happens irrespective of the initial prior probabilities. However, the number of births required for convergence can vary. The dark trajectory at the top shows the development of the posterior probability with a prior $P(H_0) = \frac{2}{3}$. This prior supports the underlying truth, so the posterior $P(H_0 \mid data)$ converges more quickly. The light trajectory at the bottom converges more slowly because it had a poorly specified prior of $P(H_0) = \frac{1}{3}$ (which placed prior belief on the wrong hypothesis). The intermediate trajectory was derived from an **uninformative prior** $P(H_0) = \frac{1}{2}$, which assigned equal prior probabilities to each of the two competing hypotheses.

Further reading

Because of its kinship with formal logic, most people find probability a challenging and often counterintuitive topic. The situation is not improved by the fact that statisticians are still bitterly debating some of the most fundamental definitions in this area. If you are interested in improving your intuitive understanding of probability, the formal but unmathematical books by Graham (2006) and Hacking (2001) are a pleasure to read. Sections on probability which cover much the same topics as this chapter can be found in statistical and mathematical textbooks. Examples include Batschelet (1979), Larson (1982), Wild and Seber (2000), Gotelli and Ellison (2004), Adler (2005), Otto and Day (2007) and Whitlock and Schluter (2009). Good introductory references for Bayesian statistics in ecology are McCarthy (2007), Bolker (2008), King *et al.* (2009) and Link and Barker (2010). A good advanced general reference for probability is Grimmett and Stirzaker (2001).

References

Adler, F.R. (2005) *Modelling the Dynamics of Life: Calculus and Probability for the Life Scientists.* Thompson Brooks/Cole, Belmont, California. 876pp.

Batschelet, E. (1979) *Introduction to Mathematics for Life Scientists.* Springer Verlag, Berlin. 643pp.

Bolker, B.M. (2008) *Ecological Models and Data in R.* Princeton. 408pp.

Gotelli, N.J. and Ellison, A.M. (2004) *A Primer of Ecological Statistics.* Sinauer Associates, Massachusetts. 510pp.

Graham, A. (2006) *Developing Thinking in Statistics.* Sage. 288pp.

Grimmett, G. and Stirzaker, D. (2001) *Probability and Random Processes.* Oxford University Press, New York. 596pp.

Hacking, I. (2001) *An Introduction to Probability and Inductive Logic.* Cambridge University Press. 320pp.

King, R., Morgan, B., Gimenez, O. and Brooks, S. (2009) *Bayesian Analysis for Population Ecology.* Chapman & Hall. 456pp.

Larson, H.J. (1982) *Introduction to Probability Theory and Statistical Inference.* John Wiley & Sons, Inc., New York. 637pp.

Larson, M., Kimura, K., Kubisch, H.M. and Roberts, R.M. (2001) Sexual dimorphism among bovine embryos in their ability to make the transition to expanded blastocyst and in the expression of the signaling molecule IFN-τ. *PNAS,* **98**, 9677–9682.

Link, W.A. and Barker, R.J. (2010) *Bayesian Inference with Ecological Applications.* Academic Press, Amsterdam. 339pp.

McCarthy, M.A. (2007) *Bayesian Methods for Ecology.* Cambridge University Press, Cambridge. 310pp.

Mills, M.G.L. and Shenk, T.M. (1992) Predator–prey relationships: the impact of lion predation on wildebeest and zebra populations. *Journal of Animal Ecology,* **61**, 693–702.

Otto, S.P. and Day, T. (2007) *A Biologist's Guide to Mathematical Modelling in Ecology and Evolution.* Princeton University Press, New Jersey. 732pp.

Trivers, R.L. and Willard, D.E. (1973) Natural selection of parental ability to vary the sex ratio of offspring. *Science,* **179**, 90–92.

Watson, A. (1985) Social class, socially-induced loss. recruitment and breeding of red grouse. *Oecologia,* **67**, 493–498.

Whitlock, M.C. and Schluter, D. (2009) *The Analysis of Biological Data.* Roberts & Co., Colorado. 700pp.

Wild, C.J. and Seber, G.A.F. (2000) *Chance Encounters: A First Course in Data Analysis and Inference.* John Wiley & Sons, Inc., New York. 611pp.

9

How to identify different kinds of randomness
(Probability distributions)

'The generation of random numbers is too important to be left to chance.'
Robert R. Coveyou (1915–1996), American physicist

Each trial of a random experiment may lead to a different outcome. However, this does not imply complete ignorance about the type of possible outcomes or their relative likelihood. In most cases, we have some idea of the kind of values a random variable can take (its event space). We may also have (or seek to obtain) a probability associated with every possible value. A **probability distribution** (Section 9.1) assigns a probability to each value in the event space. Much of the early material in this chapter focuses on the mathematical description of discrete (Section 9.2) and continuous (Section 9.3) probability distributions, and how their different statistical summaries (such as the mean and variance) can be unified by the concept of **expectation** (Section 9.4). The middle part of the chapter introduces **named probability distributions** (Section 9.5) that are routinely used by scientists to describe different types of randomness. These include distributions for experiments with **equally likely outcomes** (Section 9.6), experiments whose outcomes are **counts** of occurrences (Sections 9.7–9.10), distributions used to model the **waiting time** before one or more events occur (Section 9.11) and even distributions that can be used to describe belief in different values of probability (Section 9.12). The **normal distribution** receives particular mention (Section 9.13) because of its ability to approximate a multitude of biological and mathematical processes. Focus then shifts to more advanced issues, such as the empirical use of distributions (Section 9.14) and modelling distribution mixtures (Section 9.15). In Section 9.16 the concept of joint/conditional probability is extended to random variables, and this motivates the definition of independent random variables. Joint probability distributions are illustrated using the **bivariate normal distribution** (Section 9.17).

How to be a Quantitative Ecologist: The 'A to R' of Green Mathematics and Statistics, First Edition. Jason Matthiopoulos.
© 2011 John Wiley & Sons, Ltd. Published 2011 by John Wiley & Sons, Ltd.

The idea of multiple random variables is extended to their sums (Section 9.18 – **central limit theorem**), products (Section 9.19 – **lognormal distribution**) and residuals (Section 9.20 – **chi square distribution**). The chapter ends with a section on **simulation** (Section 9.21), which explains how to use probability distributions to generate stochasticity within computer models, hence emulating the uncertainty involved in predictions about real-world phenomena.

9.1.   Probability distributions

This section combines two terms developed independently until now: on the one hand, the frequency distribution (introduced in Chapter 7 as a graphical device for visualising data) and on the other, the probability of an event (initially defined in Chapter 8 as the relative frequency of an event after many trials). A **probability distribution** is obtained by assigning a probability to every event in an experiment's event space.

Example 9.1: Probability distribution of nominal variables

Tables 9.1 and 9.2 and Figure 9.1 show the probability distributions associated with two simple experiments.

Table 9.1: Experiment 1: Sex of next calf.

Outcome	Probability
♂	0.5
♀	0.5

Table 9.2: Experiment 2: Sex of next two calves in no particular order.

Outcome	Probability
Both ♂	0.25
Both ♀	0.25
Different sex	0.5

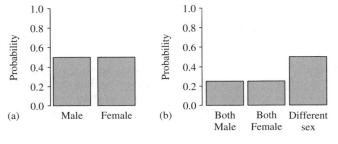

Figure 9.1: Bar charts of the probability distributions associated with two simple experiments.

Such tabular and graphical representations are good ways to specify probability distributions and, in the case of qualitative variables (Example 9.1), they are the only option. However, for quantitative variables, it may be possible to describe the probability distribution more concisely using mathematical functions. The next two sections do this for discrete and continuous random variables.

9.2.   Discrete probability distributions

A quantitative random variable is discrete if it is countable. This does not necessarily mean that its event space is finite (e.g. the set of integers is both countable and infinite).

Example 9.2: Beetle eggs per cluster: a count

Colorado potato beetle, Leptinotarsa decemlineata

The Colorado potato beetle lays its eggs in clusters, on the lower leaf surface of newly emerging potato plants. The number M of eggs in each cluster is about 30 and females of the species can deposit up to 800 eggs during their lifetime. The probability of finding a female's lifetime reproductive output in a single cluster is zero. This maximum may serve as a plausible upper limit to cluster size. Therefore, the event space for the random variable M is $\Omega = \{0, 1, 2, \ldots, 800\}$ and its probability distribution is a table of values $M = 0, \ldots, 800$ and their associated probabilities $P(M = 0), \ldots, P(M = 800)$.

In Example 9.2, using a table to fully specify the probability distribution was possible only because the event space was finite. Such assumptions are not always possible and therefore it is not always feasible to define a probability distribution by enumeration. The task can, instead, be achieved by using a function defined on the variable's event space. In the case of discrete variables, a **probability mass function** $f_X(x)$, or PMF for short, gives a probability for every value of X.

$$f_X(x) = P(X = x) \tag{9.1}$$

The notation here needs some explanation. The lower case x represents a particular value of the random variable X. In most applications, the PMF will appear with a number instead of a lower case symbol (e.g. $f_X(3)$). Hence, the name of the function (f_X), indicates which variable it refers to (X). Sometimes, when only one random variable is involved, this added clarity is unnecessary and the subscript X is dropped.

Where does the PMF of a random variable come from? Mostly, from known facts about the process generating the random variable. Things like how large the event space is, what values it contains, whether all the events are equally likely, or derived from repeated trials of simpler experiments, and so on. The values of the PMF are probabilities, so they take values between 0 and 1.

$$0 \le f(x) \le 1 \; \forall x \in \Omega \tag{9.2}$$

Also, since the PMF associates probabilities to all the events in the event space, the sum of these probabilities must be 1,

$$\sum_{All \; x \in \Omega} f(x) = 1 \tag{9.3}$$

Example 9.3: PMF for egg cluster size

I will assume that more is known about the size of the egg clusters of the potato beetle. Specifically, that clusters of 30 and 31 eggs are the most likely and that $M < 26$ and $M > 35$ have never been observed. To specify a PMF describing these facts, the event space may be split into four parts:

$$f(m) = \begin{cases} 0 & \text{if} \quad 0 \le m \le 25 \\ \frac{1}{30}(m - 25) & \text{if} \quad 26 \le m \le 30 \\ \frac{1}{30}(36 - m) & \text{if} \quad 31 \le m \le 35 \\ 0 & \text{if} \quad 36 \le m \le 800 \end{cases} \tag{9.4}$$

This PMF (an example of a **triangular distribution**) has been designed to incorporate the biological facts and obey Equations (9.2) and (9.3). It further assumes a linear increase of probability associated with cluster sizes from 26 to 30 in steps of 1/30 and a linear decline from 31 to 36 (Figure 9.2). The event space of this experiment initially comprised 801 events. Tabulating its probability distribution would require a list of 801 probabilities. By comparison, Equation (9.4) is a more concise description, using just four categories.

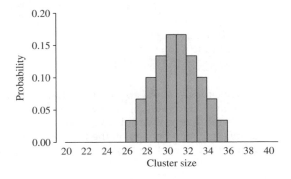

Figure 9.2: Probability mass function for the number of potato beetle eggs per cluster.

Another method to specify the distribution of a random variable is the **cumulative distribution function**, $F_X(x)$, or CDF for short. It gives the probability that the random variable X is no greater than a given value x,

$$F_X(x) = P(X \le x) \tag{9.5}$$

There is a simple relationship between the PMF and the CDF

$$F(x) = \sum_{i=-\infty}^{x} f(i) \tag{9.6}$$

Because x is now the upper limit of summation, it can no longer be used as a counter, so, instead, the counter i helps with the notation inside the sum. This is often called a **dummy** or **auxiliary variable**. Equation (9.6) considers the most general case in which the event space Ω includes all integer numbers. Hence, the starting value of the counter i is $-\infty$. If the event space does not extend that far back, the lower limit can be set to the first value in the event space.

The properties of probability imply the following properties for the CDF

$$\lim_{x \to -\infty} F(x) = 0 \tag{9.7}$$

$$\lim_{x \to \infty} F(x) = 1 \tag{9.8}$$

$$\text{If } a \le b, \text{ then } F(a) \le F(b) \tag{9.9}$$

The last property says that the CDF is a monotonically increasing function.

Example 9.4: CDF for egg cluster size

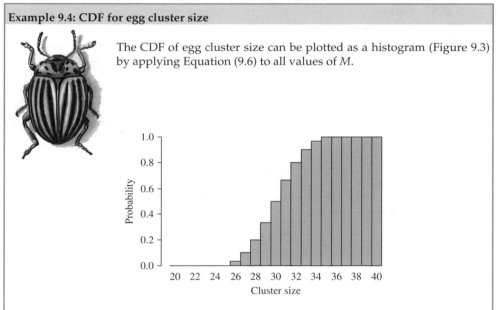

The CDF of egg cluster size can be plotted as a histogram (Figure 9.3) by applying Equation (9.6) to all values of M.

Figure 9.3: Cumulative distribution function for egg cluster size. Once the total probability has reached 1 (at $M = 35$) it no longer increases.

Just as the CDF is derived from the PMF, the PMF can be retrieved as the difference between two successive values of the CDF,

$$F(x) - F(x-1) = \sum_{i=-\infty}^{x} f(i) - \sum_{i=-\infty}^{x-1} f(i) = f(x) \tag{9.10}$$

 9.1: Cumulative probabilities

The cumulative sum of a list of values can be obtained from the command `cumsum()`. Below, this is illustrated with the PMF of the triangular distribution in Example 9.3.

```
m<-c(20:40) # Egg cluster sizes to be examined
pm<-c(rep(0,6),1/30*((26:30)-25),(36-((31:35)))/30,rep(0,5)) #PMF
cm<-cumsum(pm) # CDF
```

9.3. *Continuous probability distributions*

Whereas discrete random variables are usually counts, continuous random variables often result from measurements. Representing probabilities for continuous random variables gives rise to a conceptual problem. Think about the following paradox.

Example 9.5: Are all fern heights equally likely?

An experiment comprises measurements of the height of tree ferns. Each trial of this experiment deals with a different individual, so, in this case, the statistical population of measurements has the same size as the biological population of plants. Imagine that measurement accuracy can be increased indefinitely, making it unlikely that any two individuals will have *exactly* the same height. If no two plants have exactly the same height, then how can relative frequencies and (by implication) probabilities be associated with different values of this random variable?

The solution, as you might have guessed, lies in discretising the domain of the continuous random variable so that a probability is associated with an *interval* rather than a single value. The cumulative distribution function is a more convenient introduction point for continuous distributions, precisely because it deals with ranges of values. It is defined in the same way as for discrete variables

$$F(x) = P(X \leq x) \quad x \in \Omega \tag{9.11}$$

but it is a continuous function whose properties are a direct corollary of the laws of probability,

$$F(x) \in [0, 1] \quad \forall x \in \Omega \tag{9.12}$$

$$\lim_{x \to \infty} F(x) = 1, \quad \lim_{x \to -\infty} F(x) = 0 \tag{9.13}$$

$$\text{If } a \leq b \text{ then } F(a) \leq F(b) \tag{9.14}$$

Example 9.6: CDF for tree fern heights

Consider a continuous random variable X (such as the height of tree ferns) with the following CDF,

$$F(x) = \begin{cases} 0 & x < 0 \\ 1 - \exp(-x^2) & x \geq 0 \end{cases} \tag{9.15}$$

This function takes values close to 1 for 2 m-tall trees (Figure 9.4), meaning that trees of this species are almost certain to be shorter than 2 m. The CDF may be used to calculate probabilities associated with ranges of the random variable. For example,

$$P(X \leq 1.2) = F(1.2) \cong 0.763 \tag{9.16}$$

$$P(X > 2) = 1 - F(2) \cong 0.018 \qquad (9.17)$$

$$P(1 < X \leq 2) = F(2) - F(1) \cong 0.350 \qquad (9.18)$$

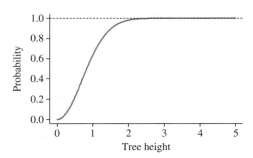

Figure 9.4: Graph of CDF in Equation (9.15).

The CDF is a continuous function and, for most continuous random variables, it is also differentiable over its entire domain (with the possible exception of a finite number of points – such as the point $X = 0$ in Example 9.6). The derivative of the CDF with respect to x is called the **probability density function**, or PDF for short

$$\frac{dF(x)}{dx} = f(x) \qquad (9.19)$$

Although this definition is straightforward, the interpretation of the PDF needs some care. The height of the graph of the PDF may be interpreted as the relative likelihood of different values of the random variable. However, it is *not* a probability. Instead, areas under the graph of the PDF (e.g. Figure 9.5) are probabilities. In general, the relationship giving the CDF in terms of the PDF is

$$F(x) = \int_{-\infty}^{x} f(s) \, ds \qquad (9.20)$$

Compare this expression with Equation (9.6) for the PMF: the sum has been replaced by an integral and the counter i has been replaced by the auxiliary variable s (remember the derivation of the integral as a limiting sum in Section 5.5). The PDF has the following properties:

$$f(x) \geq 0 \quad \forall x \in \Omega \qquad (9.21)$$

$$P(a < X \leq b) = \int_{a}^{b} f(x) \, dx \qquad (9.22)$$

$$\int_{-\infty}^{\infty} f(x) \, dx = 1 \qquad (9.23)$$

Equation (9.21) is a corollary of Equation (9.14): since the PDF is defined as the slope of an increasing function (the CDF), it should always be positive. The PDF is not bounded above by 1. However, the area under any segment of its graph cannot exceed 1.

Example 9.7: PDF for tree fern heights

The probability density function for the random variable in Example 9.6 can be found by differentiating Equation (9.15) using the chain rule

$$f(x) = \begin{cases} 0 & x < 0 \\ 2x \exp(-x^2) & x \geq 0 \end{cases} \qquad (9.24)$$

Plotting this function (Figure 9.5) indicates that short trees (around 0.5 m) are more numerous in the environment than taller ones.

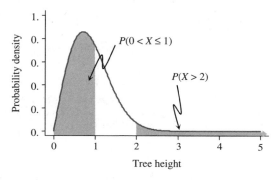

Figure 9.5: Graph of PDF in Equation (9.24) with two examples of probabilities.

9.4. Expectation

The distribution of a random variable can be summarised either by mathematical functions (the PMF and PDF – see this chapter) or by shape characteristics (the average, variance, skewness and kurtosis – see Chapter 7). These two methods can be linked by the concept of **expectation**, which I will introduce here via the mean of a discrete distribution. In Section 7.6, the average of a sample was defined in two equivalent ways:

$$\bar{x} = \frac{1}{n} \sum_{i=1}^{n} x_i = \sum_{\text{All } x} f(x)x \qquad (9.25)$$

The first sum is over all values in the population. The second is over all values in the event space. When applied to entire populations, the first version can only be used to calculate the mean if n is finite but the second can be used irrespective of population size by interpreting the relative frequency $f(x)$ as the probability of x given by the PMF.

Example 9.8: Mean egg cluster size

The mean size μ of egg clusters of potato beetles is calculated by placing Equation (9.4) into Equation (9.25)

$$\mu = \sum_{m=26}^{35} f(m)m = \sum_{m=26}^{30} f(m)m + \sum_{m=31}^{35} f(m)m$$

$$= \sum_{m=26}^{30} \frac{1}{30}(m-25)m + \sum_{m=31}^{35} \frac{1}{30}(36-m)m \qquad (9.26)$$

$$= \frac{1}{30}\left\{ \sum_{m=26}^{30} m^2 - 25 \sum_{m=26}^{30} m + 36 \sum_{m=31}^{35} m - \sum_{m=31}^{35} m^2 \right\} = 30.5$$

In general, the mean of a discrete distribution is linked to its PMF by the relationship

$$\mu = \sum_{-\infty}^{\infty} xf(x) \qquad (9.27)$$

This is also called the **expected value** or **expectation** of the random variable X. It is written as $E(X), E[X],$ or $\langle X \rangle$. The definition of expectation for a continuous random variable is analogous to the discrete case, replacing the sums by integrals and the PMF by the PDF

$$E(X) = \int_{-\infty}^{\infty} xf(x)\,dx \qquad (9.28)$$

Example 9.9: Expected tree fern heights

Assuming that the PDF of the height distribution is given by Equation (9.24), then the expected height of a tree fern (and hence the mean of the distribution) is

$$E(X) = \int_0^{\infty} f(x)x\,dx = \int_0^{\infty} 2x^2 \exp(-x^2)\,dx \qquad (9.29)$$

The lower limit of integration is 0 because the PDF is zero for negative values of X. This integral has no analytic solution but it can be solved numerically (see R5.2) to give the answer 0.886. This is the mean of the distribution in Figure 9.5.

The above establish the link between probability functions (PMF, PDF) and a distribution's mean. However, probability functions can also be linked to other statistical summaries such as the variance, skewness and kurtosis.

To establish these connections, the concept of an expectation must be generalised to *functions* of random variables. The integer powers of a random variable are particularly important for this discussion. Figure 9.6(a) represents a population of 10^5 values, symmetrically distributed with mean zero. Squaring each of these values and plotting the histogram of this new data set gives a skewed distribution (Figure 9.6(b)) comprising only non-negative values (since $X^2 \geq 0$). The mean of this new distribution is $\frac{1}{10^5} \sum_{i=1}^{10^5} x_i^2$. Plotting the histogram of X^3 gives yet another distribution (Figure 9.6(c)) with mean $\frac{1}{10^5} \sum_{i=1}^{10^5} x_i^3$. Either of these two means can also be calculated as an expectation

$$E(X^m) = \sum_{-\infty}^{\infty} x^m f(x) \quad (\text{for } X \text{ discrete}) \tag{9.30}$$

$$E(X^m) = \int_{-\infty}^{\infty} x^m f(x)\,dx \quad (\text{for } X \text{ continuous}) \tag{9.31}$$

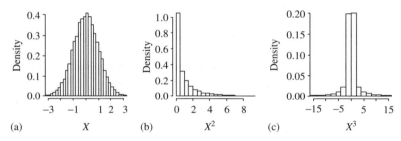

(a)   X   (b) X^2 (c) X^3

Figure 9.6: Given the distribution (a) of a particular random variable X, we can plot histograms of the distributions of its second (b) and third (c) powers.

These expectations, produced by raising the values of the variable to integer powers, are also known as the **moments of the distribution**. Now, consider the expression for the variance of a distribution from Section 7.7.

$$Var(X) = \frac{\sum_{i=1}^{n} (x_i - \mu)^2}{n} \tag{9.32}$$

where μ is the mean and n is the total number of observations that make up the population (assumed, for now, to be finite). Equation (9.32) can be manipulated as follows (but skip directly to Equation (9.34) if you are short of time)

$$Var(X) = \frac{\sum_{i=1}^{n} (x_i - \mu)^2}{n} = \frac{\sum_{i=1}^{n} (x_i^2 - 2x_i\mu + \mu^2)}{n} = \frac{\sum_{i=1}^{n} x_i^2 - \sum_{i=1}^{n} 2x_i\mu + \sum_{i=1}^{n} \mu^2}{n}$$

$$= \frac{\sum_{i=1}^{n} x_i^2}{n} - \frac{\sum_{i=1}^{n} 2x_i\mu}{n} + \frac{\sum_{i=1}^{n} \mu^2}{n} = \frac{\sum_{i=1}^{n} x_i^2}{n} - 2\mu\frac{\sum_{i=1}^{n} x_i}{n} + \mu^2 = \frac{\sum_{i=1}^{n} x_i^2}{n} - \mu^2 \tag{9.33}$$

In this, $\mu = E(X)$ and the expression $\frac{1}{n}\sum_{i=1}^{n} x^2$ is equal to $E(X^2)$, leading to the important result

$$Var(X) = E(X^2) - E(X)^2 \tag{9.34}$$

This can be shown to apply equally well to infinite populations for both discrete and continuous random variables. It establishes that variance, the second summary statistic of any distribution, can be obtained as a simple function of the first and second moments of the probability distribution. The same approach can be extended to other summary statistics. For example, skewness can be generated as a function of the first three moments and kurtosis of the first four. But why is all this important? A number of named distributions examined in this chapter are initially defined by their probability mass or density functions. Using expectations, it is possible to derive *exact* expressions for their summary statistics.

Expectation can be further generalised to any function $g(X)$ of a random variable X

$$E(g(X)) = \sum_{-\infty}^{\infty} g(x)f(x) \quad \text{(for } X \text{ discrete)} \tag{9.35}$$

$$E(g(X)) = \int_{-\infty}^{\infty} g(x)f(x)\,dx \quad \text{(for } X \text{ continuous)} \tag{9.36}$$

9.5.   Named distributions

This section is a bit abstract, so you may want to re-read it after you have had a go at Sections 9.6–9.15. Each of these sections refers to recurrent and extensively studied probability distributions that can be used to model different types of randomness. To facilitate comparison between them, presentation of each distribution follows a consistent recipe of four ingredients (although, not necessarily in the same order). The first ingredient is notation, which involves a statement of the type

$$X \sim Name(Parameters) \tag{9.37}$$

where X is the random variable and the symbol $\sim$ means 'is distributed as'. 'Name' is the full name or a shorthand symbol of the distribution and 'Parameters' is a list of values or symbols that determine the shape of the distribution. So, Equation (9.37) translates to: 'The random variable X is distributed as a (put name here) distribution, with parameters (put parameters here)'.

The second ingredient is the PMF or PDF of the distribution (depending on whether it is discrete or continuous). This will be a function $f(x)$ expressed in terms of x and the parameters of the distribution. The CDF of the distribution may also be cited. The third ingredient will either be a derivation or simple presentation of the mean and variance of the distribution, as functions of the parameters. Derivations will use the moments of the distribution (see Section 9.4). The final, and most important, ingredient will tell you where the distribution comes from and what types of randomness it can best describe.

9.2: Probability distributions

There are four distinct tools offered by R for each named probability distribution. They will be presented individually for each distribution (indeed, they are best illustrated with a particular distribution in mind) but their roles are outlined here:

❶ The PMF (for discrete distributions) and PDF (for continuous distributions). The command has the prefix d followed by the name of the distribution (e.g. `dnorm(x,mu,var)`). The arguments of the command include the parameters of the distribution and one or more values of the random variable for which the probability mass (or density) is sought. This command can be used to compare the likelihood of different values of a random variable.

❷ The CDF has the prefix p followed by the name of the distribution (e.g. `pnorm(x, mu var)`). Its arguments include the distribution's parameters and one or more values of the random variable for which cumulative probabilities are sought.

❸ The quantile function has the prefix q followed by the name of the distribution (e.g. qnorm(q, mu var)). This performs the inverse function of the cumulative. In other words, given one (or more) cumulative probability, it finds the associated value (or values) of the random variable. These values are called **quantiles**.

❹ The random numbers function has the prefix r followed by the name of the distribution (e.g. rnorm(n, mu, var)). This very important function generates n random realisations from the named distribution with the given parameter values.

9.6. Equally likely events: the uniform distribution

Example 9.10: River otter home ranges

American river otter, Lutra canadensis

River otters use stretches of river banks varying in length from 8 to 80 km. The home range of a particular individual is 10 km long. Although the animal has probably actively selected the location of its home range, its usage of different 1 km segments within the home range may be homogeneous. If so, then the probability of finding the animal in any given 1 km segment at any time would be equal for all segments. The event space of this experiment is $\Omega = \{1, \ldots, 10\}$ and the probability distribution comprises ten equal probabilities $p = 1/10$.

The **uniform distribution** is a formal restatement of an event space consisting of equiprobable events. For a discrete random variable, we first need to check that the event space Ω is finite and of known size k, e.g. $\Omega = \{x_1, \ldots, x_k\}$. The expression $X \sim U(x_1, x_k)$ states symbolically that a random variable X is uniformly distributed between x_1 and x_k. The PMF of this distribution is

$$f(x) = \frac{1}{k} \quad \forall x \in \Omega \tag{9.38}$$

If the values in the event space are consecutive integers, then the number of events in Ω is $k = x_k - x_1 + 1$ (e.g. $\Omega = \{2, 3, 4, 5\}$ contains a total of $k = 5 - 2 + 1 = 4$ numbers). Similarly, the total number of events between the starting value x_1 and any other value x is $x - x_1 + 1$. Using these facts, the CDF can be calculated as follows

$$F(x) = \sum_{i=x_1}^{x} f(i) = \sum_{i=x_1}^{x} \frac{1}{k} = \frac{x - x_1 + 1}{x_k - x_1 + 1} \tag{9.39}$$

Example 9.11: River otter home ranges

If the occurrence of the otter follows a uniform distribution $X \sim U(1, 10)$, the probability that it is observed in any given segment x is given by the constant function $f(x) = 0.1$. The probability that the otter is found at or before segment x is calculated from Equation (9.39) as $F(x) = x/10$. For example, there is a probability of 0.2 that the otter will be found inside the first two segments.

The uniform distribution for continuous variables (Figure 9.7) is motivated by the same rationale as the discrete uniform distribution, using densities instead of probabilities for each value in the event space. The distribution is denoted $X \sim U(a, b)$, where $[a, b]$ is the range of values in the continuous interval Ω. The PDF of the continuous uniform distribution is a constant function,

$$f(x) = c, x \in [a, b] \tag{9.40}$$

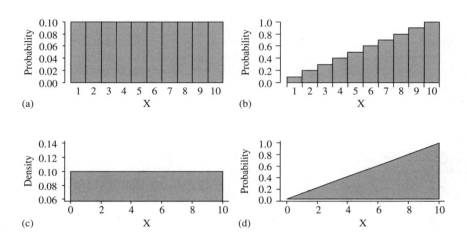

(a)

(b)

(c)

(d)

Figure 9.7: Uniform distribution for discrete and continuous variables. The PMF and PDF are shown in (a) and (c) respectively. The corresponding CDFs are shown in (b) and (d).

whose definite integral over the interval $[a, b]$ must equal 1.

$$\int_a^b f(x)\, dx = 1 \tag{9.41}$$

Combining these two relationships gives

$$\int_a^b c\, dx = 1, \therefore cx \Big|_a^b = 1, \therefore c(b - a) = 1, \therefore c = \frac{1}{b - a} \tag{9.42}$$

Therefore the PDF of a uniform distribution is

$$f(x) = \frac{1}{b - a} \tag{9.43}$$

The corresponding CDF is

$$F(x) = \int_a^x \frac{1}{b - a}\, ds = \frac{s}{b - a}\Big|_a^x = \frac{x - a}{b - a} \tag{9.44}$$

The moments of this distribution can be used to calculate its mean and variance. I illustrate using the continuous version, but similar arguments apply to the discrete. The first moment (and the distribution's mean) is

$$E(X) = \int_a^b \frac{x}{(b - a)}\, dx = \frac{1}{b - a}\int_a^b x\, dx = \frac{1}{2(b - a)}x^2\Big|_a^b = \frac{(b^2 - a^2)}{2(b - a)} = \frac{a + b}{2} \tag{9.45}$$

For the variance, the second moment is needed

$$E(X^2) = \int_a^b \frac{x^2}{(b-a)} \, dx = \frac{1}{3(b-a)} x^3 \Big|_a^b = \frac{(b^3 - a^3)}{3(b-a)} = \frac{a^2 + ab + b^2}{3} \tag{9.46}$$

This last step uses the expansion of $b^n - a^n$ from Section A1.3. Applying Equation (9.34) gives

$$Var(X) = E(X^2) - E(X)^2 = \left(\frac{a^2 + ab + b^2}{3}\right) - \left(\frac{a+b}{2}\right)^2 = \frac{(b-a)^2}{12} \tag{9.47}$$

It may seem unlikely that the uniform distribution will be able to capture many real-world phenomena, since heterogeneity abounds in nature. Indeed, the uniform distribution serves two different purposes: in some cases it formalises null hypotheses, scientific straw men offered for rejection by the real data. For example, otter home ranges may be investigated by assuming a uniform utilisation of all river segments and then checking whether the usage data deviate sufficiently from homogeneity. In other cases, uniformity quantifies ignorance about the true heterogeneities of a system. This is particularly useful for assigning equal prior probabilities to all competing hypotheses in a Bayesian framework (see Section 8.9).

ℝ   9.3: Uniform distribution

The basic implementation of the uniform distribution in R assumes a continuous random variable taking real values between `min` and `max`. The probability density of a value (or vector of values) x is given by `dunif(x, min, max)`. The cumulative distribution function for the same value(s) is given by `punif(x, min, max)`. For a given probability (or vector) of cumulative probabilities p, the command `qunif(p, min, max)` returns the associated values of the random variable. To generate n random numbers according to the uniform distribution, type `runif(n, min, max)`. For example,

```
> dunif(0,-1,1) # 0 is within the range -1 to 1
[1] 0.5
> dunif(2,-1,1) # 2 is not in the range -1 to 1
[1] 0
> punif(2,-1,1) # All the values in the range are below 2
[1] 1
> punif(c(-0.5,0,0.5), -1,1) # Returns the CDF for three values of X
[1] 0.25 0.50 0.75
> qunif(c(0.25,0.50,0.75), -1,1) # Gives three quantiles
[1] -0.5  0.0  0.5
> runif(5,-1,1) # Five random values between -1 and 1
[1] -0.8038679  0.8316444  0.1294939 -0.1554211 -0.8105164
```

To generate n random values from a discrete uniform distribution (between `min` and `max`), use any one of the following two lines:

```
round(runif(n, min-0.5, max+0.5))
sample(min:max, replace=T, n)
```

9.7. Hit or miss: the Bernoulli distribution

The simplest experiment has one of only two outcomes, often referred to as a 'success' or a 'failure', and represented numerically by 1 and 0. Each repetition of such an experiment is called a **Bernoulli trial** and the corresponding random variable is called **binary**. The Bernoulli distribution is fully specified by the table

$$
\begin{array}{c c}
x & P(X = x) \\
\hline
1 & p \\
0 & q
\end{array}
\tag{9.48}
$$

where p is the probability of success and q is the probability of failure. Since the two events are complementary, the two probabilities are related as follows

$$q = 1 - p \tag{9.49}$$

Hence, the distribution is fully specified by the parameter p. A Bernoulli-distributed variable is declared as $X \sim Bernoulli(p)$ (but see the next section for an alternative declaration). The PMF of the Bernoulli distribution is

$$f(x) = p^x q^{1-x} \text{ where } x = 0, 1 \tag{9.50}$$

Example 9.12: Bernoulli births

The probability that a particular mammal succeeds in giving birth at the age of 6 is $p = 0.4$. The occurrence (or not) of the event B ('birth') follows a Bernoulli distribution with parameter $p = 0.4$ (i.e. $q = 1 - p = 0.6$). To verify that the PMF returns the correct probabilities, simply place the values 1 and 0 into Equation (9.50) to get: $P(B) = f(1) = 0.4^1 0.6^0 = 0.4$ and $P(\bar{B}) = f(0) = 0.4^0 0.6^1 = 0.6$.

Granted, for such a simple probability distribution, the tabular definition in Equation (9.48) is more intuitive that the PMF in Equation (9.50). However, simple cases stated in a simple way may appear misleadingly isolated, when, in fact, they belong to a more general class of models, as the next section demonstrates.

9.8. Count of occurrences in a given number of trials: the binomial distribution

Example 9.13: Wildebeest reproductive histories

Female wildebeest become reproductively mature at age 3 and can give birth to a single calf each year with a constant probability $p = 0.8$. Breeding success in one year is independent of last year's outcome. We are interested in the lifetime reproductive output (X) of a female that lived to the end of its tenth year. This random variable takes values from the set

$$\Omega = \{0, 1, \dots, 8\} \tag{9.51}$$

To associate a probability with each of these possible outcomes, it is helpful to re-cast the question in terms of the event space of reproductive histories. For example, the reproductive history of a female that produced calves at age 3 and 7 is 10001000. The lifetime reproductive output of this animal would be $X = 2$. Any two reproductive histories are mutually exclusive but more than one history may lead to the same lifetime reproductive output (e.g. 11000000).

A zero outcome is the result of a string of failures to breed (00000000). Since this is the intersection of eight independent events, each with a constant probability of $q = 1 - 0.8 = 0.2$, the probability of zero offspring can be obtained as $P(X = 0) = q \times q \times q \times q \times q \times q \times q \times q = q^8 = 2.56 \times 10^{-6}$. Conversely, eight offspring can only be achieved with a string of eight successes (11111111). The probability of this event is $P(X = 8) = p^8 = 0.168$. Outcomes other than 0 and 8 will be made up of mixtures of successes and failures. One reproductive history that leads to a single offspring is the string 10000000. This has probability pq^7. However, there are another seven, equally likely, ways of achieving a single offspring (depending on which year the single birth occurs in). These strings of events are mutually exclusive so their probabilities can be added together to get $P(X = 1) = 8pq^7$. A similar approach can be applied to find $P(X = 7) = 8p^7q$. Collecting these results together,

$$
\begin{array}{c|ccccc}
x & 0 & 1 & \cdots & 7 & 8 \\
\hline
P(x) & q^8 & 8pq^7 & \cdots & 8p^7q & p^8
\end{array}
\tag{9.52}
$$

To fill in the missing entries for $x = 2,3,4,5,6$ and then come up with a PMF for X, it is necessary to discern a pattern in this table. The probability of any single reproductive history leading to a particular reproductive output x, can be written as $p^x q^{8-x}$. The total probability of x is the product of $p^x q^{8-x}$, with the number (say, m) of histories that can lead to x offspring. So, the PMF of X is of the form,

$$
f(x) = mp^x q^{8-x}
\tag{9.53}
$$

Finding the number m is easy in the cases $X = 0,1,7,8$, but for other values it can be tedious. Thankfully, **combinatorics**, a branch of mathematics that deals with counting techniques, has come up with a compact answer to the problem. The number of ways m of ordering x successes in a string of n trials is called a **binomial coefficient**. It is denoted $\dbinom{n}{x}$ (which is read: 'choose x out of n') and calculated by the formula

$$
m = \binom{n}{x} = \frac{n!}{x!(n-x)!}
\tag{9.54}
$$

where the exclamation mark indicates the factorial (e.g. $4! = 1 \times 2 \times 3 \times 4$). The factorial of zero is 1, so choosing 0 successes out of n attempts gives $\frac{n!}{0!(n-0)!} = \frac{n!}{n!} = 1$. This means that there is only one way to achieve no successes: a string of n failures.

Example 9.13 refers to the count of successes (births) achieved during a fixed and finite number of independent Bernoulli trials (the number of reproductive attempts). In general, the probability of a particular number of successes x out of n trials, each with a constant probability of success p, is given by the **binomial distribution**, with PMF

$$
f(x) = \binom{n}{x} p^x q^{n-x} \quad x = 0, \ldots, n
\tag{9.55}
$$

Example 9.14: Wildebeest reproductive output

The probability that the animal has exactly four offspring during its life (given that $n = 8, p = 0.8, q = 0.2$) is

$$f(4) = \begin{pmatrix} 8 \\ 4 \end{pmatrix} 0.8^4 0.2^{8-4} = \frac{8!}{4!4!} 0.8^4 0.2^4$$

$$= \frac{40320}{24 \times 24} 0.4096 \times 0.0016 \cong 0.046$$

(9.56)

Repeating this calculation for all values of X gives the entire probability distribution for reproductive output (Figure 9.8). Note that the x-axis only goes up to 8 (can't have more successes than trials) and that values closer to 8 are likelier than values close to 0 (with $p = 0.8$, it is likelier to reproduce than not to).

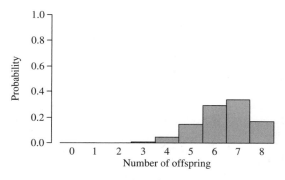

Figure 9.8: Plot of the PMF of the binomial distribution with $n = 8$ and $p = 0.8$.

The CDF of the binomial is

$$F(x) = \sum_{i=0}^{x} \begin{pmatrix} n \\ i \end{pmatrix} p^i q^{n-i} \quad x = 0, \ldots, n$$

(9.57)

Both the PMF and CDF depend only on n and p. A binomially distributed variable is declared as follows

$$X \sim B(n, p)$$

(9.58)

The binomial comprises n Bernoulli trials. Setting $n = 1$ retrieves the Bernoulli distribution. So, an alternative declaration for a Bernoulli variable is $X \sim B(1, p)$.

To decide whether a particular variable may be described by the binomial distribution, the answer to all three of the following questions should be 'yes':

❶ Can the random variable be seen as the outcome of a finite number of independent trials?
❷ Does each trial have one of only two possible outcomes?
❸ Is the probability of success in each trial constant?

The mean and variance of the binomial distribution are $E(X) = np$ and $Var(X) = npq$. The expression for the mean is intuitive: on average, the number of successes will be a fixed proportion p of the number of trials n.

ℝ 9.4: Binomial distribution and binomial coefficients

The four different commands associated with the binomial distribution are: `dbinom(x, n, p)` (probability of getting x successes in n Bernoulli trials if each has probability of success p), `pbinom(x, n, p)` (the associated CDF), `qbinom(q, n, p)` (approximate number of successes at which the CDF has value q) and `rbinom(m, n, p)` (m random numbers drawn from a binomial distribution with parameters n and p). The command `factorial()` can be used to calculate factorials directly. To calculate a binomial coefficient (i.e. the number of ways of positioning x successes in a string of n trials), use the command `choose(n, x)`.

9.9.   Counting different types of occurrences: the multinomial distribution

---

Example 9.15: Metapopulation transitions

In a five-patch Glanville fritillary metapopulation the butterflies emigrating from patch 1 arrive at another patch, or return to patch 1, with total probability 0.6. Therefore, each individual has a probability of 0.4 of getting lost during migration. If a total of 120 individuals emigrated this year from patch 1, then the number of butterflies arriving at any of the five patches is a binomial random variable, $X \sim B(120, 0.6)$. This treats each individual emigration as a Bernoulli trial with probability 0.6.

It is less straightforward to model the number of butterflies arriving at *each* of the five patches. Each emigration can have six outcomes (including loss), so it can no longer be considered a Bernoulli trial. This type of trial implies a vector of probabilities for each of the six outcomes. Here is an example in which p_0 is the probability of loss and p_i $(i = 1, \ldots, 5)$ is the probability of transiting from patch 1 to patch i : $p_0 = 0.4, p_1 = 0.2,$ $p_2 = 0.1, p_3 = 0.1, p_4 = 0.15, p_5 = 0.05$.

When there is a fixed number of independent trials but each has more than two possible outcomes, then it is necessary to use a generalisation of the binomial distribution. The **multinomial distribution** models the number of occurrences of each of k possible outcomes, assuming that the number of trials (n) is fixed and the probability (p_i) of each outcome is known. Its PMF gives the probability of a vector of counts, $\mathbf{x} = (x_1, \ldots, x_k)$ where x_i is the count referring to the ith outcome,

$$f(\mathbf{x}) = \frac{n!}{x_1! \cdots x_k!} p_1^{x_1} \cdots p_k^{x_k} \tag{9.59}$$

A multinomially distributed variable is declared $\mathbf{X} \sim \text{Multinomial}(n, \mathbf{p})$. The mean and variance for each outcome can be calculated as $\mu_i = np_i$ and $\sigma_i^2 = np_iq_i$.

Example 9.16: Metapopulation transitions

If $n = 120$ and $\mathbf{p} = (0.4, 0.2, 0.1, 0.1, 0.15, 0.05)$, then what is the probability that $x_0 = 51$ butterflies got lost and the following numbers arrived in patches 1 to 5: $x_1 = 22, x_2 = 12, x_3 = 11$, $x_4 = 19, x_5 = 5$. From Equation (9.59),

$$f(\mathbf{x}) = \frac{120!}{51!22!12!11!19!5!} 0.4^{51} 0.2^{22} 0.1^{12} 0.1^{11} 0.15^{19} 0.05^5$$

$$= 2.03 \times 10^{-5} \tag{9.60}$$

9.5: Multinomial distribution

The numbers involved in Example 9.16, as factorials and powers, are very large and very small, respectively. It is therefore advisable to use R to calculate these probabilities to avoid numerical overflow (or underflow). You can do this by using `dmultinom(x, n, p)`, where p is a vector of probabilities adding up to 1. To generate m random vectors from a multinomial distribution use the command `rmultinom(m, n, p)`. Here is a random vector of immigrants using the parameters from Example 9.16:

```
p<-c(0.4, 0.2, 0.1, 0.1, 0.15, 0.05)
> c(rmultinom(1,120,p)) # c() concatenates matrix output to one line
[1] 46 23 14  7 22  8
```

9.10. Number of occurrences in a unit of time or space: the Poisson distribution

The binomial distribution gives the probability associated with a particular number of successes out of a fixed number of Bernoulli trials. The **Poisson distribution** has a related interpretation: it gives the probability associated with a particular number of occurrences of an event within a fixed unit of space or time. Poisson random variables must satisfy the following requirements:

❶ The random variable must be a count of occurrences of an event.
❷ The count is made over a fixed unit, usually of time or space.
❸ The number of occurrences of the event within the unit of time or space could, in principle, be infinite.
❹ The average number of occurrences is the same for all intervals of time or regions in space.
❺ A single occurrence of the event must be independent of other occurrences.

For example, the number of individuals of a particular plant species counted within a quadrat may be modelled as a Poisson variable in space. The number of immigrants arriving in a local population from the outside during a single year may be thought of as a Poisson variable in time. The distribution of a Poisson random variable can be fully specified by a single **rate parameter** λ (lambda) which represents the average occurrences of the event within a unit of space or time. A Poisson random variable is declared $X \sim$ Poisson (λ) and the PMF

of the Poisson distribution is

$$f(x) = \frac{e^{-\lambda}\lambda^x}{x!} \tag{9.61}$$

Example 9.17: Data on plant abundance

A particular weed grows randomly in a field. The number N of weed rosettes occurring within a 0.25 m² quadrat can be modelled by a Poisson distribution (Figure 9.9). If the number of rosettes observed *on average* in each quadrat is $\lambda = 1.4$, then the probability of finding exactly n rosettes in any particular quadrat is

$$f(n) = \frac{e^{-1.4}1.4^n}{n!} \tag{9.62}$$

Hence, the probability of finding three rosettes is $f(3) = e^{-1.4}1.4^3/3! \cong 0.113$. The probability that the plant is present in any one quadrat is $P(N > 0) = 1 - f(0) = 1 - e^{-1.4} \cong 0.753$.

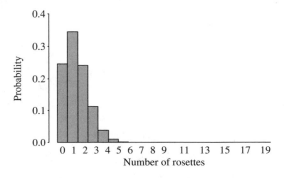

Figure 9.9: Plot of the PMF of the Poisson distribution with $\lambda = 1.4$.

The Poisson distribution is related to the binomial conceptually since they both model the total number of occurrences of an event. It turns out (see pages 183–185 in Larson, 1982, or page 100 in Mangel, 2006), that the Poisson can be derived as a limiting case of the binomial.

The mean of the Poisson is (rather predictably) the average rate of events λ. Crucially, the variance of the Poisson distribution is also λ. This offers a useful diagnostic for count data: if the sample average $\bar{x}$ and sample variance s^2 are close to each other, then the data may have originated from a Poisson distribution. Data with $s^2 > \bar{x}$ are **overdispersed** and data with $s^2 < \bar{x}$ are **underdispersed**.

Ⓡ   9.6: Poisson distribution

The PMF of a Poisson distribution with rate `lambda`, at a point x is given by `dpois(x, lambda)`. The corresponding CDF is `ppois(x, lambda)`. The function `qpois(p, lambda)` gives the value of the random variable at which the CDF is p. To generate n random values from a Poisson, write `rpois(n, lambda)`.

9.11.   The gentle art of waiting: geometric, negative binomial, exponential and gamma distributions

---

Example 9.18: Catastrophic events and species extinctions

The impending extinction of the Panamanian golden frog (Atelopus zeteki) may be the result of infection from an exotic fungus

The job of conservationists and wildlife managers is to evaluate extinction risk. Unfortunately, in the long-term, extinction is a certainty for all species, so evaluating risk focuses on when, rather than if, a species will become extinct. But why do extinctions occur? In 1984 Jared Diamond classified the ultimate causes of extinction into the 'evil quartet' of overkill (harvesting, by-catch), habitat destruction (anthropogenic or natural), introduced species (predators, competitors, parasites) and chains of extinction (incorporating indirect ecosystem effects). How do these conspire to bring a species down and how long does the process take?

Questions on waiting times are widespread in ecology and there are several probabilistic models that can help to answer them. The probability distributions presented in this section are derived from the binomial and the Poisson. The binomial fixes the number of trials and models the number of occurrences of a particular event. But the concept can be turned around to ask how many trials would be needed before observing a predetermined number of events. Similarly, instead of fixing the unit of time or space (as is done by the Poisson) we could ask how much time would need to pass (or how much distance/area/volume would need to be covered) before a predetermined number of events was encountered. Two choices need to be made before settling on an appropriate distribution: First, is there a concept of 'trial' for the particular question at hand, or is time/space considered continuous? Second, is the question about one occurrence or many?

Example 9.19: Catastrophic events and species extinctions

In trying to quantify times to extinction, facts must be established about the timing and intensity of the different causes of extinction. For example, an epidemic may invade a population at any time and be sufficiently intense to drive it to the brink. This could be modelled as a continuous process and the random variable would be the waiting time until the next major epidemic event. On the other hand, an instance of failed oviposition due to extreme, seasonal rainfall may have an equivalent effect. This process would need to be modelled in discrete time where the 'trial' is the egg-laying attempt and, rather ironically, the statistical 'success' is the biological failure to breed. Habitat destruction may require us to model several catastrophic events. For example, a species whose core range consists of five loci, each of which can be destroyed by human negligence, will only become extinct when it has been hit by five catastrophic events. This process could be modelled in continuous time (a locus can be destroyed at any time) but the random variable would be the time until all five loci have been destroyed.

Based on the answer to these questions, Table 9.3 will suggest a suitable distribution.

Table 9.3

	Only interested in the next occurrence	Interested in the next k occurrences
Time/space is discrete	**Geometric distribution** $X \sim \text{Geometric}(p)$ *Interpretation of* X: Number of trials until next success. *Parameter:* Probability of success p *PMF:* $f(x) = pq^{x-1}$ *Mean:* $1/p$ *Variance:* q/p^2	**Negative Binomial distribution** $X \sim \text{NegB}(k, p)$ *Interpretation of* X: Number of trials until k successes occur (so, $x \geq k$). *Parameters:* Probability (p) of success in each trial and number (k) of successes awaited. $PMF: f(x) = \begin{pmatrix} x-1 \\ k-1 \end{pmatrix} p^k q^{x-k}$ *Mean:* k/p *Variance:* kq/p^2
Time/space is continuous	**Exponential distribution** $X \sim M(\lambda)$ *Interpretation of* X : Waiting time (or distance) until the next occurrence of the event. *Parameter:* λ (the rate of occurrence) $PDF: f(x) = \lambda e^{-\lambda x}$ *Mean:* $1/\lambda$ *Variance:* $1/\lambda^2$	**Gamma distribution** $X \sim \text{Gamma}(k, \lambda)$ *Interpretation of* X : Waiting time (or distance) until the kth occurrence of the event. *Paramenters:* λ, k $PDF : f(x) = x^{k-1}\lambda^k \dfrac{e^{-\lambda x}}{\Gamma(k)}$ *Mean:* k/λ *Variance:* k/λ^2

Some clarifications are needed. The expression $\Gamma(k)$ that appears in the PDF of the gamma distribution is called the **gamma function**, given by

$$\Gamma(k) = \int_0^\infty s^{k-1} e^{-s} \, ds \tag{9.63}$$

where, in general, k can take any real, positive value. Although this integral has no closed-form solution, it can nevertheless be efficiently evaluated numerically by R (as well as many other packages), so, in most cases, you will not need to concern yourself with it. Note, however, that if you set $k = 1$ in Equation (9.63), you get

$$\Gamma(1) = \int_0^\infty e^{-s} \, ds = e^0 - \lim_{s \to \infty} e^{-s} = 1 \tag{9.64}$$

Therefore, setting $k = 1$ in the PDF of the gamma distribution yields the PDF of the exponential distribution. This makes sense, since k in the gamma distribution specifies the number of awaited occurrences and the exponential describes the waiting duration for a single occurrence. Similarly, the geometric distribution can be seen as a special case of the negative binomial for discrete time/space (set $k = 1$ in the PMF of the negative binomial to retrieve the PMF of the geometric).

There are alternative definitions of the geometric and negative binomial which, instead of modelling the total trials required to achieve a certain number of successes, focus on the number of *unsuccessful trials* before this number is achieved. These are the functions that R estimates by default, and they are also the ones plotted in Figure 9.10. So, the variable X now takes values from zero upwards. Under this interpretation, the PMF of the negative binomial (which contains the PMF of the geometric, $f(x) = pq^x$ for $k = 1$) is

$$f(x) = \binom{k + x - 1}{k - 1} p^k q^x \tag{9.65}$$

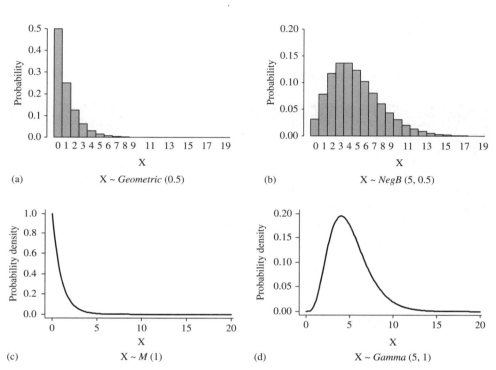

(a) $X \sim Geometric\ (0.5)$ (b) $X \sim NegB\ (5, 0.5)$

(c) $X \sim M\ (1)$ (d) $X \sim Gamma\ (5, 1)$

Figure 9.10: Example plots of (a) the geometric, (b) negative binomial, (c) exponential and (d) gamma distributions. In (a) and (b) the variable is interpreted as the number of *failures* before observing a predefined number of successes.

Example 9.20: Catastrophic events and species extinctions

A population whose size has declined because of deforestation is now restricted within a single national park. Measures are being taken to allow the population's range to be expanded but the successional period required for the regeneration of suitable habitat is ten years. In the past, the population has been affected by epidemics at an average rate of one every five years. Given its small size and concentrated distribution, the next epidemic is almost certain to drive the population to extinction. We would like to calculate the probability that the population survives long enough to be able to expand its range into the newly created habitat.

Since the impending epidemic can hit the population at any time, a continuous model is required. Since only the next epidemic is of interest, the exponential distribution is the most appropriate model. Its CDF can be found as follows

$$F(x) = \int_0^x \lambda e^{-\lambda s}\, ds = \lambda \int_0^x e^{-\lambda s}\, ds = -e^{-\lambda s}\Big|_0^x = 1 - e^{-\lambda x} \qquad (9.66)$$

The parameter λ is the rate of occurrence of epidemics, so it is equal to 1/5. The required probability therefore is

$$P(X > 10) = 1 - P(X \le 10) = 1 - F(10) = 1 - (1 - e^{-10/5}) \cong 0.135 \qquad (9.67)$$

Ⓡ 9.7: Geometric, negative binomial, exponential and gamma
R implements the alternative versions of the geometric and negative binomial in Equation (9.65). Thus (for a given probability of success p and number of successes k), `dgeom(x, p)` and `dnbinom(x, k, p)` give the probability that x failures occur before, respectively, 1 or k successes. The PDFs for the exponential and gamma are `dexp(x, lambda)` and `dgamma(x, k, lambda)`. CDFs for all four distributions are given by `pgeom(x, p)`, `pnbinom(x, k, p)`, `pexp(x, lambda)`, `pgamma(x, k, lambda)`. The values of the variables that achieve a certain cumulative probability q can be found from `qgeom(q, p)`, `qnbinom(q, k, p)`, `qexp(q, lambda)`, `qgamma(q, k, lambda)`. Finally, n random values can be drawn from each of these distributions by using the commands `rgeom(n, p)`, `rnbinom(n, k, p)`, `rexp(n, lambda)`, `rgamma(n, k, lambda)`.

9.12. Assigning probabilities to probabilities: the beta and Dirichlet distributions

The brief introduction to Bayesian concepts in Section 8.9 illustrated how to think of probability as the degree of belief in a particular scientific hypothesis (I suggest you have a quick re-read before you go on). In that section, I used the example of sex ratio determination and confined the presentation to two hypotheses: H_0: 'Probability of having a male offspring is 0.5' and H_1: 'Probability of having a male offspring is 0.4'. The subjective (or prior) probabilities associated with each of those hypotheses were $P(H_0) = \frac{2}{3}$ and $P(H_1) = \frac{1}{3}$, indicating a higher degree of belief in a proportion of 0.5 male offspring. This simple example represents a common occurrence in Bayesian statistics: often the quantity for which a prior is needed is a proportion, so it is necessary to assign a probability to a probability.

Although it is possible that in a particular scientific question there will only be two candidate answers (hypotheses), there are usually several and, often, infinite. For example, the proportion of male offspring could conceivably take any real value between 0 and 1 and each of these values could be assigned a subjective belief. Since the range [0,1] is continuous, each value in that range should be assigned a probability density by a continuous probability distribution. The standard for this purpose is the **beta distribution**, mainly because it is confined in the domain [0,1], but also because it facilitates Bayesian calculations, as we will see in Chapter 10. Its PDF is

$$f(x) = \frac{\Gamma(\alpha + \beta)}{\Gamma(\alpha)\Gamma(\beta)} x^{\alpha-1}(1 - x)^{\beta-1} \quad 0 < x < 1 \qquad (9.68)$$

This function uses two parameters α, β which enable it to take several different shapes within its domain (Figure 9.11). A beta-distributed variable is declared as $X \sim Beta(\alpha, \beta)$.

The gamma function, first introduced in Section 9.11, makes its appearance again in Equation (9.68), in the complicated expression $\frac{\Gamma(\alpha+\beta)}{\Gamma(\alpha)\Gamma(\beta)}$. The easiest way to think of it is as the normalising constant that ensures that the PDF of the beta distribution integrates to 1.

$$\int_0^1 \frac{\Gamma(\alpha + \beta)}{\Gamma(\alpha)\Gamma(\beta)} x^{\alpha-1}(1 - x)^{\beta-1}\, dx = 1 \tag{9.69}$$

Since the gamma functions in the integral do not depend on x, they can be brought out to obtain the following expression

$$\frac{\Gamma(\alpha)\Gamma(\beta)}{\Gamma(\alpha + \beta)} = \int_0^1 x^{\alpha-1}(1 - x)^{\beta-1}\, dx \tag{9.70}$$

The expression on the LHS is known as the **beta function**, written $B(\alpha, \beta)$. Equation (9.70) will come in handy in Chapter 10 in simplifying Bayesian calculations.

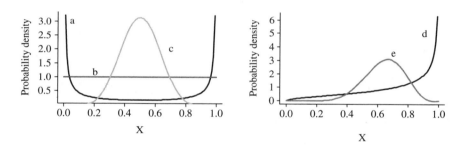

Figure 9.11: Five aspects of the beta distribution: (a) U-shaped ($\alpha = \beta = 0.1$); (b) uniform ($\alpha = \beta = 1$); (c) symmetric bell-shaped ($\alpha = \beta = 8$); (d) J-shaped ($\alpha = 1.5, \beta = 0.5$); (e) skewed bell-shaped ($\alpha = 9, \beta = 5$). Symmetric shapes such as (a), (b) and (c) indicate no prior preference for high or low values. Left-skewed shapes, such as (d) and (e) indicate belief in high values of probability. Right-skewed shapes are also possible but not shown here.

The mean and variance of the beta distribution are

$$\mu = \frac{\alpha}{\alpha + \beta} \qquad \sigma^2 = \frac{\alpha\beta}{(\alpha + \beta)^2(\alpha + \beta + 1)} \tag{9.71}$$

An extension of the beta is the **Dirichlet distribution**, which may be used to assign prior belief to the probabilities of several complementary and mutually exclusive events. For example, Example 9.15 looked at the proportions of butterflies making the transition from a starting patch to any other patch in a five-patch metapopulation. A butterfly either returned to its starting patch with probability p_0 or went to one of the other patches with probabilities $p_1, \ldots, p_5$. Each one of these probabilities ($p_0, \ldots, p_5$) could take values between 0 and 1, but not independently, since they all needed to add up to 1. A Dirichlet distribution $X \sim Dir(\mathbf{a})$ takes a vector of shape parameters $\mathbf{a}$ that has as many elements as the number of probabilities being modelled (six in the case of the butterfly metapopulation). The values of these parameters can be used to quantify belief in different values (0 to 1) of *each* of the probabilities. Random vectors $\mathbf{p}$ drawn from the Dirichlet distribution are probabilities that add up to 1.

9.8: Beta and Dirichlet

For two shape parameters a and b, the function dbeta (p, a, b) will give the beta probability density at the value p. The cumulative probability at p will be given by pbeta (p, a, b). The value p of the random variable at which the cumulative probability is q is given by qbeta (q, a, b). The function rbeta (n, a, b) will generate n random numbers from the beta distribution with parameters a and b. The package gtools provides the command rdirichlet (n, a), which, for a vector of parameters a, will generate a random vector of probabilities from the Dirichlet distribution.

9.13.   Perfect symmetry: the normal distribution

The **normal (or Gaussian) distribution** is the most frequently used probability distribution in the biological literature and elementary textbooks of statistics. It is a continuous distribution defined on the entire axis of real numbers, with a symmetric, bell-shaped PDF,

$$f(x) = \frac{1}{\sqrt{2\pi\sigma^2}} e^{-\frac{1}{2}\frac{(\mu-x)^2}{\sigma^2}} \quad x \in \mathbf{R} \tag{9.72}$$

It is useful to know your way around this expression. The quantity $(\mu - x)^2$ is zero at $x = \mu$ and positive for all other values of x. This means that the entire exponent will be negative for all values of $x \neq \mu$. An exponential function of this form decays as x takes values farther away from μ, in either direction. This generates the symmetry of the PDF around μ. The rate of decay is larger the smaller the value of the constant σ^2. Therefore, larger values of σ^2 can be used to increase the dispersion of the PDF. Finally, the constant $1/\sqrt{2\pi\sigma^2}$ ensures that the PDF integrates to 1.

A normal random variable is declared as $X \sim N(\mu, \sigma^2)$. The notation used for the two parameters is intentional since they can be shown to correspond to the mean and variance of the normal distribution.

The shape of the normal distribution is ideal for describing populations that are symmetrically clustered around a central value (the mean, mode and median of the normal distribution coincide) but it is not the only symmetric distribution. Its importance lies elsewhere: the normal distribution can be used to approximate the distributions and averages of large samples derived from non-normal populations (Section 9.18). Clearly, such a statistical chameleon deserves in-depth understanding despite the fact that, mathematically, it is not the most accommodating distribution. For example, the CDF of the normal distribution, given by

$$F(x) = \frac{1}{\sqrt{2\pi\sigma^2}} \int_{-\infty}^{x} e^{-\frac{1}{2}\frac{(\mu-s)^2}{\sigma^2}} ds \quad x \in \mathbf{R} \tag{9.73}$$

has no closed-form solution. Generations of statisticians have worked to obtain numerical approximations to this CDF for different values of the random variable. To make this task less labour-intensive, they decided to only work on a particular normal distribution, with mean zero and variance one, called the **standard normal. Standard normal variables** are declared as $Z \sim N(0, 1)$. Cumulative probabilities for any normally distributed variable $X \sim N(\mu, \sigma^2)$ can be obtained by first converting it into a standard normal variable using the following **standardising** transformation

$$z = \frac{x - \mu}{\sigma} \tag{9.74}$$

This works as follows: by subtracting μ from all values of X, its entire distribution is shifted to the left (if $\mu > 0$) or right (if $\mu < 0$) so that its new mean is zero. Then, dividing all the resulting values by σ either squeezes in (if $\sigma > 1$) or stretches out (if $0 < \sigma < 1$) the distribution around its mean so that its new standard deviation becomes 1. The cumulative probability calculated for the new standardised value z is exactly the same as the one that corresponds to the original value x.

Example 9.21: Weight distribution in voles

Weight distribution in woodland voles (*Microtus pinetorum*) is well approximated by $W \sim N(30, 16)$. The probability of finding a vole heavier than 35 g is $P(W > 35) = 1 - P(W \le 35)$. The corresponding standardised random variable is $Z = (W - 30)/4$. Therefore, $P(W > 35) = 1 - P(Z \le \frac{35-30}{4}) = 1 - F(1.25)$, where $F(x)$ is the CDF of the standard normal distribution. A comparison of actual vole weights with the standard normal distribution is shown in Figure 9.12.

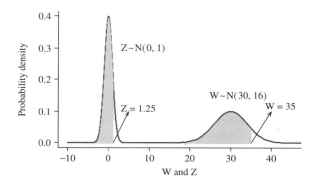

Figure 9.12: A comparison of the true distribution of vole weights with the standard normal. The shaded regions of the two distributions have exactly the same area, so that
$$P(W \le 35) = P(Z \le 1.25) \cong 0.89.$$

Since the normal distribution is defined for all real values, it associates some likelihood to negative vole weights. Therefore, using it to describe physical dimensions is only an approximation. Here, it can be calculated that the probability associated with such biologically unrealistic measurements is very small: $P(W \le 0) = 3.19 \times 10^{-14}$.

Use of the transformation in Equation (9.74) was standard practice before the advent of computers. These days, statistical packages will readily calculate probabilities over any specified interval, but the standardisation trick is still important (see chi-square distribution in Section 9.17 and confidence intervals in Section 10.5).

9.9: Normal distribution

The usual functionality is available but you need to remember that the R specification of the normal distribution requires the standard deviation instead of the variance. For a mean `mu` and a standard deviation `sd`, the functions `dnorm(x, mu, sd)` and `pnorm(x, mu, sd)`, give the probability density and cumulative probability at x. The function `qnorm(p, mu, sd)` gives the value x at which the CDF has the value p. The function `rnorm(n, mu, sd)` generates n random values from the specified normal distribution.

9.14. *Because it looks right: using probability distributions empirically*

All elementary probability distributions are derived from first principles by thinking about different experimental set-ups: counting successes per number of trials generates the Bernoulli and binomial, counting occurrences per unit space/time gives the Poisson, waiting times/distances give the geometric, negative binomial, exponential and gamma. However, there is a certain degree of liberty in how different distributions are used **empirically**. It may be that we care little about the mechanism generating the randomness in the data but would merely like to describe the shape of the resulting frequency distribution. The normal distribution is an ideal candidate for bell-shaped histograms; the negative binomial is often used to describe overdispersed count data that don't seem to follow the Poisson; the gamma distribution with its two shape parameters can be used to describe non-negative, continuous variables with skewed distributions. Empirical use of a distribution requires some fine-tuning of the distribution's parameters so that the shape of the distribution resembles the histogram of the data. Bayesian statistics makes extensive use of such 'designer' distributions in formalising prior beliefs.

Example 9.22: A beta prior for wildebeest sex at-birth

The beta distribution is a suitable prior for the probability of giving birth to a male because it is constrained between 0 and 1. Prior to collecting any data, a sensible null hypothesis might be that wildebeest have a 1:1 sex ratio. Sex ratios different from 1:1 are not unheard of in mammals, but extreme deviations are unlikely. Based on all of the above, most prior belief may be allocated to the values around 0.5, e.g. by specifying that 90% of the prior distribution will lie between the values of 0.3 and 0.7.

The mean and variance of a beta distribution with parameters α, β are given by Equations (9.71). Hence, the values of α, β corresponding to a mean $\mu = 0.5$ and a small variance σ^2 (yet to be evaluated precisely), can be calculated by solving Equations (9.71) as a system of coupled equations,

$$\alpha = \frac{\mu}{\sigma^2}(\mu - \mu^2 - \sigma^2) \quad \text{and} \quad \beta = \frac{1-\mu}{\sigma^2}(\mu - \mu^2 - \sigma^2) \tag{9.75}$$

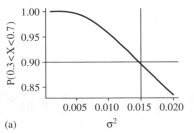

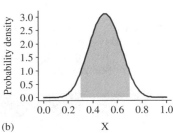

(a) (b)

Figure 9.13: (a) Plot of $P(0.3 < X \leq 0.7)$ as a function of the variance of the beta distribution; (b) the beta distribution $Beta(7.83, 7.83)$ which has mean 0.5 and contains 90% of its probability density within the values 0.3 and 0.7.

Substituting in the value $\mu = 0.5$ gives the two parameters of the beta distribution as functions of the variance: $\alpha = \beta = \frac{0.125}{\sigma^2} - 0.5$. A computer may now be used (see R9.10) to plot the probability enclosed within the values 0.3 and 0.7 as a function of σ^2 (Figure 9.13(a)). The value of σ^2 that approximately satisfies the condition $P(0.3 < X \leq 0.7) = 0.9$ is 0.015. Therefore, the required prior distribution is $Beta(7.83, 7.83)$ (Figure 9.13(b)).

Distributions aren't only used to emulate the data or our prior beliefs about parameters. Some distributions, most notably the normal, can also be used to approximate other distributions. This trick is especially useful for computational purposes and the likelihood calculations in Chapter 10. Here, I will only illustrate the normal approximation for the binomial and Poisson, but other distributions can be approximated in a similar way.

The fundamental observation is that, for large sample sizes, both the binomial and the Poisson become bell-shaped (Figure 9.14). Although they are discrete distributions, unlike the normal, the probability $P(X = x)$ of a particular (binomial or Poisson) count is nevertheless similar to the probability that the approximating normal variable is between x and $x + 1$.

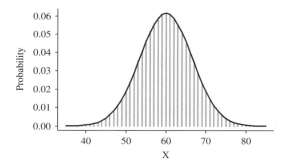

Figure 9.14: Approximation of the binomial variable $X \sim B(200, 0.3)$ (grey bars) by the normal variable $X \sim N(60, 42)$.

The approximation requires a normal distribution with the same mean and variance as the original binomial or Poisson. Hence, given a binomial distribution $B(N, p)$ with large N, the approximating normal distribution is $N(Np, Npq)$. Similarly, given the distribution $Poisson(\lambda)$, the approximating normal is $N(\lambda, \lambda)$.

9.10: Converting a range to a variance

The specification of the prior in Example 9.22 required that 90% of the beta distribution's probability density was enclosed between the values 0.3 and 0.7. For a given set of parameters α, β this probability can be calculated as $F(0.7) - F(0.3)$ where $F(x)$ is the CDF of the beta distribution. In R, this calculation is `pbeta(0.7, a, b)-pbeta(0.3, a, b))`. The parameters are given as a function of the variance $\alpha = \beta = \frac{0.125}{\sigma^2} - 0.5$. We may therefore plot $F(0.7) - F(0.3)$ for various values of σ^2 to find which value of variance brings it close to 0.9. A plot such as Figure 9.13(a) can be generated from the following code:

```
si<-seq(0.001, 0.02, 0.001) # Different values of variance
al<--0.125/si-0.5 # Corresponding parameters of the beta distr.
p<-pbeta(0.7,al,al)-pbeta(0.3,al,al) # Prob. enclosed in (0.3,0.7)
plot(si, p, xlab=expression(sigma^2), ylab="P(0.3<X<0.7)", type="l")
```

9.15.   Mixtures, outliers and the t-distribution

Example 9.23: Bimodal weight distribution in voles

A sample of weights from 2000 voles trapped in a wood-land site in the state of Indiana is bimodally distributed (Figure 9.15). Closer inspection shows that the sample comprises a mixture of woodland voles (*M. pinetorum*) and meadow voles (*M. pensylvanicus*). It is known that the weights of woodland and meadow voles can be accurately described by the distributions $W_w \sim N(30, 16)$ and $W_m \sim N(50, 25)$. The two species are equally likely to be caught (in traps or by predators) and, in the study site, woodland voles are twice as abundant as meadow voles. We wish to create a probability model for the weight distribution of potential prey of a local predator that specialises in taking small rodents. As far as the predator is concerned, a vole of a given weight w is food, irrespective of its species. The PDFs of weights of woodland and meadow voles are written $f_w(w)$ and $f_m(w)$. The PDF of weights, irrespective of species, is denoted $f(w)$. Given their relative abundance in the site, a particular prey item is a woodland vole with probability $p = \frac{2}{3}$ and a meadow vole with probability $q = 1 - p = \frac{1}{3}$. The probability of finding a vole in a very small interval of weights between w and $w + dw$ is

$$P(w < W \leq w + dw) = P(\text{meadow vole})P(w < W_m \leq w + dw)$$
$$+ P(\text{woodland vole})P(w < W_w \leq w + dw) \tag{9.76}$$

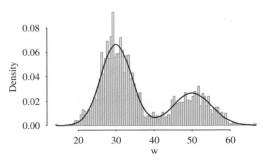

Figure 9.15: Histogram of recorded weights for the 2000 voles in the sample and the associated probability model (smooth curve).

If the probability densities remain approximately constant over the very small interval dw, then Equation (9.76) may be approximately written in terms of the PDFs,

$$f(w)\,dw = (1 - p)f_m(w)\,dw + pf_w(w)\,dw \tag{9.77}$$

The intervals dw can be cancelled out to give the final relationship

$$f(w) = (1 - p)f_m(w) + pf_w(w) \tag{9.78}$$

The particular expressions for the PDFs and the species-specific parameter values can be substituted-in to get the full PDF of vole weights

$$f(w) = \frac{1}{3\sqrt{2\pi}}\left(\frac{1}{5}e^{-\frac{1}{2}\frac{(50-w)^2}{25}} + \frac{1}{2}e^{-\frac{1}{2}\frac{(30-w)^2}{16}}\right) \tag{9.79}$$

A distribution that results from a combination of two or more populations, each with their own distributions, is called **a mixture**. It turns out that the mean and variance of a mixture distribution can be obtained as simple functions of the means and variances of its constituent parts. In the case of the mixture of two normal distributions $N(\mu_1, \sigma_1^2)$ and $N(\mu_2, \sigma_2^2)$ (see Example 9.23), the mean and variance of the mixture are

$$\mu = p\mu_1 + (1-p)\mu_2 \tag{9.80}$$
$$\sigma^2 = p(\mu_1^2 + \sigma_1^2) + (1-p)(\mu_2^2 + \sigma_2^2) - \mu^2$$

Mixtures are easier to identify when their constituent distributions do not overlap, hence presenting us with tell-tale signs such as multimodality. When there is considerable overlap between the populations, mixtures are harder to identify. A particularly important example of this deals with mixtures whose constituent populations have approximately the same mean but different variances. The resulting mixture is usually identified as a **fat-tailed distribution**. In practical terms, data generated from symmetric fat-tailed distributions may resemble a normal distribution but will tend to contain more outliers. Hence, examining the distances (the residuals) of the observations from the sample average will yield more extreme values than might be expected under a normal distribution with the estimated sample variance. The distribution most commonly used to describe such situations is the **t-distribution** with PDF

$$f(x) = \frac{\Gamma\left(\dfrac{n+1}{2}\right)}{\sqrt{n\pi}\,\Gamma\left(\dfrac{n}{2}\right)} \left(1 + \frac{x^2}{n}\right)^{-\frac{n+1}{2}} \tag{9.81}$$

A t-distributed random variable is interpreted as the residual from the sample average (so the mean of the t-distribution is always zero). The parameter n (known as the number of degrees of freedom of a particular t-distribution) takes positive values and determines the kurtosis of the distribution. For small values of n, the distribution is fat-tailed. For large values, it converges to the standard normal.

® 9.11: t-distribution

The distribution functions for the t-distribution with n degrees of freedom are: `dt(x, n)` – PDF at x; `pt(x, n)` – CDF at x; `qt(p, n)` – value of x at which CDF is p; `rt(m, n)` – m random values from the distribution.

9.16. Joint, conditional and marginal probability distributions

Before reading this section, you may find it helpful to re-read the material on functions with many independent variables in Section 1.25.

Univariate distributions characterise sample units in terms of only one variable but sample units may have several attributes (e.g. an animal has many different physical and physiological characteristics, a square metre of habitat may be described in terms of its species composition as well as its many abiotic traits). **Multivariate or joint distributions** describe how likely it is to observe a sample unit with a particular *combination* of characteristics. Sample units come in different guises. In the following example the sample unit is defined as a pair of successive foraging decisions by a seabird.

Example 9.24: Feeding site fidelity in kittiwakes

Black-legged kittiwakes, Rissa tridactyla

Black-legged kittiwakes are medium-sized gulls that spend the spring and summer months in large breeding aggregations, performing foraging trips to sea. Individuals are faithful to particular feeding sites and there is little evidence that colony members or partners share information about these sites (Irons, 1998). It therefore seems that site fidelity develops through first-hand experience, rather than communication. Consider a hypothetical group of naïve birds that have just arrived at a colony and are performing their first foraging trip. This is hypothetical because breeding birds are likely to have spent the first three years of their lives prospecting for good breeding colonies – so they are unlikely to be naïve. The distribution of food within the foraging range of the birds is shown as a radial plot in Figure 9.16, in eight directions from the colony.

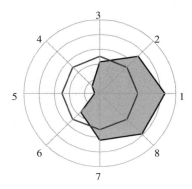

Figure 9.16: A radial plot of the distribution of food around the colony split into eight sectors in a counter-clockwise rotation. The shaded octagon shows the relative abundance of food in each sector and the unshaded octagon shows the environmental average ($= 1/8$).

In any given trip, birds stay predominantly in one sector. In their first trip, the birds are not aware of the distribution of food but have experience of the average productivity of their environment. The following is a possible foraging strategy they might use: in the absence of any knowledge, the birds head out in a uniformly random direction. If the availability of food encountered in a sector during the first foraging trip exceeds the environmental average, then that sector is twice as likely to be visited in the second trip. Conversely, sectors below the environmental average have half the probability of being visited again. We would like to model the radial distribution of foragers (i.e. proportion of visits to each sector) during the first two foraging trips of the birds. The directions of the two trips can be modelled as two random variables (X, Y) with the same event space $\Omega = \{1, 2, 3, 4, 5, 6, 7, 8\}$.

Because the first decision is naïve, it may be described by a discrete, uniform distribution $X \sim U(1, 8)$ with PMF, $f_X(x) = \frac{1}{8}$. However, the second decision will not follow the same distribution. A bird that happens to visit sectors 1, 2, 7 and 8 will encounter abundant food and will (according to the assumed decision rule) be twice as likely to return to the same sector. Conversely, birds visiting sectors 3, 4, 5 or 6 will be half as likely to return. We are therefore seeking the probabilities of events for the second decision, conditional on the first: $P(Y = 1|X), P(Y = 2|X), \ldots, P(Y = 8|X)$. If a bird visits sector 1 on its first trip, then the PMF

for the second trip can be written as

$$
\begin{array}{c|cccccccc}
y & 1 & 2 & 3 & 4 & 5 & 6 & 7 & 8 \\
\hline
f_{Y|1}(y) & \frac{2}{9} & \frac{1}{9} & \frac{1}{9} & \frac{1}{9} & \frac{1}{9} & \frac{1}{9} & \frac{1}{9} & \frac{1}{9}
\end{array}
\tag{9.82}
$$

Since there are now two random variables to consider (X and Y), it is necessary to distinguish between them, so the subscripts have reappeared in the PMF notation. This notation is flexible. In Equation (9.82) it allowed the symbol $f_{Y|1}(y)$ to represent the probability distribution for Y, conditional on $X = 1$. This is known as a **conditional probability distribution**. Conditional probabilities $P(Y|X)$ may be obtained for all values of Y given every possible value of X. These can then be converted to joint probabilities for any given combination of values (X, Y) by using the results in Chapter 8:

$$
P(Y \cap X) = P(Y|X)P(X)
\tag{9.83}
$$

Obtaining these joint probabilities for all possible pairs (X, Y) yields the joint probability distributions of the two variables, written $f_{X,Y}(x,y)$. The laws of probability introduced for single events generalise to entire random variables. Hence, the joint probability distribution is connected to the conditional distributions by the following relationship:

$$
f_{X,Y}(x,y) = f_{Y|X}(y)f_X(x)
\tag{9.84}
$$

This expression is valid for both discrete and continuous variables, although in the case of continuous variables the functions express probability densities rather than probabilities.

Example 9.25: Feeding site fidelity in kittiwakes

Conditional and joint distributions for two random variables can be represented by the two tables below. In the first table, each column represents a conditional probability distribution (column totals are all 1). The second table is obtained from the first, by applying Equation (9.84), with $f_X(x) = 1/8$. Although the values are spread across two dimensions, this is still a probability distribution (the sum of all the entries in the second table equals 1).

Conditional probability distr.: $f_{Y|X}(y)$

					X			
	1	2	3	4	5	6	7	8
1	$\frac{2}{9}$	$\frac{1}{9}$	$\frac{2}{15}$	$\frac{2}{15}$	$\frac{2}{15}$	$\frac{2}{15}$	$\frac{1}{9}$	$\frac{1}{9}$
2	$\frac{1}{9}$	$\frac{2}{9}$	$\frac{2}{15}$	$\frac{2}{15}$	$\frac{2}{15}$	$\frac{2}{15}$	$\frac{1}{9}$	$\frac{1}{9}$
3	$\frac{1}{9}$	$\frac{1}{9}$	$\frac{1}{15}$	$\frac{2}{15}$	$\frac{2}{15}$	$\frac{2}{15}$	$\frac{1}{9}$	$\frac{1}{9}$
Y 4	$\frac{1}{9}$	$\frac{1}{9}$	$\frac{2}{15}$	$\frac{1}{15}$	$\frac{2}{15}$	$\frac{2}{15}$	$\frac{1}{9}$	$\frac{1}{9}$
5	$\frac{1}{9}$	$\frac{1}{9}$	$\frac{2}{15}$	$\frac{2}{15}$	$\frac{1}{15}$	$\frac{2}{15}$	$\frac{1}{9}$	$\frac{1}{9}$
6	$\frac{1}{9}$	$\frac{1}{9}$	$\frac{2}{15}$	$\frac{2}{15}$	$\frac{2}{15}$	$\frac{1}{15}$	$\frac{1}{9}$	$\frac{1}{9}$
7	$\frac{1}{9}$	$\frac{1}{9}$	$\frac{2}{15}$	$\frac{2}{15}$	$\frac{2}{15}$	$\frac{2}{15}$	$\frac{2}{9}$	$\frac{1}{9}$
8	$\frac{1}{9}$	$\frac{1}{9}$	$\frac{2}{15}$	$\frac{2}{15}$	$\frac{2}{15}$	$\frac{2}{15}$	$\frac{1}{9}$	$\frac{2}{9}$

Joint probability distribution: $f_{X,Y}(x,y)$

					X			
	1	2	3	4	5	6	7	8
1	$\frac{1}{36}$	$\frac{1}{72}$	$\frac{1}{60}$	$\frac{1}{60}$	$\frac{1}{60}$	$\frac{1}{60}$	$\frac{1}{72}$	$\frac{1}{72}$
2	$\frac{1}{72}$	$\frac{1}{36}$	$\frac{1}{60}$	$\frac{1}{60}$	$\frac{1}{60}$	$\frac{1}{60}$	$\frac{1}{72}$	$\frac{1}{72}$
3	$\frac{1}{72}$	$\frac{1}{72}$	$\frac{1}{120}$	$\frac{1}{60}$	$\frac{1}{60}$	$\frac{1}{60}$	$\frac{1}{72}$	$\frac{1}{72}$
Y 4	$\frac{1}{72}$	$\frac{1}{72}$	$\frac{1}{60}$	$\frac{1}{120}$	$\frac{1}{60}$	$\frac{1}{60}$	$\frac{1}{72}$	$\frac{1}{72}$
5	$\frac{1}{72}$	$\frac{1}{72}$	$\frac{1}{60}$	$\frac{1}{60}$	$\frac{1}{120}$	$\frac{1}{60}$	$\frac{1}{72}$	$\frac{1}{72}$
6	$\frac{1}{72}$	$\frac{1}{72}$	$\frac{1}{60}$	$\frac{1}{60}$	$\frac{1}{60}$	$\frac{1}{120}$	$\frac{1}{72}$	$\frac{1}{72}$
7	$\frac{1}{72}$	$\frac{1}{72}$	$\frac{1}{60}$	$\frac{1}{60}$	$\frac{1}{60}$	$\frac{1}{60}$	$\frac{1}{36}$	$\frac{1}{72}$
8	$\frac{1}{72}$	$\frac{1}{72}$	$\frac{1}{60}$	$\frac{1}{60}$	$\frac{1}{60}$	$\frac{1}{60}$	$\frac{1}{72}$	$\frac{1}{36}$

Different parts of the joint probability distribution can help answer different questions. For example, the probability that a bird goes directly north on its first trip and directly south on its second is $f_{X,Y}(3,7) = \frac{1}{60}$. However, the probability that a bird first goes south and then north is lower, $f_{X,Y}(7,3) = \frac{1}{72}$. Since the 64 events in the joint probability distribution are mutually exclusive, they can be combined by addition. For example, the row sums give the unconditional probability distribution for the different destinations of the second trip:

$$
\begin{array}{c|cccccccc}
y & 1 & 2 & 3 & 4 & 5 & 6 & 7 & 8 \\
\hline
f_Y(y) & 0.14 & 0.14 & 0.11 & 0.11 & 0.11 & 0.11 & 0.14 & 0.14
\end{array}
\tag{9.85}
$$

I won't do this here but, to extend this model to the third trip of the animals, it may be convenient to assume that the animals are characterised by the **Markov property**, meaning that their next decision only depends on the previous one. This would enable the creation of a new joint distribution, in which X would represent trip 2 and Y would now stand for trip 3. The unconditional distribution of X would be obtained from Equation (9.85). This is an example of a Markov chain (see Section 6.5), whose transition matrix is given by the joint conditional distribution $f_{Y|X}(y)$, tabulated above.

The row and column totals of a two-variable joint distribution are known as the **marginal distributions**. They are, simply, the unconditional distributions $f_X(x)$ and $f_Y(y)$ of each of the two variables. This concept may be familiar from the discussion on joint frequency distributions in Example 7.19 of Section 7.13. Joint frequency distributions were summarised in terms of their sample averages and sample variance–covariance matrix. The means and variances of single-variable probability distributions are calculated using expectations (Section 9.4). There is a similar definition for the covariance, as the expectation of a function of the two variables:

$$
\text{cov}(X, Y) = E((X - \mu_X)(Y - \mu_Y))
\tag{9.86}
$$

μ_X and μ_Y are the means of X and Y respectively. For continuous random variables, this expression requires the use of double integrals (that are not covered in this book), but for discrete variables it requires double sums (covered in Section 5.4):

$$
\text{cov}(X, Y) = \sum_{All\ X} \sum_{All\ Y} (x - \mu_X)(y - \mu_Y) f_{X,Y}(x, y)
\tag{9.87}
$$

The correlation ρ between the two variables of a joint distribution is linked to the covariance with the following relationship (identical to the expression for samples in Section 7.12),

$$
\rho = \frac{\text{cov}(X, Y)}{\sigma_X \sigma_Y}
\tag{9.88}
$$

A special class of joint distributions is generated by extending the concept of independence (see Section 8.7) from single events to entire variables. Two random variables are called **independent** if the value taken by one does not affect the value taken by the other. Mathematically, this means that the conditional distribution $f_{Y|X}(y)$ is the same as the unconditional distribution $f_Y(y)$. Specifying Equation (9.84) to this case gives an expression for the joint distribution of independent variables

$$
f_{X,Y}(x, y) = f_Y(y) f_X(x)
\tag{9.89}
$$

The definition of independent variables intuitively implies that their covariance must be zero. This is indeed the case, but to demonstrate it, the following three facts about expectations are needed. The first two apply in general for any two random variables X, Y and any constant c,

$$E(X + Y) = E(X) + E(Y) \qquad (9.90)$$

$$E(cX) = cE(X) \qquad (9.91)$$

The third only applies to independent random variables (it is, in fact, a direct consequence of Equation (9.89)),

$$E(XY) = E(X)E(Y) \qquad (9.92)$$

As an illustration of how to use these identities, here is a proof that the covariance of independent variables is zero. Starting from the definition in Equation (9.86), the expression inside the expectation can be expanded,

$$\text{cov}(X, Y) = E(XY - X\mu_Y - Y\mu_X + \mu_X\mu_Y) \qquad (9.93)$$

This is the expectation of a sum, so it can be broken up using Equation (9.90)

$$\text{cov}(X, Y) = E(XY) + E(-X\mu_Y) + E(-Y\mu_X) + E(\mu_X\mu_Y) \qquad (9.94)$$

Equation (9.91) can be used to extract the constants from the expectations

$$\text{cov}(X, Y) = E(XY) - \mu_Y E(X) - \mu_X E(Y) + \mu_X\mu_Y \qquad (9.95)$$

Finally, using Equation (9.92) and the facts that $E(X) = \mu_X, E(Y) = \mu_Y$ gives $\text{cov}(X, Y) = 0$. By Equation (9.88), the implication is that the correlation of two independent variables is also zero. It should be noted that the reverse is not always true, i.e. two uncorrelated variables are not necessarily independent.

9.17. The bivariate normal distribution

All the basic concepts of joint distributions are well illustrated by the bivariate normal, which, like its univariate counterpart, is a multipurpose approximation tool. Brace yourself for its PDF:

$$f_{X,Y}(x, y) =$$

$$\frac{1}{2\pi\sigma_X\sigma_Y\sqrt{1 - \rho^2}} \exp\left\{-\frac{1}{2(1 - \rho^2)}\left(\frac{(x - \mu_X)^2}{\sigma_X^2} + \frac{(y - \mu_Y)^2}{\sigma_Y^2} - 2\rho\frac{(x - \mu_X)(y - \mu_Y)}{\sigma_X\sigma_Y}\right)\right\} \qquad (9.96)$$

where, μ_X, μ_Y are the means of the two variables, σ_X, σ_Y are their standard deviations and ρ is the correlation between them. The marginal distributions of the two variables are normally distributed: $X \sim N(\mu_X, \sigma_X^2)$ and $Y \sim N(\mu_Y, \sigma_Y^2)$.

Example 9.26: Height and root depth of tree ferns

Fern trees of a particular species are, on average, 1.9 m tall with a standard deviation of 0.22 m. Their roots are, on average, 0.57 m deep, with a standard deviation of 0.08 m. The correlation between tree height and root depth is 0.7. The joint distribution between height (H) and depth (D) may (after some numerical simplification) be described by a bivariate normal distribution with PDF

$$f_{H,D}(h, d) \cong$$

$$\frac{40}{\pi} \exp\left\{-\left(\frac{(h - 1.9)^2}{0.049} + \frac{(d - 0.57)^2}{0.006} - \frac{(h - 1.9)(d - 0.57)}{0.013}\right)\right\} \qquad (9.97)$$

This expression can be used to compare the likelihood of observing plants with particular combinations of heights and root depths. For example, a shorter-than-average tree with $H = 1.8, D = 0.4$ has $f_{H,D}(1.8, 0.4) \cong 0.31$. It may seem that an even shorter tree with $H = 1.5, D = 0.4$ should be even less likely. However, this combination has $f_{H,D}(1.5, 0.4) \cong 0.74$ because this lower height is better matched to a root depth of 0.4 m. Compare these values with the probability density associated with the average tree $f_{H,D}(1.9, 0.57) \cong 12.73$.

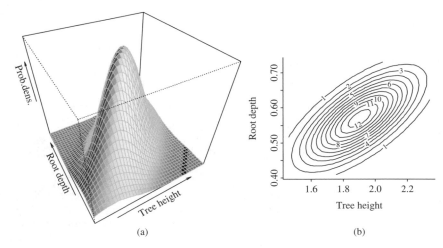

(a)   (b)

Figure 9.17: A perspective (a) and contour plot (b) of the bivariate normal distribution. The diagonal orientation of the elliptical contours in (b) is a consequence of the positive correlation between height and depth.

In the case of the bivariate normal distribution (Figure 9.17), zero correlation implies independence. This may be shown by setting $\rho = 0$ in Equation (9.96) and rearranging slightly to get

$$
\begin{aligned}
f_{X,Y}(x, y) &= \frac{1}{2\pi \sigma_X \sigma_Y} \exp\left\{ -\frac{1}{2}\left(\frac{(x - \mu_X)^2}{\sigma_X^2} + \frac{(y - \mu_Y)^2}{\sigma_Y^2} \right) \right\} \\
&= \frac{1}{\sqrt{2\pi \sigma_X^2}} \exp\left\{ -\frac{1}{2}\left(\frac{(x - \mu_X)^2}{\sigma_X^2} \right) \right\} \frac{1}{\sqrt{2\pi \sigma_Y^2}} \exp\left\{ -\frac{1}{2}\left(\frac{(y - \mu_Y)^2}{\sigma_Y^2} \right) \right\} \quad (9.98) \\
&= f_X(x) f_Y(y)
\end{aligned}
$$

which is the definition of independence given in Equation (9.89).

® 9.12: Bivariate normal

Functions for the multivariate normal distribution can be found in the package `mvtnorm`. The probability density function is `pmvnorm()` and random numbers can be drawn from the distribution with the command `rmvnorm()`. To specify the bivariate distribution you need a vector of two means and a 2×2 variance–covariance matrix. For example, to generate a contour plot of probability densities for the bivariate normal with the parameters specified in Example 9.26, you could use the following code:

```
require(mvtnorm) # Loads package for multivariate normal

# Specification of the distribution's parameters
mh<-1.9 #  Mean tree height
md<-0.57 # Mean tree depth
sh<-0.22 # St. Dev. of height
sd<-0.08 # St. Dev of depth
ro<-0.7 # Correlation between height and depth
mu<-c(mh,md) # Vector of means
cov<-ro*sh*sd # Calculates covariance from correlation
varcov<-matrix(c(sh^2,cov,cov,sd^2),nrow=2,ncol=2) #Var/covar matrix

#Plotting parameters
steps<-40 # Number of points in each dimension of the grid
hr<-c(mh-2*sh, mh+2*sh)  # Range of values for height
dr<-c(md-2*sd, md+2*sd)  # Range of values for depth

hlab<-seq(hr[1], hr[2], length.out=steps) # Height labels
dlab<-seq(dr[1], dr[2], length.out=steps) # Depth labels
h<-rep(hlab, steps)  # Height coordinates
d<-rep(dlab, 1, each=steps) # Depth coordinates
pts<-cbind(h,d) # Combined coordinates
dns<-dmvnorm(pts, mu, varcov) # List of probability densities
mat<-matrix(dns, steps, steps, byrow=T)  # Matrix of prob densities
contour(hlab, dlab, mat, xlab="Tree height", ylab="Root depth")
```

The key to how this works is in the definitions of the vectors h and d (also, have a look back at R1.15). Put together in pts, these vectors give all the combinations of height and depth in a regular 40×40 grid defined within the ranges hr and dr.

9.18.   Sums of random variables: the central limit theorem

Example 9.27: Food provisioning in starlings

In Example 4.26, the rate of provisioning of starling chicks by their parents was used as an illustration of the marginal value theorem. The chicks' fitness depends on the total weight, rather than the number of larvae offered to them, assuming that all larvae have the same weight. For this example, I assume that the weight of larvae (in grams) is a normally distributed variable $W \sim N(5, 0.2)$. If $n = 5$ is the number of larvae per trip brought back to the nest, then what can be said about the distribution of total weight (V) provided to the chicks after each trip?

This is a question about the sum of n random variables and it turns out that quite a lot can be said about it. First of all, the mean of the sum will equal the sum of the means (this is an extension of Equation (9.90) to three or more variables),

$$E(X_1 + \cdots + X_n) = E(X_1) + \cdots + E(X_n) \tag{9.99}$$

Second, *if* the variables are independent, the variance of the sum will equal the sum of the variances,

$$\text{var}(X_1 + \cdots + X_n) = \text{var}(X_1) + \cdots + \text{var}(X_n) \tag{9.100}$$

It should be noted that Equations (9.99) and (9.100) apply even if the variables $X_1, \ldots, X_n$ follow different probability distributions but not if they are dependent. There are extensions of the variance formula for correlated variables (see Larson, 1982). The assumption of independence always needs to be critiqued biologically.

Example 9.28: Food provisioning in starlings

In the case of a foraging starling, independence implies that the weight of the next larva caught is not affected by the weights of the larvae already in its beak. This may be hard to justify if the decline in foraging efficiency depends on prey mass (e.g. as a result of limited beak capacity) rather than prey number (e.g. the more larvae there are, the likelier it is for some to escape). If the independence assumption holds, then according to Equations (9.99) and (9.100), the average trip load will be 25 with variance 1. These numbers are useful, but it is still not known what type of distribution they describe.

We now come to one of the most remarkable results in statistics, called the **central limit theorem (CLT)**: if many independent random variables obey the same well-behaved distribution, then the distribution of their sum is normal with mean and variance given by Equations (9.99) and (9.100). This is remarkable because the original distributions of each of the n random variables do not, themselves, need to be normal.

The central limit theorem carries three provisos, but for most purposes they are not very restrictive. First, the original distributions must be well-behaved, meaning that their mean and variance must be finite. You may wonder whether there are any distributions that suffer from such pathology. The answer is yes, but their relevance to most ecological applications is limited. Second, the random variables must be identically distributed. Third, the number of variables being added together must be large (infinite), but as will be demonstrated in Example 9.29, even sums of bimodally distributed variables quickly converge to normal-like distributions.

Example 9.29: Mixed-diet provisioning in starlings

Consider a scenario in which adult starlings choose randomly from two types of larvae. Individual larvae of type 1 are selected with probability $p = 0.7$ and have weight $W_1 \sim N(5, 0.2)$. Type 2 larvae are selected with probability $1 - p = 0.3$ and have weight $W_2 \sim N(3, 0.1)$. This means (see Example 9.15) that the weight of any given larva in the starling's beak will follow a bimodal distribution with PDF

$$f(w) = \frac{1}{\sqrt{2\pi}} \left(1.566 e^{-\frac{(5-w)^2}{0.4}} + 0.949 e^{-\frac{(3-w)^2}{0.2}} \right) \tag{9.101}$$

Equations (9.80) can be used to calculate the distribution's mean ($\mu = 4.4$) and variance ($\sigma^2 = 1.01$). Assuming that the successive choices made by the bird are independent and that the total number of larvae (n) is large, then the load (V) brought back should be distributed as $V \sim N(4.4n, 1.01n)$. In theory, this will only be exactly true when n is infinite but, in practice, it doesn't take very large values of n for this approximation to converge to the truth (Figure 9.18).

Figure 9.18: Convergence of the sum of independent, bimodal random variables to the normal distribution. In these figures, n is the number of random variables being added together. For completeness, the case $n = 1$ is shown first to illustrate the initial discrepancy between the individual random variables and the normal approximation. For $n = 2$, an intermediate mode appears as the sum includes random draws from the two original peaks. Only traces of multimodality are retained for $n = 5$, and for $n = 10$ the distribution of sums is visually similar to its normal approximation.

The CLT is one of the cornerstones of inferential statistics (Chapter 10) and the reason why normally distributed variables are so prevalent in nature.

9.19.   Products of random variables: the log-normal distribution

Example 9.30: Stochastic exponential growth

The simple model for exponential growth in discrete time (see Example 3.3) takes the form

$$P_{t+1} = (1 + r)P_t \tag{9.102}$$

In this case, P_t will be defined as population density (rather than numbers) in a large and perfectly mixed population. To introduce environmental stochasticity on the population's growth rate, the model may be written as

$$P_{t+1} = R_t P_t \qquad (9.103)$$

where R_t is a random variable with mean $(1 + r)$ and some variance that quantifies the variability in the species' environment. There are several options available for the distribution of R_t. The normal is one possibility but it may need to be truncated below zero to make sure the population does not become negative. Alternatively, a zero-truncated t-distribution might be an appealing model for environmental stochasticity because its fat tails have the tendency to produce a mixture of small perturbations and rare catastrophic events. The gamma is also a suitable candidate because it is designed to give non-negative values. In any case, starting from an initial density P_0, the density at time t will be

$$P_t = (R_{t-1} \times \cdots \times R_0) P_0 \qquad (9.104)$$

The density of the final population relative to the initial value will be

$$\frac{P_t}{P_0} = R_{t-1} \times \cdots \times R_0 \qquad (9.105)$$

Because the terms on the RHS are random variables, the final outcome of this model will differ each time it is run. Figure 9.19 shows several such realisations. A particularly important question is whether the population will exceed or drop below a particular threshold density. For example, what is the probability that a threatened population will be half its current size in 50 years, or, alternatively, the probability that a pest population will double in 50 years? These questions can be answered by calculating the areas under the left- or right-hand tail of the probability distribution of the random variable P_t/P_0. The frequency distribution of the final ratios from 1000 simulations (see histogram in Figure 9.19) is skewed, ranging from 0 to infinity.

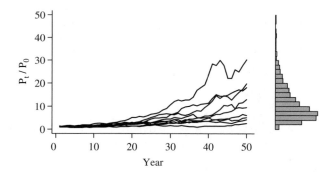

Figure 9.19: The time series plot shows ten different realisations of the stochastic model in Equation (9.104). Each trajectory was generated by running the model for 50 time steps (years). The y-axis is standardised by the initial population density, indicating the size of the final population relative to the initial size. The histogram on the right shows the distribution of these final ratios generated from 1000 trajectories.

This example illustrates the biological relevance of the probability distribution of the product of many independent, identically distributed random variables

$$Y = X_1 \times \cdots \times X_n \tag{9.106}$$

It is convenient to take logs on both sides so that the product becomes a sum

$$\ln(Y) = \sum_{i=1}^{n} \ln(X_i) \tag{9.107}$$

Here, the RHS is the sum of identically distributed random variables, so the central limit theorem ensures that $\ln(Y)$ will be normally distributed. When the logarithm of a random variable is normally distributed (say with mean μ and variance σ^2), the variable itself is said to be **log-normally distributed**,

$$Y \sim LogN(\mu, \sigma^2) \tag{9.108}$$

The log-normal is only defined for non-negative values and has PDF

$$f(y) = \frac{1}{y\sqrt{2\pi\sigma^2}} e^{-\frac{(\ln y - \mu)^2}{2\sigma^2}} \tag{9.109}$$

The mean and variance of the log-normal distribution are related to the mean and variance of the log-transformed normal variable as follows

$$E(Y) = e^{\mu + \frac{1}{2}\sigma^2} \quad var(Y) = (e^{\sigma^2} - 1)e^{2\mu + \sigma^2} \tag{9.110}$$

The log-normal distribution has found wide application in the natural sciences as an empirical description of observed frequency distributions (Limpert *et al.*, 2001) and appears to be a particularly good model of the relative abundance of species in ecological communities (see May *et al.*, 2007 for an intuitive explanation).

⊛ 9.13: Stochastic growth and the Log-normal distribution

The following code will generate a single realisation of the model in Equation (9.105) in which R_t comes from a zero-truncated normal distribution with mean 1.05 and variance 0.01.

```
T<-50     # Total number of years
r<-0.05   # Mean intrinsic growth rate
sd<-0.1   # Standard deviation for population change
h<-pmax(0, rnorm(T,1+r,sd)) # Values from truncated normal distr.
PT<-cumprod(h) # Ratio of final over initial population density
plot(PT, type="l", xlab="Year", ylab=expression(P[t]/P[0]))
```

Here, the truncation is achieved by the command pmax() which selects the pairwise maxima by comparing the vector of random values with zero (the command can also compare two vectors). The time series of products is obtained by the command cumprod(), which does for products what cumsum() does for sums (see R9.1).

The R functions pertaining to the log-normal are dlnorm(), plnorm(), qlnorm() and rlnorm(). The parameters required by all of them are the mean and standard deviation of the normally distributed variable whose logarithm has the log-normal distribution.

9.20. Modelling residuals: the chi-square distribution

Example 9.31: Position of limpets relative to water

Common limpet, Patella vulgata

Regular, long-term observations of a small group of intertidal limpets has shown that their distance (Z) from the wave is almost normal, with mean 0 and standard deviation 1 m. The zero mean implies that these animals spend 50% of their time out of the water. The site-specific value of the standard deviation results from the inclination of the coast, exposure to tide and the limited extent of limpet foraging movement when submerged. It is desirable to repeat the data collection at other sites with less effort. Assuming that a mean of zero applies to all limpets of this species, we might ask what would be the variance calculated from a sample of n measurements of Z. In general, the variance of a particular sample from a population with known mean can be written (see Section 7.7)

$$v(X) = \frac{\sum_{i=1}^{n} (x_i - \mu)^2}{n} \tag{9.111}$$

Here we are interested in the distribution of the variances of all samples from a population with mean zero,

$$v(Z) = \frac{1}{n} \sum_{i=1}^{n} Z_i^2 \tag{9.112}$$

If the sum of random variables on the RHS is denoted by W_n, then the sample variance is $\frac{1}{n} W_n$. To quantify how different the sample variances will be from the true underlying variance, it is useful to describe the distribution of W_n. Treating the variance of a sample as a random variable is a novel concept about which I will have more to say in Chapter 10. Since it is the sum of identically distributed random variables Z^2, the central limit theorem states that, for large n, it will be normal. However, as in this example, the distribution of W_n is required for small n.

You may have noticed that I artificially chose the mean and variance of Z in Example 9.30 to give a standard normal distribution. The sum of squares (W_n) of n standard normal variables

$$W_n = \sum_{i=1}^{n} Z_i^2 \tag{9.113}$$

is said to have a **chi-square distribution with n degrees of freedom**, $W_n \sim \chi^2(n)$. In the more general case of nonstandard normal variables $X \sim N(\mu, \sigma^2)$, the standardising transformation of Equation (9.74) can give

$$W_n = \sum_{i=1}^{n} \left(\frac{X_i - \mu}{\sigma} \right)^2 \sim \chi^2(n) \tag{9.114}$$

It turns out that the chi-square distribution is a special case of the gamma. Specifically, $\chi^2(n)$ is the same as $Gamma(n/2, 1/2)$, so this can be used to calculate densities, probabilities and summary statistics.

There is also a useful link between the chi-square, standard normal and t-distributions. Given three variables $Z \sim N(0, 1)$, $W \sim \chi^2(n)$ and $T \sim t(n)$, then

$$T = \frac{Z}{\sqrt{W/n}} \tag{9.115}$$

This fact will be useful in constructing confidence intervals in Chapter 10.

9.14: Chi-square

The functions associated with the chi-square distribution of n degrees of freedom are `dchisq(x, n)` (PDF), `pchisq(x, n)` (CDF), `qchisq(p, n)` (inverse CDF) and `rchisq(m, n)` (m random draws). The same results can be achieved by using the functions of the gamma distribution with parameters $n/2$ and $1/2$.

9.21. Stochastic simulation

Many simple phenomena can be modelled using the named distributions of this chapter. The approach is to identify a distribution whose assumptions are satisfied by the research question and then obtain the values for the parameters of that distribution. But what happens when the parameters are themselves random variables? For example, given a cohort of P_0 animals, a binomial distribution might be used to decide how many will be alive one month from today: $P_1 \sim B(P_0, s)$, where s is the per capita probability of survival. For any given month t, the survivors can be modelled as $P_{t+1} \sim B(P_t, s)$ but any one of the two parameters of this distribution can be random. Certainly P_t, because it depends on the number of survivors from previous months, but also s, which may vary from month to month due to chance environmental effects. When the parameters of one random variable have their own distribution, then the variable is called **compound**. The distributions of some compound variables are tractable mathematically, but in most cases they are not. Then, the only recourse is **stochastic simulation**, i.e. the implementation of the system in a computer using random draws from the constituent distributions of the compound variable. Repeatedly running the simulation to obtain a large number of final outcomes leads to an empirical description of the compound distribution. The population trajectories in Example 9.30 are a good example of this process.

Dynamic iteration of the type studied in Chapter 3 (e.g. going from P_t to P_{t+1}) is one of the main mechanisms that lead to compounding. The term **random process** is often used to describe stochastic systems in which the parameters of random variables vary, either systematically with time or randomly according to their own distribution. Perhaps you can now see a whole new field of research (let's call it stochastic dynamic modelling) which takes the deterministic versions of the dynamic models in Chapters 3, 4, 5 and 6 and examines how the behaviour of these systems changes when they are affected by different types and amounts of randomness.

Example 9.32: Population viability analysis

Questions that concern the viability of threatened biological populations usually take the form: 'What is the probability p that, in n years from now, the population's size P_t will be below a critical level P_{crit}?'. The simulation approach to answering such questions can be illustrated with the simple model of unrestricted growth (see Example 3.2). Because in many instances we are interested in mitigation, it is important to understand the demographic drivers of population dynamics, so a good starting point is the version of the deterministic model that distinguishes between recruitment (b) and mortality (d).

$$P_{t+1} = P_t + bP_t - dP_t \tag{9.116}$$

To start with, the values of b and d may be assumed constant. Therefore, the expected numbers of new recruits and fatalities will be bP_t and dP_t. Even so, there will be stochastic variations around these expectations that are purely the result of chance (not to mention that animals are individuals, so the numbers recruited and dying in a small population need to be modelled as integers). These chance variations are collectively known as **demographic stochasticity**. A version of Equation (9.116) with demographic stochasticity is

$$P_{t+1} = P_t + B_t - D_t$$
$$B_t \sim Poisson(bP_t) \quad D_t \sim B(P_t, d) \tag{9.117}$$

Here, the Poisson distribution is used to model recruitment because each animal may yield more than one offspring, and mortality is modelled by a binomial distribution because the number of animals currently available sets the upper limit to how many fatalities can occur in the year. Both of these distributions are discrete, so they will generate integer numbers of recruits and deaths. Figure 9.20 shows the results of simulations based on Equation (9.117), from which the probability of dropping below P_{crit} was estimated as 0.015.

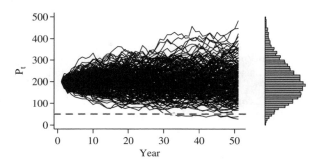

Figure 9.20: Simulation of Equation (9.117) with the values $b = d = 0.5, n = 50 P_0 = 200$ and $P_{crit} = P_0/4$. The left part shows 200 runs of the simulation. The dashed line corresponds to P_{crit}. The histogram of final populations on the right was generated from 10 000 runs of the simulation. Approximately 1.5% of these populations were below the critical level at the end of the 50-year period.

As a further illustration of compound random processes, I will introduce **environmental stochasticity** to the survival and recruitment parameters. A possible extension of the model to include these features is the following:

$$P_{t+1} = P_t + B_t - D_t$$
$$B_t \sim Poisson(b_t P_t) \quad D_t \sim B(P_t, d_t) \tag{9.118}$$
$$b_t \sim Gamma(k, \lambda) \quad d_t \sim Beta(\alpha, \beta)$$

Here, the beta is used to assign randomness to a probability (per capita mortality) and the gamma is used to model a non-negative rate (recruitment). Adding small amounts of environmental stochasticity to the demographic rates increases the risk for the population to 0.11 (Figure 9.21).

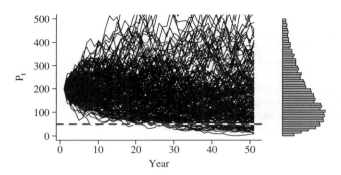

Figure 9.21: Simulation of Equation (9.118) with the values $n = 50$ $P_0 = 200$ and $P_{crit} = P_0/4$. The left part shows 200 runs of the simulation. The dashed line is P_{crit}. The histogram of final populations on the right was generated from 10 000 runs of the simulation. Approximately 11% of these populations were below the critical level at the end of the 50-year period. The demographic parameters were $b \sim Gamma(100, 200)$ and $d \sim Beta(20, 20)$.

9.15: Unrestricted growth with demographic stochasticity

The following piece of code will simulate 1000 population trajectories from Equations (9.117) and report the probability that the final population is smaller than a quarter of its original size.

```
n<-50                 # Number of years
b<-0.5                # Recruitment rate
d<-0.5                # Mortality probability
p0<-200               # Initial pop size
trials<-1000          # Number of simulation runs
fin<-rep(0,trials)    # Vector to store final populations

 for(tr in 1:trials)  # Loop for simulation runs
   {
   p<-rep(p0,n)  # Initialises population vector for each run
   for(t in 1:(n-1)) # Time loop for each simulation run
      {
      B<-rpois(1,p[t]*b)    # New recruits
      D<-rbinom(1,p[t], d)  # Deaths
      p[t+1]<-p[t]+B-D      # Updates current population size
      }
   fin[tr]<- p[n] # Stores final population size from current run
   }
 print(length(fin[fin<(p0/4)])/trials)  # Prints required probability
```

Here, the probability $P(P_{50} < P_{crit})$ is approximated by the relative frequency of simulation runs that yielded populations smaller than p0/4. Can you see how the last line of the code achieves this?

Further reading

General sections on probability distributions can be found in Larson (1982), Gotelli and Ellison (2004), Adler (2005) and Otto and Day (2007). The ecological applications of probability distributions are emphasised by Hilborn and Mangel (1997), Mangel (2006) and Bolker (2008). Of those, Mangel (2006) presents several useful derivations rarely found in textbooks. Larson (1982) has tractable sections with good examples on multivariate distributions. Stochastic models and simulation are extensively covered by Lande *et al.* (2003), Mangel (2006) and Otto and Day (2007). More details on population viability analysis can be found in Beissinger and McCullough (2002). Crawley (2007) has a helpful chapter on writing simulations in R.

References

Adler, F.R. (2005) *Modelling the Dynamics of Life: Calculus and Probability for the Life Scientists.* Thompson Brooks/Cole, Belmont, California. 876pp.

Beissinger, S.R. and McCullough, D.R. (2002) *Population Viability Analysis.* Chicago University Press, Chicago. 593pp.

Bolker, B.M. (2008) *Ecological Models and Data in R.* Princeton. 408pp.

Crawley, M.J. (2007) *The R Book.* John Wiley & Sons, Ltd, Chichester. 942pp.

Diamond, J.M. (1984) 'Normal' extinctions of isolated populations. In *Extinctions* (Ed. Nitecki, M.H.). University of Chicago Press, Chicago, IL. pp. 191–246.

Gotelli, N.J. and Ellison, A.M. (2004) *A Primer of Ecological Statistics.* Sinauer Associates, Massachusetts. 510pp.

Grimmett, G. and Stirzaker, D. (2001) *Probability and Random Processes.* Oxford University Press, New York. 596pp.

Hilborn, R. and Mangel, M. (1997) *The Ecological Detective: Confronting Models with Data.* Princeton University Press, Princeton. 315pp.

Irons, D.B. (1998) Foraging area fidelity of individual seabirds in relation to tidal cycles and flock feeding. *Ecology,* **79**, 647–655.

Lande, R., Engen, S. and Sæther, B.-E. (2003) *Stochastic Population Dynamics in Ecology and Conservation.* Oxford University Press, Oxford. 212pp.

Larson, H.J. (1982) *Introduction to Probability Theory and Statistical Inference.* John Wiley & Sons, Inc., New York. 637pp.

Limpert, E., Stahel, W.A. and Abbt, M. (2001) Log-normal distributions across the sciences: Keys and clues. *Bioscience,* **51**, 341–352.

Mangel, M. (2006) *The Theoretical Biologist's Toolbox: Quantitative Methods for Ecology and Evolutionary Biology.* Cambridge University Press, Cambridge. 375pp.

May, R.M., Crawley, M.J. and Sugihara, G. (2007) Communities: Patterns. In *Theoretical Ecology: Principles and Applications.* (Eds May, R.M. and McLean, A.). Oxford University Press, Oxford. pp. 111–131.

Otto, S.P. and Day, T. (2007) *A Biologist's Guide to Mathematical Modelling in Ecology and Evolution.* Princeton University Press, New Jersey. 732pp.

10

How to see the forest from the trees
(Estimation and testing)

'All you need in this life is ignorance and confidence, then success is sure.'
Mark Twain (1835–1910)

In Chapter 9, random values were generated from distributions with known parameters (hence going from population parameters to data). This chapter presents the fundamental concepts and methods of **statistical inference**: the process of using samples to guess the unknown parameters of entire populations. It deals with three related questions: what is the best estimate of an unknown population parameter (**estimation**, covered in Sections 10.1–10.4 and then, in more depth, in Sections 10.7–10.11)? How far is this estimate likely to be from the true value (**confidence intervals**, covered in Sections 10.5–10.6)? Is the estimated value significantly different from some postulated value (**hypothesis testing**, covered in Sections 10.12–10.17)?

10.1. Estimators and their properties

The reason people turn to statistics is uncertainty, and one of the main causes of uncertainty is sampling. A sample contains some, but not all, of the information about a population. So, with a sample in our hands, we are neither completely ignorant nor totally knowledgeable. The challenge is to use whatever information is available in the sample to describe the population and then quantify the remaining uncertainty in this description. Statistical populations are random variables, with associated probability distributions, so 'describing a population'

How to be a Quantitative Ecologist: The 'A to R' of Green Mathematics and Statistics, First Edition. Jason Matthiopoulos.
© 2011 John Wiley & Sons, Ltd. Published 2011 by John Wiley & Sons, Ltd.

means specifying the underlying, unknown distribution of all possible measurements, of which we may have only a few. In Chapter 7, distributions were described by summary statistics (mean, mode, variance, etc.). Chapter 9 achieved this using probability functions (PDFs and PMFs, each with their own parameters). In some cases, such as the normal and Poisson distributions, the summary statistics and the parameters of the probability function coincided. In others, such as the binomial, the summary statistics needed to be expressed as functions of the distribution's parameters.

We can attempt to **estimate** any one of these descriptors from the data. For example, the average $\bar{x}$, calculated from a particular sample, is an estimate of the population mean μ. The formula used to calculate an estimate is called an **estimator** (hence $\frac{1}{n}\sum_{i=1}^{n} x_i$ is an estimator for the population mean).

Sampling uncertainty raises its ugly head as soon as a second sample is collected from the same population (with the same mean) and *its* average is found to be different from the first. The first crucial breakthrough of inferential statistics was to treat the estimator as a random variable, thus expressing sampling uncertainty in familiar distributional terms: If the estimator is a random variable, then any given estimate is a realisation of that variable. The **sampling distribution of an estimator** is a hypothetical distribution generated from a large (infinite) number of estimates obtained from independent samples of the same size. Hence, the variance of the sampling distribution of an estimator quantifies the uncertainty in any one estimate. The second breakthrough of inferential statistics was to estimate this uncertainty from a single sample, but more on that later.

Example 10.1: Sampling gannet egg weights

Figure 10.1(a) shows the hypothetical probability density for gannet egg weight. The true, but unknown, mean of this distribution is $\mu = 106$. To estimate it, we might collect a sample of egg weights and calculate its average. If we obtained 1000 such samples and plotted the 1000 averages as a histogram, we would get a good approximation of the sampling distribution of the estimator (Figures 10.1(a) and 10.1(b)). The distribution of averages is narrower than the original distribution (compare Figure 10.1(a) with 10.1(b)) and becomes narrower still as the sample size increases (compare Figure 10.1(b) with 10.1(c)). This makes sense: the bigger the sample, the less its average will deviate from the true mean.

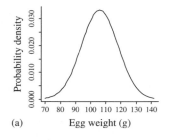 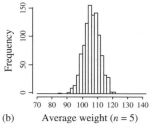 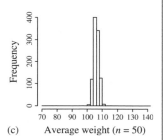

(a)   Egg weight (g)   (b)   Average weight ($n = 5$)   (c)   Average weight ($n = 50$)

Figure 10.1: (a) The true, but unknown, distribution of gannet egg weights; (b) distribution of averages from 1000 samples of size 5; (c) distribution of averages from 1000 samples of size 50.

To introduce the notation for estimators, consider the random variable X with distribution function $f(x)$. As we saw in the previous chapter, this function will contain parameters: The

Poisson has rate λ, the binomial has probability p for each of N trials, etc. The dependence of the distribution on any given parameter θ can be emphasised by using the notation $f(x; \theta)$. Any estimate of the parameter obtained from a sample of data is written $\hat{\theta}$. The estimator of the parameter, being a random variable, is written in upper case $\hat{\Theta}$.

Many estimators can be intuited. For example, the average is a **natural estimator** of the population mean and a sample's median, variance, kurtosis and skewness are natural estimators of the corresponding population statistics. However, as we will see in later sections, they are neither the only, nor necessarily the best, estimators for population statistics/parameters.

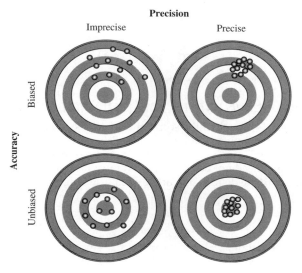

Figure 10.2: A visual depiction of the trade-off between two criteria for the quality of an estimator. The bull's eye symbolises the population parameter being estimated.

What makes a good estimator depends on intended use, but is generally decided on the basis of the following criteria:

❶ **Accuracy:** If the mean of the distribution of an estimator coincides with the true parameter value (i.e. $E(\hat{\Theta}) - \theta = 0$), then the estimator is said to be **unbiased**. The bias of an estimator is quantified as $B(\hat{\Theta}) = E(\hat{\Theta}) - \theta$. Negative values of bias imply **underestimates** of θ and positive values imply **overestimates**. Values of bias close to zero indicate high accuracy.
❷ **Precision:** An estimator is called **precise** or **efficient** if its distribution has a low variance.
❸ **Consistency:** An estimator is called **consistent** if both its variance and bias go to zero as the sample size increases.

The classic 'target practice' illustration in Figure 10.2 is a useful memory aid, but it belies the difficulty of obtaining a good estimator: in most cases, accuracy is obtained at the cost of precision. Of course, if an estimator is consistent and the sample is large, bias and imprecision are both small.

10.2. Normal theory

Bear with me through this and the next three sections while I present the topics in a somehow abstract manner with no examples. You should read them together, try to understand their links and then confirm your understanding with the example at the end of Section 10.5.

The most important aspect of estimation is determining what the distribution of the estimator is. When the estimator involves the sum of values from a large sample, then the central limit theorem (CLT – Section 9.18) can be used to approximate it by a normal distribution. There is a body of theory that uses this, and similar, approximations to derive compact analytical expressions for the population estimates and their associated uncertainty. We will see how this is done in the next three sections, but it is first necessary to introduce the following two results regarding the mean and variance of the product between a number and a random variable (X) (The first of these, was already seen in Section 9.16).

$$E(aX) = aE(X) \tag{10.1}$$

$$Var(aX) = a^2 Var(X) \tag{10.2}$$

10.3. Estimating the population mean

It is important to remind yourself of the scope of the CLT (Section 9.18): it applies to the sum of n identically distributed, independent random variables. From a sampling perspective, it is even more intuitive to think of this as the sum of n random draws from the same distribution. The original distribution need not be normal, but it must have a finite mean (μ) and variance (σ^2). We are interested in the distribution of the sample average,

$$\bar{X} = \frac{1}{n} \sum_{i=1}^{n} X_i \tag{10.3}$$

(Using the estimator notation, I might have written any particular sample average as $\hat{\mu}$, implying the rather unconventional notation $\hat{M}$ for the average, instead of $\bar{X}$). We know from the CLT that, for large samples, the distribution of the sum in Equation (10.3) will be normal with mean $n\mu$ and variance $n\sigma^2$. But what about the mean of the distribution of sample averages? It can be calculated as a sum of expectations with the aid of Equation (10.1)

$$E(\bar{X}) = E\left(\frac{1}{n} \sum_{i=1}^{n} X_i\right) = \frac{1}{n} \sum_{i=1}^{n} E(X_i) = \frac{1}{n} \sum_{i=1}^{n} \mu = \mu \tag{10.4}$$

This important result says that the average is an *unbiased* estimator of the population mean (because $E(\bar{X}) - \mu = 0$).

The variance of the sampling distribution of averages is written $Var(\bar{X})$ or $\sigma_{\bar{X}}^2$. To calculate it, we initially make the (unrealistic) assumption that, although the mean μ is unknown, the population variance σ^2 is, somehow, known exactly. Then, with the help of Equation (10.2), we get,

$$\sigma_{\bar{X}}^2 = Var(\bar{X}) = Var\left(\frac{1}{n} \sum_{i=1}^{n} X_i\right) = \frac{1}{n^2} \sum_{i=1}^{n} Var(X_i) = \frac{\sigma^2}{n} \tag{10.5}$$

This confirms that the variance of the distribution of averages decreases with sample size (see Example 10.1, above).

In most situations, we do not know the value of σ^2 so we need to estimate it. The problem of variance estimation is tackled in the next section, but assuming that we have an estimator S^2 of the population's variance, the variance of the distribution of the mean can be written

$$\hat{\sigma}_{\bar{X}}^2 = \widehat{Var(\bar{X})} = \frac{S^2}{n} \tag{10.6}$$

The resulting estimate of the standard deviation of $\bar{X}$ is $S/\sqrt{n}$. This is called the **standard error** for the estimator of the population mean (μ). It gives an *indication* of how precisely we can estimate the mean of the true distribution from the sample average. In general, the standard error is the estimated standard deviation of the distribution of *any* estimator, so we could define the standard error for any other statistic, such as the median or the variance of the population.

A note of caution: the above calculations assumed that the sample size was sufficiently large for the CLT to ensure a good approximation of the sampling distribution of $\bar{X}$. When sample size is small (the usual informal definition for a small sample is $n<30$), then the sampling distribution of the mean tends to produce outlying measurements more often than would be expected under the normal. It turns out that the t-distribution is a great multipurpose model that can capture such outliers. More on this in Section 10.5.

10.4.   Estimating the variance of a normal population

To derive an estimator for the population variance, we may start from the definition of sample variance

$$v = \frac{\sum\limits_{i=1}^{n} (x_i - \bar{x})^2}{n} \tag{10.7}$$

This describes the variability in any given sample (Section 7.7). It may be extended to make inferences about the population by replacing $\bar{x}$ by μ, assuming for the moment that the population mean is known,

$$S^2 = \frac{\sum\limits_{i=1}^{n} (X_i - \mu)^2}{n} \tag{10.8}$$

In most cases, the mean of the population will not be known, making it impossible to calculate the estimator in Equation (10.8). We may, instead, use $\bar{X}$ (the estimator of the population mean) to obtain

$$S^2 = \frac{\sum\limits_{i=1}^{n} (X_i - \bar{X})^2}{n} \tag{10.9}$$

This deceptively simple step leads to an underestimate of the true variance because the denominator is larger than it should be. To see why, consider a sample of n measurements. If you were given the sample average and the values of all but the nth measurement, then you could calculate the missing value as follows:

$$\bar{x} = \frac{1}{n} \sum_{i=1}^{n} x_i$$

$$\bar{x} = \frac{1}{n} \left(\sum_{i=1}^{n-1} x_i + x_n \right) \tag{10.10}$$

$$x_n = n\bar{x} - \sum_{i=1}^{n-1} x_i$$

This means that there is just as much information in the combination of $n-1$ measurements and the sample average as there is in the sample of n measurements. An alternative interpretation is that, once you have used the sample to get an estimate of the population's mean, the effective sample size that you have left for estimating the population's variance is one smaller, because the last observation is no longer free to take any value. We call this the **loss of one degree of freedom**, and taking it into account gives an unbiased estimator of σ^2

$$S^2 = \frac{\sum_{i=1}^{n} (X_i - \bar{X})^2}{n-1} \qquad (10.11)$$

Although we cannot determine the distribution of this estimator directly, *if* the X_i come from a normal population, rearranging the expression and dividing both sides by the unknown population variance gives

$$\frac{(n-1)S^2}{\sigma^2} = \sum_{i=1}^{n} \frac{(X_i - \bar{X})^2}{\sigma^2} \qquad (10.12)$$

The RHS of this expression is a chi-squared variable (check Equation (9.114) in Section 9.20). Here, the sample average has already been used to estimate the mean, so although n variables are being added together, only $n - 1$ are independent of each other. Therefore, the sum on the RHS has a chi-squared distribution with $n - 1$ degrees of freedom. Looking now to the LHS, it is the product between a random variable (S^2) and a constant $(n - 1)/\sigma^2$. Therefore, S^2 must be a rescaling of the variable $W_{n-1} \sim \chi^2(n-1)$. This observation will prove useful in the following section.

10.5. Confidence intervals

Standard errors provide an index of uncertainty that can be used in comparative statements about estimates from different populations and samples of different size. A more tangible question is: how far is a particular estimate ($\hat{\theta}$) likely to be from the true population parameter (θ) ? Such questions can be addressed by defining a **confidence interval (CI)**, i.e. a range of values that is highly likely to bracket the true population parameter. The probability with which the interval succeeds in bracketing the mean (known as the **confidence level**) is a subjective decision, but convention dictates confidence levels of 0.95 or 0.99. A 95% CI with lower value l and upper value u implies that if 100 intervals were calculated by the same method from 100 different samples, 95 of them would contain θ (a 100% CI is uninformative because it includes all the candidate values of the parameter, no matter how unlikely).

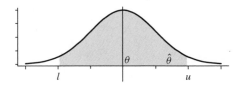

Figure 10.3: An estimator $\hat{\Theta}$ is used to make inferences about the unknown parameter θ. The curve shows the sampling distribution of $\hat{\Theta}$ and the shaded region encompasses a probability of 0.95. So, 95% of estimates ($\hat{\theta}$) will fall within the interval at the base of the shaded region.

If the estimator $\hat{\Theta}$ is unbiased, then its mean is θ. The interval (l, u) encompasses 95% of the sampling distribution of $\hat{\Theta}$ (Figure 10.3).

$$P(l \leq \hat{\Theta} \leq u) = 0.95 \qquad (10.13)$$

Although we don't know what l and u are in general, for some estimators, this expression can be turned into a statement about the population parameter. As a first example, consider the

CI for the mean of a population whose variance σ^2 is known. The estimator of the mean is the average of n observations. According to Section 10.3 it will be distributed as

$$\bar{X} \sim N\left(\mu, \frac{\sigma^2}{n}\right) \tag{10.14}$$

As we saw in Equation (9.74), this can be converted into a standard normal variable by the transformation

$$Z = \sqrt{n}\frac{\bar{X} - \mu}{\sigma} \tag{10.15}$$

The 95% CI (l, u) for the mean of a standard normal random variable satisfies the expression

$$P(l \leq Z \leq u) = 0.95 \tag{10.16}$$

A computer can find the values of l and u for a standard normal distribution.

R 10.1: 95% CI for standard normal variable

```
> p<-(1-0.95)/2 # Probability contained in lower/upper tail
> qnorm(p, 0,1)               # Lower limit
[1] -1.959964
> qnorm(p, 0, 1, lower.tail=F) # Upper limit
[1] 1.959964
```

So,

$$P(-1.96 \leq Z \leq 1.96) = 0.95 \tag{10.17}$$

I focus on the inequality inside the brackets, substitute-in according to Equation (10.15) and manipulate as follows

$$
\begin{aligned}
-1.96 &\leq \sqrt{n}\frac{\bar{X} - \mu}{\sigma} \leq 1.96 \\
-1.96\frac{\sigma}{\sqrt{n}} &\leq \bar{X} - \mu \leq 1.96\frac{\sigma}{\sqrt{n}} \\
-\bar{X} - 1.96\frac{\sigma}{\sqrt{n}} &\leq -\mu \leq -\bar{X} + 1.96\frac{\sigma}{\sqrt{n}} \\
\bar{X} - 1.96\frac{\sigma}{\sqrt{n}} &\leq \mu \leq \bar{X} + 1.96\frac{\sigma}{\sqrt{n}}
\end{aligned}
\tag{10.18}
$$

The last step involved changing the signs of the three parts of the inequality and therefore reversing the direction of the inequality (Section 1.15). This interval is highly likely (with a probability of 0.95) to contain the true population mean. So, given a particular average $\bar{x}$ from a single sample, we can write a realisation of this interval as $\bar{X} \pm 1.96\frac{\sigma}{\sqrt{n}}$. This interval is stochastic, in the sense that $\bar{X}$ will be different for different samples. In general, the CI will be larger for smaller samples drawn from highly variable populations.

We may now use the above approach to devise a CI for a population whose variance σ^2 is unknown. In Equation (10.15), the estimate S^2 may be used instead of the variance to obtain a new variable,

$$\sqrt{n}\frac{\bar{X} - \mu}{S} \tag{10.19}$$

We can show that this has a t-distribution (skip the next five equations unless, like me, you always wondered where the need for t-values comes from) by using two facts. First, from the definition of Z, we have

$$Z = \sqrt{n}\frac{\bar{X} - \mu}{\sigma} \sim N(0, 1) \tag{10.20}$$

Second, from Equation (10.12) we have

$$W = \frac{(n - 1)S^2}{\sigma^2} \sim \chi^2(n - 1) \tag{10.21}$$

A standard normal and a chi-square variable with $n - 1$ degrees of freedom can be linked by Equation (9.115) to give a t-distributed variable (again, with $n - 1$ degrees of freedom),

$$T = \frac{Z}{\sqrt{W/(n - 1)}} \sim t(n - 1) \tag{10.22}$$

Substituting the definitions of Z and W from Equations (10.20) and (10.21) into Equation (10.22) and simplifying gives

$$T = \frac{\sqrt{n}(\bar{X} - \mu)/\sigma}{\sqrt{[(n - 1)S^2]/[\sigma^2(n - 1)]}} = \sqrt{n}\frac{\bar{X} - \mu}{S} \tag{10.23}$$

This mathematical result has an intuitive interpretation. Replacing the fixed value σ by the stochastic estimator S in Equation (10.19) must inject an additional amount of uncertainty in the resulting CI. We would therefore expect the 95% interval of T

$$P(l \leq T \leq u) = 0.95 \tag{10.24}$$

to be wider than the 95% CI of Z. This is indeed the case, as the following calculation shows.

® 10.2: 95% CI for t-variable

Compare the following 95% CIs for a t-variable with the CI for a normal variable in R10.1. First, a 95%CI for a small sample:

```
> n<-10             # Sample size
> p<-(1-0.95)/2     # Probability contained in lower/upper tail
> qt(p, df=n-1)     # Lower limit
[1] -2.262157
> qt(p, df=n-1, lower.tail=F) # Upper limit
[1] 2.262157
```

Then, calculation for a large sample:

```
> n<-500  # Sample size
> qt(p, n-1)                   # Lower limit
[1] -1.964729
> qt(p, n-1, lower.tail=F) # Upper limit
[1] 1.964729
```

This shows that the CI for T shrinks towards the length of the CI for Z as the sample size increases and the uncertainty in the estimate of σ vanishes. This dependence of the CI on sample size can be written

$$P(t_{0.025(n-1)} \leq T \leq t_{0.975(n-1)}) = 0.95 \tag{10.25}$$

where $t_{p(n)}$ is the p-quantile of the t-distribution with n degrees of freedom. Equation (10.25) can now be manipulated in a way similar to Equations (10.18) to give the CI for the mean

$$\bar{X} - t_{0.975(n-1)}\frac{S}{\sqrt{n}} \leq \mu \leq \bar{X} + t_{0.025(n-1)}\frac{S}{\sqrt{n}} \tag{10.26}$$

It may also be possible to calculate CIs for other parameters. For example, a 95% CI for the population variance is a short step away from what we have already discussed. Using Equation (10.21), we can write

$$P\left(w_{0.025(n-1)} \leq \frac{(n-1)S^2}{\sigma^2} \leq w_{0.975(n-1)}\right) = 0.95 \tag{10.27}$$

where $w_{p(n)}$ is the p-quantile of the χ^2-distribution with n degrees of freedom. This can then be rearranged (try it) to give

$$\frac{(n-1)S^2}{w_{0.975(n-1)}} \leq \sigma^2 \leq \frac{(n-1)S^2}{w_{0.025(n-1)}} \tag{10.28}$$

Example 10.2: Estimating the unknown parameters of the gannet egg weight distribution

Assume that gannet egg weight X is normally distributed with $\mu = 106$ and $\sigma^2 = 144$. Neither of these numbers are known to us, so we will attempt to draw inferences about them from the following sample of 20 weights: {120.0, 88.6, 110.8, 125.9, 120.0, 100.9, 108.1, 110.9, 93.2, 124.4, 110.5, 103.2, 97.2, 127.9, 112.2, 114.8, 112.5, 98.7, 95.8, 81.2}. The estimate of the mean is $\bar{x} = 107.84$. The unbiased estimate of the variance $S^2 = 164.14$ can be calculated from Equation (10.11). The sample size is $n = 20$, so, according to Equation (10.6), the standard error (the estimated standard deviation of the distribution of averages) is $\hat{\sigma}_{\bar{X}} = 2.87$. Since this is a normal population with an unknown variance, the 95% CI for the mean can be calculated from Equation (10.26). It is: (101.84, 113.84). Also, given normality of the population of egg weights, the 95% CI for the population variance can be found from Equation (10.28). It is: (94.93, 350.14). In this particular example, the sample yielded CIs that bracketed both the population mean and variance, but this is only going to be the case in 95 out of 100 samples.

10.3: CIs for the mean and variance of a normal population

Below is an R function (`normInf`) that will draw inferences about the mean and variance of a normal population given a sample `dat` and a required confidence level `clevel`. Note that the built-in R command `sd()` calculates the standard deviation with a denominator $n - 1$, so, here, it is the appropriate one to use for the standard error (contrast with R7.5).

```
normInf<-function(dat, clevel)
  {
  n<-length(dat) # Sample size
  av<-mean(dat)    # Sample average
  s<-sd(dat)    # Estimate of population standard deviation
  se<-s/sqrt(n)    # Standard error
  p<-(1-clevel)/2 # Probability contained in lower/upper tail
```

```
t1<-qt(p, n-1)                    # Lower t-value for CI of mean
t2<-qt(p, n-1, lower.tail=F)   # Upper t-value for CI of mean
chi1<-qchisq(p, n-1)               # Lower chisq for var CI
chi2<-qchisq(p, n-1, lower.tail=F)  # Upper chisq for var CI


# Calculates CIs
muCI<-round(c(av-t2*se, av-t1*se),2)
varCI<-round(c((n-1)*s^2/chi2, (n-1)*s^2/chi1),2)


# Generates printed output
print(paste("Mean is estimated as ",round(av,2), "with ",clevel*100,"%CI
(", muCI[1],",", muCI[2],")"))
print(paste("Variance is estimated as ",round(s^2,2), " with ",clevel*100,
"%CI (", varCI[1],",", varCI[2],")"))
}
```

A call to this function using the data from Example 10.2 looks like this:

```
> dat<-c(120.0,88.6,110.8,125.9,120.0,100.9,108.1,110.9,93.2,
+ 124.4,110.5,103.2,97.2,127.9,112.2,114.8,112.5,98.7,95.8,81.2)
> normInf(dat, 0.95)
[1] "Mean estimated as 107.84 with 95 %CI ( 101.84, 113.84 )"
[1] "Variance estimated as 164.14 with 95 %CI ( 94.93, 350.14 )"
```

10.6. Inference by bootstrapping

The analytical CIs derived in Equations (10.26) and (10.28) are, sadly, restricted to normal populations. More punishing mathematics is required to make inferences about other parameters and non-normal distributions, and even this is not always possible. If the type of population distribution is unknown, or known to be non-normal, a computer-based technique known as the **bootstrap** can be used for inference.

Statistical inference relies on the concept of the sampling distribution. This can be thought of as a hypothetical histogram of parameter estimates constructed from infinitely many samples drawn independently from the population. However, in most studies, we only have a single sample. Instead of just using the data to obtain a single parameter estimate, we can think of the frequency distribution of the sample as a (poor) approximation of the population distribution. This frequency distribution can be used to draw a large number of independent samples, hence approximating the sampling distribution of the parameter. The key idea here is **re-sampling**. Given a data set of n measurements, it is possible to generate many new samples of the same size by re-sampling with replacement, so that in each new sample some data may be included more than once while others are completely omitted. It turns out that this approximation of the sampling distribution is very good, even if the original sample is small. Having obtained the sampling distribution, it is then relatively straightforward to obtain estimates, standard errors and CIs. For example, a 95% CI can be obtained by generating 10 000 bootstrapped estimates, sorting them in increasing order and noting the 250th and 9750th values (i.e. the 2.5th and 97.5th percentiles).

Does this sound like it shouldn't work? Does it appear that we are getting more out of the data than they have to give? As its name suggests, bootstrapping is reminiscent of a person lifting themselves up by pulling on the laces of their own shoes. There are good theoretical arguments for why bootstrapping works, but here, I will simply illustrate its performance with an example.

Example 10.3: Bootstrap inference for gannet age distribution

A small number ($n = 20$) of breeding gannets on the colony of Bass Rock are aged. The true (but unknown) age distribution of breeders on the colony is heavily skewed (Figure 10.4(a)). Unsurprisingly, the frequency distribution of ages obtained by plotting the sample of 20 birds (Figure 10.4(b)) is a poor approximation for the true distribution in Figure 10.4(a). Inferences about the mean age of gannets at the colony invoke the concept of the sampling distribution. Figure 10.4(c) shows the true sampling distribution of averages that would be obtained if a large number of 20-bird samples were drawn repeatedly from the population. Compare this with Figure 10.4(d), which shows the bootstrapped sampling distribution of averages, obtained by re-sampling several thousand samples of 20 ages (with replacement) from the data set of 20 age measurements. Hopefully, you agree that the similarity between the true (Figure 10.4(c)) and bootstrapped (Figure 10.4(d)) sampling distributions is striking.

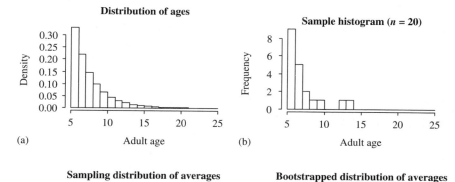

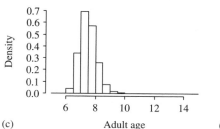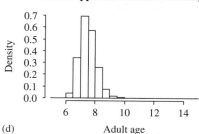

Figure 10.4: The true distribution of ages of breeding gannets on the colony (a) is sampled once ($n = 20$) to generate a histogram of ages (b).To make inferences about the mean of the age distribution, we require the sampling distribution of the mean shown in (c). Although this is not available, a convincing approximation (d) can be obtained by bootstrapping from the single available sample in (b).

10.4: Bootstrap

The generic tool used for re-sampling is the command `sample()` (see R7.2). Given a data set (`dat`) comprising 20 observations, a single re-sampling of 20 numbers with replacement can be achieved as follows:

```
> dat
[1]  6 10  6  6  7  7  6  8  6 10  8  6 11  7  7  5  5  5  6 19
> sample(dat, 20, replace=T)
[1]  6  6  7 10  6  7  6 10 11  6 19  6  7  7 11  6  5 19  6  7
```

This could be used to obtain a single bootstrap estimate of the parameter of interest and then placed in a loop to yield an approximate sampling distribution for this parameter. However, R offers more advanced functionality for bootstrapped inference through the package `boot`. Load the package by `require(boot)`. The command `boot()` does all the hard work for you. For basic usage, you need to specify the name of the data set, the estimator (mean, median, variance or any user-defined summary) and the number of bootstrap iterations that are to be performed. The only tricky bit is that the estimator first needs to be defined as a function. For example, the function for the average could be declared as follows:

```
av<-function(dat, i) {sum(dat[i])/length(dat[i])}
```

Note the peculiar requirement for the iterator `i` in this. This is for the benefit of more advanced uses of `boot()` that will not concern us here. The following creates and analyses 1000 bootstrap replicates from a given sample

```
> bs<-boot(dat, av,R=1000)
> bs


ORDINARY NONPARAMETRIC BOOTSTRAP


Call:
boot(data = dat, statistic = av, R = 1000)


Bootstrap Statistics :
   original  bias   std. error
t1*    7.55  -0.0446  0.6806167
```

In this output, `original` is the value of the estimate obtained directly from the original sample. A better estimate can be obtained by calculating the expectation of the bootstrapped estimates. The difference between this more robust estimate (the average of averages in the above example) and `original` is reported as `bias`. The standard error is also estimated directly as the standard deviation of the bootstrapped estimate. To obtain a variety of CIs for your specified statistic, use the command `boot.ci(bs)`. Of the four types of CIs reported, `Normal` is based on normality assumptions and `Percentile` is based on the 2.5th and 97.5th percentiles.

10.7. More general estimation methods

The general estimation problem can be stated as follows: we have a set of observations from the real world (the data) and make certain assumptions about the stochastic process that generated them. These assumptions are structured into a model containing parameters and (in more advanced cases) covariates of the data. An **estimation method** is then used to answer the question: which values of the parameters are most plausible in the light of the data?

The preceding sections assumed that the data were independent observations coming from the same underlying distribution and the objective was to estimate one population parameter at a time. It was therefore relatively easy to propose natural estimators and evaluate them. However, all of these assumptions may need to be relaxed for some problems. In Sections 10.8–10.10, I will present three more general estimation methods and illustrate them with simple examples. In the meantime, here is a challenging problem, intended merely to illustrate what levels of complexity can eventually be handled by these more general methods.

Example 10.4: Estimating parameters for animal movement

The data are observations of position collected at 100 sec intervals by time-lapse photography. At this timescale, the animal may be assumed to be performing a random walk in two dimensions (see Example 2.9) with a variable step length. Its speed depends on the temperature of the liquid that it is moving through. We would like to estimate the quantitative characteristics of the animal's movement from the data set of positions, but to do this, these characteristics must first take the form of parameters in a model. Given the animal's current position $\mathbf{X}_t$, its next position can be modelled as

$$\mathbf{X}_{t+1} = \mathbf{X}_t + \mathbf{Z}_t \tag{10.29}$$

where $\mathbf{Z}_t$ describes the displacement of the animal in a single 100 sec interval. The notation here needs some explanation: $\mathbf{X}_t$ and $\mathbf{Z}_t$ are uppercase because they are random variables. They are also in boldface because they are positional vectors, each containing two spatial coordinates, $\mathbf{X}_t = (X_t, Y_t)$, $\mathbf{Z}_t = (\Delta X_t, \Delta Y_t)$. The displacement variable $\mathbf{Z}_t$ can be modelled by a bivariate normal distribution (see Section 9.17) with the following (unknown) parameters:

$$\mathbf{m} = (\mu_X, \mu_Y) \text{ (mean)} \tag{10.30}$$

$$\mathbf{v} = \begin{pmatrix} \sigma^2_{\Delta X} & \text{cov}(\Delta X, \Delta Y) \\ \text{cov}(\Delta X, \Delta Y) & \sigma^2_{\Delta Y} \end{pmatrix} \text{ (var-covar matrix)} \tag{10.31}$$

This is a complicated model containing many parameters that need to be estimated simultaneously from the same data set. Take a moment to consider its biological interpretation. Equation (10.29) clearly builds a dependence between successive observations of position (so, out of the window goes the assumption of independence between the data). Use of the bivariate normal to model displacement implies that very long jumps are unlikely. The values of the parameters in Equations (10.30) and (10.31) are informative about the behaviour of the animal. For example, if μ_X and μ_Y are different from zero, then this implies a tendency to move in a certain direction (e.g. $\mu_Y > 0$ implies a northward moving tendency). Similarly, the variances $\sigma^2_{\Delta X}, \sigma^2_{\Delta Y}$ relate to the speed of movement (high values of these imply a more dispersed bivariate normal and, hence, a higher likelihood for long jumps). If the speed of the animal is affected by the medium in which movement occurs, then these variances would need to be written as functions of the environmental conditions at the current location of the animal (implying that displacements will come from a different distribution each time). Having written down this model and with the aid of some positional observations, the task would be to guess all of the model's parameters and their associated uncertainty. Can you begin to see the challenges involved in doing this?

10.8. Estimation by least squares

Least squares estimation is a highly efficient method, both numerically and statistically, that relies on few assumptions. It is particularly useful for estimating regression models, as we will see in the next chapter. Consider a set of observations $X_1, \ldots, X_n$ that may originate from distributions with different means. If there is no pattern in the means of those distributions, then there is little else that can be done. However, it may be possible to write a function $f(\mathbf{y}_i; \theta)$ that describes the expected value of each X_i in terms of known conditions $\mathbf{y}_i = (y_1, \ldots, y_m)_i$, pertaining to the ith observation, and unknown parameters $\theta = (\theta_1, \ldots, \theta_k)$.

Example 10.5: Weight distribution in voles of different ages

Weight (W) in voles is strongly correlated with age (Y). We may therefore decide that expected weight can be written as a linear function of age: $W = bY + a$ (where a is the mean weight at birth). Given a random sample of voles of known weight and age, we may be interested in estimating the parameters a and b. Here, the function f is linear with age and the question of estimating the mean weight translates to a question of estimating the parameters a and b. The utility of this is that given the age of any particular cohort of voles, we can write down an estimate for their expected weight.

The general least squares approach is the following. First, write down the expression

$$Q = \sum_{i=1}^{n} (X_i - f(\mathbf{y}_i; \theta))^2 \tag{10.32}$$

This expression, the **sum of squared residuals**, quantifies the agreement between the data and any given parameterisation of the model for the means. Then, use calculus or a computer to find the values of θ that minimise this quantity. These values are the required estimates $\hat{\theta}$ of the unknown parameters. The estimate of the mean of X_i under particular conditions $\mathbf{y}_i$ is then $f(\mathbf{y}_i; \hat{\theta})$. The estimator of the variance around that mean will be

$$S^2 = \frac{1}{n-k} \sum_{i=1}^{n} (X_i - f(\mathbf{y}_i; \hat{\theta}))^2 \tag{10.33}$$

where k is the number of parameters involved in the model for the mean. This is all quite abstract, so we can try to apply the method to the most trivial case of estimating a single population mean. We have already developed a natural estimator of the mean for this scenario (the average of all the data), so it would be good if the least squares approach led to the same result.

Example 10.6: Mean weight in a cohort of voles

In a single cohort, all the voles are assumed to have the same age (say y), so we expect the weight measurements to originate from the same distribution with mean μ_y. Therefore, the expectation is given by the constant function $f(y; \mu_y) = \mu_y$ which has a single parameter, μ_y. The quantity in Equation (10.32) is written

$$Q = \sum_{i=1}^{n} (W_i - \mu_y)^2 \tag{10.34}$$

The next step is the trickiest to digest: Although in truth, the value of μ_y is a constant, we are ignorant of its value. While we search for it, we need to temporarily think of μ_y as a variable θ. This enables us to minimise Q with respect to θ. Brush up on optimisation (Section 4.10) if you need to before reading on. $Q_{\min}$ will satisfy

$$\frac{dQ}{d\theta} = 0 \tag{10.35}$$

The derivative on the LHS can be expanded and simplified considerably

$$\frac{d}{d\theta}\left\{\sum_{i=1}^{n}(W_i - \theta)^2\right\} =$$
$$\sum_{i=1}^{n}\frac{d}{d\theta}(W_i - \theta)^2 =$$
$$\sum_{i=1}^{n}\frac{d}{d\theta}(W_i^2 - 2W_i\theta + \theta^2) = \tag{10.36}$$
$$\sum_{i=1}^{n}(-2W_i + 2\theta) =$$
$$-2\sum_{i=1}^{n}W_i + 2\sum_{i=1}^{n}\theta$$

When this quantity becomes zero, then Q is minimised and θ becomes equal to the required estimate $\hat{\mu}_y$ of the parameter μ_y. Hence, we can write

$$-2\sum_{i=1}^{n}W_i + 2\sum_{i=1}^{n}\hat{\mu}_y = 0 \tag{10.37}$$

Rearranging this expression gives

$$\hat{\mu}_y = \frac{1}{n}\sum_{i=1}^{n}W_i = \bar{W} \tag{10.38}$$

Reassuringly, this confirms that, under least squares estimation, the average is still the best estimator of the mean. According to Equation (10.33), the estimated variance around the mean is

$$S^2 = \frac{1}{n-1}\sum_{i=1}^{n}(W_i - \bar{W})^2 \tag{10.39}$$

ⓡ 10.5: Minimising the sum of squared residuals

Coming up with an analytical expression for the estimator is ideal, but if this isn't possible, then numerical optimisation must be used. Have a look at R4.9 to see how to minimise Q numerically in the case where a single parameter is sought. For more than one parameter, the optimisation methods of R10.6 can be used.

10.9. Estimation by maximum likelihood

The method of **maximum likelihood** is broader and more flexible than least squares. Also, whenever its estimates differ from those of least squares, they tend to be more precise. The fundamental ingredients of estimation are the same: we have data that were generated by an assumed biological process. The stochastic model representing this process contains parameters that we want to estimate. The method focuses on the probability of obtaining the observations from a given parameterisation of the model

$$f(data|\theta) \tag{10.40}$$

Such probabilities can be readily calculated using the distributions in Chapter 9. Maximum likelihood assumes that this quantity is proportional to the probability of a particular parameterisation θ given the data

$$g(\theta|data) \propto f(data|\theta) \tag{10.41}$$

This function may be a probability if the parameters take discrete values or a probability density if the parameters are continuous. Nevertheless, as was discussed in Chapter 9, probabilities and probability densities both convey likelihood (although one is bounded above by 1 and the other is not). Therefore, we can introduce a function called the **likelihood** as an index of the plausibility of a given parameterisation under the data, i.e. $L(\theta) \propto g(\theta|data)$. Because of the proportionality in Equation (10.41), the likelihood can be defined as

$$L(\theta) = f(data|\theta) \tag{10.42}$$

The most likely estimates of the parameters under this method are the ones that maximise this function. These can be found by any available method (calculus or computers). The method can deal with more than one type of parameter simultaneously, as the following example shows.

Example 10.7: Maximum likelihood estimation of mean and variance of vole weight distribution

 A set **w** of weight measurements from the same cohort of voles can be treated as independent random variables $W_1, \ldots, W_n$. Assuming that the weights in the cohort are normally distributed, the probability density of a single measurement is given by

$$f(W|\mu, \sigma^2) = \frac{1}{\sqrt{2\pi\sigma^2}} e^{-\frac{1}{2}\frac{(\mu-W)^2}{\sigma^2}} \tag{10.43}$$

Here, the parameters to be estimated from the data are μ and σ. Since the weights in the sample are independent, the probability density of the entire sample can be written as the product of the densities of individual points.

$$f(W_1, \ldots, W_n|\mu, \sigma^2) = f(W_1|\mu, \sigma^2) \times \ldots \times f(W_n|\mu, \sigma^2) \tag{10.44}$$

This is the likelihood of the parameters under the data. Such products are written in the shorthand notation (analogous to the summation notation of Section 5.4)

$$L(\mu, \sigma^2) = \prod_{i=1}^{n} f(W_i|\mu, \sigma^2) \tag{10.45}$$

Because the members of such products are often smaller than 1, the resulting likelihoods can be very small indeed, exceeding the numerical capabilities of most computers. A simple trick is used to work around this: since we are not interested in the actual value of the likelihood but, rather, where in parameter space this maximum occurs, we can log-transform the likelihood. The position of the maximum is unaffected by this transformation and the so-called log-likelihood is a numerically more convenient sum (remember, the log of a product is the sum of the logs)

$$l(\mu, \sigma^2) = \sum_{i=1}^{n} \ln f(W_i | \mu, \sigma^2) \tag{10.46}$$

Any base log will do for the job but, in this example, the natural logarithm simplifies the expanded log-likelihood considerably. This can be seen by substituting Equation (10.43) into Equation (10.46):

$$l(\mu, \sigma^2) = \sum_{i=1}^{n} \left\{ \ln \frac{1}{\sqrt{2\pi\sigma^2}} - \frac{1}{2} \frac{(\mu - W_i)^2}{\sigma^2} \right\} \tag{10.47}$$

The values of μ and σ^2 that maximise the log-likelihood are the required estimates. To distinguish them from the true population values and the final maximum likelihood estimates, I will use the symbols m and v for the candidate values of the mean and variance. Because we now have two parameters, we must maximise the likelihood with respect to both of them. This involves the use of partial derivatives (Section 4.9). The maximum must satisfy the following two conditions:

$$\frac{\partial l(m, v)}{\partial m} = 0 \quad and \quad \frac{\partial l(m, v)}{\partial v} = 0 \tag{10.48}$$

The first of these derivatives is

$$\begin{aligned}
\frac{\partial l(m, v)}{\partial m} &= \frac{\partial}{\partial m} \sum_{i=1}^{n} \left\{ \ln \frac{1}{\sqrt{2\pi v}} - \frac{1}{2} \frac{(m - W_i)^2}{v} \right\} \\
&= \sum_{i=1}^{n} \left\{ \frac{\partial}{\partial m} \left(\ln \frac{1}{\sqrt{2\pi v}} \right) + \frac{\partial}{\partial m} \left(-\frac{1}{2} \frac{(m - W_i)^2}{v} \right) \right\} \\
&= -\frac{1}{2v} \sum_{i=1}^{n} \left\{ \frac{\partial (m - W_i)^2}{\partial m} \right\} \\
&= -\frac{1}{v} \sum_{i=1}^{n} (m - W_i)
\end{aligned} \tag{10.49}$$

The second derivative (calculations require the chain rule in Section 4.7) is

$$\begin{aligned}
\frac{\partial l(m, v)}{\partial v} &= \frac{\partial}{\partial v} \sum_{i=1}^{n} \left\{ \ln \frac{1}{\sqrt{2\pi v}} - \frac{1}{2} \frac{(m - W_i)^2}{v} \right\} \\
&= \sum_{i=1}^{n} \left\{ \frac{\partial}{\partial v} \left(\ln \frac{1}{\sqrt{2\pi v}} \right) + \frac{\partial}{\partial v} \left(-\frac{1}{2} \frac{(m - W_i)^2}{v} \right) \right\} \\
&= \sum_{i=1}^{n} \left\{ -\frac{1}{2v} + \frac{1}{2} \frac{(m - W_i)^2}{v^2} \right\} \\
&= -\frac{n}{2v} + \frac{1}{2} \sum_{i=1}^{n} \frac{(m - W_i)^2}{v^2}
\end{aligned} \tag{10.50}$$

Setting both of these derivatives to zero and substituting m and v by $\hat{\mu}$ and $\hat{\sigma}^2$ gives us a system of two equations in two unknowns

$$-\frac{1}{\hat{\sigma}^2} \sum_{i=1}^{n} (\hat{\mu} - W_i) = 0$$

$$-\frac{n}{2\hat{\sigma}^2} + \frac{1}{2} \sum_{i=1}^{n} \frac{(\hat{\mu} - W_i)^2}{\hat{\sigma}^4} = 0 \tag{10.51}$$

The first of these can be solved directly to find that the estimate for the mean is, once again, the average.

$$\hat{\mu} = \frac{1}{n}\sum_{i=1}^{n}W_i = \bar{W} \tag{10.52}$$

Solving the second equation gives

$$\hat{\sigma}^2 = \frac{1}{n}\sum_{i=1}^{n}(\bar{W} - W_i)^2 \tag{10.53}$$

Interestingly, the denominator here is n, not $n - 1$ (as obtained with least squares in Equation (10.39)). It seems that the method of maximum likelihood suggests a biased estimator for population variance (compare with Section 10.4). This is indeed the case but, as discussed in Section 10.1, estimator accuracy is neither the only, nor necessarily the most important priority of inference.

Below is a simpler example of maximum likelihood applied to something other than normal densities.

Example 10.8: Maximum likelihood estimator for wildebeest sex ratio

In a total of N offspring produced during the breeding season in a given population of wildebeest, there were n male calves. We seek to derive an estimator for p, the sex ratio at birth. Intuitively, this must be the proportion of males n/N in the cohort of calves since this proportion will tend to the true value of p as $N \to \infty$ (Example 8.5). It would be good if maximum likelihood confirmed this intuitive result. We first need to write the likelihood. According to Section 9.8, the likelihood for the number of n occurrences given N trials and a probability of occurrence p is binomial, therefore the likelihood for p can be written

$$L(p) = \binom{N}{n} p^n (1 - p)^{N-n} \tag{10.54}$$

The corresponding log-likelihood is

$$l(p) = \ln\binom{N}{n} + n\ln p + (N - n)\ln(1 - p) \tag{10.55}$$

To avoid confusion with the true value (p) and its estimator ($\hat{p}$) while this likelihood is being maximised, I will write it in terms of the variable θ. The derivative of the log-likelihood is

$$\frac{dl(\theta)}{d\theta} = \frac{d}{d\theta}\left[\ln\binom{N}{n}\right] + \frac{d}{d\theta}[n\ln\theta] + \frac{d}{d\theta}[(N - n)\ln(1 - \theta)] = \frac{n}{\theta} - \frac{(N - n)}{(1 - \theta)} \tag{10.56}$$

Setting this to zero and rearranging gives

$$\hat{p} = n/N \tag{10.57}$$

A word about confidence intervals: both of the above examples have focused on maximum likelihood estimators; some more work is required to derive CIs for the required parameters. Hilborn and Mangel (1997) and Bolker (2008) provide entry points to CI estimation with maximum likelihood.

10.6: Maximising likelihood numerically

Use of algebra to find a general expression for the estimators is not always possible because the equations resulting from setting the derivatives to zero are not always solvable. This necessitates the use of numerical techniques. The important thing to remember here is that the results of numerical techniques apply to particular data sets, so they do not have the generality of analytical results. The techniques that we covered in R4.9 will maximise log-likelihood functions in one parameter. I illustrate with the wildebeest sex ratio in Example 10.8, so that we can compare the numerical with the analytical estimates. The first step is to input the data and the statement of the likelihood

```
N<-10   # Number of trials
n<-4    # Number of male calves observed
f<-function(N,n,p) {log(dbinom(n,N,p))} # statement of log-lik.
```

Plotting the log-likelihood (Figure 10.5) is informative about the existence of peaks

```
ps<-seq(0,1,0.01) # Candidate values for p
plot(ps, f(N,n,ps), type="l", ylab="log-likelihood", xlab="p")
abline(v=0.4, lty="dashed")# Theoretical maximum from Example 10.8
```

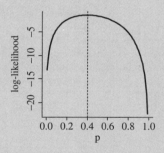

Figure 10.5: Plot of log-likelihood function in Equation (10.55).

The maximum likelihood estimate of p is found as follows:

```
> g<-optimize(f, lower=0, upper=1, N=10, n=4, maximum=T)
> g$maximum
[1] 0.4000015
```

This is close to, although not exactly the same as, the theoretical estimate ($\hat{p} = n/N = 0.4$).

When more than one parameter needs to be estimated, as was the case in Example 10.7, then the more general command `optim()` must be used. This can be illustrated with a small sample of vole weights and the log-likelihood of Equation (10.47).

```
w<-c(32.5, 38.3, 33.7, 32.3, 31.4, 24.8, 26.2, 26.2, 27.8, 26.0)
f<-function(p,w) {
 mu=p[1]
 si=p[2]
 -sum(log(dnorm(w,mu,si)))
 }
```

For the function f to be usable by `optim()`, the above declaration of the log-likelihood has two peculiarities: first, the parameters to be estimated (here, the mean and variance) are passed to f as the vector p, which is then expanded inside the function. Second, because the default option for optim() is to perform minimisation, instead of maximising the log-likelihood, we will minimise the negative log-likelihood.

A plot of the log-likelihood surface in the parameter space of mean and standard deviation is shown in Figure 10.6 (the cross-lines indicate the position of the theoretical maximum as derived in Example 10.7).

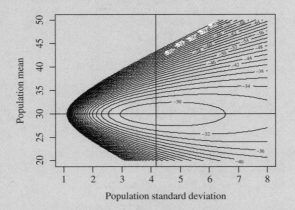

Figure 10.6: Plot of log-likelihood function in Equation (10.47).

To optimise this log-likelihood numerically and obtain parameter estimates, the following call is used

```
> g<-optim(fn=f,par=c(mu=30, si=4), w=w)
> g$par
        mu          si
29.920338 4.156209
```

Here, the option par offers to the optimisation algorithm initial values for the parameters mu and si.

10.10.   Bayesian estimation

In Section 8.9 we saw the alternative (Bayesian) interpretation of probability as the degree of belief in different candidate explanations of reality. Bayesians use data to inform their prior opinions about the truth of scientific hypotheses. In the context of statistical estimation, a hypothesis is just a particular candidate value for the parameter θ being estimated. Consider the Venn diagram in Figure 10.7. The prior belief associated with a particular value of the parameter, say θ_2, is $P(\theta_2)$. As we saw in the previous section, it is usually possible to construct a model (called the likelihood) that gives the probability of obtaining our data if this parameter value is true. In Figure 10.7, this model gives the probability $P(data|\theta_2)$. As we saw in Section 8.6, the product of these two probabilities gives the joint probability of θ_2 being true and obtaining the observed data. However, since we have already observed the data, we would like to obtain an updated (or posterior) estimate of belief for the parameter value conditional on these data. Bayes's law (Section 8.6) is used for this purpose because it is ideal for reversing the direction of conditional probabilities,

$$P(\theta_2|data) = \frac{P(data|\theta_2)P(\theta_2)}{P(data)}$$

(10.58)

Given the prior $P(\theta_2)$ and the likelihood $P(data|\theta_2)$, the only missing component is the unconditional probability of the data occurring $P(data)$. We can get this from the law of total probability (Section 8.8)

$$P(data) = \sum_{i=1}^{7} P(data|\theta_i)P(\theta_i) \qquad (10.59)$$

This sum is nothing more than the sum of the pieces forming the shaded ellipse in Figure 10.7. If you are having any problems following these arguments, go back to Section 8.9 and make sure you understand the calculations in Example 8.17.

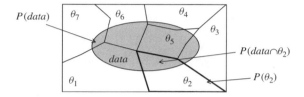

Figure 10.7: Visualising the components of Bayes's law as stated in Equation (10.58).

This approach can be extended in two directions. First, it can be stated generally with respect to all possible values of the parameter θ

$$P(\theta|data) = \frac{P(data|\theta)P(\theta)}{P(data)} \qquad (10.60)$$

This expression is subtly different from Equation (10.58) because the subscripts have been dropped. Now, the expression $P(\theta)$ and $P(data|\theta)$ are no longer probabilities but entire probability distributions. In Example 9.22, we saw how a probability distribution can be used to assign prior Bayesian beliefs to all candidate values of a parameter. Here, Bayes's rule helps update this **prior distribution**, using data, into the **posterior distribution** for θ.

Second, it turns out that Equation (10.60) remains valid, even if the parameter or the data take continuous values, by replacing the probabilities by probability densities and the sum by an integral

$$f(\theta|data) = \frac{f(data|\theta)f(\theta)}{f(data)} = \frac{f(data|\theta)f(\theta)}{\int_{All\theta} f(data|\theta)f(\theta)\,d\theta} \qquad (10.61)$$

The following example illustrates these concepts, pulling together several strands of material from Chapters 8, 9 and 10. The maths looks scary so you may prefer to skip the main body of this example during your first reading. But do read its conclusion.

Example 10.9: Bayesian estimator for wildebeest sex ratio

As in Example 10.9, we seek to derive an estimator for p, the sex ratio at birth. The data are n, the number of male calves in a cohort of size N. Using the terminology of this example, Bayes's law can be written as follows

$$f(p|n) = \frac{f(n|p)f(p)}{f(n)} \qquad (10.62)$$

The likelihood model in this case is the binomial distribution (see also Example 10.8).

$$f(n|p) = \binom{N}{n} p^n (1-p)^{N-n} \tag{10.63}$$

The prior distribution can incorporate our belief that, for animals such as wildebeest, the sex ratio is likely to be 1:1. In anticipation of the present example, such a prior was generated in Example 9.22. It is beta-distributed (see Section 9.12) with pdf

$$f(p) = \frac{\Gamma(\alpha+\beta)}{\Gamma(\alpha)\Gamma(\beta)} p^{\alpha-1}(1-p)^{\beta-1} \tag{10.64}$$

and has parameter values $\alpha = \beta = 7.83$. According to the general result in Equation (10.61), the denominator of Equation (10.62) can be calculated as the definite integral of the numerator over the interval [0,1]

$$f(n) = \int_0^1 \binom{N}{n} p^n (1-p)^{N-n} \frac{\Gamma(\alpha+\beta)}{\Gamma(\alpha)\Gamma(\beta)} p^{\alpha-1}(1-p)^{\beta-1}\, dp \tag{10.65}$$

The most intimidating parts of this expression are the binomial coefficient and the fraction of gamma functions. Note, however, that neither depends on p, so they can be pulled out of the integral, which then simplifies to

$$f(n) = \binom{N}{n} \frac{\Gamma(\alpha+\beta)}{\Gamma(\alpha)\Gamma(\beta)} \int_0^1 p^{n+\alpha-1}(1-p)^{N-n+\beta-1}\, dp \tag{10.66}$$

We are now ready to tackle Equation (10.62) by substituting-in the expressions from Equations (10.63), (10.64) and (10.66)

$$f(p|n) = \frac{\binom{N}{n} p^n (1-p)^{N-n} \frac{\Gamma(\alpha+\beta)}{\Gamma(\alpha)\Gamma(\beta)} p^{\alpha-1}(1-p)^{\beta-1}}{\binom{N}{n} \frac{\Gamma(\alpha+\beta)}{\Gamma(\alpha)\Gamma(\beta)} \int_0^1 p^{n+\alpha-1}(1-p)^{N-n+\beta-1}\, dp} \tag{10.67}$$

Rather conveniently, the binomial coefficients and gamma functions at the top and bottom cancel out and we are left with

$$f(p|n) = \frac{p^{n+\alpha-1}(1-p)^{N-n+\beta-1}}{\int_0^1 p^{n+\alpha-1}(1-p)^{N-n+\beta-1}\, dp} \tag{10.68}$$

This may seem like the end of the road because the integral in the denominator looks impossible. However, we have encountered something similar before, in Equation (9.70) of Section 9.12. Using that equation, the integral can be rewritten as

$$\int_0^1 p^{n+\alpha-1}(1-p)^{N-n+\beta-1}\, dp = \frac{\Gamma(n+\alpha)\Gamma(N-n+\beta)}{\Gamma(N+\alpha+\beta)} \tag{10.69}$$

So, here are the gamma functions making another appearance. Puting together Equations (10.68) and (10.69) yields the following:

$$f(p|n) = \frac{\Gamma(N+\alpha+\beta)}{\Gamma(N-n+\beta)\Gamma(n+\alpha)} p^{(n+\alpha)-1}(1-p)^{(N-n+\beta)-1} \tag{10.70}$$

It may take you a while to recognise, but this is the pdf of a beta distribution with parameters $\alpha' = n + \alpha$ and $\beta' = N - n + \beta$. Therefore, the posterior distribution for the proportion of male offspring is beta-distributed with mean (see Equation (9.71))

$$\mu = \frac{\alpha'}{\alpha' + \beta'} = \frac{n + \alpha}{N + \alpha + \beta} \tag{10.71}$$

Note how the mean of this posterior is a combination of data (n, N) and prior beliefs (α, β). The posterior mean can be seen as the Bayesian estimator $\hat{p}$ of the parameter p.

Here is a numerical example of how all of this works together. Assume that, from a total of $N = 10$ calves, $n = 4$ were male. The prior distribution $Beta(7.83, 7.83)$ was designed in Example 9.22 to be concentrated around a mean of 0.5 (shaded density in Figure 10.8). The posterior distribution (dark curve in Figure 10.8) can be generated directly from Equation (10.70). The posterior estimate of the sex ratio is then obtained from Equation (10.71) as $\hat{p} = 0.46$.

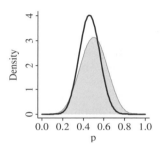

Figure 10.8: Comparison of prior (shaded curve) and posterior distribution (thick curve) in this example.

The reason it was possible to make progress analytically with this example was the mathematical relatedness between the binomial and beta distributions. These two distributions are called **conjugate**, because when used as the likelihood and prior in Bayes's law, they yield a posterior that is of the same type as the prior. Other examples of such conjugate pairs are a Poisson likelihood with a gamma prior and a normal likelihood with a normal prior. Why is this useful? As we saw in Section 8.9, Bayes's law can be used iteratively to update our beliefs as new data are collected. The posterior beliefs generated on the basis of the first batch of data can be used as the priors for the next update, using new data. If the prior is of the same type as the posterior, we can perform as many updates as we want, analytically.

10.7: Markov Chain Monte Carlo methods

Although the analytical approach presented in Example 10.9 is immensely efficient and, to some extent, generalisable, it has its limits. More complicated problems require the use of numerical methods. One such method that has taken pride of place in modern ecological literature is **Markov Chain Monte Carlo (MCMC)**, a class of computer algorithms that generates a sample of values from the posterior distribution, given a likelihood function and posterior distributions for the parameters of the likelihood. This sample can then be used to approximate the posterior distribution and derive things like parameter estimates and parameter intervals (called **credible intervals** in Bayesian terminology). Currently, the R library MCMC offers basic functionality for Bayesian estimation, but a more flexible alternative is the open source package openBugs (obtainable from http://mathstat.helsinki.fi/openbugs/). Good introductions to MCMC and openBugs for ecologists are given by McCarthy (2007), Bolker (2008) and Link and Barker (2009).

10.11.   Link between maximum likelihood and Bayesian estimation

Maximum likelihood and Bayesian estimation represent two philosophically distinct viewpoints of statistics. However, in practical terms, the two approaches are intimately linked.

Example 10.10: Comparing ML and Bayes estimates of sex ratio at birth

In Example 10.8 we derived the maximum likelihood estimate of sex ratio at birth as $\hat{p}_{ML} = n/N$. In Example 10.9, the Bayesian estimate of the same parameter was obtained as $\hat{p}_B = (n + \alpha)/(N + \alpha + \beta)$. For the particular numerical values of $n = 4, N = 10, \alpha = \beta = 7.83$, the estimates obtained are $\hat{p}_{ML} = 0.4$ and $\hat{p}_B = 0.46$. The Bayesian estimate is closer to 0.5 than the maximum likelihood estimate because we specified a prior that was concentrated around 0.5. Hence, for the small sample size of $N = 10$, prior beliefs have an effect on the posterior estimate. But what happens for large samples? Dividing numerator and denominator in the Bayesian estimator by N and taking the limit as sample size increases gives

$$\lim_{N \to \infty} \hat{p}_B = \lim_{N \to \infty} \left(\frac{\alpha/N + n/N}{1 + \alpha/N + \beta/N} \right) = \frac{n}{N} = \hat{p}_{ML} \tag{10.72}$$

So, as the weight of evidence increases, prior beliefs become irrelevant and the Bayesian estimate coincides with the ML estimate.

It seems, therefore, that maximum likelihood yields estimates in which prior beliefs do not matter. This can be shown mathematically, starting from the definition of Bayesian estimation.

$$f(\theta|data) = \frac{f(data|\theta)f(\theta)}{\int\limits_{All\,\theta} f(data|\theta)f(\theta)\,d\theta} \tag{10.73}$$

Here, I have used the continuous version of the denominator, but the arguments apply equally well to a sum. We can declare ignorance about θ by giving it a uniform prior, $f(\theta) = c$, where c is just a constant. Then, the posterior probability in Equation (10.73) simplifies to

$$f(\theta|data) = Kf(data|\theta) \tag{10.74}$$

where $K = \left(\int\limits_{All\,\theta} f(data|\theta)d\theta \right)^{-1}$ is a constant. Therefore, in the absence of prior beliefs about θ, its posterior is proportional to the likelihood of the data. This is exactly the definition of likelihood in ML methods, as stated in Equation (10.41). Therefore, one way to describe frequentists is as Bayesians who refuse to have a personal opinion. Conversely, Bayesians could be criticised for lack of objectivity. Which viewpoint suits you more as a scientist?

10.12.   Hypothesis testing: rationale

Hypotheses are at the heart of the scientific method and constructing them is one of the most pleasurable and skilled tasks for any scientist. The process of statistical hypothesis testing is rather less glamorous: it usually tries to determine whether a parameter estimate is different from what we think it might be. Most frequently, the hypothesis that the two quantities are the same is called a **null hypothesis** (written H_0). The **alternative hypothesis** (written H_1) says that they differ.

Because estimators are random variables, it is impossible to say with certainty whether the unknown parameter is exactly the same as the value it is being compared with. So, the next best question is to ask whether they are **significantly different**. Deciding on what amounts to a significant difference is a subjective decision. The overall approach has the following steps (summarised in Figure 10.9):

❶ Define the null and alternative hypotheses. The null hypothesis is usually the one that represents the less complicated explanation of the real world (see discussion on parsimony in Example 2.1). Under the nebulous notion that everything would be simpler if all things were identical, the null hypothesis is conventionally chosen to represent no deviation from the norm.
❷ Construct a sampling distribution for the estimator on the basis of the null hypothesis and the size of the sample.
❸ The sampling distribution will assign a likelihood (no matter how small) to every possible value of the estimator. However, very small values of likelihood (at the extremes of the sampling distribution) can be taken as evidence that the population generating the data has a parameter different to the one postulated by H_0. A probability value α is chosen to represent the level of significance required of the result. For example, $\alpha = 0.01$ means that, under H_0, the estimator will be found in the extreme regions of parameter space only once every one hundred samples.
❹ Given the sampling distribution arising out of H_0, the significance level α must map onto one or two **critical values** for the estimator. These values tell us if a particular estimate is significantly different from what would be expected under the null hypothesis.
❺ The estimate ($\hat{\theta}$) is calculated from the data and compared with the critical value(s).
❻ If the estimate falls in the region of extreme values, then the null hypothesis is rejected, otherwise we say that there is not enough evidence to reject it.

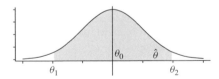

Figure 10.9: In this illustration, the hypothesis test focuses on the value of an unknown population parameter θ. The value θ_0 represents the null hypothesis (H_0), as formulated in step ❶. The curve is the sampling distribution for the estimator Θ constructed in step ❷ on the assumption that H_0 is true. The tails of the sampling distribution contain a probability α decided upon in step ❸. Therefore, the shaded area encloses a probability $1 - \alpha$. The critical values θ_1, θ_2 defining the shaded region are the boundaries between ordinary and extraordinary values of the estimator (step ❹).

To express how likely it is to obtain the observed estimate under the null hypothesis, we can calculate the probability that this value, or a value even more extreme, is observed. For example, if the estimate $\hat{\theta}$ is on the right-hand side of the sampling distribution, the probability under the right-hand tail is $P(\Theta \geq \hat{\theta})$. This is called a **p-value**.

In constructing the sampling distribution (step ❷), it is necessary to decide whether we can assume a distribution for the population from which the sample is drawn. If we are happy to do that, then the test is called **parametric**. If not, then we need to use a **nonparametric test**. Because of the central limit theorem and related results, nonparametric tests can help us derive a sampling distribution for the estimator, even though the distribution of the population is not specified.

One sensitive aspect of hypothesis testing comes in step ❻. The critical values, which are themselves a direct consequence of an arbitrary decision for the value of α, are treated as hard boundaries between what should be considered ordinary and extraordinary. This means that the conclusion in step ❻ can commit two types of error. We may reject the null hypothesis when it is, in fact, true (**Type I error**), or we may accept the null hypothesis when it is, in fact, false (**Type II error**). The next four sections briefly discuss various often-encountered tests and use them as examples of how the six steps (above) are put to practice.

10.13.   Tests for the population mean

We saw in Section 10.3 that, for large samples, the sampling distribution of the average will usually be normal. If the variance of the population is known, then the sampling distribution of the normalised average $(\mu - \bar{X})/\sigma$ is standard normal (Section 10.3), but if the variance is estimated from the sample, then the quantity $(\mu - \bar{X})/s$ is t-distributed (Section 10.5). This suggests a clear way forward for testing hypotheses about the mean of a particular population:

❶ *Formulate hypotheses*
   H_0: The mean of the population is equal to (some value) μ
   H_1: The mean of the population is not equal to μ
❷ *Construct sampling distribution*
   If the variance is known, then, under the null hypothesis, $Z = (\mu - \bar{X})/\sigma \sim N(0,1)$. In the more likely case where the variance is estimated from a sample, then $T = (\mu - \bar{X})/s \sim t(n-1)$.
❸ *Select significance level α*
❹ *Find critical values*
   These will be the quantiles of either the standard normal or the t–distribution with $n-1$ degrees of freedom, at the points $\alpha/2$ and $1 - \alpha/2$. For example, for $\alpha = 0.05$, we are looking for the points $t_{0.025}, t_{0.975}$, such that $P(T < t_{0.025}) = 0.025$ and $P(T > t_{0.975}) = 0.975$.
❺ *Calculate test statistic*
   This will be $(\mu - \bar{X})/\sigma$ if the population standard deviation is known and $(\mu - \bar{X})/s$ if the sample standard deviation is used instead.
❻ *State result*
   If the test statistic is outside the range $(t_{\alpha/2}, t_{1-\alpha/2})$, then H_0 can be rejected (implying that the mean of the population is not μ).

    10.8: t test for the mean

R will carry out the above steps using the command t.test(). To illustrate, we can first create a sample of data ($n = 20$) from a normal distribution with mean 2 and standard deviation 3.

```
d<-rnorm(20,2,3)
```

Assuming that the values of these parameters are unknown, R can decide if there is significant evidence (at $\alpha = 0.05$) that the mean of the population is different from zero:

```
> t.test(data, mu=0, conf.level=0.95)

    One Sample t-test

data: data
t = 3.335, df = 19, p-value = 0.00348
alternative hypothesis: true mean is not equal to 0
95 percent CI:
 0.652504 2.851769
sample estimates:
mean of x
 1.752136
```

The resulting p-value (0.00348) is (considerably) smaller than the significance level of 0.05, so the null hypothesis can be rejected.

The test distribution usually hides some assumptions. In this case, we either assume that the sample is large enough for the sampling distribution to be normal, or, if the sample is small, that the underlying population is normal. These assumptions need to be checked before the t test is applied.

ⓇR   10.9: Examining normality by a QQ plot

A simple graphical tool, qqnorm(), can be used to visually assess whether a sample of data comes from a normal distribution. The method plots successive quantiles from the theoretical cumulative normal distribution (with mean and variance estimated from the data) against the corresponding quantiles from the cumulative frequency distribution of the data. If the data come from a normal distribution, then the paired quantiles should be aligned along a 45° line. Here are two examples using a simulated sample from a normal and non-normal distribution (their output is in Figure 10.10).

```
# Normal distribution
data<-rnorm(30, 2,2)
qqnorm(data)
qqline(data)

# Non-normal distribution
data<-runif(30, -4,4)
qqnorm(data)
qqline(data)
```

Useful though this is, it is not entirely reliable, especially for samples with $n < 40$: often, small samples from normal populations deviate from the line and samples from non-normal populations arrange themselves closely along its length. There is even a hypothesis test (implemented in shapiro.test()) that can be used to examine normality. Here are two examples using the above data:

```
# Normal distribution - This should pass the test (p-value>0.05)
> data<-rnorm(30, 2,2) #
> shapiro.test(data)

    Shapiro-Wilk normality test
```

```
data: data
W = 0.9537, p-value = 0.2124

>
> # Non-normal distr. - This should fail the test (p-value<0.05)
> data<-runif(30, -4,4)
> shapiro.test(data)

    Shapiro-Wilk normality test

data: data
W = 0.935, p-value = 0.04658
```

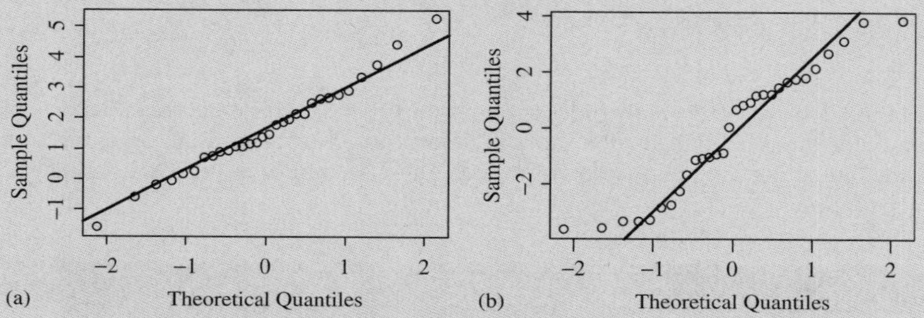

Figure 10.10: QQ plots for (a) normal and (b) non-normal samples.

So, what do we do if the sample is small and we are not confident that it comes from a normal population? If the distribution of the population is symmetric, then the **sign test** for the mean is a nonparametric alternative to the t-test that sidesteps the assumption of normality. It simply examines which side of the hypothesised mean each observation is on. Observations to the right are given a positive sign and vice versa, hence the name of the method. If the total number of pluses is significantly different from the total number of minuses, then the null hypothesis is rejected. For the sign test, the six steps of hypothesis testing are as follows:

❶ *Formulate hypotheses*
   H_0: The mean of the population is equal to (some value) μ
   H_1: The mean of the population is not equal to μ
❷ *Construct sampling distribution*
   If H_0 is true, then the probability of getting an observation to the right of μ is the same as the probability of it falling to the left (this is the definition of the mean for symmetric distributions). Therefore, given a sample of size n, the total number X of observations on the right of μ is the binomial variable $X \sim B(n, \frac{1}{2})$.
❸ *Select significance level* α
❹ *Find critical values*
   These will be the quantiles $x_{\alpha/2}, x_{1-\alpha/2}$ of the binomial at the points $\alpha/2$ and $1 - \alpha/2$.
❺ *Calculate test statistic*
   This is just the number of observations to the right of μ.
❻ *State result*
   If test statistic is outside the range $(x_{a/2}, x_{1-a/2})$, then null hypothesis can be rejected.

The sign test does not require the data to be normal and can be applied to small samples. Why then can it not be used all the time in favour of the t test? By only examining whether the data are to the right or left of the mean, the sign test throws away information about their exact position. It therefore fails to use the data with maximum efficiency. This exemplifies a fundamental trade-off: *parametric methods are more efficient and powerful while nonparametric methods are more robust and general.*

10.10: Sign test for the mean

This can be carried out with the command `binom.test()`. Here is an example using data from $U(-1, 5)$ (i.e. $\mu = 2$). The null hypothesis is $H_0 : \mu = 0$, so we would expect it to be rejected as, indeed, happens (check p-value).

```
> n<-30 # Sample size
> data<-runif(n, -1,5) # Simulated data
> p<-0.5 # Probability of getting a plus if Ho is true
> mu<-0 # Hypothesised distribution mean
> difs<-mu-data # Differences from mean
> pls<-length(difs[difs>0]) # Number of pluses
> binom.test(pls, n, p, conf.level=0.95) # Carries out the test


    Exact binomial test

data: pls and n
number of successes = 5, number of trials = 30, p-value = 0.0003249
alternative hypothesis: true probability of success is not equal to 0.5
95 percent CI:
 0.0564217 0.3472117
sample estimates:
probability of success
          0.1666667
```

10.14. Tests comparing two different means

Another interesting question is whether the means of two samples are significantly different. The simplest situation arises with paired samples.

Example 10.11: Paired samples of gannet condition

A group of 30 breeding female gannets is tagged before the breeding season of 2010 and found again at the same time in 2011. In both years, measurements are taken of some index I of condition (e.g. ratio of size over weight). The biological question asked is whether the mean condition of breeders in the population has declined from one year to the next. We are therefore interested in the quantity $\Delta I = I_{2011} - I_{2010}$ and would like to know if, on average, it is positive or negative across the wider population of breeders.

The trick with paired samples is to consider the differences between values in each pair. This results in a new variable $\Delta X = X_2 - X_1$. If, as in Example 10.11, we are interested in an

H_0 of no change, then we want to see if ΔX deviates significantly from zero. This reduces the paired samples problem to a single sample problem, so we can apply to it the tools from the previous section (i.e. test for normality and then decide if a t-test or a sign test is appropriate).

A different approach is required if the samples are collected independently (in Example 10.11, above, this would happen if gannets were not tagged and hence the two samples of condition came from different individuals). If we can assume that the distribution of the population is normal in both sampling instances, then we are interested in the variables $X_1 \sim N(\mu_1, \sigma_1^2), X_2 \sim N(\mu_2, \sigma_2^2)$. If we can additionally assume that $\sigma_1 = \sigma_2$, then there is a t-test for the purpose. It goes as follows:

❶ *Formulate hypotheses*
  $H_0 : \mu_1 = \mu_2 \quad H_1 : \mu_1 \neq \mu_2$
❷ *Construct sampling distribution*
  The test statistic (see step ❺, below) has a t-distribution with $df = n_1 + n_2 - 2$ (note, therefore, that the test allows the sample sizes to be different).
❸ *Select significance level α*
❹ *Find critical values*
  The quantiles of the t-distribution at the points $\alpha/2$ and $1 - \alpha/2$.
❺ *Calculate test statistic*
  The test statistic is

$$T = \frac{\bar{X}_1 - \bar{X}_2}{s} \tag{10.75}$$

Here, the denominator is the standard error of the distribution of differences. It can be calculated from the following expression (where s_1 and s_2 are the standard errors of the individual samples):

$$s = \frac{(n_1 - 1)s_1^2 + (n_2 - 1)s_2^2}{n_1 + n_2 - 2} \sqrt{\frac{1}{n_1} + \frac{1}{n_2}} \tag{10.76}$$

❻ *State result*
  If value of the test statistic is outside the range $(t_{a/2}, t_{1-a/2})$, then H_0 can be rejected.

Ⓡ 10.11: t test for unpaired means comparison

Now that the hypotheses are getting complicated and the test statistics more unwieldy, it is useful to know how to carry out the tests using built-in R commands. The command `t.test()` that was examined in R10.8 can also deal with the general case of comparison between means. The following example examines whether the two unpaired samples of unequal size come from distributions with the same means (i.e. the mean of the differences is zero). The option (`var.equal = F`) implies that we can also relax the assumption of equal variances.

```
> # Simulated data with different means, variances and sample sizes
> x1<-rnorm(20, 10, 5)
> x2<-rnorm(25, 20, 6)
> t.test(x1, x2, mu=0, paired=F, var.equal=F, conf.level=0.95)

    Welch Two Sample t-test

data: x1 and x2
t = -4.9488, df = 42.858, p-value = 1.205e-05
alternative hypothesis: true difference in means is not equal to 0
```

```
95 percent CI:
 -13.132923 -5.527753
sample estimates:
mean of x mean of y
 11.68898 21.01932
```

If it is not possible to justify the assumption of normality for both populations, then a non-parametric alternative, such as the Wilcoxon rank sum test, is required. This makes use of **ranking** as follows: the two samples are merged and the pooled data are sorted in increasing order. Ranks are assigned to the data, appropriately dealing with tied (i.e. identical) observations, as shown in Example 10.12. The test statistic is based on the sum of ranks for each of the two samples and then compared to a purpose-specific test distribution. The main idea is to see if one of the two samples dominates the higher end of the ranks.

Example 10.12: How to rank observations

Here are two samples: $A = \{1, 3, 3, 2, 4\}, B = \{5, 5, 3, 2, 5, 4, 11, 5\}$. We can rearrange these into a table of sorted values (Table 10.1), also noting their membership (sample A or B), order and rank. Notice that when two values are tied, they share the same rank (calculated as the average of orders).

Table 10.1

Membership	A	A	B	A	A	B	A	B	B	B	B	B	B
Value	1	2	2	3	3	3	4	4	5	5	5	5	11
Order	1	2	3	4	5	6	7	8	9	10	11	12	13
Rank	1	2.5	2.5	4	5.5	5.5	7.5	7.5	10.5	10.5	10.5	10.5	13

R **10.12: Wilcoxon rank sum test**

If required, the ranks of the members of a data set can be calculated with the command `rank()`. However, the entire procedure of testing can be carried out by R. Here is an example using simulated data (unequal sample sizes) from two different Poisson distributions (so, out of this, we should expect to get a p value smaller than 0.05).

```
> x1<-rpois(20, 2.8)
> x2<-rpois(35, 5)
> wilcox.test(x1,x2)

    Wilcoxon rank sum test with continuity correction

data: x1 and x2
W = 181, p-value = 0.002749
alternative hypothesis: true location shift is not equal to 0

Warning message:
In wilcox.test.default(x1, x2) : cannot compute exact p-value with ties
```

10.15. Hypotheses about qualitative data

Many legitimate biological questions refer to categorical data (e.g. species, habitat type, sex, life stage, etc.). Although the specific tests examined so far are not much use for these questions, the overall philosophy of hypothesis testing can be carried over, intact.

Example 10.13: Tree selection in Peruvian ants

Carpenter ant, genus Camponotus

We consider the study system of Yu and Davidson (1997) who looked for associations between ant species and seven members of the *Cercopia* family of tropical trees in Madre de Dios, Peru. Table 10.2 lists the numbers (n) of trees of each species that might have been encountered in the survey and the number (x) of those trees on which the ant *C. balzani* was found

Table 10.2

Species	1	2	3	4	5	6	7
n_i	53	6	7	31	17	17	24
x_i	16	0	5	3	2	1	12

This data may be used to ask if *C. balzani* use one tree species more than another. Certainly, species 1 has the highest count of trees with *C. balzani*, but it is also the most abundant *Cercopia* in the study area, so we need to account for host availability. We also need to allow for chance variations in the occurrence of ants. Are the observed variations sufficient evidence for selectivity by the ants?

These data are counts, associated with different values of a categorical variable (tree species). In Section 7.11 these were called contingency tables. Questions pertaining to categorical data are most often dealt with by **chi-square tests**. These fundamentally involve a comparison between observed (O_i) and expected (E_i) frequencies for the ith value of the categorical variable. Because they examine the closeness of the data to some preconceived idea of the process generating them, they are also known as **goodness-of-fit** tests. A typical chi-square test has the following structure:

❶ *Formulate hypotheses*
$H_0 : O_i = E_i$ $H_1 : O_i \neq E_i$. This is a bit cryptic. The alternative hypothesis expresses significant deviation from an expected outcome but some creative thinking may be required in constructing the expected frequencies. In the simplest case, if we are testing the hypothesis that all m outcomes are equally likely and the sample size is n, then the expected frequencies are n/m, for all m outcomes.
❷ *Construct sampling distribution*
The test statistic (see step ❺, below) has a chi-square distribution whose degrees of freedom depend on the number of categories in the contingency table and the experimental design.
❸ *Select significance level α*
❹ *Find critical value*
A single quantile is obtained at the desired level α

❺ *Calculate test statistic*
The test statistic is

$$\chi_0^2 = \sum_{i=1}^{m} \frac{(O_i - E_i)^2}{E_i} \tag{10.77}$$

A requirement of the test is that none of the categories have a count smaller than 5. If any categories suffer from data sparsity, they can either be pooled with other categories or dropped from the test completely, with an appropriate modification of the null hypothesis.

❻ *State result*
If test statistic is to the right of the critical value, the null hypothesis is rejected.

The particulars of this test seem to come out of nowhere. So, some illustration and explanations are needed.

Example 10.14: Tree selection in Peruvian ants

The ants had low or zero counts in five of the tree species. As it happened, *Cercopia* species 1 and 2 were closely related, and so were species 4, 5 and 6. These categories can therefore be pooled to give the results shown in Table 10.3.

Table 10.3

Species	(1+2)	3	(4+5+6)	7
n_i	59	7	65	24
x_i	16	5	6	12

If the ants were not selective in their use of different *Cercopia* species, then we should expect their frequency of occurrence on different trees to be proportional to the frequency of occurrence of the trees themselves. Therefore, the expected number of ant occurrences E_i for the ith tree species can be obtained as $E_i = \left(\frac{n_i}{155}\right) \times 39$, where $n_i/155$ is the proportion of trees of the ith species and 39 is the total number of trees with *C. balzani* on them. This calculation yields Table 10.4 of observed and expected frequencies.

Table 10.4

Species	(1+2)	3	(4+5+6)	7
E_i	14.85	1.76	16.35	6.04
O_i	16	5	6	12

The test statistic calculated from these, with the aid of Equation (10.77), is $\chi_0^2 = 18.49$. This needs to be compared with a chi-square distribution but we must first decide on the degrees of freedom. Of the six species of tree (categories) initially in the data, only four categories remain. Also the total number of ant occurrences was used in the calculation of expected frequencies. This is the equivalent of losing one data point (a degree of freedom). Therefore, there are three degrees of freedom left. The quantile of $\chi^2(3)$ at the 95% level is 7.91 (you can calculate this by typing `qchisq(0.95, 3)` in R, see R9.14), providing evidence for selectivity by ants.

The test statistic in Equation (10.77) appears to do something sensible insofar as it contains the sum of squared differences between expectations and observations. Clearly, the smaller these differences are, the closer the whole thing will be to zero and the less likely it will be to exceed the quantile of the chi-square distribution. Note that, since the differences are squared, extreme values can only be positive, hence the need for only a single quantile for this test. So far so good. But why is this particular test statistic chi-square distributed, and why do we need to have frequencies in excess of 5 in all categories? The reasoning is as follows: since the content of each category i is a count, its observed value could be modelled as a Poisson variable with rate E_i. Hence, $O_i \sim Poisson(E_i)$. A memorable characteristic of the Poisson is that its mean and variance are the same (Section 9.10). In this case, $\mu_i = \sigma_i^2 = E_i$. Hence, the statistic in Equation (10.77) can be rewritten as

$$\chi_0^2 = \sum_{i=1}^{m} \frac{(O_i - \mu_i)^2}{\sigma_i^2} \qquad (10.78)$$

Now, if the number of occurrences in each category is large, then the Poisson variable in each category can be approximated by a normal variable with the same mean and variance (Section 9.14), so, $O_i \sim N(\mu_i, \sigma_i^2)$ for $\mu_i = \sigma_i^2 = E_i$. This last statement converts Equation (10.78) into the sum of squares of m standard normal variables, which is the definition of a chi-square distributed variable (compare Equation (10.78) with Equation (9.114) in Section 9.20).

The analysis of categorical data can be extended to more than one dimension. Such analyses on multi-way contingency tables may aim to reveal associations between two or more categorical variables, as in the following example.

Example 10.15: Niche partitioning in Peruvian ants

The main thrust of Yu and Davidson's (1997) paper was to reveal specialisation in the use of different *Cercopia* trees by different ant species. Four ant species were considered in total, yielding the joint frequency distribution of two categorical variables (ant species vs *Cercopia* species). The null hypothesis here was one of **homogeneity**: different species were assumed to respond to the set of trees in the same way. So, under the null hypothesis, even if the ants don't use the tree species in proportion to their availability, different ant species use the same tree species equally. The biological interpretation of the alternative hypothesis is that ants have specialised on different trees to avoid competition.

ℝ 10.13: Chi-square tests

Both simple and complicated tests for contingency tables can be carried out by the single command `chisq.test()`. To carry out the chi-square test for Example 10.14, we need to provide a vector of observations (o) and a vector (p) of expected proportions,

```
> n<-c(59,7,65,24)    # Counts for tree species
> o<-c(16,5,6,12)    # Observed occurrences of ants
> p<-n/sum(n)    # Expected proportion of ants by tree
> chisq.test(o, p=p)

        Chi-squared test for given probabilities

data:  o
```

```
X-squared = 18.4862, df = 3, p-value = 0.0003491

Warning message:
In chisq.test(o, p = p) : Chi-squared approximation may be incorrect
```

10.16.   Hypothesis testing debunked

The reign of hypothesis testing over applied statistics saw the flourishing of an industry of tests (both parametric and nonparametric), addressing different scientific questions. There are now methods to test if two variables are correlated, if a proportion is significantly different from a postulated value, if the variances of two populations are equal, if two multivariate samples come from the same distributions, and so on. Hypothesis testing also dominated the ecological literature for decades and reading historical reports requires at least a rudimentary knowledge of the logic behind the tests. The previous four sections have tried to instil just such an understanding. However, the approach suffers from several weaknesses, particularly when used blindly and indiscriminately:

❶ *Significance levels are arbitrary*: Changing the significance level magically turns an ordinary result into something worth reporting.
❷ *Results are only qualitative*: We get an idea of whether the null hypothesis is true but not how well supported it is by the data. The use of *p*-values as measures of evidence has also received criticism by many authors.
❸ *The dichotomy between null/alternative hypotheses is limiting*: Why not look at several candidate values at the same time?
❹ *Null hypotheses are guaranteed to be false*: In the sense that all models are wrong, no population parameter will ever be exactly the same as our expectations.
❺ *A significant result is guaranteed if the sample size is large enough*: This issue is related to problem ❶ above. The significance level must be appropriately chosen in relation to sample size. There are methods for doing this, under the broader area of **power analysis**.

These weaknesses have resulted in something of a backlash against hypothesis testing: a spirited critique is presented in McCarthy (2007) and a more detailed review of the arguments, for and against, in Quinn and Keough (2002). The methodological void created by the gradual abandonment of hypothesis tests (Eberhardt, 2003) is now getting filled by statistical modelling approaches, the subject of the next chapter.

Further reading

The bare essentials of this chapter are concisely covered in Siegel (1988), Rowntree (2000), Cann (2003) and Harris *et al.* (2005). Like this chapter, Larson (1982) and Bolker (2008) provide deeper glimpses into the mechanics and methodology of inference, by covering maximum likelihood and Bayesian statistics. Bolker's book, in particular, is a strong manual on how to do maximum likelihood in ecology. There are now an increasing number of good introductory books for Bayesian estimation for ecologists (McCarthy, 2007; King *et al.*, 2009; Link and Barker, 2009). More classically styled textbooks such as Sokal and Rohlf (1995), Wild and Seber (2000), Quinn and Keough (2002) and Whitlock and Schluter (2009) cover statistical tests more exhaustively. Nonparametric tests are comprehensively covered by Conover (1999). Some of the more philosophical aspects of inference, clumsily sidestepped by me, are broached by Gotelli and Ellison (2004) and flairfully confronted by Jaynes (2003).

References

Bolker, B.M. (2008) *Ecological Models and Data in R*. Princeton. 408pp.

Cann, A.J. (2003) *Maths from Scratch for Biologists*. John Wiley & Sons, Ltd, Chichester.

Conover, W.J. (1999) *Practical Nonparametric Statistics*. John Wiley & Sons, Inc., New York. 584pp.

Eberhardt, L.L. (2003) What should we do about hypothesis testing ? *The Journal of Wildlife Management*, **67**, 241–242.

Gotelli, N.J. and Ellison, A.M. (2004) *A Primer of Ecological Statistics*. Sinauer Associates, Massachusetts. 510pp.

Harris. M, Taylor, G. and Taylor, J. (2005) *CatchUp Maths & Stats For the Life and Medical Sciences*. Scion, Kent.

Hilborn, R. and Mangel, M. (1997) *The Ecological Detective: Confronting Models with Data*. Princeton University Press, New Jersey. 330pp.

Jaynes, E.T. (2003) *Probability Theory: The Logic of Science*. Cambridge University Press, Cambridge. 727pp.

King, R., Morgan, B.J.T., Gimenez, O. and Brooks, S.P. (2009) *Bayesian Analysis for Population Ecology*. CRC Press. 456pp.

Larson, H.J. (1982) *Introduction to Probability Theory and Statistical Inference*. John Wiley & Sons, Inc., New York. 637pp.

Link, W. and Barker, R. (2009) *Bayesian Inference: With Ecological Applications*. Academic Press, London. 400pp.

McCarthy, M.A. (2007) *Bayesian Methods for Ecology*. Cambridge University Press, Cambridge. 310pp.

Quinn, G.P. and Keough, M.J. (2002) *Experimental Design and Data Analysis for Biologists*. Cambridge University Press. 537pp.

Rowntree, D. (2000) Statistics Without Tears: An Introduction for Non-Mathematicians. Penguin, London. 195pp.

Siegel, A.F. (1988) *Statistics and Data Analysis*. John Wiley & Sons, Inc., New York. 518pp.

Sokal, R.R. and Rohlf, F.J. (1995) *Biometry*, 3rd edition. Freeman, New York. 887pp.

Whitlock, M.C. and Schluter, D. (2009) *The Analysis of Biological Data*. Roberts & Co., Colorado. 700pp.

Wild, C.J. and Seber, G.A.F. (2000) *Chance Encounters: A First Course in Data Analysis and Inference*. John Wiley & Sons, Inc., New York. 611pp.

Yu, D.W. and Davidson, D.W. (1997) Experimental studies of species-specificity in *Cercopia*-ant relationships. *Ecological Monographs*, **67**, 273–294.

11

How to separate the signal from the noise
(Statistical modelling)

'Questions were thrown at him: How did he explain the ''life-causing property'' of the signal? How did it originate? Was it, according to him, a ''pure accident''? And, most of all – where did we get Frog Eggs from?'
From His Master's Voice *by Polish writer Stanislaw Lem (1921–2006)*

A primary task of statistics is to separate systematic from random patterns. Systematic patterns may offer scientific insights. Randomness is just a nuisance. Patterns unfold in space and time, so to detect them we need to collect repeated observations, possibly under different circumstances. Picking up the thread from Chapter 10, this chapter first asks whether hypothesis testing can be used to compare the means of multiple populations and, assuming differences between them are detected, whether they can be explained by particular attributes of the populations. The first question can be addressed by an extension of the t-test called **analysis of variance** (**ANOVA** – Section 11.1). The second question can be examined by empirically modelling the replicate observations as a function of the conditions in which they were made. This gives rise to **regression**, a statistical approach to estimating the parameters and associated uncertainty of these empirical models. The simplest such model, **linear regression**, estimates the slope and intercept of a linear relationship between the observations and a single explanatory variable (Section 11.2). Compared to ANOVA, the regression approach brings gains in predictive and inferential ability (Section 11.3), assuming that its assumptions of linearity, normality and independence are satisfied.

The rest of this chapter discusses what can be done if the **fit** of the linear model to the data is poor (Section 11.4). For example, it may be that more than one explanatory variable is required to account for the variation in the data (**multiple linear regression models** – Section 11.5).

How to be a Quantitative Ecologist: The 'A to R' of Green Mathematics and Statistics, First Edition. Jason Matthiopoulos.

Deciding how many and which explanatory variables should be included in a multiple regression model is known as **model selection**, a problem that can be examined as a trade-off between a model's quality of fit and its predictive ability (Section 11.6).

In many cases, the variation of the observations around the mean cannot be assumed to be normal. Using distributions other than the normal gives rise to a broader class of regression models known as **generalised linear models** (**GLMs** – Section 11.7). The particularly important cases of **binomial** and **Poisson GLMs** are discussed in Sections 11.7 and 11.8. The method of **quasilikelihood** is briefly introduced as a solution to **overdispersed** and **heteroscedastic** data (Section 11.9). If the assumption of linearity is violated and the more flexible forms offered by the GLMs are still not sufficient to describe the patterns in the data, then a functional form can be designed from biological first principles and fitted to the data by **nonlinear regression** (Section 11.10). Alternatively, if such biological information is unavailable, the data may be allowed to suggest an appropriate functional form through the use of **smoothing functions** (Section 11.11).

Variability around the mean prediction may be due to different identifiable sources. Real ecological data are hierarchical, often grouped according to nested or overlapping sample units (e.g. metapopulation, population, individual). The components of variability in the data can be accounted for by an extension of the linear model known as the **mixed effects model** (Section 11.12).

11.1.   Comparing the means of several populations

Example 11.1: Samples of gannet condition

Change in the condition of breeding female gannets during a particular time period of the year is given by $\Delta I = I_{After} - I_{Before}$. To investigate the spatial patterns in this variable, we may track groups of females from several different colonies (Figure 11.1(a)) This will yield multiple samples of ΔI that we can use to obtain estimates of mean change in condition for each colony.

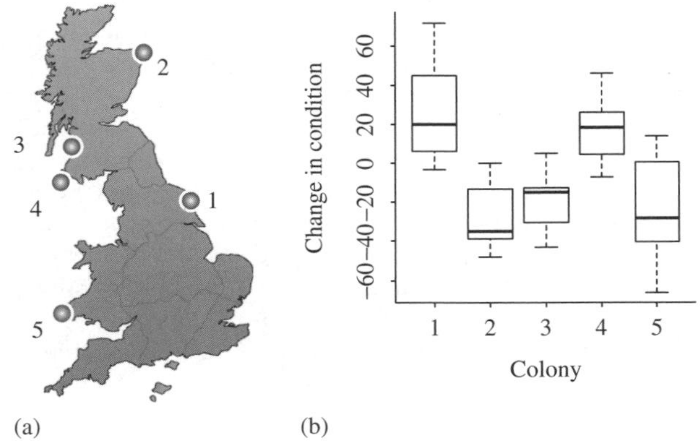

(a)   (b)

Figure 11.1: (a) Positions of five gannet colonies on mainland UK; (b) boxplots of the change in condition, incorporating ten females from each of the five colonies.

The first question is whether a spatial pattern exists at all, i.e. if mean change in condition differs between colonies. Comparisons between any two colonies can be carried out by using the t-test in Section 10.14. But how can we compare all of them at the same time? Could we, perhaps, carry out all pair-wise tests until we encounter evidence of dissimilarity? That is not a great idea, because the probability of finding a difference by chance increases with the number of tests performed (see critique of hypothesis testing in Section 10.16 and pp. 345–348 in Gotelli and Ellison, 2004).

The formal alternative to the t-test when more than two means are involved is called **analysis of variance** (**ANOVA** for short). The test examines the following two hypotheses: H_0: The means of all k populations are the same ($\mu_1 = \mu_2 = \cdots = \mu_k$) and H_1: At least two means are different. It works by comparing the variability between populations with the variability within populations. Its test statistic takes the form:

$$f = \frac{\text{Variability between populations}}{\text{Variability with in populations}} \tag{11.1}$$

So, larger values of f make it more likely that H_0 will be rejected. The numerator in Equation (11.1) is the following expression

$$\text{Variability between populations} = \frac{1}{k-1} \sum_{i=1}^{N} n_i (\hat{\mu}_i - \hat{\mu})^2 \tag{11.2}$$

where k is the number of populations whose means are being compared, N is the sample size of the pooled data set, n_i is the sample size from the ith population, $\hat{\mu}$ is the average of the pooled data and $\hat{\mu}_i$ is the average of data from the ith population. This quantity, known as the **error mean square**, first carries out a comparison $(\hat{\mu}_i - \hat{\mu})^2$ between the pooled average ($\hat{\mu}$) and each of the individual averages ($\hat{\mu}_i$). The multiplication with n_i inside the sum implies that individual means that come from larger samples carry a greater weight in the calculation. The error mean square has $k-1$ degrees of freedom because, given all but one of the individual means together with the pooled mean, we can calculate the missing average.

The denominator in Equation (11.1) is given by

$$\text{Variability within populations} = \frac{1}{N-k} \sum_{i=1}^{N} (n_i - 1) s_i^2 \tag{11.3}$$

where s_i^2 is the estimated variance for the ith population. This quantity, known as the **group mean square**, adds together the sample sums of squares $(n_i - 1)s_i^2 = \sum_{j=i}^{n_i} (x_{i,j} - \hat{\mu}_i)^2$ and then divides by the degrees of freedom. Here, we have $df = N - k$ because each sum of squares uses one degree of freedom in calculating the individual average $\hat{\mu}_i$ from the raw data.

If variabilities within and between populations are normally distributed, then the sampling distribution for this statistic is known as the **F-distribution**. Its parameters are the degrees of freedom $df_1 = k - 1$ and $df_2 = N - k$.

11.1: Analysis of variance

The data presented in Figure 11.1(b) are a data frame (called `dat`) with 50 rows (ten individuals for each of the five colonies). Each row contains information about the change in individual condition (column `dat$Condition`) and colony membership (column `dat$Colony`, encoded as a factor – see R1.1 on factors). To test the H_0 that ΔI is not dissimilar between colonies, we need to define a **linear model object**.

```
mod<- lm(Condition~Colony,data=dat)
```

The reason and syntax for this will become clear in the following section. The model object, here called mod, is created by calling the command lm(). This call contains a **formula** (Condition~Colony) which tells R that we want to examine the effect of Colony on Condition. It also contains the name of the data frame, so that R knows where to find the data for Condition and Colony. Once this object has been created, an ANOVA can be performed very simply

```
> anova(mod)
Analysis of Variance Table

Response: Condition
          Df Sum Sq Mean Sq F value  Pr(>F)
Colony     4 24638.0 6159.5 15.893 3.508e-08 ***
Residuals 45 17439.9  387.6
---
Signif. codes: 0 '***' 0.001 '**' 0.01 '*' 0.05 '.' 0.1 ' ' 1
```

Here, we have five colonies (so, $k = 5$) and a total of 50 animals ($N = 50$). The degrees of freedom listed in the column Df of the output are $df_1 = k - 1 = 4$ and $df_2 = N - k = 45$. The entries in column Mean Sq give the quantities in Equations (11.2) and (11.3) respectively. The ratio of these quantities (Equation (11.1)) is listed under F value. The p value for the null hypothesis is listed under Pr(>F). In this example it is highly significant, implying that H_0 should be rejected. This test therefore suggests emphatically that there are differences in ΔI between gannets in at least two of the five colonies.

The ANOVA test makes several assumptions whose violation could annul the test result.

❶ *Independence:* In the case of the gannet example, this means that the measured change in condition of one female from a given colony is not affected by the change in condition of any other female (from the same, or another, colony).

❷ *Normal residuals:* For example plotting $(\Delta I - \overline{\Delta I})$ for females from a given colony should give a distribution close to a normal.

❸ *Equal standard deviations between sampling instances* (**homoscedasticity assumption**): For example, the distributions of residuals $(\Delta I - \overline{\Delta I})$ for different gannet colonies should have the same standard deviations.

If there are significant differences between populations, then perhaps a statistical summary can be estimated for each population individually, using the estimation methods of Chapter 10. You may find Example 11.2 below pedantic because it uses a lot of maths to arrive at the common-sense result that the best summary for each population is its average. However, it is useful in demonstrating a straightforward case of simultaneous estimation of many parameters and in proving the equivalence between maximum likelihood and least squares under the assumptions of ANOVA. Feel free to skip this, but come back if you have difficulties understanding how LSE and MLE are used to estimate the parameters of linear regression models in Section 11.2.

Example 11.2: Estimates of multiple population averages

Given a data set from 50 female gannets (ten from each of five colonies) we would like to find the best estimate of ΔI for each population. Intuitively, this should be the average value of ΔI for each population. Let's see if this is confirmed by least squares estimation (LSE – see Section 10.8). If the required LSE estimate for the jth population is written θ_j, then the sum of squared residuals for this data set is

$$SSR = \sum_{j=1}^{5} \sum_{i=1}^{10} (\theta_j - \Delta I_{ij})^2 \tag{11.4}$$

Here, j counts the gannet colonies and i counts the individuals in the sample from each colony. Hence, ΔI_{ij} is the observed change in condition in the ith animal of the jth colony. The LSE values of the parameters θ_i are those that collectively minimise the SSR. We are seeking a total of five parameters ($\theta_1, \ldots, \theta_5$), so we need to minimise the quantity in Equation (11.4) with respect to all five of them. Have a quick look back at Section 4.9 and Section 4.10: To minimise a function with respect to several variables, we need to minimise it with respect to every single one of them. Hence, here we need to calculate and solve the following system of equations:

$$\frac{\partial SSR}{\partial \theta_1} = 0, \ldots, \frac{\partial SSR}{\partial \theta_j} = 0, \ldots, \frac{\partial SSR}{\partial \theta_5} = 0 \tag{11.5}$$

Let's calculate the first of these (the others are done in exactly the same way):

$$\frac{\partial SSR}{\partial \theta_1} = \frac{\partial}{\partial \theta_1} \left[\sum_{j=1}^{5} \sum_{i=1}^{10} (\theta_j - \Delta I_{ij})^2 \right] \tag{11.6}$$

The quantity inside the square brackets only depends on θ_1 when $j = 1$. Therefore, the partial derivatives will be zero for all $j \neq 1$. This simplifies the expression considerably.

$$\frac{\partial SSR}{\partial \theta_1} = \frac{\partial}{\partial \theta_1} \left[\sum_{i=1}^{10} (\theta_1 - \Delta I_{i1})^2 \right] = \sum_{i=1}^{10} \frac{\partial}{\partial \theta_1} (\theta_1 - \Delta I_{i1})^2$$
$$= \sum_{i=1}^{10} 2(\theta_1 - \Delta I_{i1}) = 20(\theta_1 - \overline{\Delta I_1}) \tag{11.7}$$

This will only be zero if $\theta_1 = \overline{\Delta I_1}$, confirming that the best estimate for each population is its average.

The same result can be reached via MLE (Section 10.9). Since we are assuming a normal distribution of independent residuals around the estimates, the likelihood can be written as a product of normal densities:

$$L(\theta_1, \ldots, \theta_5) = \prod_{j=1}^{5} \prod_{i=1}^{10} \frac{1}{\sqrt{2\pi\sigma^2}} e^{-\frac{1}{2} \frac{(\theta_j - \Delta I_{ij})^2}{\sigma^2}} \tag{11.8}$$

Note here that the variances are the same across populations (homoscedasticity assumption). The corresponding log-likelihood is

$$l(\theta_1, \ldots, \theta_5) = \sum_{j=1}^{5} \sum_{i=1}^{10} \left[\ln\left(\frac{1}{\sqrt{2\pi\sigma^2}}\right) - \frac{1}{2} \frac{(\theta_j - \Delta I_{ij})^2}{\sigma^2} \right] \tag{11.9}$$

The MLE values of the θ_i are obtainable by maximising this quantity. A few observations can quickly simplify this expression: of the two parts of the expression in the square brackets, only the second depends on the θ_i. Also, this second part is negative. Therefore, maximising Equation (11.9) is equivalent to minimising

$$l(\theta_1, \ldots, \theta_5) = \frac{1}{2\sigma^2} \sum_{j=1}^{5} \sum_{i=1}^{10} (\theta_j - \Delta I_{ij})^2 \qquad (11.10)$$

The constant $1/2\sigma^2$ that has appeared in front of this expression simply scales the likelihood, so it can be removed. Having made all these simplifications, the resulting expression is exactly the same as the LSE criterion in Equation (11.4). Therefore, the assumptions of normality and homoscedasticity have ensured that the parameter estimates obtained from LSE coincide with MLE.

Having detected differences between populations, and having obtained estimates for the expected within-population values, the next question is what causes these differences. In the gannet example, what is it about some colonies that results in a deterioration of condition (negative $\overline{\Delta I}$) when in others, the condition of gannets is improving? The ecologist's response to this question is to list possible explanations (e.g. variable availability of food, variable predation risk, agonistic effects due to crowding, etc.). The statistician's response is to take these ecological insights and incorporate them in the modelling. If these **explanatory variables** (or **covariates**) are qualitative (e.g. factors such as the type of regional conservation plan for the species, or whether the colony is facing the Atlantic, North Sea or Irish Sea) then the approach continues to be called ANOVA. However, if the explanatory variables are quantitative, the approach is called **regression**.

11.2.   Simple linear regression

Example 11.3: Density dependence of gannet condition

It is postulated that differences (ΔI) in gannet condition between colonies may be explained by density dependence. Therefore, data are collected on the density (P) of birds at the five colonies. The simplest possible relationship between these two quantities is linear, with slope a_1 and intercept a_0

$$\Delta I = a_1 P + a_0 \qquad (11.11)$$

The values of the parameters a_1 a_0 are biologically interesting. For example, we might expect that when population density is zero, the change in condition should be positive $\Delta I = a_0 > 0$. A negative value for a_1 implies that higher densities reduce the ability of birds to improve their condition.

More generally, **simple linear regression** involves a collection of statistical methods for evaluating a linear relationship between a response (or dependent) variable Y and explanatory (or independent) variable X,

$$Y = a_1 X + a_0 \qquad (11.12)$$

In the real world, such relationships are stochastic: the same value of X can give different values of Y. Hence, a **stochastic component** is often added to the end of Equation (11.12),

$$Y = a_1 X + a_0 + \varepsilon \tag{11.13}$$

In simple regression, like in ANOVA, this stochastic component is assumed to be normally distributed with a zero mean and a fixed variance for all values of $X : \varepsilon \sim N(0, \sigma^2)$. An alternative way to think of Equation (11.13) is as a model for the mean of the data. The ith observation in the data set can be thought of as coming from the following distribution

$$Y_i \sim N(a_1 X_i + a_0, \sigma^2) \tag{11.14}$$

There are two fundamental benefits in using this approach over ANOVA: ANOVA requires us to estimate as many expectations as the number of subpopulations in the data (five parameters in the gannet example). In contrast, linear regression can model the expectations of all populations using only two parameters (the slope and intercept). Furthermore, the estimates from ANOVA only refer to the specific populations in the data. Linear regression can be used to predict the expected value of the response variable for unobserved values of the explanatory variable (see Section 11.3). These increases in estimation efficiency and predictive ability are achieved at a price: the assumption that the two variables are *linearly* related. In general, the process of estimating the best parameters for a particular curve on the basis of data is called **model-fitting**. The rationale behind the estimation of the two parameters in linear regression can again be illustrated with the specific example of gannet condition.

Example 11.4: Modelling density dependence of gannet condition

The objective is to obtain values for the parameters a_0, a_1 in Equation (11.11). This can be done by LSE or MLE. For LSE, we need to minimise the same criterion as Equation (11.6), but here the expected population-specific change in condition is replaced by the linear density-dependent model

$$SSR = \sum_{j=1}^{5} \sum_{i=1}^{10} (a_1 P_j + a_0 - \Delta I_{ij})^2 \tag{11.15}$$

Here, the change in condition is subscripted by both individual and population, and the density is subscripted by population. Equation (11.15) can be simplified by pooling all the individuals together so that change in condition (ΔI_i) for the ith individual is contrasted with the density (P_i) experienced by that animal

$$SSR = \sum_{i=1}^{50} (a_1 P_i + a_0 - \Delta I_i)^2 \tag{11.16}$$

The required parameter values a_0, a_1 are those that minimise this expression. MLE gives a similar result. We seek to maximise the following likelihood

$$L(a_0, a_1) = \prod_{i=1}^{50} \frac{1}{\sqrt{2\pi\sigma^2}} e^{-\frac{1}{2} \frac{(a_1 P_i + a_0 - \Delta I_i)^2}{\sigma^2}} \tag{11.17}$$

Using arguments similar to those in Example 11.2, this corresponds to minimising the log-likelihood

$$l(a_0, a_1) = \frac{1}{2\sigma^2} \sum_{i=1}^{50} (a_1 P_i + a_0 - \Delta I_i)^2 \qquad (11.18)$$

This expression is minimised at the same values of a_1, a_0 as Equation (11.16).

In general, therefore, estimating the parameters of the linear regression model in Equation (11.13) requires us to minimise the following quantity

$$f(a_0, a_1) = \sum_{i=1}^{n} (a_1 x_i + a_0 - y_i)^2 \qquad (11.19)$$

where n is sample size and y_i, x_i are the response and explanatory data for the ith sample unit. The partial derivatives of this expression with respect to the two parameters are

$$\frac{\partial f(a_0, a_1)}{\partial a_0} = 2 \sum_{i=1}^{n} (a_1 x_i + a_0 - y_i) \qquad \frac{\partial f(a_0, a_1)}{\partial a_1} = 2 \sum_{i=1}^{n} (a_1 x_i + a_0 - y_i) x_i \qquad (11.20)$$

The MLE and LSE estimates of the two parameters $(\hat{a}_1, \hat{a}_0)$ satisfy the coupled system of equations

$$\sum_{i=1}^{n} (\hat{a}_1 x_i + \hat{a}_0 - y_i) = 0 \qquad \sum_{i=1}^{n} (\hat{a}_1 x_i + \hat{a}_0 - y_i) x_i = 0 \qquad (11.21)$$

Carrying out the algebra (try it) gives the following estimates for slope and intercept:

$$\hat{a}_1 = \frac{\sum_{1}^{n} x_i y_i - n\overline{xy}}{\sum_{1}^{n} x_i^2 - n\overline{x}^2} \qquad (11.22)$$

$$\hat{a}_0 = \overline{y} - \hat{a}_1 \overline{x}$$

where $\overline{x}, \overline{y}$ are the sample averages for the explanatory and response data. Further calculations (not presented here) can be carried out to obtain standard errors and confidence intervals for these parameters. These calculations rely heavily on the underlying normality assumption about the distribution of points around the line. Standard errors can be used to construct t-tests for the slope and the intercept. A t-test for the slope is particularly informative because it tests the H_0 that $a_1 = 0$. Rejection of the H_0 implies a nonzero slope, and therefore a linear relationship between the explanatory and response variables. Thankfully, all of the above tasks can be carried out by R.

ℝ 11.2: Fitting a linear model to data

ANOVA is really a special version of linear regression and that is why the command `lm()` used for fitting a linear model was also used in R11.1. The syntax is the same as before: a formula is declared within `lm()` which specifies the names of the response and explanatory variables. These names are taken from the column headings of the data frame. For example, a model of gannet condition change as a function of colony density can be created with the following line:

```
mod<- lm(Condition~Density,dat)
```

The resulting model object is stored in `mod` (or pick any other name that suits you) and the properties of this object can be explored as follows:

```
> summary(mod)

Call:
lm(formula = Condition ~ Density, data = dat)

Residuals:
    Min      1Q  Median      3Q     Max
-40.612 -14.240  -3.571  15.169  55.241

Coefficients:
            Estimate Std. Error t value Pr(>|t| )
(Intercept) 36.48166    6.17323   5.910 3.44e-07 ***
Density     -0.72160    0.09389  -7.686 6.59e-10 ***
---
Signif. codes: 0 '***' 0.001 '**' 0.01 '*' 0.05 '.' 0.1 ' ' 1

Residual standard error: 19.82 on 48 degrees of freedom
Multiple R-squared: 0.5517,    Adjusted R-squared: 0.5424
F-statistic: 59.07 on 1 and 48 DF, p-value: 6.59e-10
```

This rich output needs to be interpreted, one piece at a time:

- `Call`: confirms the formula of the model.
- `Residuals`: gives information about the distribution of the difference between observed and fitted values. Sometimes the linear model will be above the observations (+ve residual) and sometimes it will be below (-ve residual). In this example, the greatest overestimate of ΔI is 55.241 units over the observation for that particular bird. The inter-quartile range of the residuals can be obtained as $3Q-1Q$ $(= 29.409)$.
- `Coefficients`: a table whose rows carry information about each of the model's parameters (the model intercept and the coefficient of gannet density). The column `Estimate` gives the values of the parameters as obtained from Equations (11.22). Here, the linear model suggests a negative relationship between density and change in condition. The next four columns in this table deal with hypothesis testing about the two parameters. The estimated standard errors for each of the two parameters are used to calculate t values and then compared with a t-distribution to find p values. Both parameters have highly significant p values (as indicated by the star significance codes). Biologically, the conclusion of a negative coefficient a_1 means that breeding gannets will be worse off at crowded colonies. Also, because we are confident that the intercept is positive, we can say that at low gannet densities, birds will gain condition. The conclusion that density is negatively related with condition can be interpreted in two ways: either high densities diminish condition (via crowding and competition) or declining colonies are those whose members are failing in their effort to accumulate condition (and therefore suffer lower fecundity and survival).

The remaining output of `summary()` details how well the model fits the data. We will return to this issue in Section 11.4.

11.3. Prediction

Plotting the linear model with the estimated parameters gives a **best-fit line** through a scatter plot of the data. This line is defined for all values of the explanatory variable and can therefore be used to predict the expected value of the response for values of the explanatory variable that were not originally in the data. If prediction is carried out for values within the observed range of the explanatory variable, then it is called **interpolation**. Predicting the response outside the range of observed explanatory data is called **extrapolation**. For a given value of the explanatory variable, these predictions will be subject to two types of uncertainty: first, uncertainty in mean response, as conveyed by the uncertainty in the estimated values of the regression parameters; second, variability in the actual values observed around this mean. We can visualise different types of uncertainty using **confidence and prediction intervals**. These are bands around the estimated mean, running the range of the explanatory variable.

Example 11.5: Predicting density-dependent changes in gannet condition

There are two ways to plot uncertainty in the predictions of the linear model for gannet condition changes. Confidence intervals (Figure 11.2(a)) represent uncertainty in the estimates of the model's parameters. If the regression line were to be perturbed according to this uncertainty (i.e. shifted up and down by nudging the intercept, and pivoted around its midpoint by altering the slope), the movements would chart a band that is narrower towards the middle of the observations and wider towards the extremes.

Prediction intervals (Figure 11.2(b)) represent parameter uncertainty combined with the additional stochasticity around the mean due to chance individual variation. Because variables other than density may act to determine the changes in condition observed for any one gannet, we would still get some variation around the trend even if this was known precisely. Since they account for both sources of uncertainty, prediction intervals are always wider than confidence intervals. This is also the reason why 95% confidence intervals do not generally encompass 95% of the data points (see Figure 11.2(a)).

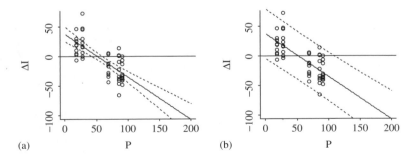

Figure 11.2: (a) Confidence intervals for the estimated mean response; (b) prediction intervals for particular observations around the estimated mean.

Therefore, the two parts of Figure 11.2 are tackling two different questions: if you were to visit a colony of known density, and measured ΔI from a large sample of individuals, there is a probability of 0.95 that the average of this sample would be within the band shown in

---

Figure 11.2(a). However, if you were to take a random animal from a colony having a given density, then the value of ΔI for that animal would fall within the band of Figure 11.2(b) with a probability of 0.95.

 11.3: Prediction and confidence intervals

Generating predictions from a model object requires the values of the independent variables for which the predictions are sought. These values must be passed as new explanatory data to the command `predict()` as follows:

```
dens<-seq(0,200,1) # Generates a vector of densities from 0 to 200

# Create data frame with the new explanatory data
datNew<-data.frame("Density"=dens)

# Create predictions from model object mod at the positions
# of datNew and store them in the list preds
preds<-predict(mod, datNew)
```

To generate a simple plot of the data with the best fit line use the following:

```
plot(dat$Density, dat$Condition) # Scatter plot of data
lines(dens, preds) # Plots best fit line
```

To create a plot with confidence intervals (such as the one in Figure 11.2(a)), use the interval option inside `predict()`. This generates a matrix of three columns, one for the best fit line and two for its lower and upper confidence curves.

```
preds<-predict(mod, datNew, interval="confidence")
plot(dat$Density, dat$Condition)
lines(preds[,1], lty=1)
lines(preds[,2], lty=2)
lines(preds[,3], lty=2)
```

To generate a plot with prediction intervals (like the one in Figure 11.2(b)), specify the option `interval="prediction"`.

11.4.   How good is the best-fit line?

Linear regression assumes that the relationship between the explanatory and the response variables is linear, that the residuals around this line are independent, normally and identically distributed. If there is a strong linear signal in the data and all other assumptions are satisfied, then the estimated parameters and confidence intervals can in good conscience be used for prediction. However, when the simple linear model fails for one or more reasons, then it may need to be extended. Therefore, this section can be treated as a springboard to the rest of the chapter because it reviews the different ways in which a linear model can fail and refers you to appropriate extensions introduced in later sections.

The first step in the diagnostic process is to decide if the model is any good at all. A measure of the goodness of fit is provided by the **coefficient of determination**, the squared correlation between the observed and estimated response values $r^2 = corr(y, \hat{y})^2$. This takes values between 0 and 1, with values closer to 1 indicating a good fit. Alternatively, r^2 can be interpreted as the **proportion of explained variance** (e.g. a linear model with $r^2 = 0.93$ accounts for 93% of the variance in the raw data).

 11.4: Coefficient of determination

When fitting a linear model the summary of the output contains the value of the coefficient of determination under the entry `Multiple R-squared`. Have a look at the output in R11.2. The r^2 value is a relatively low 0.55, indicating that there may be more that can be done with this model.

If the r^2 value is low, then something is certainly amiss with the model, but, even if it is high, there could be problems lurking. Hence, the following diagnostics should be carried out regardless. Most of the diagnostics make use of residuals obtained by subtracting the estimated response value from the corresponding observation: $e_i = y_i - \hat{y}_i$. Any disagreements between the model and the data will be manifested in the residuals but it is hard to know what constitutes a large disagreement (and, hence, one worth worrying about). Therefore, raw residuals are often replaced by **standardised residuals**, which are designed to look like an independent sample from $N(0, 1)$ if all the assumptions of linear regression are satisfied. This offers a good basis for diagnosing problems with each assumption:

- *Linearity:* In a plot of raw residuals against the estimated values (e.g. Figure 11.3(a)), deviations from linearity appear as consistent patterns of overestimation or underestimation. These patterns may even indicate how the linear model should be modified to better describe the data. If you think that you should be fitting a nonlinear model to your data, then Sections 11.7–11.11 are of particular relevance.
- *Homoscedasticity:* The assumption of equal variance can be examined by inspecting a plot of the residuals (raw or standardised) against fitted values (Figure 11.3(a)). If, for example, it appears that there are bottlenecks in the scatter of residuals around the mean, then this could indicate non-constant variance. A more formal diagnostic is provided by an **ncv test** (see R11.5) of the null hypothesis of homoscedasticity. If you conclude that your model suffers from this pathology, there are extensions that enable you to explicitly model changes in the variance jointly with the mean. You can find more in Section 11.9 and Faraway (2006, Section 7.3).
- *Normality:* This can be examined using the Q-Q plot of the residuals (Figure 11.3(b)). The Q-Q plot was introduced in R10.9. A Shapiro–Wilk test can be used to get a p value for the hypothesis of normality (again, see R10.9). If you know that your residuals are non-normal, then Sections 11.7–11.8 offer useful generalisations to the linear model.
- *Outliers:* Sometimes, data collection or data entry go wrong, generating outliers that have a disproportionate effect on the parameters of the resulting model. Cook's distance is a metric which measures the effect on the resulting fit of removing each point in the data. The rule of thumb is that Cook's distance values exceeding 1 should make you go back to the data and re-examine the circumstances in which that observation was collected. Plotting Cook's distances for all points (Figure 11.3(c)) is sufficient as a first measure. If outliers are present but no suspicion lies with the data collection process, then perhaps normality is also violated (see above) or the residuals are overdispersed (see Section 11.9).
- *Independence:* Dependence between observations can take many different forms. For example, spatial autocorrelation (see Example 2.4 and R2.1) may occur because neighbouring gannets in a given colony tend to perform similarly to each other. If the sample

is obtained from a localised group of birds, then it may give a misleading impression about changes in condition across the colony. Similarly, **temporal** or **serial autocorrelation** occurs when successive observations are more similar (or dissimilar) to each other than the average similarity in the sample. For example, the position of a moving animal after 5 min will not be totally independent of its present position. If you have information on the spatial or temporal structure of the data, then plot the residuals in order of spatial or temporal proximity. Better still, plot a correlogram of the residuals as a function of time or distance. If there is evidence of serial correlation in your residuals, then Section 11.13 has some relevant material.

Example 11.6: Diagnostics for the density dependence model

Figure 11.3 shows three diagnostic plots derived from the model of R11.2.

- *Linearity:* The residuals in Figure 11.3(a) do not indicate systematic biases (i.e. consistently low or high residuals in particular regions of the fitted values axis).
- *Homoscedasticity:* It is hard to tell with only five values of population density, but the scatter of residuals in Figure 11.3(a) does not appear to vary as we move from left to right.
- *Normality:* The Q-Q plot and superimposed Q-Q line in Figure 11.3(b) are typical of a sample of normally distributed residuals.
- *Outliers:* All data points have Cook's distances well below 1 (Figure 11.3(c)).

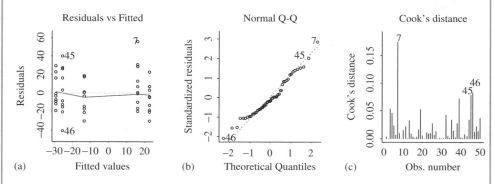

Figure 11.3: (a) Raw residuals $(y_i - \hat{y}_i)$ against the fitted values $(\hat{y}_i)$; (b) Q-Q plot of standardised residuals against the corresponding quantiles from a normal distribution $N(0, 1)$; (c) Cook's distances for all 50 observations.

As Example 11.6 illustrates, it is entirely possible for the diagnostics to reveal no problem with the assumptions and yet have poor model fit (as evidenced by r^2). This may mean that we have not offered enough information to describe the observed patterns. Specifically, it may indicate that one or more important explanatory variable has been missed out. This is the subject of the next section.

11.5: Linear model diagnostics

For a given model object `mod`, the standardised residuals are given by `rstandard(mod)`. The homoscedasticity test is carried out by the command `ncv.test(mod)`, found in the library car. The command `plot(mod, which=c(1,2,4))` will generate a sequence of plots like the ones shown in Figure 11.3.

11.5.   Multiple linear regression

Three influences determine the value of a particular measurement (Section 8.1): conditions, covariates and stochasticity. Conditions are considered known and fixed. Once a covariate has been used to explain away some of the variability in the data, the remaining variability may be called stochasticity. Alternatively, it may be attributed to another covariate that operates together with the first.

Example 11.7: Combined effects of density and the environment

In addition to population density, there are a great number of other, colony-specific characteristics that may explain variations in the ability of gannets to improve their condition. Some of these may be quantitative (e.g. density of competitors in the region), others qualitative (e.g. immediate proximity to Atlantic Ocean, Irish Sea, North Sea). There is no *a priori* reason to assume that changes in condition are only affected by one covariate, so we need to model their combined effect.

The simplest form of multiple linear regression examines the effects of n different covariates $(X_1, X_2, \ldots, X_n)$ in an additive fashion

$$Y = a_0 + a_1 X_1 + a_2 X_2 + \cdots a_n X_n + \varepsilon \tag{11.23}$$

So, now the signal is represented by the sum of terms, each involving a covariate and its coefficient. Stochasticity is represented by the stochastic component ε, a random variable with distribution $N(0, \sigma^2)$. An alternative interpretation is that the response data come from the following distribution

$$Y \sim N(\mu, \sigma^2) \text{ where } \mu = a_0 + a_1 X_1 + a_2 X_2 + \cdots a_n X_n \tag{11.24}$$

All of the assumptions of simple linear regression are carried over and estimation follows the familiar path of ❶ writing the likelihood or least squares criterion, ❷ setting to zero all partial derivatives with respect to the parameters $a_0, a_1, a_2, \ldots, a_n$ and ❸ solving the resulting system of linear equations.

Example 11.8: Combined effects of density and the environment

To get a better idea of what affects changes (ΔI) in the condition of gannets, data are needed from more colonies. A total of 100 birds are sampled (ten birds from each of ten colonies). For each colony, there are records of gannet density (P) and the density (M) of marine mammals (White-beaked dolphins *Lagenorhynchus*

albirostris and Harbour porpoises *Phocoena phocoena*) in the feeding grounds associated with each colony. Both of these explanatory variables are included in the following multiple regression model

$$Y = a_0 + a_1 P + a_2 M + \varepsilon \tag{11.25}$$

Fitting the model to this particular data gave the parameters $a_0 = 25.96, a_1 = -0.81, a_2 = 0.23$.

Equation (11.25) describes a plane in three dimensions that can be drawn using the parameters estimated for the model (Figure 11.4) and visually compared with the data for the ten colonies. The inclination of this plane can be interpreted biologically: The declining condition with increasing conspecific density points to the detrimental effect of intra-specific competition. The positive slope with marine mammal density is more intriguing. It is explained by the observation that gannets have a commensalistic relationship with marine mammals. Porpoises and dolphins operate as 'beaters', bringing prey closer to the surface (Camphuysen and Webb, 1999). Hence, according to this data set, gannets are expected to gain condition most rapidly when they live in uncrowded colonies that are close to the foraging grounds of marine mammals.

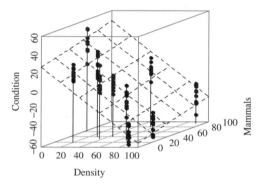

Figure 11.4: The response and covariate data plotted as dots in 3D space. The tilted plane is the graph of the estimated model which describes how condition is expected to change in response to prevailing gannet and marine mammal densities.

11.6: Fitting multiple regression models

The material presented in R11.2 is enough to get started with fitting multiple regression models. For example, if you have a data frame (`dat`) which contains columns for bird condition (`Condition`), colony density (`Density`) and the density of marine mammals (`Mammals`), then the model can be estimated, and detailed output produced, with two lines of code:

```
> mod<- lm(Condition~Density+Mammals,dat)
> summary(mod)

Call:
lm(formula = Condition ~ Density + Mammals, data = dat)

Residuals:
      Min        1Q    Median        3Q       Max
-25.75748  -5.39267   0.07778   5.36553  23.71727
```

```
Coefficients:
            Estimate Std. Error t value Pr(>|t|)
(Intercept) 25.95908    2.36954   10.955 < 2e-16 ***
Density     -0.81328    0.03056  -26.617 < 2e-16 ***
Mammals      0.22949    0.03502    6.553 2.71e-09 ***
---
Signif. codes: 0 '***' 0.001 '**' 0.01 '*' 0.05 '.' 0.1 ' ' 1

Residual standard error: 9.163 on 97 degrees of freedom
Multiple R-squared: 0.8846,   Adjusted R-squared: 0.8822
F-statistic: 371.6 on 2 and 97 DF, p-value: < 2.2e-16
```

Note how the addition of some more data and a second covariate has led to an improved r^2 value (0.88 compared to the 0.55 of the single covariate model in R11.2). The p values listed under the heading `Pr(>|t|)`, test the null hypothesis of no relationship between the response and each explanatory variable.

To obtain mean predictions from this model for new data, use the approach of R11.3. Plots of these predictions with respect to two covariates (e.g. Figure 11.4) can be generated using the library `scatterplot3d` as follows:

```
require(scatterplot3d)
attach(dat)
s3d <-scatterplot3d(Density,Mammals,Condition, type="h")
s3d$plane3d(mod)
detach(dat)
```

Finally, never forget to run diagnostics for your model (see Section 11.4). The code of R11.5 can be used intact, with models of more than one covariate.

11.6. Model selection

When trying to account for patterns in a set of observations, a good biologist will come up with a wealth of possible explanatory variables. A good modeller will try to prune these back to a minimal set.

Example 11.9: Should the beating of a butterfly's wings be used to explain gannet condition?

In Example 11.6, we found that conspecific and heterospecific density can account for 88% of the observed variability in gannet condition change. The remaining 12% could be due to any list of plausible characteristics, such as the various biotic and abiotic attributes of each colony and their surrounding habitats, or the individual characteristics (age, size, genetic makeup, etc.) of each gannet in the sample. This should help the model account for more of the variability in the data. Carrying on in this way, we may try some less plausible possibilities, such as the indirect ecosystem effects of species that are separated from gannets by two or more trophic levels. In the end, we might consider far-fetched ideas such as the effect of particular pollutants that are not toxic to seabirds. Eventually, we are guaranteed to come up with a model that accounts for 100% of the variability in the data. Surely, this is a good result?

The answer to this question is an emphatic 'no'. Philosophically, it is the wrong thing to do because it goes against the principle of parsimony (look back at Example 2.1). In the context of regression models, adding all conceivable covariates is a practical concern because most data sets carry only a limited amount of information about the study system. It is foolhardy to believe that we can perfectly explain changes in the condition of a sea bird species by observing 100 individuals from ten colonies.

Why, then, will the r^2 value continue increasing with the inclusion of covariates of ever-decreasing plausibility? The crucial fact here is that regression models are empirical: they describe patterns, not mechanisms (perhaps you have heard the saying 'correlation is not causation'?). With each additional covariate, the regression model simply becomes more flexible, and by playing one covariate against the other it can fit the data exactly, but this does not mean that all these variables are biologically important or even relevant. Such models are called **overparameterised** or **overfitted** and they have one lethal flaw: they fit the observations really well but predict appallingly in new situations.

We must, therefore, come to terms with the fact that some of the residual variability after fitting a model is due to ❶ variables that we haven't thought of, ❷ variables we don't have data for, ❸ measurement errors or ❹ violations of the model's assumptions not picked up by the diagnostics of Section 11.4. Deciding which variables to keep and which to drop is a major aspect of **model selection**.

The traditional approach to this question was to use the p values generated automatically by model-fitting software (e.g. see summary output in R11.6). If, in the summary of the model, all covariates had low p values (usually annotated with one to three stars), then the model would remain intact. If one or more covariates appeared with high p values, then one of them would be dropped and the model re-fitted. The process would continue until all covariates had stars (one, two or three depending on the significance level desired by the analyst).

This approach is a poor basis for model selection (Burnham and Anderson, 2002) because the levels of significance are arbitrary, and due to the problems associated with multiple nonindependent hypothesis tests (see Section 11.1). For these reasons, model selection by p values is liable to retain irrelevant covariates or drop important ones. So, the star system in computer output (e.g. R11.6) should be used as a rough guide during the exploratory stages of the analysis, not for model selection.

When fitting a model by LSE, a better approach looks at the balance between sample size (n) and the number of coefficients (K). It is generally poor practice to use many covariates relative to the number of observations in the data. This is easy to understand by thinking about simple linear regression: if the data have a single observation, then a linear model that contains both a slope and an intercept cannot be specified because infinite lines can be drawn through a single point. Thus, a good model is one that achieves a high r^2 using few covariates relative to the sample size. This gives rise to the **adjusted coefficient of determination**

$$\text{adjusted } r^2 = 1 - (1 - r^2)\frac{n-1}{n-K} \tag{11.26}$$

Given two models with the same r^2, the adjusted r^2 takes higher values for the model with fewer parameters. This quantity is estimated automatically by software packages (see output in R11.2 and R11.6). To employ it for model selection you need to estimate all the models you would like to consider and select the one with the highest adjusted r^2.

This idea is carried over to a much broader class of model selection criteria by determining quality of fit on the basis of likelihood rather than residuals. The approach leads to various criteria that balance the likelihood against the number of parameters (K) in a particular model. The best known of these is the rather cryptically named **Akaike Information Criterion** or **AIC**,

$$AIC = -\{2l(\text{model} \mid \text{data}) - mK\} \tag{11.27}$$

In which $m = 2$ and the term $l(\text{model} \,|\, \text{data})$ is the maximum log-likelihood of a model under the observations. As this quantity increases, the fitted model passes closer to the data. However, if this is achieved at the price of an increased numbers of parameters, the value of the criterion is penalised by the number of parameters (K) in the model. The negative sign in front of Equation (11.27) means that the best model in the set is the one with the smallest value of AIC. The name of the criterion is in honour of its proposer (Hirotugu Akaike) and the branch of mathematics (**information theory**) used to prove that, asymptotically (i.e. in the long run), this expression offers the best balance between model fit and predictive ability. Other information criteria, using slightly different derivations, result in different values for m.

Using information criteria is preferable to adjusted r^2 values because they are more general and reliable. This is particularly useful to remember when the two methods disagree.

Example 11.10: Model selection by adjusted r^2 and AIC

Four different models are fit to the data to examine the possible improvements in model quality brought by introducing weather variables (annual averages of temperature and rainfall) for each colony. The covariates in each model and their associated adjusted r^2 and AIC are listed in Table 11.1 (the best models according to each criterion are shown in boldface).

Table 11.1

Model	adj r^2	AIC
1 Gannet dens+Mar mam dens	0.882	**731.776**
2 Gannet dens+Mar mam dens+Rain	**0.883**	732.083
3 Gannet dens+Mar mam dens+temp	0.881	733.361
4 Gannet dens+Mar man dens+Rain+Temp	0.882	733.988

We are therefore faced with the (not atypical) situation of two model selection techniques that do not entirely agree with each other. As it happens, this example is based on made-up data in which values for temperature and rain are random numbers with no effect on condition, so AIC has got closer to the underlying truth.

Information criteria rely on asymptotic arguments, meaning that they are guaranteed to work in the majority of studies and as sample size increases. It is, however, important to check, for any particular study, whether the model selection process has yielded the best predictive model. This can be achieved by re-sampling techniques such as **cross-validation**. The idea of cross-validation is to keep aside a part of the data (e.g. data from one of the ten gannet colonies), fit the candidate models to the rest of the data and then use them to predict the missing part. Since there is no particular reason for selecting one part of the data over the others for this purpose, the whole process needs to be repeated for all parts (e.g. all ten colonies, one at a time). The model that best manages to predict the hidden data is the one that should be chosen for fitting to the entire data set. This approach certainly gets to the heart of the problem by trying to maximise the predictive power of the chosen model but it is computer-intensive and thus impractical for some studies. One possibility is to use an AIC-derived ranking to arrive at a confidence set of models (a small set of highly promising models), and then use cross-validation to select between them. A more robust alternative is to keep the best of all worlds by generating **model-averaged predictions** from this confidence set of models (see Burnham and Anderson, 2002).

One particularly thorny problem often resolved by model selection methods is **collinearity**. Imagine that you have two candidate covariates that, within the range of observed values, are closely and linearly related to each other.

Example 11.11: Collinearity in covariates of gannet condition

The ability of different gannets to acquire condition will almost certainly depend on their individual characteristics, such as age and body size. For any given bird, these are not independent of each other. Furthermore, different measures of body size such as length, girth and weight may be exactly or approximately proportional to each other. These characteristics contain similar information, so using them all together in a regression model uses up valuable parameters.

Models with collinear covariates tend to be less robust than models with linearly independent ones: consider a model $Y = a_0 + a_1X_1 + a_2X_2$ with two covariates (X_1, X_2) that are approximately proportional to each other (i.e. $X_2 \cong cX_1$). Instead of being arranged across an area in the X_1, X_2 plane, the explanatory data are arranged closely around the line $X_2 = cX_1$. We are therefore trying to fit a model with two covariates (a plane) to a set of points that are, in fact, one-dimensional. Metaphorically, it is easier to lay a flat sheet of metal on top of a bed of nails rather than trying to balance it on the teeth of a rake.

Model selection will often manage to yield a reduced set of covariates, keeping the ones with the most explanatory power. However, if you are worried that collinearity may still be a problem for your model, you may want to read more about detection with **variance inflation factors** (Fox, 2002) which can also deal with the more general problem of **multicollinearity** (i.e. one covariate being a linear function of two or more covariates).

11.7: Manual and automated model selection

Given a small set of candidate models, manual model selection is feasible. The four models in Example 11.10 can be estimated and compared (on the basis of AIC) as follows:

```
> mod1<-lm(Condition~Density+Mammals,dat)
> mod2<-lm(Condition~Density+Mammals+Rainfall,dat)
> mod3<-lm(Condition~Density+Mammals+Temperature,dat)
> mod4<-lm(Condition~Density+Mammals+Temperature+Rainfall,dat)
> AIC(mod1, mod2, mod3, mod4)
     df      AIC
mod1  4 731.7764
mod2  5 732.0837
mod3  5 733.3605
mod4  6 733.9877
```

Automated model selection, an exploratory search through a large set of models, may be carried out using any of the following three methods: **forward selection** starts with an intercept-only model and examines if any of the covariates (added one at a time) improves the AIC. If so, then that model is used to construct all two-covariate models by adding the remaining covariates one at a time, and so on until none of the new models can improve on the AIC. **Backward elimination** begins with a model that contains all covariates and drops them, one at a time, using AIC to decide if these changes result in model improvement. **Stepwise selection** is a combination of the other two methods. It drops variables one at a time but, with every step, it tries to re-introduce some of the variables that were rejected in previous iterations of elimination, just in case this yields an even better (lower) AIC. To carry out stepwise selection in R, first estimate the full model and use it as the input to the command `step()`

```
mod<-lm(Condition~Density+Mammals+Temperature+Rainfall,dat)
step(mod)
```

The output of `step()` can be quite verbose because it prints a summary of every model that it has tried out. The very last model described is the one selected.

11.7.   Generalised linear models

Example 11.12: Many response variables are constrained

Loggerhead turtle, Caretta caretta

So far, all the examples in this chapter have modelled variations in the animals' ability to improve their condition. The ultimate goal of many population studies is to link short-term behavioural or physiological responses with long-term performance; i.e. fitness. The fitness of an individual can be broken down into the demographic components of survival and reproduction. Extended further, the concept of inclusive fitness also examines the ability of an individual's offspring to be recruited into the breeding population. However, unlike changes in condition, which can theoretically take any values from $(-\infty, \infty)$, the component variables of fitness are constrained: a turtle has a probability of survival in the range [0,1], its reproductive output cannot be less than zero and the number of its offspring recruited can only range between zero and the number of offspring it produced. The values taken by these variables within their allowed ranges will depend on various environmental influences, so it would be interesting to create regression models linking any one attribute of fitness to its environmental covariates. However, linear regression is patently inappropriate for the task because linear functions are unconstrained.

Transformations of the response data are a rather neat trick, historically used to shoe-horn nonlinear data into the constraints of the linear regression framework. If you look back at Figure 3.19 in Example 3.16, you will notice how data on population growth (a non-negative variable in that example, defined as the ratio of population size in two successive years) were converted into a linear arrangement by a log-transformation.

Example 11.13: Log-transforming fecundity data

How does a turtle's ability to reproduce at the end of a year depend on the availability of forage in the months leading up to the breeding season? If Y is a random variable describing per capita fecundity (number of eggs produced in breeding season)

and $\mathbf{y} = \{y_1, \ldots, y_n\}$ is a data set of fecundity measurements from n turtles, then a plot of these data on a log-scale may reveal a linear pattern (Figure 11.5).

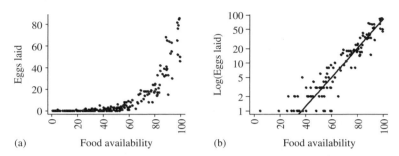

Figure 11.5: (a) Counts of eggs laid by 200 Loggerhead turtles plotted against an index of food availability and (b) the same data plotted on a log scale for the eggs axis, with an indication of the emergent linear pattern.

This example reveals two important problems with log-transforming the data. The first is that real data often contain zeros and log(0) is not defined mathematically. The simulated example in Figure 11.5 contained 76 zeroes, meaning that 38% of the data were thrown away simply by going from Figure 11.5(a) to Figure 11.5(b). The second problem is that fitting a line to the scatter plot in Figure 11.5(b) violates the homoscedasticity assumption, one of the cardinal requirements of linear regression.

One clue to the solution of the problem is to understand why zero counts occur in the first place. Is it impossible for an undernourished turtle to have any eggs at all, or is it just unlikely? Reductions in the availability of forage will diminish the expected number of eggs produced. Even though this expectation will not itself be zero, it will lead to several turtles not producing eggs. In other words, instead of transforming the data, we may transform their expected value and then model the data as arising from some random process around this expectation.

Example 11.14: Modelling count data

Y and $\overline{Y}$ are, respectively, the actual and expected number of eggs laid by a turtle living under certain conditions. In this example, where the modelled variable is a count, the Poisson distribution may be used to represent the stochasticity of Y around $\overline{Y}$. Furthermore, if X is a covariate of Y (e.g. forage availability), we may model the dependence of the mean number of eggs on X by a suitably transformed linear model. In short,

$$Y \sim Poisson(\overline{Y})$$
$$\overline{Y} = \exp(a_0 + a_1 X) \qquad (11.28)$$

Here, the exponential transformation of the linear model ensures that the mean number of eggs does not become zero or negative. Equivalently, the transformation can be seen as a log-transform of the mean number of eggs ($\ln(\overline{Y}) = a_0 + a_1 X$).

This is an example of a **generalised linear model (GLM)**. It is 'generalised' in the sense that it can deal with both constrained and unconstrained response variables. It is 'linear' because the expectation of the response variable is modelled as a transformation of a linear model. GLMs usually appear in the following form

$$Y \sim \text{Distribution}(\overline{Y}, \theta)$$
$$\overline{Y} = h(a_0 + a_1 X_1 + a_2 X_2 + \ldots)$$

(11.29)

The GLM therefore has three components, the distribution (or **stochastic component**) which describes the stochasticity of the data, the **linear predictor** $(a_0 + a_1 X_1 + a_2 X_2 + \ldots)$ which introduces the effect of covariates on the mean of the distribution and some **link function** (h) which transforms the mean of the distribution so that it can be modelled by a linear predictor. Apart from the mean ($\overline{Y}$), the distribution of the GLM may also depend on additional parameters (θ). In the case of a count variable such as the number of eggs, the Poisson is a suitable distribution and the appropriate link function is ln(). The resulting GLM is called **log-linear**.

It is helpful to acquaint yourself with the three components of the GLM (stochastic component, linear predictor, link) by recognising them in the familiar linear regression model.

Example 11.15: Seeing the linear model as a special case of the GLM

Under simple regression, we model the mean of the normal distribution directly by a linear model (see also Equation (11.14))

$$Y \sim N(\overline{Y}, \sigma^2)$$
$$\overline{Y} = a_0 + a_1 X_1$$

(11.30)

Here, the additional parameter is the variance σ^2, and the mean is untransformed. In GLM terminology, the link of this model is the **identity function**, i.e. the mean of the stochastic component is the same as the linear predictor.

As with all statistical modelling, the objective is to estimate the parameters of the linear predictor from a sequence of response observations. In this task, estimation by maximum likelihood really proves its usefulness.

Example 11.16: Likelihood for a log-linear GLM

For a particular expected value $\overline{Y}$, the probability of y eggs being laid is given by the PMF of the Poisson distribution (see Section 9.10)

$$f(y \mid \overline{Y}) = \frac{e^{-\overline{Y}} \overline{Y}^y}{y!}$$

(11.31)

If y_i is the number of eggs laid by the ith turtle in the sample, and $\overline{Y}_i = \exp(a_0 + a_1 X_i)$ is the expected value for that particular turtle, considering its access to food (X_i), then the probability associated with the observation of that turtle is

$$f(y_i \mid a_0, a_1) = \frac{e^{-\overline{Y}_i} \overline{Y}_i^{y_i}}{y_i!}$$

$$\overline{Y}_i = \exp(a_0 + a_1 X_i)$$

(11.32)

Notice that now the probability is conditional on the parameters of the regression model. Assuming that the fecundity of different turtles in the sample is independent, the probability of the data under a set of regression coefficients is

$$f(y_1, \ldots, y_n \mid a_0, a_1) = \prod_{i=1}^{n} \frac{e^{-\overline{Y}_i} \overline{Y}_i^{y_i}}{y_i!}$$

$$\overline{Y}_i = \exp(a_0 + a_1 X_i)$$

(11.33)

We can now turn this around (see Section 10.9) to obtain the likelihood of any pair of parameter values being true, given the data

$$L(a_0, a_1 \mid y_1, \ldots, y_n) = \prod_{i=1}^{n} \frac{e^{-\overline{Y}_i} \overline{Y}_i^{y_i}}{y_i!}$$

$$\overline{Y}_i = \exp(a_0 + a_1 X_i)$$

(11.34)

The task is now to estimate the parameters a_0, a_1 by maximising the following log-likelihood

$$l(a_0, a_1 \mid y_1, \ldots, y_n) = \sum_{i=1}^{n} \{-\overline{Y}_i + y_i \ln \overline{Y}_i - \ln(y_i!)\}$$

$$\overline{Y}_i = \exp(a_0 + a_1 X_i)$$

(11.35)

The awkward term $-\ln(y_i!)$ inside the sum does not depend on the parameters a_0, a_1 and can therefore be dropped from the log-likelihood because it will not affect the position of the maximum. Note that the same likelihood function can be written for more than one covariate. So, if we wanted to extend the investigation to environmental influences other than food (e.g. human disturbance near the beach, pollutants, etc.) we could just extend the linear predictor to include these additional covariates, just like we did for multiple linear regression (Section 11.5).

Maximising the likelihood (see R11.8) will yield MLE parameter values. We can place these back into the model for the mean number of eggs $\overline{Y} = \exp(\hat{a}_0 + \hat{a}_1 X)$ and plot it along with the data (Figure 11.6).

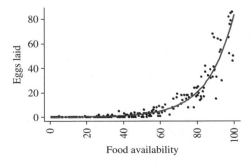

Figure 11.6: Raw data on turtle egg production and the associated best-fit log-linear model.

11.8: Fitting a and predicting from a log-linear GLM

The central command for fitting GLMs of all types is `glm()`. At a minimum, it requires the model formula, the name of the data frame from which the data are to be taken (specified via the option `data`) and the stochastic component of the GLM (specified via the option `family`). For example, given the two vectors `eggs` and `food`, containing the data in Figure 11.6, the GLM can be estimated as follows

```
dat<-data.frame(food, eggs)
mod<-glm(eggs~food, family=poisson, data=dat)
```

The name `mod` now holds the GLM object. You can get more information about the estimated coefficients by typing `summary(mod)`. To find out what are the fitted values for the response variable (i.e. the expected value of the response for the observed values of the explanatory variable), then type `fitted(mod)`. To predict for new values of the explanatory variable(s), use the command `predict(mod, newdata)`, where `newdata` is a data frame containing the explanatory values for which predictions are required. As an illustration, the curve in Figure 11.6 can be produced as follows

```
newdat<-data.frame("food"=seq(1,100)) # Food availabilities
preds<-predict(mod, newdat, type = "response")
plot(food, eggs, xlab="Food availability", ylab="Eggs laid")
lines(preds)
```

The option `type` inside the command `predict` tells R that predictions are required at the scale of the response variable, not the linear predictor. The alternative option, `type="link"`, generates predictions on the scale of the linear predictor. This is useful if untransformed predictions are required (e.g. this is how I produced the trend line shown in Figure 11.5(b)).

Fecundity is an example of a demographic variable that is constrained below by 0. Modelling response variables that are constrained both above and below is also possible by choosing the appropriate stochastic component and link function. In the simplest case, the response (Y) may be a binary variable, taking only the values 1 and 0, representing success or failure (see Section 9.7).

Example 11.17: Modelling senescence in turtles

 We might be interested in examining how the probability of survival of adult turtles changes with their age. The data comprise a random sample of 200 turtles whose age can be established at the beginning of the year and whose death can also be ascertained with little error.

Even if there is no senescence in the species, the frequency distribution of ages in the sample will not be uniform (Figure 11.7(a)): even if the annual probability of survival is the same for all ages (say, $p < 1$), the proportion of animals surviving to 10 years is larger than the proportion surviving to 50 ($p^{50} < p^{10}$).

The response data will be binary, taking the value 1 if a turtle survives to the end of the year and 0 if it doesn't. The explanatory data are the ages of the turtles at the start of the year. It might be anticipated that, in the presence of senescence, we would observe more deaths at higher ages but if we plot the response against the explanatory data, we get an incredibly uninformative plot (Figure 11.7(b)). The fact that sample size declines with age means that the dots in Figure 11.7(b) become sparser as we move from left to right. This pattern applies to the survivals and deaths equally, making it hard to decide by eye whether the density of zeroes is higher than the density of ones towards the right of the plot. A better graphical approach might be to bin the observations into age categories and plot the proportion of survivors in each class as a bar plot (Figure 11.7(c)). This gives an indication of a declining trend in survival but suffers from two problems: the picture is sensitive to the arbitrary choice of bin width and the bars on the right are based on ever-decreasing sample sizes.

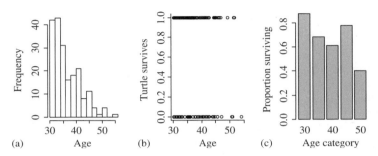

Figure 11.7: (a) Histogram of turtle ages in the sample; (b) plot of survivals (1) and deaths (0) against the age of turtles; (c) proportions of surviving turtles calculated in 5 yr age classes.

A binary data set like this can be modelled as follows

$$Y \sim \text{Bernoulli}(p) \tag{11.36}$$

A sigmoidal link function known as the **logit** can be used to write the success probability (p) of each trial as a function of covariates (turtle age, in Example 11.17). The logit link function enables a probability to be transformed so that it can be modelled by a linear predictor (this is written $\text{logit}(p) = a_0 + a_1 X$). Mathematically, the transformation has the form

$$p = \frac{\exp(a_0 + a_1 X)}{1 + \exp(a_0 + a_1 X)} \tag{11.37}$$

Depending on the values of the parameters a_0, a_1, this sigmoid function either declines from 1 to 0 or it increases from 0 to 1. More covariates can be added to the linear predictor to obtain a multiple logistic regression model. You can check, by starting from the PMF of the Bernoulli (Section 9.7) that the log-likelihood of this model is

$$l(a_0, a_1 \mid y_1, \ldots, y_n) = \sum_{i=1}^{n} \{y_i \ln p_i + (1 - y_i) \ln(1 - p_i)\} \tag{11.38}$$

where the data y_i are either 1 or 0 and p_i is given by Equation (11.37) for any particular observation. The combination of binary data, Bernoulli stochastic component and logit link is a specific version of a **logistic GLM**. More generally, logistic GLMs can deal with response data that are constrained by 0 and 1 (i.e. binary responses or proportions) by modelling the associated probability with a sigmoidal link, such as the logit. As example (11.18) shows, if the data are in the form of a proportion of successes from a given number of attempts, the likelihood is derived from the binomial distribution.

Example 11.18: Estimating survival as a function of age

If we have detailed and precise data on single animals, then the analysis may be based on individual survivorship and the parameters of Equation (11.37) can be estimated by maximising the likelihood in Equation (11.38). Finally, the survival model can be plotted as a function of age (Figure 11.8(a)).

However, it may be necessary to treat animals in groups. For example, our data on turtle age may not be very precise. We may, for instance, only be able to classify age within five-year bins. In that case, the data would come in the form of Figure 11.7(c): for the ith age bin we would know the number of turtles (n_i) surviving out of the initial number (N_i) of turtles of that age. Hence, using the binomial PMF (Section 9.8) for the ith bin, the probability of the observations is

$$f(n_i \mid p_i, N_i) = \binom{N_i}{n_i} p_i^{n_i} (1 - p_i)^{N_i - n_i} \tag{11.39}$$

The age-specific probability of survival (p_i) is modelled by the logit function in Equation (11.37). A little algebra can generate the log-likelihood of the parameters for all of the m age bins combined.

$$l(a_0, a_1 \mid n_1, \ldots, n_m, N_1, \ldots, N_m) = \sum_{i=1}^{m} \{n_i \ln p_i + (N_i - n_i) \ln(1 - p_i)\} \tag{11.40}$$

This likelihood depends on the absolute frequencies n_i of survivors, but also on the number of available animals N_i. In the terminology of logistic GLMs, this second set of numbers are called **weights**. The explanation for their name is the following: different age classes can yield the same observed proportions of survivors, even if they are populated by different numbers of individuals to start with. However, the estimated proportion obtained from an age class with more animals should be more precise. Therefore, the numbers N_i enter the likelihood so that they can correctly weight the maximum likelihood parameter estimates according to the distribution of the sample size across different age classes.

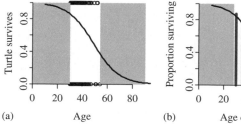

Figure 11.8: Logistic regression on binary (a) and proportional (b) response data. The solid curve shows the fitted model. Its interpretation is subtly different in the two cases. In (a) it can be thought of as the probability of survival of an animal of that age. In (b) it represents the proportion of animals of that age that are expected to survive into the next year. To illustrate the sigmoidal nature of the logistic curve, both models are extrapolated outside the range of the data (into the shaded regions).

A second point about Equation (11.40) is that it no longer contains the binomial coefficient of Equation (11.39) because, fortunately, that part of the expression does not depend on p_i and the parameters of the model (so it does not affect the parameter estimates). Estimating the parameters of the logistic via this likelihood should give similar results to the binary case, assuming that the age classes are not too coarse (Figure 11.8(b)).

11.9: Fitting a logistic GLM to binary and proportion data

Using the specific details of Examples 11.17 and 11.18, I will deal with the binary case first. We have two vectors, the first (y) refers to the event of survival and contains only 1s and 0s. The second (age) contains individuals' ages. The model can be estimated in just two lines of code

```
dat<-data.frame(y, age) # Data frame with binary data
mod<-glm(y~age, family=binomial, data=dat)
```

Now, assume that the data come in the form of proportions (a vector f of relative frequencies of survival) grouped into age bins (the vector age). We also need to know the vector (N) of the initial numbers of animals in each age bin.

```
datp<-data.frame(f, N, age) # Data frame with proportions
mod<-glm(f~age, family=binomial, data=datp, weights=N)
```

Note the use of the vector N as the weights in this version of logistic regression. The commands summary(), fitted() and predict() can be used in the same way as in R11.8. For example, the curve in Figures 11.8(a) and (b) can be plotted as follows:

```
ager<-seq(10,120) #Creates new vector of age values for prediction
newdat<-data.frame("age"=ager) #Specifies single-vector data frame
plot(ager, predict(mod, newdat, type="response"), type="l")
```

11.8.   Evaluation, diagnostics and model selection for GLMs

Thanks to its normality assumptions, the linear regression model was traditionally estimated by least squares (minimising the residuals of the model from the data) and so, most of its evaluation and diagnostic criteria are based on residuals. This approach does not work for GLMs which rely heavily on likelihood. We therefore need to develop an alternative criterion for goodness of fit. To do this, we require the concept of a **saturated model**, one which uses up all the information in the data by having as many parameter values as there are observations. Remember that the central aim of regression is to tease apart the signal from the noise. A saturated model is one that insists that all the information in the data is signal and, therefore, requires a different parameter to describe each observation. Such a model is guaranteed to be overfit (because real data always contain noise) but it is also guaranteed to have the closest possible fit to the data. Therefore, the maximised log-likelihood of the saturated model (say l_S) will be greater than the maximised log likelihood (say l) of any other model that might be considered. A simple comparison between these two quantities, known as the **residual deviance** (or **deviance** for short) can tell us how close our model is to achieving the closest fit

$$Deviance = 2(l_S - l) \tag{11.41}$$

For all models, the deviance will be greater than zero (equal to zero if the model considered is the saturated model). Closer-fitting models have low values of deviance. But how low is low enough? Unfortunately, pseudo-r^2 criteria for GLMs, based on deviance (e.g. $1 - l/l_s$) are not very reliable, but if you want to report on goodness-of-fit for a particular GLM, use the χ^2 test described below.

ⓡ 11.10: Null, residual deviance and diagnostic plots for a GLM

Consider the egg production case (Example 11.16). The model summary looks like this

```
Call:
glm(formula = eggs ~ food, family = poisson, data = dat)

Deviance Residuals:
    Min      1Q   Median      3Q      Max
-4.4730  -0.9846  -0.3440   0.7510   5.1735

Coefficients:
            Estimate Std. Error z value Pr(>|z|)
(Intercept) -3.115797   0.135532  -22.99   <2e-16 ***
food         0.077182   0.001552   49.72   <2e-16 ***
---
Signif. codes: 0 '***' 0.001 '**' 0.01 '*' 0.05 '.' 0.1 ' ' 1

(Dispersion parameter for poisson family taken to be 1)

    Null deviance: 5589.98 on 199 degrees of freedom
Residual deviance: 447.37 on 198 degrees of freedom
AIC: 1007.0

Number of Fisher Scoring iterations: 5
```

The residual deviance of the model is reported close to the bottom. The null deviance is the maximised log-likelihood of the saturated model. There is nothing particularly mystical about a saturated model. You can specify such a model by asking R to treat each observation as a unique occurrence with no similarity to any of the others. In this example, we can achieve this by treating food availability as a qualitative variable:

```
glm(eggs~as.factor(food), family=poisson, data=dat)
```

In the summary of this model, the residual deviance is practically zero.

For goodness-of-fit, the following approximate chi-square test can be used, in which a small value (e.g. <0.05) indicates lack of model fit.

```
1-pchisq(mod$deviance, mod$df.residual)
```

The idea of deviance as a criterion of fit can be extended to single observations in the data. The **deviance residual** is the measure of deviance contributed from each observation. This can be used to generate diagnostic plots similar to the ones in Figure 11.3. Using the turtle egg production data, the command `plot(mod, which=c(1,2,4))` will produce the plots in Figure 11.9. Plotting the residuals against the predicted values of the fitted GLM (Figure 11.9(a)) has a slightly different interpretation to the linear model. Here we are dealing with a Poisson stochastic component, so the variance of the model increases with its mean.

Hence the increasing spread of points from left to right. The striations observed towards the left are due to the discreteness of the count data (at low expectations, the Poisson can only yield a few distinct values). Other than that, the residuals are neither consistently over nor under the x-axis, so this is a perfectly healthy plot, indicating that the log-linear model is appropriate for these data.

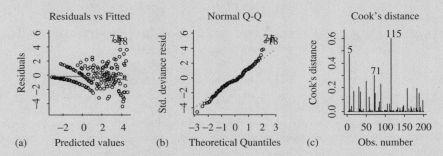

Figure 11.9: Diagnostic plots for log-linear GLM. (a) Simple residuals plotted against the predicted values; (b) Q-Q plot using the standardised deviance residuals; (c) an approximate Cook's distance.

The modified Q-Q plot (Figure 11.9(b)) uses the fact that deviance residuals are approximately normally distributed to evaluate the appropriateness of the stochastic component of the GLM. The alignment of the central bulk of the data with the Q-Q line indicates that the Poisson is an appropriate distribution for these data. Finally, the plot of Cook's distances in Figure 11.9(c) reveals no outliers (all distances are smaller than 1).

The GLMs described above are estimated on the basis of Poisson or binomial likelihoods. It is therefore possible to define for each model a value of the Akaike Information Criterion (Section 11.6) and carry out manual or automatic model selection as outlined in R11.7. In this way, particularly when dealing with many candidate covariates at the same time, it is possible to arrive at a parsimonious model that combines quality of fit and predictive ability.

11.9. Modelling dispersion

The models in the previous sections assume a stochastic component (normal, Poisson, binomial) and use the data to estimate a trend (linear, log-linear, logistic). This approach only estimates the expected value of the response variable. It deals with the variance around this estimated expectation by making some rather restrictive assumptions. For example, in the linear model, the variance of the normal stochastic component was assumed to be constant across the range of the explanatory variables (homoscedasticity). In the case of log-linear regression, the Poisson distribution implicitly imposed equality between mean and variance. Similar implicit constraints were placed on the variance by the use of the binomial in logistic regression. All these assumptions will be violated if the data contain extra signal or extra noise than what is being accounted for by the model. Hence, omitting an important covariate or a source of stochasticity from the likelihood will usually lead to increased dispersion around the fitted model. The first step to a cure is to detect such deviations from the assumed likelihood.

11.11: Investigating violations of the dispersion assumptions

The `ncv.test()` (see R11.5) will detect heteroscedasticity in linear models. Within the GLM framework, the main diagnostic for detecting deviations from the assumed variance is the ratio of the deviance over the residual degrees of freedom. Given a model `mod`, the **dispersion coefficient** ϕ is given by

```
phi<-mod$deviance/mod$df.residual
```

If this is much greater than 1 (e.g. 1.5 or more), the model is overdispersed. Values of ϕ smaller than 1 indicate the (rarer) condition of underdispersion.

Heteroscedasticity may be due to model mis-specification. For example, an important covariate may have been omitted or a more flexible model may be required for the mean (see Sections 11.10 and 11.11 below). If there are no such improvements possible, then it may be useful to model the variance of the model together with the mean. For example, heteroscedasticity may be modelled by allowing the variance to change as a linear or power function of the mean. Alternatively, overdispersion may be modelled by inflating the model's variance. All of these tricks can be achieved by the approach of **quasi-likelihood**.

 11.12: Specifying quasi-likelihood models

Consider the simple case of a Poisson likelihood with a log-linear model in which the dispersion coefficient is much greater than 1. The same model can be estimated using quasi-likelihood in a way that allows the stochastic component of the GLM to have a variance higher than its mean. Here is how,

```
mod<-glm(y~x, data=dat, family=quasipoisson)
```

For a binomial model, the option would become `family=quasibinomial`.

In both of these cases the variance is a simple scaling of what was previously stipulated by the Poisson or binomial family. The more complicated case, in which the variance increases or decreases in a nonstandard way, can also be accommodated. For example, a heteroscedastic normal model in which the variance appears to increase as the square of the mean can be specified as follows

```
mod<-glm(y~x, data=dat, family=quasi(link="identity", variance="mu^2"))
```

A difficulty with quasi-likelihood models is that they do not yield a true likelihood and therefore do not come with an associated AIC. This requires an alternative route for model selection. A manual approach using successive testing is described by Faraway (2006) pp. 149–150.

11.10. Fitting more complicated models to data: polynomials, interactions, nonlinear regression

Historically, most regression models used simple monotonic (i.e. either increasing or decreasing) functions. However, many relationships are nonlinear and nonmonotonic. There are now several different options for fitting more complicated curves to data and, to choose between them, you must first ask yourself if you have *a priori* reasons (e.g. theory, first principles or previous analyses) that suggest a particular shape for the curve to be fit. If you do, then the methods in this section will be useful. If you suspect that a nonmonotonic curve is needed but don't quite know what shape it should have, then look at the methods in Section 11.11.

Example 11.19: Density dependence and polynomial terms

Consider a biological population with a small generation time, such as a passerine bird. We have annual data on population size (P_t), an index of immigration from nearby populations (I_t) and accurate information about the number (N_t) of predators in the study area (Figure 11.10). We would like to determine which of these three (density, immigrants, predators) influence population size.

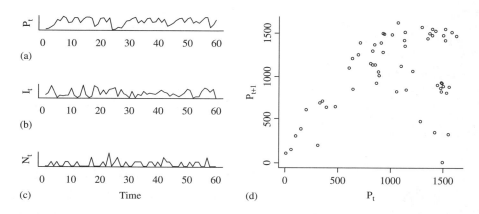

Figure 11.10: (a) Time series of population size for a passerine bird; (b) index of immigrants from nearby populations; (c) number of predators foraging in the study area; (d) population size for next year plotted against this year's population.

It is first necessary to decide on the models to be fit. The simplest possible model in discrete time is unrestricted growth $P_{t+1} = (1 + r)P_t$ (Example 3.2). This is a linear model in P_t so the corresponding regression model is $Y = aX + \varepsilon$ (where ε is the stochastic component. The intercept must be zero because no new individuals can be produced when population size is zero). However, Figure 11.10(d) indicates that the relationship between P_t and P_{t+1} is not linear. P_{t+1} decreases for high values of P_t, probably indicating the existence of density dependence. We can capture this biological feature with a model of constrained growth, such as the discrete logistic model (Example 3.3).

$$P_{t+1} = P_t + r_{max}\left(1 - \frac{P_t}{K}\right)P_t \tag{11.42}$$

This can be expanded to give a second order polynomial in P_t

$$P_{t+1} = \left(-\frac{r_{max}}{K}\right)P_t^2 + (1 + r_{max})P_t \tag{11.43}$$

The general form of the corresponding regression model can be obtained by setting $Y = P_{t+1}$, $P_t = X$, $a_0 = 0$, $a_1 = (1 + r_{max})$ and $a_2 = -r_{max}/K$

$$Y = a_2X^2 + a_1X + a_0 + \varepsilon \tag{11.44}$$

The need to fit nonlinear models to ecological data arises often. In some situations (as in the above example), the assumed ecological process gives rise to a polynomial. In other cases, a polynomial merely offers a convenient approximation to the shape of the data. Polynomial terms can be added to linear models or GLMs to increase their flexibility, but it is good practice during model selection to drop the highest order terms of each variable first.

11.13: Fitting models with polynomial terms

In the formula definition, a higher order term needs to be passed inside the expression `I()`. Below are the specifications for the linear and quadratic model in Example 11.19.

```
# Create a data frame with current (N) and
# previous (Npr) population size
dat<-data.frame("N"=n[2:tmax],"Npr"=n[1:(tmax-1)])
# Fit the models
mod0_EX<-lm(N~ -1+Npr, dat) # Linear model
mod1_DD<-lm(N~ -1+Npr+I(Npr^2), dat) # Quadratic model
```

The presence of -1 in both formulae forces the models to have a zero intercept.

Example 11.20: Predation, immigration and interaction terms

The remaining information in the data may now be used to further explain the observed pattern in Figure 11.10(d). A model containing immigration and predation can be written as

$$P_{t+1} = P_t + r_{max}\left(1 - \frac{P_t}{K}\right)P_t + \beta I_t + \alpha P_t N_t \qquad (11.45)$$

Here, I have assumed that the number of immigrants is proportional to the index of immigration (I_t) and that the predators consume their prey according to a Type I functional response (see Example 4.14) with parameter α. This model now has three explanatory variables (P_t, I_t, N_t), it is quadratic in (P_t) and has a multiplication between two of its variables as part of the functional response. Setting $X_1 = P_t$, $X_2 = N_t$, $X_3 = I_t$, $a_0 = 0$, $a_2 = 0$ gives the following regression model

$$Y = a_0 + a_1 X_1 + a_2 X_2 + a_3 X_3 + a_4 X_1^2 + a_5 X_1 X_2 \qquad (11.46)$$

This latest model includes a second order polynomial term, multiple explanatory variables and a multiplicative term between two explanatory variables. In regression models, such multiplicative terms are known as **interactions**. This name makes good sense in the above example since it refers to the trophic interaction between the predator and prey populations. More generally, an interaction term in a regression model implies that two explanatory variables do not act additively on the response (i.e. the slope of the response to one variable changes with the value of the other variable). Similar to polynomial terms, it is accepted practice that when using an interaction term, the linear terms participating in it must also be in the model (even if their coefficients are estimated close to zero).

11.14: Fitting models with interaction terms

Models with interaction terms are easily specified, although it is generally harder to interpret their meaning. So, before you introduce an interaction, make sure you have a biological reason. Here are three different models examining different biological explanations for the data in Example 11.19.

```
# Model with density dependence and immigration
mod2_IM<-lm(N~ -1+Npr+I(Npr^2)+I, dat)
# Model with density dependence and predation
mod3_PR<-lm(N~ -1+Npr+I(Npr^2)+P+Npr*P, dat)
# Model with density dependence, immigration and predation
mod4_All<-lm(N~ -1+Npr+I(Npr^2)+I+P+Npr*P, dat)
```

A comparison of the AIC values between the five models in R11.12 and R11.13 indicates that the model with density dependence, immigration and predation offers the best explanation for the observed patterns.

```
> AIC(mod0_EX, mod1_DD, mod2_IM, mod3_PR, mod4_All)
         df      AIC
mod0_EX   2  898.2920
mod1_DD   3  858.8015
mod2_IM   4  859.9041
mod3_PR   5  725.4281
mod4_All  6  709.0039
```

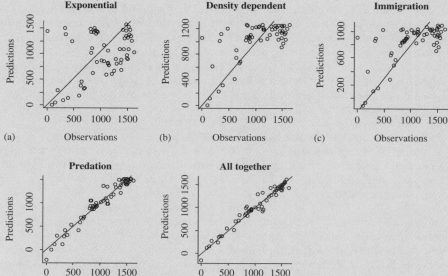

Figure 11.11: Plots of fitted against observed values for the five models examined here: (a) exponential growth; (b) density dependence; (c) density dependence with the effect of immigration; (d) density dependence with the effect of predation; (e) density dependence with combined effect of predation and immigration.

We can visualise the quality of fit by plotting fitted against observed values for all five models (Figure 11.11).

```
par(mfrow=c(2,3)) # Splits graphics device into 2 rows and 3 columns
plot(dat$N, fitted(mod0_EX), main="a.Exponential", xlab="Observations",
                                               ylab="Predictions")
abline(0,1)
plot(dat$N, fitted(mod1_DD), main="b.Density dependent",
              xlab="Observations", ylab="Predictions")
abline(0,1)
plot(dat$N, fitted(mod2_IM), main="c.Immigration", xlab="Observations",
                                               ylab="Predictions")
abline(0,1)
plot(dat$N, fitted(mod3_PR), main="d.Predation", xlab="Observations",
                                            ylab="Predictions")
abline(0,1)
plot(dat$N, fitted(mod4_All), main="e.All together", xlab="Observations",
                                               ylab="Predictions")
abline(0,1)
par(mfrow=c(1,1)) # Returns graphics device to original
```

This latest R box illustrates the power of model selection as a method for testing scientific hypotheses. In this data set, it has been able to identify the mechanisms that drive the population dynamics of this population.

I have chosen to motivate the different types of empirical models in this section from a mechanistic viewpoint. In the above discussion, quadratic and interaction terms arise naturally from the population dynamics models examined in Chapter 3. This approach allows you to appreciate the necessity and meaning of such nonlinear terms from an ecological perspective but it is also somewhat misleading. We rarely motivate empirical models from specific mechanistic models. We may have a vague notion that predators interact multiplicatively with prey, and a population responds nonlinearly to its past density but we rarely express the regression model as a reparameterisation of a mechanistic model – it is simply too much work and not always possible.

However, if you have a mechanistic model that you feel very strongly is a good representation of reality then, assuming that it cannot be written as a linear combination of polynomial and interaction terms, there are increasingly good methods for fitting it directly to data. Such methods belong to the broader area of **nonlinear regression** that can be implemented via three routes:

❶ *Nonlinear least squares*: If the residuals around the nonlinear model can be assumed to be normal, then least squares is the quickest method. Crawley (2007) and Bolker (2008) offer more details with ecological examples and R implementation.

❷ *Maximum likelihood*: A custom stochastic component can be specified around the model from the selection of the distributions presented in Chapter 9. If the likelihood for this model can be written, then it may also be optimised using the methods discussed in Chapter 10. Bolker (2008) ventures quite deeply into this area.

❸ *Bayesian methods*: It makes similar requirements as the maximum likelihood approach, but with the added advantages of being able to specify priors for the parameters. McCarthy (2008) and Bolker (2008) offer good introductions.

11.11.   Letting the data suggest more complicated models: smoothing

---

Example 11.21: Distribution along a linear habitat

A particular study maps the presence of otters along a river. Researchers visit 100 m-long segments of the river bank and count the number of otter scats they find. The time intervals between visits are larger than the time it takes for scats to decompose, to ensure no double counting. Because of accessibility constraints, the sampling effort differs for different river segments. The resulting data are shown in Figure 11.12.

Assuming that the number of scats is an unbiased index of usage, these data will still be affected by sampling variation (i.e. even if usage does not change, repeat visits to the same 100 m segment will encounter different numbers of scats). So, there is certainly noise in this data, but is there also a signal? There appear to be two peaks at around the 100- and 200-segment points. Given just these data, there is no *a priori* reason to assume a curve of a particular form (as was done in Section 11.10), but it is still useful to draw a curve through the cloud, to help generate some hypotheses.

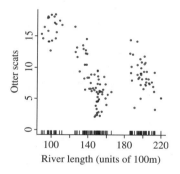

Figure 11.12: Number of otter scats found on different visits plotted against their position on the river banks. Sampling effort is indicated by the 'rug' of small vertical lines on the x-axis.

In situations like this, when both signal and noise are present but there is no clue for how to describe the signal mathematically, **smoothing** is a handy tool. The idea behind it is simple: if there is signal in the data, then points that are close together in the x-axis must also have similar y values. In other words, there must be some autocorrelation in the response data (see Section 2.4).

The easiest way to iron out the sampling variation is to obtain an estimate of the response variable as the average of y values in the vicinity of a particular x value ($x \pm \frac{1}{2}\Delta x$). Doing this for all x values in the range of the data yields a **moving average**, the simplest type of smoother. Defining the 'vicinity of a particular x value' sounds simple, but in reality the shape of the resulting curve will depend on this **smoothing window** or **bandwidth** (Δx). Furthermore, a moving average is problematic because, for any particular point x it uses all points inside the range $x \pm \frac{1}{2}\Delta x$ and ignores all points outside it. Ideally, the similarity between observations should decay with their separation along the x-axis.

There are several generalisations of the moving average that can account for these drawbacks. For example, a technique known as **kernel smoothing** weights the influence of neighbouring points on each-other according to their proximity along the x-axis. The 'kernel' in the name of the method is the weighting function that achieves this. Mathematically, for a sample of n observations, the kernel-smoothed estimate $f(x)$ at a value x of the independent variable looks just like a weighted average

$$f(x) = \sum_{i=1}^{n} w_i y_i \bigg/ \sum_{i=1}^{n} w_i \tag{11.47}$$

In this expression, the kernel is w_i. It works by allocating a portion of the observation y_i to the estimate at x. Many functions can serve as kernels, but we usually want functions that decay away from x, such as the following bell-shaped function that is related to the normal (Gaussian) PDF.

$$f(x) = e^{-\frac{(x-x_i)^2}{\lambda}} \tag{11.48}$$

This has two desirable properties: it attains its maximum value at x, so that observations close to the position being estimated drive the estimate. It also has a parameter λ which operates analogously to the variance in the normal distribution. By increasing the value of λ, the smoothing window is extended and hence imposes a greater degree of smoothness on the estimated function. Deciding just how smooth a function should be is tricky. If you have some information about this, then by all means decide for yourself what the value of λ should be. There are also automatic methods for choosing the smoothing window, based on cross-validation (see Section 11.6). They work by omitting one observation at a time, trying a range of values for λ and measuring how well each missing observation is predicted by the curves created from each λ.

Example 11.22: Otter density estimation by kernel smoothing

The data in Example 11.21 are simulated, so the true distribution of otter scats is known to us but not to the smoother. The performance of the normal kernel with automatic bandwidth selection (via cross-validation) is shown in Figure 11.13.

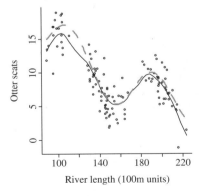

Figure 11.13: Kernel-smoothed estimate of otter scat density (solid line) and the underlying true density (dashed grey line).

11.15: Estimation by kernel smoothing

The built-in command `ksmooth()` will readily produce a kernel-smoothed curve for a data set with two variables x and y.

```
plot(x,y) # Scatter plot
rug(x) # Helps visualise sampling effort along x-axis

# Estimates & plots smooth
lines(ksmooth(x,y, kernel="normal", bandwidth=10))
```

Here, the options of `ksmooth` specify a normal kernel (see Equation (11.48)) and an arbitrary bandwidth of 10. The bandwidth can be determined automatically by cross-validation, via the library `sm`.

```
require(sm) # Loads library
bandw<-hcv(x, y) # Applies cross validation to estimate bandwidth
sm.regression(x, y, h=bandw) # Generates plot and adds smooth
```

Kernel smoothing can be applied with more than one explanatory variable, but requires a multivariate kernel (e.g. one based on the multivariate normal distribution, see Section 9.17). A similar method that I will not cover here is **locally weighted regression scatterplot smoothing** (**LOESS**). Both of these methods are equally flexible (or inflexible, depending on the choice of bandwidth) across the entire range of values for the independent variable. But, as we saw in Example 11.21, unbalanced sampling effort may result in more information being available for particular segments of the x-axis. Also, their application to more than two explanatory variables is cumbersome.

A more advanced technique, the **generalised additive model** (**GAM**), takes the concept of smoothing further by applying it in a piece-wise manner. A GAM first splits the range of x values into segments whose cut-off points are called **knots**. It then fits a flexible curve (usually a polynomial) to the data within each segment. The parameters of each of these piece-wise polynomials are selected in a way that satisfies two criteria: ❶ the polynomials must fit the data in each segment well and ❷ the polynomials in adjacent segments must join smoothly at the knots. By choosing the position of the knots according to the availability of the data, the GAM can describe the response with higher detail in data-rich parts of the x-axis.

GAMs have the full functionality of a GLM (variety of stochastic components, ability to deal with large numbers of explanatory variables, use of interaction terms, estimation of confidence intervals for predictions) combined with the flexibility of a smoother. This flexibility is both a blessing and a curse – GAMs have often been accused of overfitting the data. In the case of multiple explanatory variables in particular, a GAM needs to be parsimonious both in terms of the variables it contains and the 'wiggliness' with which it describes the response to each of them. A recent development in GAMs called a **shrinkage smoother** achieves both of these objectives simultaneously: by using cross-validation, it allows the wiggliness for each covariate to go to zero as required by the data. Such a flat-lined covariate nominally remains in the model but, practically, has no effect on model predictions (it is, in effect, selected out).

Example 11.23: Otter density estimation by GAM

One of the design features of the generalised additive model is its smoothness. By construction, GAMs are best suited to modelling response variables that change gradually with changes in the explanatory variables. In this sense, the simulated otter scat density is captured extremely well by the GAM in Figure 11.14.

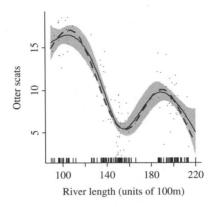

Figure 11.14: GAM fit to the simulated otter density data. The thin solid line represents expected scat density as estimated by the GAM. The dashed line is the real underlying density. The confidence limits around the GAM predictions are shown as a grey band.

11.16: Estimation by GAMs

There are two different packages in R for fitting GAMs, gam and mgcv, the second being the most actively developed, documented and user-friendly. The syntax for invoking a GAM is similar to the GLM. The following example fits a GAM to the otter scat data and plots the results in a graph similar to Figure 11.14.

```
require(mgcv) # Loads library
dat<-data.frame(x,y) # Creates data frame with two variables
mod<-gam(y~s(x),data=dat, family=gaussian) # Fits the gam
plot(mod, xlab="River length (units of 100m)", ylab="Otter scats",
                    shade=T, residuals=T, shift=mean(y))
```

Notice how the explanatory variable is now enclosed in s(x) in the formula, indicating that the response to this variable is to be smoothed by the GAM. The option family here isn't strictly necessary since the default for this is gaussian anyway, but it does show you where to specify a Poisson or binomial stochastic component if your response data are counts or proportions. Some of the plotting options need further explanation. The confidence limits around the mean prediction can be plotted as a pair of curves or a shaded zone. The option shade=T gives a result like the one in Figure 11.14. The option residuals=T shows (as dots) the residuals of the raw data from the predictions. Also, by default, the y-axis for a GAM plot shows the residuals on the scale of the linear predictor. This being a Gaussian model, I have simply shifted the mean of the predictions up to the mean of the response data by specifying `shift=mean(y)`. As a result, here the residual dots are the raw data.

The implementation of GAMs in the `mgcv` library automates the selection of knots and the flexibility of the smooths, so all the complicated choices are transparent to the user.

11.12.   Partitioning variation: mixed effects models

I started this chapter by asserting that the observed variation in real data comprises both signal and noise (although, of course, not all data sets contain detectable signal and some can be assumed to be almost noise-free). In the models examined, the signal took the form of an expectation (a point, a curve or a surface) and the noise was the remaining variation around it. All of the sections so far have looked at methods for constructing this expectation by trying out different transformations and combinations of one or more explanatory variables. In the process, much of the original variation in the data was explained away. However, in Section 11.6, while discussing model selection, I mentioned that we must resign ourselves to having some residual noise because not all of the variation in the data can be explained with a deterministic signal. This is true, but we may still have some idea of what causes the residual variation in the data and may yet be able to attribute *portions* of this variation to different properties of the sample units. This final section can be motivated by the same example that introduced the chapter.

Example 11.24: Samples of gannet condition

Change in the condition of breeding female gannets during a particular time period of the year is $\Delta I = I_{After} - I_{Before}$. This index represents the ability of birds to acquire condition. Values close to zero indicate no change, negative values indicate deterioration and positive values mean that resources are being acquired faster than they are being expended. Through slight differences in their genetic makeup and experience, different individuals will be more or less efficient at the tasks of foraging, maintenance and breeding. Additionally, through slight differences in local environmental conditions, the birds living in different colonies will be differentially efficient. We are aware that the residual variation in a model for ΔI will comprise both individual and colony variation, so how can we account for these two sources in a regression approach?

Consider the linear model in one covariate

$$Y = a_1 X + a_0 + \varepsilon \qquad (11.49)$$

Bundled together in the stochastic component ε are two, or perhaps more, different types of stochasticity, originating from different processes. If two such processes can be identified (stochastic components are conventionally lower-case Greek letters), then perhaps the model can be rewritten as

$$Y = a_1 X + a_0 + (\phi + \psi) \qquad (11.50)$$

where $\varepsilon = \phi + \psi$. If the two errors ϕ, ψ apply equally and independently to all sample units, then this latter version of the linear model gains us nothing. However, if there are asymmetries or similarities between sample units that can be attributed to group membership, then Equation (11.50) can extract more information from the data than can Equation (11.49).

Example 11.25: Modelling individual and colony variation

In Example 11.3, the change in condition was written as a linear function of local density for 50 birds (ten from each of five colonies). I will here use subscripts to represent the ith bird in the jth colony.

$$\Delta I_{i,j} = a_1 P_j + a_0 + \varepsilon_{i,j} \tag{11.51}$$

Here, the change in condition (ΔI) and the residual (ε), refer to a particular individual so they are subscripted by both i and j. Density measurements (P) are the same for all birds living in the same colony, so colony density is subscripted only by j. Following the rationale of Equation (11.50), the stochastic component can be broken up into individual-specific ($\psi_{i,j}$) and colony-specific (ϕ_j) terms

$$\Delta I_{i,j} = a_1 P_j + (a_0 + \phi_j) + \psi_{i,j} \tag{11.52}$$

The brackets inserted in this latest equation hint at an alternative interpretation of this model. A quick comparison between Equations (11.51) and (11.52) shows them to be structurally identical, apart from the fact that the second equation has an intercept that changes randomly from one colony to the next. Biologically, this **random-intercept** model says that the baseline ability of animals to improve their condition is affected by a number of unknown, colony-specific influences.

This model can be further extended to represent the idea that the effect of density on animals is also colony specific. Mathematically, we can achieve this by introducing a colony-specific random effect (ω_j) to the slope of the response to density, to obtain a **random slope and random intercept model**.

$$\Delta I_{i,j} = (a_1 + \omega_j) P_j + (a_0 + \phi_j) + \psi_{i,j} \tag{11.53}$$

Although the terms $\omega_j, \phi_j, \psi_{i,j}$ in Equation (11.53) resemble model coefficients, they are, in fact, random variables with mean zero. So, what we seek to estimate for a model such as Equation (11.53) are the expectations of the **fixed effects** (a_0, a_1) and the variances of the **random effects** ($\omega_j, \phi_j, \psi_{i,j}$). Biologically, this quantifies the between-colony variation in the ability of birds to acquire condition and the impact of colony density. The mathematics used for estimating mixed effects models is beyond the scope of this textbook. It is more important for you to know when to seek the assistance of this type of model. If your data can be subdivided into discrete classes and if there is a possibility that common class membership leads to similarity between sample units, then a mixed effects model may be what you want, *assuming* you are interested in making population-level inferences that reach beyond the specific classes in your data.

Example 11.26: Why not use colony as a factor?

Consider the random intercept model in Equation (11.52). One entirely legitimate approach to dealing with colony effects would be to incorporate the colony as a factor in the model. We would then have one coefficient for the effect of density and one coefficient for each of the five levels of the factor colony. The ability of such a model to describe the data would be similar to Equation (11.52) but it would have two distinct disadvantages: first, the number of parameters in the model would increase linearly with the number of colonies in the data; second, it would only model the response of animals at the particular colonies included in the data, without giving us an estimate of between-colony variation.

11.17: Fitting a mixed effects model in `lme4`

Much of the original published output for mixed models with R used the `nlme` package (described in detail in Pinheiro and Bates, 2000). Users are now gradually shifting to the newer `lme4`. The command for fitting a mixed model is `lmer()`. Inside it, you have to specify similar options to what was used in `lm()` and `glm()`. The only new element is the declaration of the random effects in the model formula. A typical mixed effects formula in `lme4` looks like this:

```
response variable ~ fixed effects+(random effects | grouping factor)
```

The data frame `dat` is the same as was used in R11.2, containing a row for each individual gannet and columns for colony membership (Colony), colony density (Density) and change in condition (Condition). Below is the code for estimating the two models in Equations (11.52) and (11.53)

```
require(lme4)
# Random intercept model
mod0<-lmer(Condition~Density+(1|Colony), data=dat, family=gaussian)

# Random slope and random intercept model
mod1<-lmer(Condition~Density+(Density|Colony), data=dat, family=gaussian)
```

and here is the output generated for `mod1`

```
> summary(mod1)
Linear mixed model fit by REML
Formula: Condition ~ Density + (Density | Colony)
   Data: dat
   AIC   BIC logLik deviance REMLdev
 448.8 460.3 -218.4    438.7   436.8
Random effects:
 Groups   Name        Variance   Std.Dev. Corr
 Colony   (Intercept) 125.476633 11.20164
          Density       0.016649  0.12903 -1.000
 Residual             377.198612 19.42160
Number of obs: 50, groups: Colony, 5

Fixed effects:
            Estimate Std. Error t value
(Intercept)  36.1525     9.6958   3.729
Density      -0.7189     0.1291  -5.567

Correlation of Fixed Effects:
        (Intr)
Density -0.943
```

This report is split between two main parts, referring to random and fixed effects. The output for the fixed effects is similar to what we would get out of `lm()` or `glm()` (i.e. coefficient estimates, their standard errors and significance). The output for the random effects reports their estimated variances. Given the very small value of the random effect for density (0.0166), it would appear that the estimated slope of the animals' response to density (−0.7189) does not change between colonies. It would also appear that individual variation is three times greater than between-colony variation (compare 377.199 with 125.477).

There are three important extensions of the basic mixed effects model:

❶ The response variable can be modelled with a non-normal stochastic component (giving rise to Generalised Linear Mixed Models – GLMMs) or using flexible smoothers (giving rise to Generalised Additive Mixed Models – GAMMs).

❷ The structure of the groupings can become much more elaborate. For example, sources of variation can be **nested** in multilevel hierarchies (e.g. individual, subpopulation, metapopulation) or using **crossed** classifications (e.g. male and female individuals within different populations).

❸ Mixed effects models are ideal for analysing **longitudinal data**, i.e. repeat measurements collected from the same subject at different points in time. Examples include the collection of multiple physiological and physical measurements from individuals during the course of their lives, replicate measurements of biodiversity at predetermined spatial locations, positional data on the movement of satellite-tagged animals, etc.

Longitudinal data are correlated in two different ways: first, any two observations are likely to be similar when they originate from the same subject (**within-subject correlation**) and second, two successive observations are likely to be similar because any one subject changes gradually through time (**serial correlation**). Both of these types of autocorrelation are responsible for the ailment of **pseudoreplication**, affecting all longitudinal data sets. The problem can be summarised as follows: when data are autocorrelated, the apparent sample size (i.e. the number of rows in the data frame) is misleadingly large because nonindependent data contain less information than independent ones. Assuming independence (as most elementary regression methods do) for fitting and model selection generally leads to overfitted models. In contrast, within-subject correlation is automatically taken care of by the standard mixed effects model and the addition of extra components called **autoregressive structures** can account for serial correlation.

Further reading

ANOVA is covered extensively by Sokal and Rohlf (1995), Quinn and Keough (2002), Gotelli and Ellison (2004) and Whitlock and Schluter (2009). Perhaps the most useful books on the linear model and its various extensions are by Faraway (2004, 2006), but a wealth of information can also be found in the all-purpose textbooks by Fox (1997, 2002) and Krzanowski (1998). The definitive monograph on model selection is Burnham and Anderson (2002) but the topic is covered by every textbook on multiple regression. Density estimation by smoothing is covered by Faraway (2006) and a good theoretical reference is Silverman (1986). The most comprehensive reference for GAMs is Wood (2006). Mixed effects models are covered by Pinheiro and Bates (2000) and Zuur *et al.* (2009). Most modern books on regression now offer introductory chapters to GLMs, GAMs and mixed models.

References

Bolker, B.M. (2008) *Ecological Models and Data in R*. Princeton. 408pp.

Burnham, K.P. and Anderson, D.R. (2002) *Model Selection and Multimodel Inference: A Practical Information-theoretic Approach*. Springer-Verlag, New York. 488pp.

Camphuysen, K. and Webb, A. (1999) Multi-species feeding associations in North Sea seabirds: Jointly exploiting a patchy environment. *ARDEA*, **87**, 177–198.

Crawley, M.J. (2007) *The R Book*. John Wiley & Sons, Ltd, Chichester. 942pp.

Faraway, J.J. (2004) *Linear Models with R*. Chapman & Hall, Boca Raton, Florida. 240pp.

Faraway, J.J. (2006) *Extending the Linear Model with R: Generalised Linear, Mixed Effects and Nonparametric Regression Models*. Chapman & Hall, Boca Raton, Florida. 301pp.

Fox, J. (1997) *Applied Regression and Analysis, Linear Models and Related Methods*. Sage, Thousand Oaks, California. 597pp.

Fox, J. (2002) *An R and S-plus Companion to Applied Regression*. Sage, Thousand Oaks, California. 311pp.

Gotelli, N.J. and Ellison, A.M. (2004) *A Primer of Ecological Statistics*. Sinauer Associates, Massachusetts. 510pp.

Krzanowski, W.J. (1998) *An Introduction to Statistical Modelling*. Arnold, London. 252pp.

Pinheiro, J.C. and Bates, D.M. (2000) *Mixed-effects Models in S and S-plus*. Springer, New York. 528pp.

Quinn, G.P. and Keough, M.J. (2002) *Experimental Design and Data Analysis for Biologists*. Cambridge University Press. 537pp.

Silverman, B.W. (1986) *Density Estimation for Statistics and Data Analysis*. Chapman & Hall, Boca Raton, Florida. 176pp.

Sokal, R.R. and Rohlf, F.J. (1995) *Biometry*, 3rd edition. Freeman, New York. 887pp.

Whitlock, M.C. and Schluter, D. (2009) *The Analysis of Biological Data*. Roberts & Co., Colorado. 700pp.

Wood, S. (2006) *Generalized Additive Models: An Introduction with R*. Chapman & Hall/CRC, Boca Raton, Florida. 391pp.

Zuur, A.F., Ieno, E.N., Walker, N.J., Saveliev, A.A. and Smith, G.M. (2009) *Mixed Effects Models and Extensions in Ecology with R*. Springer, New York. 574pp.

12

How to measure similarity
(Multivariate methods)

'It has long been known that one horse can run faster than the other. But which one?
Differences are crucial'
From Time Enough for Love by American writer Robert A. Heinlein (1907–1988)

Seeing patterns of similarity in an assemblage of objects involves two decisions: inspecting the assemblage from the correct viewpoint and measuring dissimilarity with the right yardstick. This chapter offers an introduction to a broad class of exploratory and predictive methods that try to organise objects according to their similarities and differences. I give a brief outline of the challenges confronted by multivariate statistics (Section 12.1), outline its three different objectives and illustrate each objective with one or two representative methods. The first objective, called **ordination**, seeks to rearrange the sample units in a way that highlights their similarities using as few descriptors as possible (Section 12.2). The main example of ordination is **principal components analysis** (Section 12.3), a technique that seeks the primary gradients of variability within the data. If, instead of continuous gradients, the objective of the analysis is to group sample units into classes, then the favoured approach depends on whether these classes can be defined *a priori*. **Clustering** methods (Section 12.4) can simultaneously group the sample units and decide on the definition of the classes. Two examples are **hierarchical clustering** (Section 12.5) and **k-means clustering** (Section 12.6). If, on the other hand, the classes are known and specimen class members are at-hand, then the task, known as **classification** (Section 12.7), is to decide which attributes of these objects can be used to define class membership for new sample units. I illustrate the use of binomial logistic regression for the two-class problem (Section 12.8) and then extend the approach to **multinomial logistic regression** to deal with the many-class problem (Section 12.9).

How to be a Quantitative Ecologist: The 'A to R' of Green Mathematics and Statistics, First Edition. Jason Matthiopoulos.
© 2011 John Wiley & Sons, Ltd. Published 2011 by John Wiley & Sons, Ltd.

12.1.   The problem with multivariate data

Many debates have been won by the side that could first accuse its opposition of 'comparing apples and oranges'. This argument has traditionally signalled instant death for questions that required the mixing of qualitatively dissimilar attributes in quantitative comparisons. Yet, these are exactly the sort of questions that, as scientists, we often have to address.

Example 12.1: Characterising environmental similarity

The ecology of species distributions enquires whether a given species is more or less likely to be found in a particular type of environment. This assumes that similar locations can be identified on the basis of a potentially large list of attributes. For example, the types of data listed in Table 12.1 may be useful in modelling the distribution of a particular species of fern.

Table 12.1

Geomorphology	Soil-related
1 Altitude	7 pH
2 Slope	8 Porosity
3 Roughness	9 Salinity
Climate	**Biotic**
4 Average precipitation	10 Density of grazers
5 Wind exposure	11 Density of competitors
6 Average temperature	

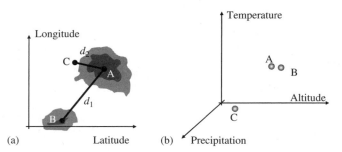

Figure 12.1: (a) Geographical distance is not the best indicator of similarity. The centres A and B of these two distant habitat patches are more like each other than the nearby points just outside each patch; (b) visualising points in environmental space gives a better idea of their similarity.

Consider first the similarity between geographical locations. It is likely that points close-by in geographical space will have similar environmental attributes, but geographical proximity is no guarantee of environmental similarity, just like geographical separation does not necessarily imply dissimilarity (Figure 12.1(a)). It is preferable to map points in **environmental space**, a multidimensional system of axes, each representing an environmental attribute (Figure 12.1(b)). A good (but naïve) first approach is to represent dissimilarity as the Euclidean distance (Section 2.4) between two points in environmental space. This has two problems: first, the arbitrary choice of measurement units (e.g. metres vs feet) for the different axes of environmental space can affect the final measure of similarity; second, many of these variables are likely to be correlated (e.g. altitude vs temperature). Including in the analysis many variables that contain similar information inadvertently increases their collective importance in the measure of similarity.

The objects (sample units) being compared in this example are the spatial locations A, B and C. A typical data frame for such a study would contain a row for each sample unit and a column for each variable measured. Other examples of sample units might include individual organisms, subpopulations or entire species. Note that, unlike the data sets considered in Chapter 11, analyses of similarity do not contain a response variable. But this is not always true for multivariate questions, as the following example illustrates.

Example 12.2: Characterising patterns of occurrence

Consider the issue of habitat suitability for the fern species. To make things simpler, we may focus on occurrence rather than abundance so that a particular location in space is given the value 1 if a fern is present and 0 if it is not. The research question is: 'What are the environmental attributes of a particular location that make it more or less likely to be occupied by the organism of interest?'. For such a question, our subjective evaluation of similarity may be irrelevant. Locations that appear similar to us because of their proximity in environmental space may be nothing like each other from the point of view of the study organism. Conversely, what appear as different environments to us may be indistinguishable to the study species. This inconsistency arises because our definition of environmental space may contain a number of irrelevant dimensions and our definition of similarity may weight equally environmental variables of unequal importance.

12.2.   Ordination in general

Example 12.3: Correlations represent redundancy

The plot in Figure 12.2(a) shows the strong correlation between altitude (Y) and temperature (T) in different sampling locations. The dissimilarity between any two locations, in terms of these two variables, can be measured as

$$d = \sqrt{(T_1 - T_2)^2 + (Y_1 - Y_2)^2} \tag{12.1}$$

The two axes have different units (degrees Celsius and metres), but the following transformations lead to similarly scaled data with mean zero (Figure 12.2(b))

$$Y' = \frac{Y - \bar{Y}}{\sigma_Y}, \ T' = \frac{T - \bar{T}}{\sigma_T} \tag{12.2}$$

Although the data contain information about two different variables and hence the environmental space in Figure 12.2 is two-dimensional, most of the variability between points occurs along a one-dimensional gradient (the regression line relating altitude to temperature). By rotating the plotting axes (Figure 12.2(b)) the x-axis becomes aligned with this gradient. The variables represented in the new axes are a mixture of altitude and temperature and have no correlation. The dissimilarity between any two points can now be well approximated by a one-dimensional measure, the distance along the new x-axis.

$$d = |X_1 - X_2| \tag{12.3}$$

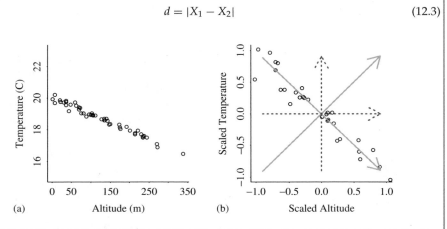

(a)   Altitude (m)

(b)   Scaled Altitude

Figure 12.2: (a) Temperature plotted against altitude for different sampling locations in the fern data set; (b) plot of the same information along scaled axes. The dashed arrows indicate the orientation of the axes in the original environmental space. The solid grey arrows indicate the new rotated axes.

Ordination is a category of statistical techniques that shift, re-scale and rotate the data to reveal similarities between them. They are mainly used for their ability to approximate high-dimensional similarity by lower-dimensional indexes, as was done in Example 12.3. This is useful for visualisation (reducing multidimensional spaces into graphs with one, two or three axes) and for applications that require uncorrelated variables. In addition to **principal components analysis** (examined in the next section), the names of some other techniques in this category are **factor analysis, correspondence analysis** and **principal coordinates analysis**.

12.3. Principal components analysis

Given a data set in n dimensions $(X_1, \ldots, X_n)$, principal components analysis (PCA) generates a new set of n axes $(Y_1, \ldots, Y_n)$ as linear combinations of the originals

$$Y_1 = a_{11}X_1 + \cdots + a_{1n}X_n$$

$$\vdots \tag{12.4}$$

$$Y_n = a_{n1}X_1 + \cdots + a_{nn}X_n$$

or, in matrix notation,

$$\mathbf{Y} = \mathbf{Ax} \tag{12.5}$$

This transformation corresponds to a rotation of the original axes (just like the one shown in Figure 12.2(b)). Since there are an infinite number of possible rotations, there are also an infinite number of possible values for the coefficients a_{ij} in Equations (12.4). However, the matrix of coefficients **A** (known as the PCA **loadings**) is special: imagine a cloud of data plotted in n dimensions. The first new axis (known as the **first principal component**) is the line along which most of the observed variation occurs. The next axis (the **second principal component**) is constrained to be perpendicular to the first. When $n = 2$, as was the case in Example 12.3, the orientation of the second component is completely defined by the first component, but when $n \geq 3$, the second component may be rotated to lie along the line of maximum *residual* variation. The third principal component is constrained to be perpendicular to the first two, and so on. The matrix mathematics of how this is achieved is well within the remit of Chapter 6, but is perhaps less important than the interpretation of the technique's output.

Example 12.4: Collinearities between four environmental variables

Consider a data set that contains information about temperature (T), altitude (A), slope (S) and wind exposure (W). A cursory look at pairwise correlations (Figure 12.3) indicates that PCA may be able to reduce the dimensionality of the data set.

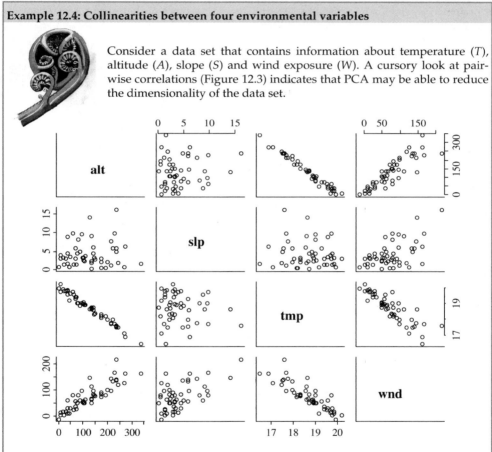

Figure 12.3: Pair-wise scatter plots of the four environmental variables in the data set, altitude (alt), slope (slp), temperature (tmp) and wind exposure (wnd).

Three outputs of PCA are of interest: the composition of each principal component (as conveyed by its loadings), the proportion of the variance explained by each principal component and an **ordination plot** of the data.

Example 12.5: PCA for four environmental variables

Since there are four environmental variables in the data of Figure 12.3, PCA will generate four principal components. Here are the first two:

$$PC1 = -0.558A - 0.230S + 0.556T - 0.572\,W$$
$$PC2 = -0.284A - 0.893S + 0.289T - 0.198\,W \qquad (12.6)$$

PC1 comprises altitude, temperature and wind exposure in equal measures. This is a direct result of strong correlations between these three variables. Therefore, PC1 can be loosely interpreted as a 'climate' variable. PC2 is easier to interpret since it is dominated by the loading for slope. The relative importance of the four principal components can be visualised by a bar chart known as the **screeplot** (Figure 12.4). The variability accounted for by PC1 and PC2 is almost 100%, so PC3 and PC4 can be dropped from any further analysis.

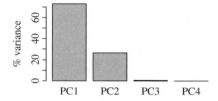

Figure 12.4: Screeplot indicating the proportion of variance accounted for by each of the principal components.

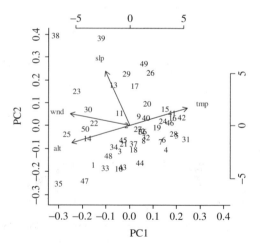

Figure 12.5: Ordination plot of the results of PCA on the four environmental variables. The secondary axes (top and right) indicate the scale of the original variables.

The resulting 2D environmental space can be visualised in an ordination plot which is simply a scatterplot of the data in the transformed axes PC1 and PC2 (Figure 12.5). The position of sample units in this plot highlights their similarities. The annotated arrows superimposed on the cloud of points represent the relative loadings of the original variables in the first two principal components. Hence, temperature, altitude and wind exposure are more aligned with PC1 and slope is more aligned with PC2. Also, the negative correlation between altitude and temperature shows up as the opposing direction of the corresponding arrows.

This type of analysis carries several caveats. It is only sensible to apply PCA when there are collinearities between the data. Even then, it is unlikely that the variability in the data can be captured by just two principal components (as in Example 12.5). Furthermore, the gains in dimensionality offered by PCA have a rather high price, because it is generally difficult to interpret the principal components ecologically, in terms of the original variables.

 12.1: Principal components analysis

For a given data frame `dat` containing information on the four variables of Example 12.4, PCA can be applied as follows:

```
pr<-prcomp(dat, scale=T)
```

The option `scale = T`, specifies that, prior to analysis, the raw data are to be transformed according to Equation (12.2). The loadings for the principal components are listed under the heading `rotation` in the resulting object:

```
> pr
Standard deviations:
[1]  1.71017730  1.02954347  0.12293497  0.01486053

Rotation:
            PC1          PC2          PC3          PC4
alt  -0.5579066  -0.2835636   0.5368908   0.56575632
slp  -0.2301698   0.8927044  -0.1458949   0.35890860
tmp   0.5554498   0.2889520   0.7783689  -0.04608652
wnd  -0.5720468   0.1979326   0.2908689  -0.74093214
```

A report of the relative importance of the PCs can also be generated,

```
> summary(pr)
Importance of components:
                          PC1    PC2     PC3      PC4
Standard deviation       1.710  1.030  0.12293  0.01486
Proportion of Variance   0.731  0.265  0.00378  0.00006
Cumulative Proportion    0.731  0.996  0.99994  1.00000
```

A rough-and-ready screeplot can be generated from `screeplot(pr)` but a slightly more presentable one (with labels and bar heights adding to 100%) can be built from the summary output

```
barplot(100*summary(pr)$importance[2,], ylab="% variance")
```

Finally, the ordination plot of Figure 12.5 can be generated from `biplot(pr)`.

12.4.   Clustering in general

Example 12.6: Identifying functional groups in ecological communities

Ecological communities are complex, but so are the dynamics of any member species. Hence, whole-ecosystem models need to make compromises between the degree of realism used to represent single-species ecology and community interactions. An abstraction that can help deal with this dilemma is to model **functional groups**, rather than individual species (e.g. Petchey and Gaston, 2002). Although ecologists talk about functional groups all the time (e.g. primary producers, apex predators, detritivores, etc.) it is hard to identify these groups objectively. To do this, we need to list the salient functional characteristics for each species, rank them according to their importance and then use them to cluster different species into groups according to their overall similarity.

Clustering methods attempt to group sample units objectively according to a measure of their similarity derived from several of their traits. This is harder than it first appears: in the above example, no sooner did I mention objectivity than I had to invoke two subjective decisions on 'salience' and 'ranking'. Which traits should be included? Which ones are important for the application at hand? How many groups should we end up with? Should this number be dictated by the data or by the intended application? There is more than one answer to these questions, and that is why clustering is a research field in its own right. Broadly speaking, however, all clustering methods have the same three components:

❶ *The data frame:* Similar to most data frames, each row represents a sample unit and each column contains trait data.
❷ *The clustering algorithm:* There are two distinct ways to create clusters. The first, known as **hierarchical clustering** (see Section 12.5) works by successively merging the sample units into ever-larger groups. **Nonhierarchical clustering** (see Section 12.6) starts from a random arrangement of units in a predefined number of groups which is then optimised according to certain criteria.
❸ *Decision on the number of groups:* Depending on what clustering method is used, this decision may need to be made before/after clustering, or not at all.

Example 12.7: Clustering data frame for Antarctic species

Consider the (already rather simplified) animal community of Figure 1.11(a). Each species in that system becomes a row in the analysis data frame. The columns represent species traits (Figure 12.6). There are three types of traits that may be useful for deciding functional similarity in the community: morphology (including physical and physiological measures), demography (age to maturity, annual survival and fecundity) and trophic connectivity.

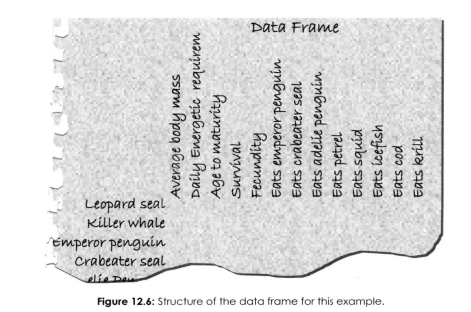

Figure 12.6: Structure of the data frame for this example.

12.5. Agglomerative hierarchical clustering

Hierarchical clustering operates on a distance matrix (**D**): this is a square $n \times n$ matrix, where n is the number of sample units in the data. A particular entry d_{ij} in that matrix represents the dissimilarity between the ith and jth sample unit, quantified by any of the distance measures discussed in Section 2.4.

Agglomerative hierarchical clustering starts by assuming that every sample unit is the sole member of a group. It selects the two most similar units from the distance matrix **D** and merges them into one group. It recalculates the data frame and distance matrix for this new grouping and goes on to form the next grouping. The method continues until all units are in the same group. The agglomeration process can be represented graphically by a **dendrogram**, an upside-down tree diagram whose trunk represents a single group containing all sample units and whose n leaves represent single-unit groups. These two extremes of the clustering process are of little interest compared to the intermediate branchings.

Example 12.8: Dendrogram for Antarctic species

The dendrogram resulting from the application of hierarchical clustering to the Antarctic community data is shown in Figure 12.7. Onto this, I have placed a horizontal dashed line to indicate what seems (to me) a biologically sensible number of groups. The dashed line intersects the tree at five branches which correspond to the following groups: ❶ large filter feeders (Humpback), ❷ large predators (Killer whale), ❸ fish (Antarctic cod and icefish), ❹ invertebrates (squid and krill) and ❺ medium predators (pinnipeds and seabirds).

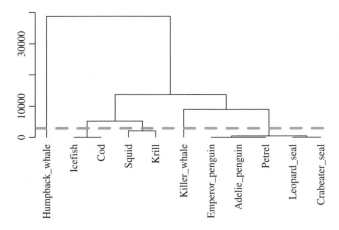

Figure 12.7: Dendrogram for dissimilarity matrix of animals belonging to the Antarctic community.

There is a degree of arbitrariness in deciding how many groups to choose. Several criteria have been devised to try and decide on the optimal number of groups, but these tend to yield different results. In some applications, the decision is not always needed. For example, studies in ecological genetics use the proximity with which the organisms occur at the bottom of the dendrogram to determine relatedness between individuals (e.g. Blouin *et al.*, 1996).

Ⓡ   12.2: Hierachical clustering

Given a data frame `dat` with the structure shown in Figure 12.6, the first step is to create a dissimilarity (or distance) matrix using the command `dist()` which offers several methods for calculating distance between sample units (the default is `euclidean`). Clustering is implemented in `hclust()` which, again, has several different options for determining how to cluster groups and how to recalculate the dissimilarity matrix. In the simplest case, a dendrogram like the one in Figure 12.7 can be generated with the following three lines:

```
di<-dist(dat)
cl<-hclust(di)
plot(cl)
```

12.6.   Nonhierarchical clustering: k means analysis

The problem with hierarchical techniques is that they cannot fully explore the space of possibilities: once a grouping is made, it is carried through until the end. There is no guarantee that merging the groups with the highest similarity at every single step of the clustering will yield the greatest possible similarity within the eventual groups. Sometimes, a mountain peak is reached only by going through a gorge (similar problems were encountered in the discussions about maximisation techniques in Section 4.10 and automated model selection in Section 11.6).

Ideally, the configuration of clusters should maximise within-group similarity and between-group dissimilarity. This is achieved by the method of **k means** in which the user must first specify the number (k) of clusters (this is the main operational difference with hierarchical clustering). To see how the method works, imagine a data frame containing n sample units, each characterised by m traits. Each sample unit can be visualised as a point in m-dimensional space. Each cluster of units is a cloud of points in m-space and can be characterised by the position of its mean – a point whose coordinates in m–space are the averages of the coordinates of the cluster members. The algorithm tries to minimise the aggregate distance of cluster points from their corresponding means. It has the following steps:

❶ The k means are randomly positioned (or randomly shifted from their previous positions) in m-space.
❷ The sample units in the data are assigned to their closest mean.
❸ The sum of squared distances of sample units from their means is calculated and stored.
❹ If the sum of distances is smaller than that obtained for the previous configuration of the means, then the new configuration is retained and the process starts again from ❶.
❺ The process ends when the positions of the means can no longer be improved.

 12.3: k means clustering

For the data frame used in R12.2, a five-group clustering by k means can be achieved as follows:

```
km<-kmeans(dat, 5)
```

The resulting object contains a lot of diagnostic information (such as the positions of the means in m–space and the within-cluster sum of squares for all clusters). We are mostly interested in the cluster membership, the first element of the object km:

```
> km[1]
$cluster
   Leopard_seal      Killer_whale Emperor_penguin Crabeater_seal
              2                 4               1              2
Adelie_penguin            Petrel           Squid        Icefish
              1                 1               3              3
           Cod  Humpback_whale           Krill
              3                 5               3
```

It therefore seems that k means clustering has settled on an (apparently) equally sensible biological definition of functional groups that are almost the same as those generated by hierarchical clustering apart from the fact that invertebrates are grouped together with fish, and pinnipeds are kept separate from seabirds.

12.7. Classification in general

Classification is the process of allocating objects (sample units) to classes. It has two differences from clustering: first, the classes are known *a priori* and second, there is a **training set** of objects, of known class membership. Some authors treat clustering (Sections 12.4-12.6) as a form of **unsupervised classification** (in the sense that it classifies objects without previously having encountered class members). There are many classification techniques, particularly in the computer sciences (e.g. **neural networks**). Of those, the most widely applied category in ecology is **discriminant analysis**. You can find information about it in Crawley (2007) and Maindonald and Braun (2007). However, here, I will apply to the classification problem the statistical modelling tools introduced in Chapter 11. These tools offer considerably more flexibility and expandability than the traditional discriminant analyses (see discussion on pp. 128–130 in Fielding, 2007) and you should already be familiar with them from the previous chapter.

12.8.   Logistic regression: two classes

Example 12.9: Characterising fern habitat

We return to the problem of habitat suitability for the fern species. Given the properties of a spatial location we would like to be able to decide if it is suitable for the species or not. This can be treated as a problem of classification in which the sample unit is the spatial location and the two classes correspond to fern presence and absence. The problem can be approached in the following five stages:

➊ *Model fitting:* The data frame contains two types of information for every spatial location: the response variable (taking the values 1 if fern was present, 0 if absent) and the explanatory variables (altitude, slope, wind exposure, temperature). The following logistic regression model can be fit to these data (see Section 11.7)

$$Y \sim \text{Bernoulli}(p)$$

$$p = \frac{\exp(a_0 + a_1 alt + a_2 slp + a_3 tmp + a_4 wnd)}{1 + \exp(a_0 + a_1 alt + a_2 slp + a_3 tmp + a_4 wnd)} \tag{12.7}$$

All the usual diagnostic and model selection methods (Section 11.8) can be applied to Equation (12.7) to arrive at a more parsimonious model.

➋ *Prediction:* From this fitted model, predictions can be derived for new locations (given the environmental data for a new location, the model can be used to generate a probability of encountering ferns there).

➌ *Definition of fern habitat:* Some thought needs to be put into what should be defined as 'fern habitat'. The model generates probabilities, which are, arguably, more informative than binary output, but sometimes when asked about the suitability of a particular location we need to be able to give a 'yes', or 'no' answer. The most obvious solution is to define fern habitat as the locations with probabilities over 0.5, but this is not always correct. It assumes that the fern is abundant enough in its core habitat and that the size of the sampling quadrat is sufficiently large to ensure that at least one fern is found in suitable habitat. There are yet more considerations about transient dynamics resulting from local extinction and recolonisation. Organisms may be temporarily present in unsuitable habitat or absent from as-yet uncolonised but favourable regions.

➍ *Validation:* A part of the training data set can be put aside to validate the methodology. This can be done with the aid of a **confusion matrix**, tabulating the number of times the model predicted presence and absence correctly, and the times it failed to predict the validation data.

➎ *Wider application:* Once the model is validated and fitted to the entire data set, it can be rolled out for larger-scale application. For example, environmental GIS layers can be used to map suitable habitat.

12.4: Classification by binomial logistic regression

The analysis outlined in Example 12.9 can be carried out as follows. First, the model fitting:

```
>dat<-data.frame(alt, slp, tmp, wnd, rdat) #Create data frame
># Specify model with all four covariates
>mod<-glm(rdat~alt+slp+tmp+wnd, data=dat, family=binomial)
>mod<-step(mod) # Carry out automatic model selection
> mod

Call: glm(formula = rdat ~ slp + wnd, family = binomial, data = dat)

Coefficients:
  (Intercept)        slp        wnd
      -0.4223     0.8950    -0.1373

Degrees of Freedom: 199 Total (i.e. Null); 197 Residual
Null Deviance:       154.6
Residual Deviance: 74.1      AIC: 80.1
> 1-pchisq(mod$deviance, mod$df.residual)
[1] 1
```

Here, the `step()` function has retained the effects of slope and wind, and the approximate chi-square test gives no evidence for lack of fit. Given a validation data set `datp` of the same format (i.e. columns have the same names as the data frame `dat`, which was used for fitting), predictions can be generated as follows:

```
pr<-predict(mod, newdata=datp, type="response")
```

These probabilistic predictions can be thresholded at 0.5 (so that predictions larger than 0.5 are given the value 1 and lower probabilities are assigned the value 0). The easiest way to do this is:

```
pr<-round(pr)
```

More generally, a different threshold (say, 0.7) could be applied as follows:

```
pr<-ceiling(pr-0.7)
```

Finally, given the response values from the validation data `rdat` and the thresholded predictions `pr`, the confusion matrix can be generated by using the cross-tabulation command

```
> table(pr, rdat)
    rdat
pr   0 1
  0 40 1
  1  4 5
```

So, in this example, the classification got it right in 45 out of 50 cases. In one case it predicted an absence when the fern was actually there, and in four cases it predicted presence in quadrats in which the fern was not found.

12.9. Logistic regression: many classes

The problem of classifying objects in more than two categories can be dealt with by an extension of binomial logistic regression. The two-class problem models the probability p that the object belongs to one of the two classes. This implicitly models the complementary probability $q = 1 - p$ that the object belongs to the other class. For more than two classes, the probabilities $p_1, p_2, \ldots, p_n$ must be modelled, where $p_n = 1 - (p_1 + \cdots + p_{n-1})$. We have already encountered a named probability distribution, the multinomial (Section 9.9), that can serve as the likelihood function for each sample unit in multi-class data sets. Such a model is known as **multinomial logistic regression** (but can often be found listed under **polytomous** or **polychotomous regression**). It takes the form

$$Y \sim \text{Multinomial}(1, p_1, \ldots, p_n)$$

$$p_i = \frac{\exp(\eta_i)}{1 + \sum\limits_{j=1}^{n} \exp(\eta_j)} \tag{12.8}$$

Notice the similarity between the second of these expressions and the logit link function of the binomial model (Section 11.7).

Example 12.10: Classification of whale vocalisations

Sperm whale, Physeter macrocephalus

The abundance and distribution of cetaceans is increasingly being investigated with the aid of passive acoustic methods (Moore *et al.*, 2006). For example, input waveforms may be segmented into regular time windows and the distribution of sound energy across different frequencies in each time window can be described by a spectral diagram (see Section 2.16). Different statistical characteristics of the spectral diagram (e.g. mean frequency, standard deviation of frequencies, dynamic range) may be used to classify the calls by species.

Since whale calls will be absent from most time windows, a binary logistic regression (see Section 12.8) can be applied first to classify windows as 'whale' and 'no whale'. The analysis can then proceed on the subset of time windows believed to contain whale calls. By treating the acoustic segment as the sample unit and the different species as the classes, the probability that a call belongs to a particular species can be calculated by multinomial regression. The approach would require a training data set of calls from visually identified whales. Once fitted to these data, the model would then be asked to recognise the species of whale vocalising in the vicinity of the hydrophone. Different whales are also capable of emitting different types of calls. This can be handled equally well by this approach if each class is subscripted by whale species and call type.

℞ 12.5: Classification by multinomial logistic regression

The data set comprises 303 time windows containing whale calls from visually identified individuals belonging to six different species (listed in the `species` column in the data frame). The explanatory variables are mean frequency (`muf`), standard deviation of frequencies (`sif`) and dynamic range (`dyn`). The model will be estimated using the command `multinom()` from the `nnet` library. I will use the first 253 observations for fitting and the remaining 50 for validation.

```
dat<-data.frame(species, muf, sif, dyn)
datf<-dat[1:253,] # Data for fitting
datv<-dat[254:303,]# Data for validation
require(nnet) # Loads library
mod<-multinom(species~muf+sif+dyn, data=datf) # Estimates model
mod<-step(mod) # Performs automatic model selection
```

I have omitted all of the automatic output generated by this code, but here is the final model, which has retained only the variables muf and dyn

```
> mod
Call:
multinom(formula = species ~ muf + dyn, data = datf)

Coefficients:
   (Intercept)        muf        dyn
2    -4.312751 0.03716399 0.1219503
3   -13.959216 0.05198608 0.2900206
4   -40.518367 0.06018279 0.6026860
5   -65.186294 0.06526570 0.8155116
6  -304.661447 0.04865704 2.4439845

Residual Deviance: 156.3877
```

The validation part is done as follows:

```
pr<-predict(mod, newdata=datv, type="probs")
prs<-rep(0,50)
for(i in 1:50) {prs[i]<-match(max(pr[i,]), pr[i,])}
```

Here, the first line creates predictions for each of the new sound clips (note the specification type = "probs" used to get probabilistic predictions). The resulting object pr is a matrix in which each row represents a sample unit (a sound clip). The rows contain six probabilities specifying how likely it is for that sound clip to have come from each of the six whale species. The second line of code creates an empty list prs of length 50. The third line sweeps through the rows of the matrix pr and decides on the likeliest species for that sound clip. It does this via the command match(u, v) which finds the position(s) of the value u in the vector v. The result for each sample unit is stored in the vector prs. Finally, the confusion matrix for this classification attempt can be viewed as follows

```
> #Confusion matrix
> table(prs, datv$species)

prs 1 2  3 4  5 6
  1 4 0  0 0  0 0
  2 0 6  1 0  0 0
  3 0 3 12 0  0 0
  4 0 0  0 9  1 0
  5 0 0  0 1 10 0
  6 0 0  0 0  1 2
```

Any entries found off the diagonal indicate misclassification. Here, the correct classifications amount to 86%. You can quickly calculate this by typing

```
100*sum(diag(table(prs, datv$species)))/50
```

Further reading

There are few dedicated books for multivariate techniques in ecology. A near-comprehensive list is Jongman *et al.* (1995); others include Legendre and Legendre (1998) and Fielding (2007). Quantitative ecology books with some multivariate material are Quinn and Keough (2002), Gotelli and Ellison (2004) and Zuur *et al.* (2007). References using R are Everitt and Hothorn (2006), Maindonald and Braun (2007) and Crawley (2007). More general references are Johnson (1998) and Manly (2005). Further details about multinomial regression can be found in the excellent toolkit of techniques assembled in Faraway (2006).

References

Blouin, M.S., Parsons, M., Lacaille, V. and Lotz, S. (1996) Use of microsatellite loci to classify individuals by relatedness. *Molecular Ecology*, **5**, 393–401.

Crawley, M.J. (2007) *The R Book.* John Wiley & Sons, Ltd, Chichester. 942pp.

Everitt, B.S. and Hothorn, T. (2006) *A Handbook of Statistical Analyses using R.* Chapman & Hall, Boca Raton, Florida. 275pp.

Faraway, J.J. (2006) *Extending the Linear Model with R: Generalized Linear, Mixed Effects and Nonparametric Regression Models.* Chapman & Hall, Boca Raton, Florida. 301pp.

Fielding A.H. (2007) *Cluster and Classification Techniques for the Biosciences.* Cambridge University Press, New York. 246pp.

Gotelli, N.J. and Ellison, A.M. (2004) *A Primer of Ecological Statistics.* Sinauer Associates, Massachusetts. 510pp.

Johnson, D.E. (1998) *Applied Multivariate Methods for Data Analysis.* Wadsworth. 425pp.

Jongman, R.H.G., Ter Braak, C.J.F. and van Tongeren, O.F.R. (1995) *Data Analysis in Community and Landscape Ecology.* Cambridge University Press, Cambridge. 324pp.

Legendre, P. and Legendre, L. (1998) *Numerical Ecology.* Elsevier, Amsterdam. 870pp.

Maindonald, J. and Braun, J. (2007) *Data Analysis and Graphics using R: An Example-based Approach*, 2nd edition. Cambridge University Press, New York. 509pp.

Manly, B.F.J. (2005) *Multivariate Statistical Methods: A Primer.* Chapman & Hall, Boca Raton, Florida. 208pp.

Moore, S.E., Stafford, K.M., Mellinger, D.K. and Hildebrand, J.A. (2006) Listening for large whales in the offshore waters of Alaska. *Bioscience*, **56**, 49–55.

Petchey, O. and Gaston, K.J. (2002) Functional diversity (FD), species richness and community composition. *Ecology Letters*, **5**, 402–411.

Quinn, G.P. and Keough, M.J. (2002) *Experimental Design and Data Analysis for Biologists.* Cambridge University Press. 537pp.

Zuur, A., Ieno, E.N. and Smith, G.M. (2007) *Analysing Ecological Data.* Springer, New York. 672pp.

Appendix: Formulae

1. Algebra

1 The properties of powers

(i) $c^0 = 1$

(ii) $c^{-x} = \frac{1}{c^x}$

(iii) $c^{\frac{x}{y}} = \sqrt[y]{c^x} = (\sqrt[y]{c})^x$

(iv) $(c^x)^y = c^{xy} = (c^y)^x$

(v) $c^x c^y = c^{x+y}$

(vi) $\frac{c^x}{c^y} = c^{x-y}$

2 Manipulating expressions.

$a + b = b + a$

$ab = ba$

$(a+b)+c = a+(b+c) = (a+c)+b$

$(ab)c = a(bc) = b(ac)$

3 Some useful algebraic identities.

$a^2 - b^2 = (a-b)(a+b)$ or, more generally,

$a^n - b^n = (a-b)(a^{n-1} + a^{n-2}b + \cdots + ab^{n-2} + b^{n-1})$

$(a \pm b)^2 = a^2 \pm 2ab + b^2$

4 Solution of second order polynomial equation of the form $a_2 x^2 + a_1 x + a_0 = 0$. The 'quadratic formula' is

$$x_1, x_2 = \frac{-a_1 \pm \sqrt{a_1^2 - 4a_2 a_0}}{2a_2}$$

5 Complex numbers are of the form $z = a + bi$ where $i^2 = -1$ (or, $i = \sqrt{-1}$). Algebra with complex numbers uses the following identities:

$z_1 \pm z_2 = (a_1 \pm a_2) + (b_1 \pm b_2)i$

$z_1 z_2 = a_1 a_2 + (a_1 b_2 + a_2 b_1)i - b_1 b_2$

$\frac{z_1}{z_2} = \frac{z_1 \bar{z}_2}{z_2 \bar{z}_2}$, where $\bar{z}$ is the conjugate of z, such that

$z\bar{z} = (a + bi)(a - bi) = a^2 + b^2$.

How to be a Quantitative Ecologist: The 'A to R' of Green Mathematics and Statistics, First Edition. Jason Matthiopoulos.
© 2011 John Wiley & Sons, Ltd. Published 2011 by John Wiley & Sons, Ltd.

2. Geometry

1☞ Pythagoras's theorem:

$$a^2 = c^2 + b^2$$

Distance between two points in 2D space:

$$d = \sqrt{(x_1 - x_2)^2 + (y_1 - y_2)^2}$$

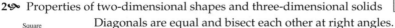

2☞ Properties of two-dimensional shapes and three-dimensional solids

Square

Diagonals are equal and bisect each other at right angles.
Perimeter: $4a$
Area: a^2

Rectangle

Diagonals are equal and bisect each other
Perimeter: $2(a + b)$
Area: ab

Parallelogram

Diagonals bisect each other
Perimeter: $2(a + b)$
Area: hb

Trapezium

Two sides are parallel
Perimeter: $a + b + c + d$
Area: $h\dfrac{a + b}{2}$

Circle

All points on the perimeter (circumference) are the same distance (radius) from a point (the centre) internal to the circle.
Perimeter: $2\pi r$
Area: πr^2

Ellipse

An ellipse is the set of points on a plane whose sum of distances $(a + b)$ from two given points is the same.
Perimeter (approximate):

$$\frac{h + r}{2}\left(1 + \frac{3\left(\dfrac{r - h}{r + h}\right)^2}{10 + \sqrt{4 - 3\left(\dfrac{r - h}{r + h}\right)^2}}\right)$$

Area: πrh

Triangle

The same area can be calculated using any side of the triangle with its corresponding height.

Perimeter: $a + b + c$

Area: $\dfrac{ha}{2}$

Regular hexagon

All the triangles formed by drawing the three diagonals of the hexagon are equilateral.

Perimeter: $6a$

Area: $3ha = \dfrac{3\sqrt{3}}{2}a^2$

Cube

All sides are squares

Surface area: $6a^2$

Volume: a^3

Cuboid

All sides are rectangles

Surface area:

$2ab + 2bc + 2ac$

Volume: abc

Sphere

Surface area: $4\pi r^2$

Volume: $\dfrac{4}{3}\pi r^3$

Cylinder

Surface area: $2\pi r(r + h)$

Volume: $\pi r^2 h$

3. Trigonometry

1⟐ Notation. Angles are usually denoted by the Greek letters θ, ϕ, φ

2⟐ Angle conversions between degrees (d) and radians (ϕ): $360/d = 2\pi/\phi$. Some useful angle conversions are

Degrees	Radians	Degrees	Radians
0	0	135	$3\pi/4$
30	$\pi/6$	150	$5\pi/6$
45	$\pi/4$	180	π
60	$\pi/3$	270	$3\pi/2$
90	$\pi/2$	360	2π
120	$2\pi/3$		

3✎ Definitions of trigonometric functions $\cos(\theta) = x/r$, $\sin(\theta) = y/r$, $\tan(\theta) = y/x$, $\cot(\theta) = x/y$

4✎ Numerical approximation of basic trigonometric functions: $\sin(\theta) = \theta - \frac{\theta^3}{3!} + \frac{\theta^5}{5!} - \frac{\theta^7}{7!} + \cdots$ $\cos(\theta) = 1 - \frac{\theta^2}{2!} + \frac{\theta^4}{4!} - \frac{\theta^6}{6!} + \cdots$

5✎ Converting polar to Cartesian coordinates: $(x, y) = (r\cos(\theta), r\sin(\theta))$

6✎ Basic trigonometric identities:

$\tan\theta = \sin\theta / \cos\theta$ $\cos^2(\theta) + \sin^2(\theta) = 1$

$\cos(-\theta) = \cos\theta$ $\sin(-\theta) = -\sin\theta$

$\cos(\theta + n2\pi) = \cos\theta$ $\sin(\theta + n2\pi) = \sin\theta$

$\cos(\pi - \theta) = -\cos\theta$ $\sin(\pi - \theta) = \sin\theta$

7✎ Trigonometric identities for sums of angles:

$\sin(\theta \pm \phi) = \sin\theta\cos\phi \pm \cos\theta\sin\phi$

$\cos(\theta \pm \phi) = \cos\theta\cos\phi \mp \sin\theta\sin\phi$

$\tan(\theta \pm \phi) = \dfrac{\tan\theta \pm \tan\phi}{1 \mp \tan\theta\tan\phi}$

8✎ Trigonometric identities for double angles:

$\sin(2\theta) = 2\sin\theta\cos\theta$

$\cos(2\theta) = \cos^2\theta - \sin^2\theta$

$\tan(\theta \pm \phi) = \dfrac{2\tan\theta}{1 - \tan^2\theta}$

9✎ Identities for the product of trigonometric functions:

$\sin\theta\cos\phi = \frac{1}{2}(\sin(\theta + \phi) + \sin(\theta - \phi))$

$\cos\theta\sin\phi = \frac{1}{2}(\sin(\theta + \phi) - \sin(\theta - \phi))$

$\cos\theta\cos\phi = \frac{1}{2}(\cos(\theta + \phi) + \cos(\theta - \phi))$

$\sin\theta\sin\phi = \frac{1}{2}(\cos(\theta + \phi) - \cos(\theta - \phi))$

10✎ Identities for the sum of trigonometric functions:

$\sin\theta + \sin\phi = 2\sin\left(\dfrac{\theta + \phi}{2}\right)\cos\left(\dfrac{\theta - \phi}{2}\right)$

$\sin\theta - \sin\phi = 2\cos\left(\dfrac{\theta + \phi}{2}\right)\sin\left(\dfrac{\theta - \phi}{2}\right)$

$\cos\theta + \cos\phi = 2\cos\left(\dfrac{\theta + \phi}{2}\right)\cos\left(\dfrac{\theta - \phi}{2}\right)$

$\cos\theta - \cos\phi = -2\sin\left(\dfrac{\theta + \phi}{2}\right)\sin\left(\dfrac{\theta - \phi}{2}\right)$

11✎ Identities for the angles and sides of an oblique triangle (i.e. one that does not have a right angle).

The law of sines : $\dfrac{A}{\sin a} = \dfrac{B}{\sin b} = \dfrac{C}{\sin c}$

The law of cosines : $A^2 = B^2 + C^2 - 2BC\cos a$
$B^2 = A^2 + C^2 - 2AC\cos b$
$C^2 = A^2 + B^2 - 2AB\cos c$

12. General transformation of cos into a new periodic function:

$$f(t) = \frac{f_{max} + f_{min}}{2} + \frac{f_{max} - f_{min}}{2} \cos\left(\frac{2\pi}{T}(t - t_0)\right)$$

where f_{min} and f_{max} are the minimum and maximum of the new function (replacing -1 and 1), T is the new period (replacing 2π) and t_0 is the location of the new peak (replacing 0).

13. Superposition of several periodic components is described by a trigonometric polynomial

$$f(t) = a + (\beta_1 \cos \omega t + \beta_2 \cos 2\omega t + \cdots) + (\gamma_1 \sin \omega t + \gamma_2 \sin 2\omega t + \cdots)$$

4. Sequences

1. Arithmetic progression. The iterative definition is $a_{n+1} = a_n + m$ and the corresponding general definition is $a_n = a_0 + nm$.

2. Geometric progression. The iterative definition is $a_{n+1} = ma_n$ and the corresponding general definition is $a_n = m^n a_0$.

3. A taxonomy of difference equations presented using generic examples: c_1, c_2 and c_3 are constants and $c_1(t), c_2(t)$ and $c_3(t)$ are functions of time. The entries describe the right-hand side of the difference equation, so a homogeneous, autonomous, linear, affine equation is $a_{t+1} = c_1 a_t$

	Homogeneous	
	Autonomous	Nonautonomous
Linear affine	$c_1 a_t$	$c_1(t)a_t$
Linear	$c_1 a_t + c_0$	$c_1(t)a_t + c_0$
Nonlinear affine	$c_2 a_t^2 + c_1 a_t$	$c_2(t)a_t^2 + c_1(t)a_t$
Nonlinear	$c_2 a_t^2 + c_1 a_t + c_0$	$c_2(t)a_t^2 + c_1(t)a_t + c_0$

	Nonhomogeneous	
	Autonomous	Nonautonomous
Linear	$c_1 a_t + c_0(t)$	$c_1(t)a_t + c_0(t)$
Nonlinear	$c_2 a_t^2 + c_1 a_t + c_0(t)$	$c_2(t)a_t^2 + c_1(t)a_t + c_0(t)$

5. Logarithms

1. Properties of the logarithmic function

(i) $\log_c 1 = 0$

(iv) $\log_c(xy) = \log_c x + \log_c y$

(ii) $\log_c c = 1$

(v) $\log_c(x/y) = \log_c x - \log_c y$

(iii) $c^{\log_c x} = x$

(vi) $\log_c x^k = k \log_c x$

2. Relationship between natural and common logarithms: $\ln x = \ln 10 \log x$. This implies that the two are proportional to each other.

6. Limits

1. Properties of limits.
If $\lim_{x \to a} f(x) = c_1$ and $\lim_{x \to a} g(x) = c_2$, then
(i) The limit of the sum equals the sum of the limits

$\lim_{x \to a}(f(x) \pm g(x)) = c_1 \pm c_2$

(ii) The limit of the product equals the product of the limits

$\lim_{x \to a}(f(x)g(x)) = c_1 c_2$

(iii) The limit of the ratio equals the ratio of the limits, assuming that $c_2 \neq 0$

$\lim_{x \to a}\left(\frac{f(x)}{g(x)}\right) = \frac{c_1}{c_2}$ given $c_2 \neq 0$

(iv) The limit of the product between a constant and a function equals the product of the constant and the function's limit

$\lim_{x \to a}(cf(x)) = cc_1$

2🪶 Limits involving infinity.

(i) $\lim_{x \to 0} \frac{1}{x} = \infty$ (ii) $\lim_{x \to \infty} \frac{1}{x} = 0$

7. Derivatives

1🪶 Definition of the derivative

$\frac{df}{dx} = \lim_{x \to x_1} \frac{f(x) - f(x_1)}{x - x_1}$

or, alternatively,

$\frac{df}{dx} = \lim_{\Delta x \to 0} \frac{f(x + \Delta x) - f(x)}{\Delta x}$

There are five alternative notations for the derivative of a function:

$\frac{df}{dx} = \frac{df(x)}{dx} = \frac{d}{dx}f(x) = f'(x) = \dot{f}(x)$

2🪶 Rules of differentiation

	Function	Derivative
1	$f(x) = c$	$f' = 0$
2	$f(x) = x^n$	$f' = nx^{n-1}$
3	$f(x) = cg(x)$	$f' = cg'$
4	$f(x) = g(x) + h(x)$	$f' = g' + h'$
5 Product rule	$f(x) = g(x)h(x)$	$f' = g'h + h'g$
6	$f(x) = g(x)^n$	$f' = ng^{n-1}g'$
7 Quotient rule	$f(x) = \frac{g(x)}{h(x)}$	$f' = \frac{g'h - h'g}{h^2}$
8	$f(x) = g(x)^{\frac{n}{m}}$	$f' = \frac{n}{m}g^{\frac{n}{m}-1}g'$
9	$f(x) = \ln x$	$f' = \frac{1}{x}$
10	$f(x) = e^x$	$f' = e^x$
11	$f(x) = \sin x$	$f' = \cos x$
12	$f(x) = \cos x$	$f' = -\sin x$
13	$f(x) = \tan x$	$f' = \sec^2 x$

3🪶 The chain rule

For a function $f(x) = f(g(x))$,

$$\frac{df}{dx} = \frac{df}{dg}\frac{dg}{dx}$$

More generally, for $f(x) = f(g_1(g_2(\cdots g_n(x))))$,

$$\frac{df}{dx} = \frac{df}{dg_1}\frac{dg_1}{dg_2}\cdots\frac{dg_n}{dx}$$

4 Taylor expansion

$$f(x) \cong f(x_1) + \sum_{i=1}^{n} \frac{1}{i!}f^{(i)}(x_1)(x - x_1)^i$$

8. Integrals

1 Basic rules of integration

	Function $(F(x))$	Indefinite integral $(\int F(x)dx)$
1	a	$ax + c$
2	x^n	$\frac{1}{n+1}x^{n+1} + c$
3	$cG(x)$	$c\int G(x)dx$
4	$G(x) + H(x)$	$\int G(x)dx + \int H(x)dx$
5	$G(x)^n\frac{dG}{dx}$	$\frac{G(x)^{n+1}}{n+1} + c$
6	$G(x)^{\frac{n}{m}}\frac{dG}{dx}$	$m\frac{G(x)^{\frac{n}{m}+1}}{n+m} + c$
7	$\frac{1}{x}$	$\ln x + c$
8	e^x	$e^x + c$
9	$\sin x$	$\cos x$
10	$\cos x$	$-\sin x$
11	$\tan x$	$\sec^2 x$

2 Manipulating definite integrals

(i) Given an interval $[a, b]$ that comprises the two subintervals $[a, c]$ and $[c, b]$ then

$$\int_a^b f(x)\,dx = \int_a^c f(x)\,dx + \int_c^b f(x)\,dx$$

(ii) $\int_a^b f(x)dx = -\int_b^a f(x)dx$

9. Matrices

1 Properties of matrix arithmetic. For any three matrices $\mathbf{A}, \mathbf{B}, \mathbf{C}$ and scalars a, b, the following are true

1. $\mathbf{A} + \mathbf{B} = \mathbf{B} + \mathbf{A}$
2. $(\mathbf{A} + \mathbf{B}) + \mathbf{C} = \mathbf{A} + (\mathbf{B} + \mathbf{C})$
3. $a(\mathbf{A} + \mathbf{B}) = a\mathbf{A} + a\mathbf{B}$
4. $(a + b)\mathbf{A} = a\mathbf{A} + b\mathbf{A}$
5. $\mathbf{A}(\mathbf{BC}) = (\mathbf{AB})\mathbf{C}$
6. $\mathbf{IA} = \mathbf{A}, \mathbf{AI} = \mathbf{A}$
7. $\mathbf{A}(\mathbf{B} + \mathbf{C}) = \mathbf{AB} + \mathbf{AC}$

2 Properties of the transpose

1. $(\mathbf{A}^T)^T = \mathbf{A}$
2. $(\mathbf{A} + \mathbf{B})^T = \mathbf{A}^T + \mathbf{B}^T$
3. $(\mathbf{AB})^T = \mathbf{B}^T\mathbf{A}^T$

3♐ Properties of the determinant
1 $\det(\mathbf{AB}) = \det(\mathbf{A})\det(\mathbf{B})$
2 $\det(A^T) = \det(A)$
3 $\det(\mathbf{I}_n) = 1$
and, if **A** is invertible,
4 $\det(\mathbf{A}^{-1}) = \dfrac{1}{\det(\mathbf{A})}$

4♐ Nonlinear dynamical systems. The Jacobian:

$$\mathbf{M} = \begin{pmatrix} \dfrac{\partial f_1}{\partial x_1} & \cdots & \dfrac{\partial f_1}{\partial x_n} \\ \vdots & \ddots & \vdots \\ \dfrac{\partial f_n}{\partial x_1} & \cdots & \dfrac{\partial f_n}{\partial x_n} \end{pmatrix}$$

10. Descriptive statistics

1♐ The average. For a sample of observations $x_1, \ldots, x_n$, the average is: $\bar{x} = \frac{1}{n}\sum_{i=1}^{n} x_i$. If $f(x)$ is the relative frequency with which a value x occurs in the sample, then the average is also written $\bar{x} = \sum_{\text{All } x} f(x)x$. The two expressions will be identical if no binning is used to calculate the relative frequencies (e.g. as is the case in discrete variables).

2♐ One random variable

Variance $\qquad\qquad v(x) = \frac{1}{n}\sum_{i=1}^{n}(x_i - \bar{x})^2$

Standard deviation $\quad s(x) = \sqrt{v(x)}$

Skewness $\qquad\qquad \dfrac{1}{ns^3}\sum_{i=1}^{n}(x_i - \bar{x})^3$

Kurtosis $\qquad\qquad \dfrac{1}{ns^4}\sum_{i=1}^{n}(x_i - \bar{x})^4 - 3$

3♐ Two random variables

Covariance $\quad \text{cov}(x, y) = \frac{1}{n}\sum_{i=1}^{n}(x_i - \bar{x})(y_i - \bar{y})$

Correlation $\qquad r = \dfrac{\text{cov}(x, y)}{s_x s_y}$

11. Sets and events

1♐ The empty set is written $\emptyset$ and the event space of an experiment is conventionally written Ω.

2♐ Comparisons between two sets
$A = B$ The set A is identical to the set B
$A \subset B$ The set A is a subset of the set B
$A \subseteq B$ Set A is a subset of, or the same as, set B
$A \not\subseteq B$ Set A is a not a subset of B

3♐ Set operators
$\bar{A}$ Negation: 'not A'
$A \cup B$ Union of two sets: 'A or B'
$A \cap B$ Intersection between two sets: 'A and B'

4☙ Important relationships between events
$B = \bar{A}$: Complementary events
$A \cup B = \Omega$: Collectively exhaustive events
$A \cap B = \emptyset$: Mutually exclusive events

12. Rules of probability

1☙ 1 For a set of mutually exclusive and collectively exhaustive events $E_1, E_2, E_3, \ldots, E_n$, it is always true that $\sum_{i=1}^{n} P(E_i) = 1$. If these events are equally likely, each with probability p, then $p = \frac{1}{n}$. Also, if two events are complementary, then $P(E) = 1 - P(\bar{E})$.

2☙ The probability of the union of two mutually exclusive events is $P(E_1 \cup E_2) = P(E_1) + P(E_2)$. The probability of the union of any two events is $P(E_1 \cup E_2) = P(E_1) + P(E_2) - P(E_1 \cap E_2)$

3☙ Rules dealing with conditional probability

1. $P(E_2|E_1) = \dfrac{P(E_1 \cap E_2)}{P(E_1)}$

2. $P(E_2|E_1) = \dfrac{P(E_1|E_2)P(E_2)}{P(E_1)}$ (Bayes's law)

3. $P(E_1 \cap E_2) = P(E_1)P(E_2)$ (Independent events)

4. Total probability. Given the probabilities $P(E_1), \ldots, P(E_n)$ of n mutually exclusive and collectively exhaustive events, and the conditional probabilities $P(G|E_1), \ldots, P(G|E_n)$ referring to some other event G, then the total probability of G is $P(G) = \sum_{i=1}^{n} P(G|E_i)P(E_i)$

4☙ Bayesian probability.

$$P(H|data) = \frac{P(data|H)P(H)}{P(data)}$$

13. Probability distributions

1☙ Discrete random variables
Probability mass function (PMF): $f_X(x) = P(X = x)$
Properties of the PMF:

(i) $0 \leq f(x) \leq 1 \ \forall x \in \Omega$

(ii) $\sum_{All \ x \in \Omega} f(x) = 1$

Cumulative distribution function (CDF): $F_X(x) = P(X \leq x)$
Properties of the CDF:

(i) $\lim_{x \to -\infty} F(x) = 0$

(ii) $\lim_{x \to \infty} F(x) = 1$

(iii) If $a \leq b$, then $F(a) \leq F(b)$

Relationship between the PMF and CDF:

$$f(x) = F(x) - F(x-1) \quad \text{and} \quad F(x) = \sum_{i=-\infty}^{x} f(i)$$

2✎ Continuous random variables

Definition and properties of CDF: same as in discrete case.

Probability density function (PDF): defined as the derivative of the CDF. Properties of the PDF:

(i) $f(x) \geq 0 \ \forall x \in \Omega$

(ii) $P(a < X \leq b) = \int_a^b f(x)dx,$

(iii) $\int_{-\infty}^{\infty} f(x)dx = 1$

Relationship between PDF and CDF:

$$\frac{dF(x)}{dx} = f(x) \quad \text{and} \quad F(x) = \int_{-\infty}^{x} f(s)\,ds$$

14. Expectation

1✎ Expectation of a random variable:

If X is discrete: $E(X) = \sum_{-\infty}^{\infty} xf(x)$

If X is continuous: $E(X) = \int_{-\infty}^{\infty} xf(x)dx$

2✎ Moments of a distribution:

If X is discrete: $E(X^m) = \sum_{-\infty}^{\infty} x^m f(x)$

If X is continuous: $E(X^m) = \int_{-\infty}^{\infty} x^m f(x)dx$

3✎ Expectation of a general function $g(X)$

If X is discrete: $E(g(X)) = \sum_{-\infty}^{\infty} g(x)f(x)$

If X is continuous: $E(g(X)) = \int_{-\infty}^{\infty} g(x)f(x)dx$

4✎ Relationship between moments and descriptive statistics

$\mu = E(X), Var(X) = E(X^2) - E(X)^2, \ \text{cov}(X, Y) = E((X - \mu_X)(Y - \mu_Y))$

5✎ Manipulating expectations and variances

$E(X + Y) = E(X) + E(Y)$
$E(cX) = cE(X)$
$Var(cX) = a^2 Var(X)$
$E(XY) = E(X)E(Y)$(independent variables)
$Var(X + Y) = Var(X) + Var(Y)$(independent variables)

15. Combinatorics

1✎ The binomial coefficient counts the number of ways m of ordering x successes in a string of n trials: $m = \begin{pmatrix} n \\ x \end{pmatrix} = \dfrac{n!}{x!(n-x)!}$

16. Discrete probability distributions

Name	PMF	Mean	Variance
Uniform $U(x_1, x_k)$	$\dfrac{1}{x_k - x_1 + 1}$	$\dfrac{x_1 + x_2}{2}$	$\dfrac{(x_k - x_1 + 1)^2 - 1}{12}$
Binomial $B(n, p)$	$\dbinom{n}{x} p^x q^{n-x}$	np	npq
Poisson Poisson(λ)	$\dfrac{e^{-\lambda} \lambda^x}{x!}$	λ	λ
Geometric (x is no of trials) Geometric(p)	$f(x) = pq^{x-1}$	$1/p$	q/p^2
Geometric (x is no of failures) Geometric(p)	$f(x) = pq^{x-1}$	q/p	q/p^2
Negative binomial if x is no of trials NegB(k, p)	$f\dbinom{x-1}{k-1} p^k q^{x-k}$	k/p	kq/p^2
Negative binomial if x is no of failures $X \sim$ NegB(k, p)	$\dbinom{k+x-1}{k-1} p^k q^x$	kp/q	kp/q^2
Multinomial Multinom$(n, \mathbf{p})$	$\dfrac{n!}{x_1! \cdots x_k!} p_1^{x_1} \cdots p_k^{x_k}$	$\mu_i = np_i$	$\sigma_i^2 = np_i q_i.$

17. Continuous probability distributions

Name	PMF	Mean	Variance
Uniform $U(a, b)$	$\dfrac{1}{b-a}$	$\dfrac{a+b}{2}$	$\dfrac{(b-a)^2}{12}$
Exponential $M(\lambda)$	$\lambda e^{-\lambda x}$	$1/\lambda$	$1/\lambda^2$
Gamma $Gamma(k, \lambda)$	$x^{k-1} \lambda^k \dfrac{e^{-\lambda x}}{\Gamma(k)}$	k/λ	k/λ^2
Beta $Beta(\alpha, \beta)$	$\dfrac{\Gamma(\alpha + \beta)}{\Gamma(\alpha)\Gamma(\beta)} x^{\alpha-1}(1-x)^{\beta-1}$	$\dfrac{\alpha}{\alpha + \beta}$	$\dfrac{\alpha\beta}{(\alpha+\beta)^2(\alpha+\beta+1)}$
Normal $N(\mu, \sigma^2)$	$\dfrac{1}{\sqrt{2\pi\sigma^2}} e^{-\frac{1}{2}\frac{(\mu-x)^2}{\sigma^2}}$	μ	σ^2
t-distribution $t(n)$	$\dfrac{\Gamma(\frac{n+1}{2})}{\sqrt{n\pi}\,\Gamma(\frac{n}{2})} \left(1 + \dfrac{x^2}{n}\right)^{-\frac{n+1}{2}}$	0	$\begin{cases} \infty & \text{if } 1 < n \le 2 \\ \dfrac{n}{n-2} & \text{if } n > 2 \end{cases}$
Lognormal $LogN(\mu, \sigma^2)$	$\dfrac{1}{y\sqrt{2\pi\sigma^2}} e^{-\frac{(\ln y - \mu)^2}{2\sigma^2}}$	$e^{\mu + \frac{1}{2}\sigma^2}$	$(e^{\sigma^2} - 1)e^{2\mu+\sigma^2}$
Chi-square $\chi^2(n)$	$\dfrac{x^{n/2-1}}{2^{n/2}} \dfrac{e^{-x/2}}{\Gamma(n/2)}$	n	$2n$

R Index

Index

How to be a Quantitative Ecologist: The 'A to R' of Green Mathematics and Statistics, First Edition. Jason Matthiopoulos.
© 2011 John Wiley & Sons, Ltd. Published 2011 by John Wiley & Sons, Ltd.